Forest Management and Planning

Forest Management and Planning

Edited by Lester Jacob

MURPHY & MOORE
www.murphy-moorepublishing.com

Published by Murphy & Moore Publishing,
1 Rockefeller Plaza,
New York City, NY 10020, USA
www.murphy-moorepublishing.com

Forest Management and Planning
Edited by Lester Jacob

International Standard Book Number: 978-1-63987-236-7 (Hardback)

Cataloging-in-Publication Data

Forest management and planning / edited by Lester Jacob.
p. cm.
Includes bibliographical references and index.
ISBN 978-1-63987-236-7
1. Forest management. 2. Forest plants--Control. 3. Forests and forestry. I. Jacob, Lester.

SD431 .F67 2022
634.92--dc23

TABLE OF CONTENTS

PREFACE

The branch of forestry concerned with the overall administration of forests is known as forest management. It includes economic, legal as well as social aspects. Scientific and technical aspects such as forest regulation, silviculture, forest protection, etc. also fall under this domain. Some of the focus areas of forest management are forest genetic resources, management of water, urban values, management for aesthetics, deriving wood products, etc. Conservation and economics are the two pillars of forest management. Some of its varied techniques include planting and replanting of different species, fire prevention, timber extraction and cutting roads and pathways through forests. The aim of such management is either economical, like increasing yields and forest products or ecological like carbon sequestration, species recovery, etc. This book presents the complex subject of forest management in the most comprehensible and easy to understand language. From theories to research to practical applications, case studies related to all contemporary topics of relevance to this field have been included herein. This book includes contributions of experts and scientists which will provide innovative insights into this discipline.

The information contained in this book is the result of intensive hard work done by researchers in this field. All due efforts have been made to make this book serve as a complete guiding source for students and researchers. The topics in this book have been comprehensively explained to help readers understand the growing trends in the field.

I would like to thank the entire group of writers who made sincere efforts in this book and my family who supported me in my efforts of working on this book. I take this opportunity to thank all those who have been a guiding force throughout my life.

Editor

1

Technical assessment of forest road network using Backmund and surface distribution algorithm in a hardwood forest of Hyrcanian zone

Aidin Parsakhoo

Assistant Professor, Department of Forestry, Faculty of Forest Sciences, Gorgan University of Agricultural Sciences and Natural Resources, Gorgan, Iran

Abstract

Aim of study: Corrected Backmund and Surface Distribution Algorithms (SDA) for analysis of forest road network are introduced and presented in this study. Research was carried out to compare road network performance between two districts in a hardwood forest.

Area of study: Shast Kalateh forests, Iran.

Materials and methods: In uncorrected Backmund algorithm, skidding distance was determined by calculating road density and spacing and then it was designed as Potential Area for Skidding Operations (PASO) in ArcGIS software. To correct this procedure, the skidding constraint areas were taken using GPS and then removed from PASO. In SDA, shortest perpendicular distance from geometrical center of timber compartments to road was measured at both districts.

Main results: In corrected Backmund, forest openness in district I and II were 70.3% and 69.5%, respectively. Therefore, there was little difference in forest openness in the districts based on the uncorrected Backmund. In SDA, the mean distance from geometrical center of timber compartments to the roads of districts I and II were 199.45 and 149.31 meters, respectively. Forest road network distribution in district II was better than that of district I relating to SDA.

Research highlights: It was concluded that uncorrected Backmund was not precise enough to assess forest road network, while corrected Backmund could exhibit a real PASO by removing skidding constraints. According to presented algorithms, forest road network performance in district II was better than district I.

Keywords: technical assessment; forest road network; corrected Backmund; skidding distance; surface distribution algorithm.

Funding: The authors received no specific funding for this work.

Competing interests: The authors have declared that no competing interests exist.

Correspondence should be addressed to Aidin Parsakhoo: aidinparsakhoo@yahoo.com

Introduction

Forest road networks are designed in forest for a variety of reasons, including harvesting, forest conservation and recreational programs (Ismail, 2009). The analysis of different alternatives of forest road network especially in hard mountainous condition has received great attention as an appropriate tool to select the best alternative (Eastaugh & Molina, 2012). Assessment of road network is a key step to provide appropriate transportation planning (Heralt, 2002; Aruga *et al.*, 2006). By technical assessment of forest road network, analyzers will generate maps and data that show management opportunities for changing current road network to address future needs, budgets, and harvesting concerns in a better way (Dean, 1997; Gumus *et al.*, 2008). There are many algorithms to assess the performances of road alternatives based on environmental, economic and traffic safety point of views (Klassen, 2006; Ismail, 2009). The advantages and potential limitations of using this algorithm are different (Aruga *et al.*, 2005; Sessions, 2007).

Backmund algorithm is the traditional theoretical technique in Iranian forestry which can determine the degree of forest openness through forest roads. The openness of a commercial forest area had to be calculated for determining the skidding distance and accessible area. The size of openness is an indicator of forest transport network performance (Pentek *et al.*, 2005).

Road density, road spacing and skidding distance parameters are used to calculate Backmund algorithm. In this technique, maximum distances of skidding so-called Potential Area for Skidding Operations (PASO) is calculated around the road network and then overlaps is determined. One of the most important limitations of Backmund algorithm is that the constraints of ground-based skidding are not considered in all the PASO (Puya *et al.* 2009). Ground-based skidding is a system of moving wood on the ground to the depot (temporary timber storage areas). Severe constraints on the skidding operations and resources harvesting lead to decreased forest accessibility. Puya *et al.* (2009) recorded using GPS and mapped the position of stream network, soil condition and steep terrain as skidding constraints.

Contreras & Chung (2007) used grid cells and shortest path algorithm in raster-based GIS data to show constraints and obstacles of ground-based timber harvesting. This information was suitable for locating log depot. The existence of these constraints forces the shortest skid trails to be re-routed, which lead to increase skidding distances (Jaafari *et al.*, 2015). Another limitation of Backmund algorithm is that it is often ineffective in comparing road network alternatives with different density (Pentek *et al.*, 2005; Sessions & Boston, 2006). Hayati *et al.* (2012) showed that despite a very good forest openness (90%), the efficiency coefficient of the studied road networks were not satisfactory (32% only) due to overlapping or falling outside the district area. Overlap areas are created when PASO contact together. These areas are opened twice so overlapping is unsuitable for forest openness (Pentek *et al.*, 2005).

Surface Distribution Algorithm (SDA) is another method for analyzing road network. In this algorithm the distribution of road network on forest land is assessed and the proximity of roads to the center of timber compartments is determined (Wang & Crowcroft, 1992; Rogers, 2005; Xie & Levinson, 2006). The network with lower distance to the centers is better due to skidding distance can be shorter (Anderson & Nelson, 2004). Skidding distance can be determined by SDA, but forest openness and overlaps cannot be measured by this algorithm (Musliman *et al.*, 2008).

Single-tree selection cutting which is a closes-to-nature silviculture system is planned for harvesting hardwood forests of Hyrcanian zone. The aim is to achieve an uneven-aged and mixed forest. Tree marking in this method needs to precise statistical data about tree distribution in diameter classes (Eshaghi Rad & Seyyedi, 2012). This work is difficult because of the government regulations on skid trail construction and maximum skidding distance limit (Hayati *et al.*, 2012). There are many constraints (for example steep terrain, wetlands, erodible area, rocky area, deep trench, instable and marshy soil, and etc) in forest that a harvesting planner should consider with regards to skidding operations and machine selection. Another important harvesting challenge is the density and distribution of forest road. Road with higher density and better distribution make harvesting cost lower. Skidding distance for rubber-tire skidders may reach 3 kilometers and maximum allowable longitudinal slope on skid trails is 25%. Nowadays, many studies are conducted in hardwood forests of Hyrcanian zone to determine optimal road density as well as analysis of road performance (Ghaffariyan & Sobhani, 2008; Najafi *et al.*, 2008; Naghdi & Mohammadi Limaei, 2009). Previously applied methodology in the literature review has mostly failed to assess accurate PASO and forest road network distribution because the constraints and obstacles of ground-based skidding were not located and measured and this is a research gap to be solved. The novelty of our proposed methodology is correction of Backmund algorithm by considering these areas to plan skidding operations in a better way. The main objectives of this work were to:

A) determine constraints and obstacles of ground-based skidding in PASO of study area.
B) correct Backmund algorithm to better design new routes in unopened areas
C) use corrected Backmund and SDA to compare the performance of existing forest road network in two districts of Shast Kalate Educational and Research Forests, northern Iran.

Material and methods

Study area

Shast Kalateh forest is located in Golestan province and in watershed number of 85. This forest has been divided in two districts. District I with an area of 1,713.3 ha in extended from 36° 43' 27" to 36° 45' 6" N and 54° 21' 26" to 54° 24' 57" E. District II with an area of 1,992 ha is located from 36° 42' 30" to 36° 43' 30" N and 54° 21' 6" to 54° 23' 30" E. The total length of forest roads in districts I and II were 30.3 and 28 km (21 km is located in timber compartments with area of 995 ha), respectively. Stream network, deep trench, erodible area, rocky area, nursery, steep terrain, botany garden, wetland, forest park and camp are the main skidding constraints in our study area. In our study area, single-tree selection method is done in timber compartments at 10-year intervals. Two cycles of operations has already been done in district I. District II has not been harvested yet. The trees species of a case study are *Parrotia persica*, *Carpinus betulus*, *Fagus orientalis*, *Quercus cas-*

taneifolia and *Zelkova carpinifolia*. The mean of trees density per hectare was 214.92 and the canopy cover was 75-85% (Mohammadi *et al.*, 2014).

The roads in district I were constructed in 1989, but the roads of district II has not been built yet. The bedrock is lime and sand stone with altitude ranging from 100 to 1,935 m above sea level. The forest is mixed deciduous which has been established on brown forest soil with mostly sandstone as bedrock Clay-loam-silty texture and worn stones are spread around the region. The climate is moderate and moist. The mean annual precipitation is varying from 528 mm to 817 mm which the lowest is in July and August (Fig. 1).

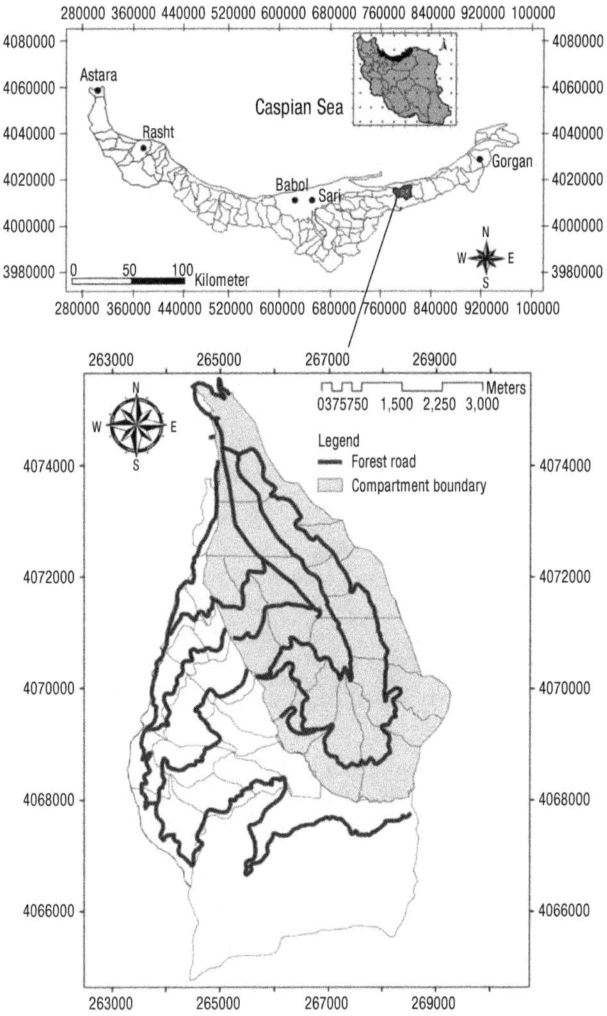

Figure 1. The geographical position of the study area.

Backmund algorithm

In this study the density of forest roads in harvestable area and the distances among roads were calculated using Equation 1 and 2, respectively.

$$R_{density} = \frac{R_{Length}}{A} \tag{1}$$

$$R_{spacing} = \frac{10000}{R_{density}} \tag{2}$$

Where $R_{density}$ is road network density (m ha^{-1}), R_{Length} is total length of forest roads (m), A is total harvestable area (ha) and $R_{spacing}$ is distance among roads (m). $S_{distance}$ is maximum skidding distance (m), which was calculated using Equation 3.

$$S_{distance} = \frac{R_{spacing}}{2} \tag{3}$$

Then PASO with a width of maximum skidding distance for rubber-tire skidder or mean distance among road was designed at both sides of the road in ArcMap software using the proximity buffer tool. In uncorrected Backmund algorithm, total area of PASO was calculated to determine openness degree of forest by forest road network (Equation 4). Overlaps among roads were detected using the overlay intersect tool. In another algorithm it can be possible to calculate unopened areas in forest and consequently the forest openness areas.

$$E = \frac{S}{A} \times 100 \tag{4}$$

Where S is PASO (ha) and E is forest openness (%). The coordinates and polygons of skidding constraints were taken in field survey using GPS. In the next stage the data was transmitted from GPS to ArcMap and these areas were removed from PASO map. The new PASO shows the openness percentage of forest after application of corrected Backmund algorithm (Backmund, 1968).

Surface Distribution Algorithm (SDA)

In SDA, the model builds a network consisting of a set of points and links. Points represent geometrical center of timber compartments and depots, and links represent the skid trails. The geometrical center of timber compartments was determined using the data management features tool and then the shortest perpendicular distance of the center of timber compartments to roads was measured using the distance shortest path tool. Equation 5 is used to analyze different alternatives of forest roads and determine the best distribution of road network in the field.

$$Y = \frac{\sum D_i}{n} \tag{5}$$

Where Y is mean distances from the center of timber compartments to road (m). D_i is shortest perpendicular distance from geometrical center of timber compartment i to road (m) and n is the number of compartments (Fig. 2).

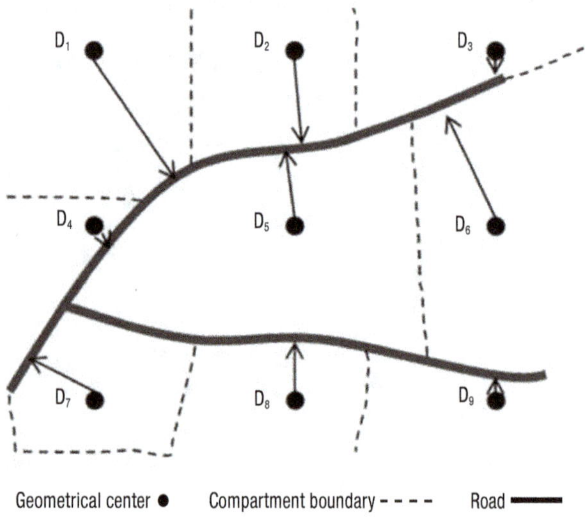

Figure 2. Schematic of surface distribution algorithm.

Results

Network analysis using corrected Backmund

Skidding constraints and obstacles fragmented PASO of two districts (Fig. 3). PASO map exhibit the extent of permissible areas which the ground-based skidding can be done for rubber-tire skidders. It was produced on the basis of uncorrected Backmund and road spacing value 565.6 m (Fig. 4). 75.5% of district I had potential areas for skidding operations. 6.94% of these areas (PASO) were unsuitable for skid trail construction due to skidding constraints and so were removed from PASO (Table 1). New PASO was prepared based on corrected Backmund algorithm (Fig. 5). In our case study forest openness with an area of 1,713.3 ha decreased to 70.3% after use of corrected Backmund (Table 2).

According to uncorrected Backmund, 75.9% of district II had potential areas for skidding operations (Fig. 6). But, 8.48% of this area were unsuitable for skid trail construction and so were removed from PASO (Table 3).

In district II with an area of 995 ha, forest openness decreased to 69.5% after the use of corrected Backmund algorithm (Fig. 7). Table 4 shows the technical parameters of forest road network according to corrected and uncorrected Backmund algorithms.

Network analysis using SDA

In SDA, the mean distance from the geometrical center of the timber compartments of district I to the road was 199.45 m (Fig. 8). The mean distance from the geometrical center of the timber compartments of

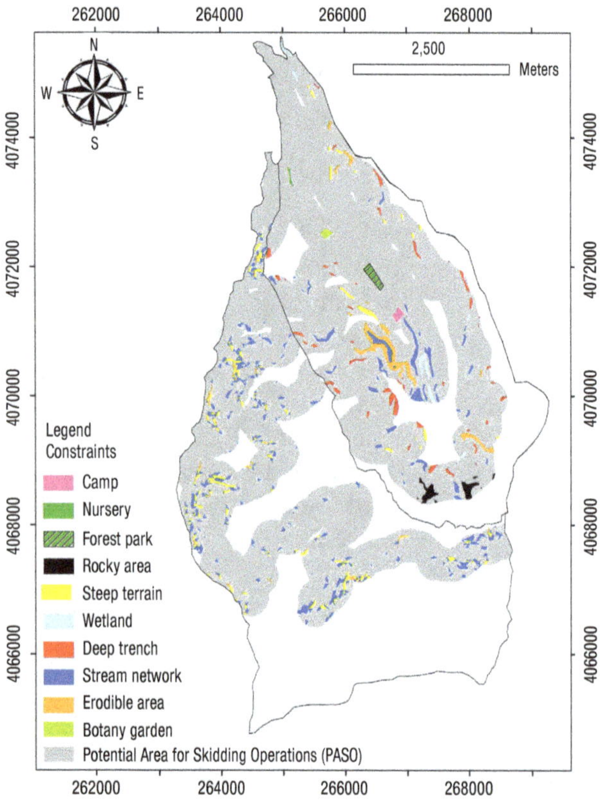

Figure 3. Distribution of the skidding constraints in both districts.

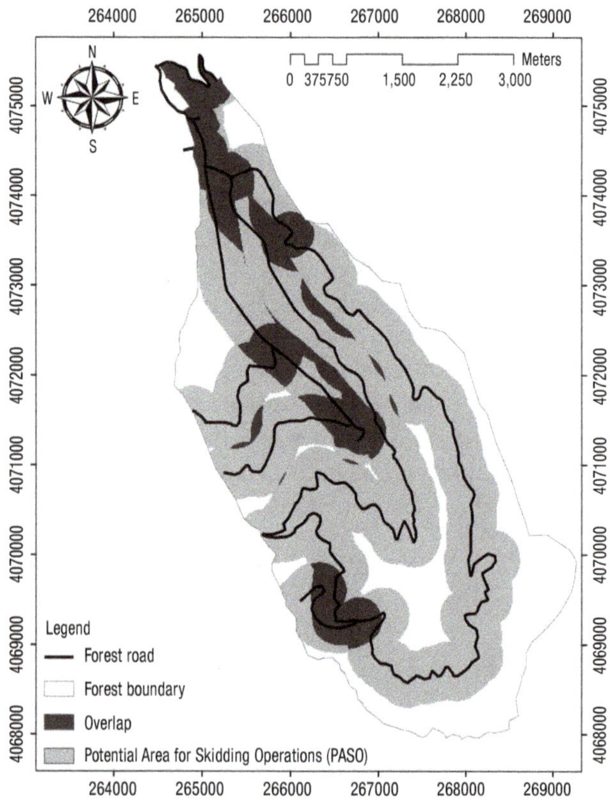

Figure 4. PASO using uncorrected Backmund algorithm in district I.

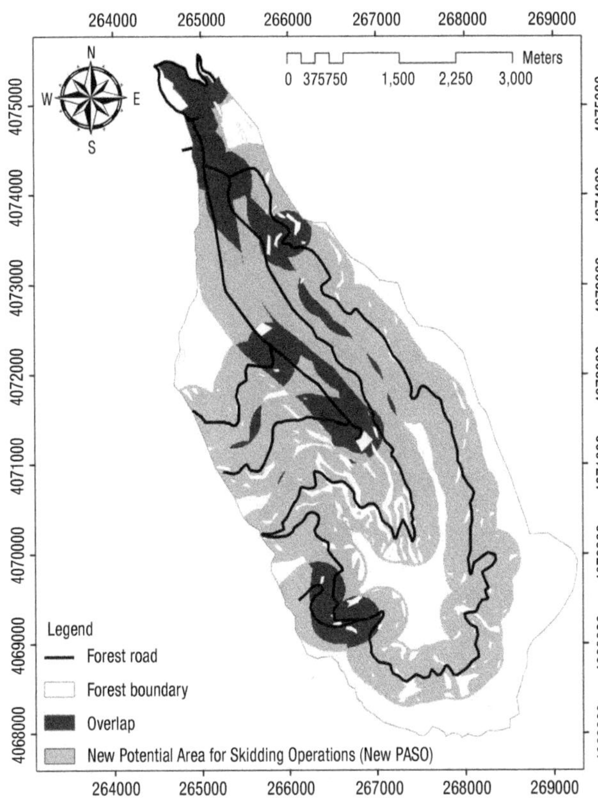

Figure 5. New PASO after removing skidding constraints from district I.

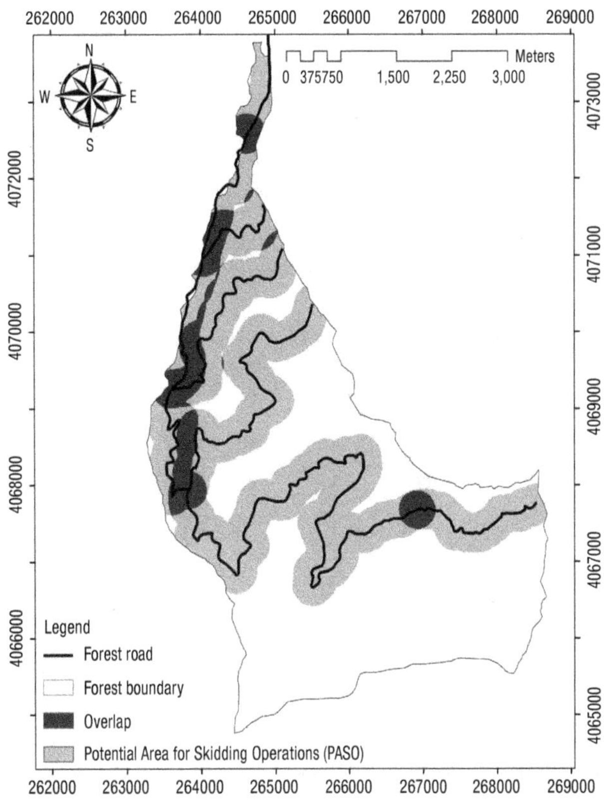

Figure 6. PASO using uncorrected Backmund algorithm in district II.

Table 1. Relative percentage of skidding constraint areas to the whole area of PASO in district I.

Constraints	Area (%)	Constraints	Area (%)	Constraints	Area (%)
Stream network	3.23	Nursery	0.0005	Forest park	0.36
Deep trench	0.98	Steep terrain	0.29	Camp	0.17
Erodible area	0.76	Botany garden	0.09	Total	6.94
Rocky area	0.33	Wetland	0.71		

Table 2. Technical parameters of forest road network in district I.

Algorithms	Road density (m ha^{-1})	Road spacing (m)	PASO & New PASO (ha)	Forest openness (%)
Uncorrected Backmund	17.68	565.6	1,294.19	75.5
Corrected Backmund	17.68	565.6	1,204.41	70.3

Table 3. Relative percentage of skidding constraint areas to the whole area of PASO in district II.

Constraints	Area (%)	Constraints	Area (%)	Constraints	Area (%)
Stream network	3.84	Wetland	0.07	Deep trench	0.41
Steep terrain	3.44	Rocky area	0.06	Erodible area	0.66
Total	8.48				

Table 4. Technical parameters of forest road network in district II.

Algorithms	Road density (m ha^{-1})	Road spacing (m)	PASO & New PASO (ha)	Forest openness (%)
Uncorrected Backmund	21.1	473.8	755.25	75.9
Corrected Backmund	21.1	473.8	691.17	69.5

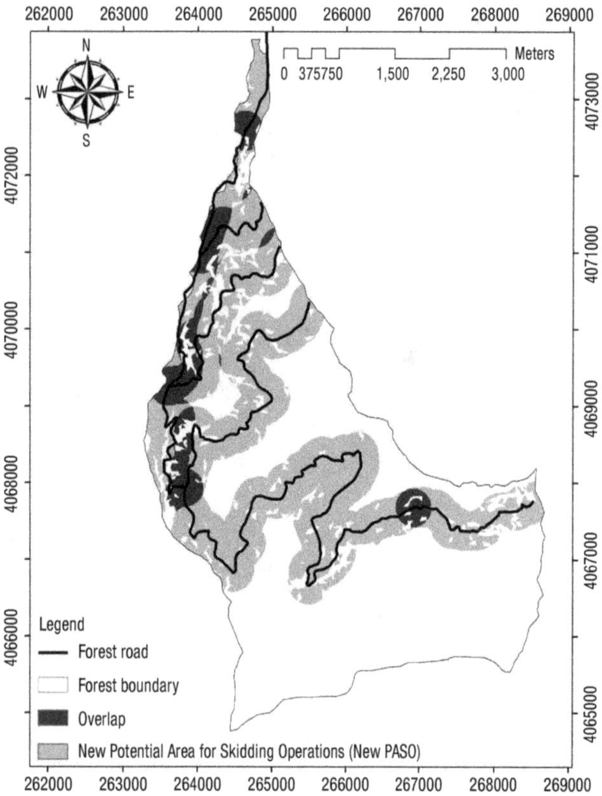

Figure 7. New PASO after removing skidding constraints area from district II.

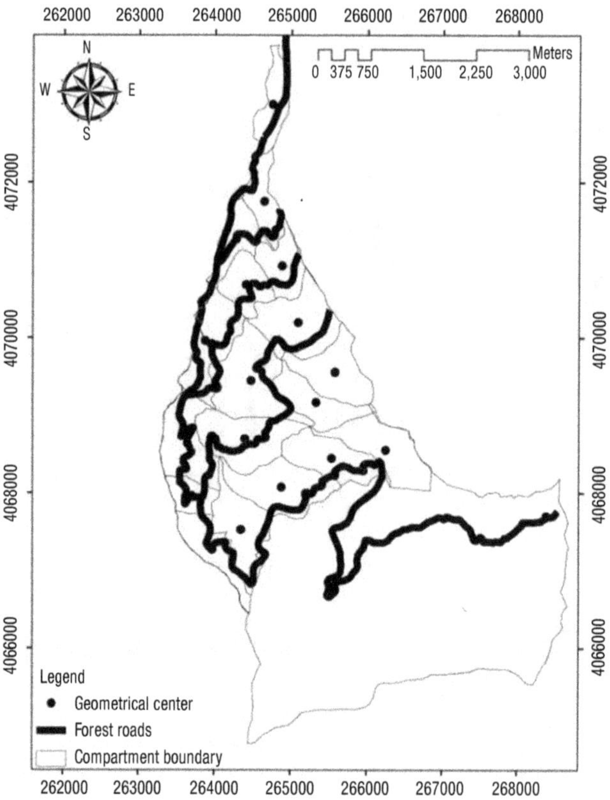

Figure 9. Geometrical center of timber compartments in district II of Shast Kalate forests.

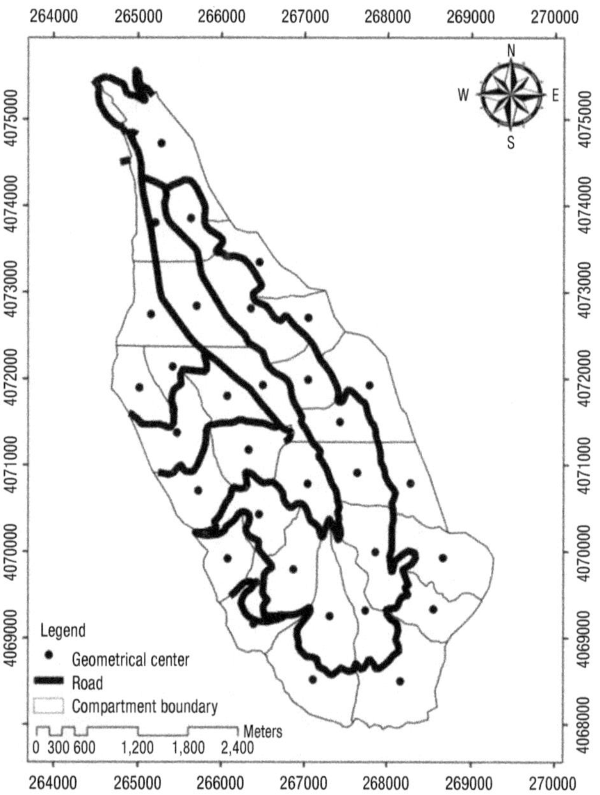

Figure 8. Geometrical center of timber compartments in district I of Shast Kalate forests.

Table 5. Mean shortest distance from geometrical center of timber compartments to road.

District	Number of timber compartment	Mean	Std. Deviation
I	33	199.45	84.103
II	19	149.31	132.421

district II to the road was 149.31 m (Fig. 9). Therefore, according to SDA the forest road network distribution in district II was better than that of district I. Means shortest distance presented differences between districts, with the maximum and the minimum path ranging between 199.45 and 149.31 m for district I and II, respectively (Table 5).

Discussion

The overall large investment for road network construction in the case study area reflects the importance of precise network analysis in the forest area, which suggests that the traditional algorithms such as uncorrected Backmund in this area has not accurately resulted in roads being located in mountainous areas. This paper has presented two metrics (Corrected

Backmund and SDA) to analysis forest road networks in a way meaningful to road network managers. It was concluded that uncorrected Backmund was not precise enough to evaluate forest road network (Fig. 4 and 6). In this algorithm it is assumed that all the forest areas around the road network are opened by skid trails and/or has potential for skidding operations up to maximum skidding distances, while it is not possible due to the existence of constraints and obstacles. Therefore, it is necessary to determine the forest openness based on the constraints of depot and skid trail construction. The position of constraints and obstacles should be located and then removed from potential area for skidding operations to achieve corrected Backmund. South of study area with high slope and sensitive geological and hydrographic situations was not an appropriate location to design forest roads. In the current study, geographical position of skidding constraints was the important information layer for analyzing road performance. Naghdi & Babapour (2009) stated that the slope, soil texture, bed rock and hydrographic layers were most important layers that determine skidding constraints and possibility of road construction. Their results are similar to those of the current study.

The existing forest road network analysis showed that the road density and spacing is acceptable but it is possible to carry on with the opening of unopened areas. Calculating average skidding distance is an important parameter for the determination of the optimum road density and spacing. Skidding distances in districts I and II were 282.8 m and 236.9 m, respectively. Therefore, the width of areas surrounding of roads in two districts I and II were equal to 565.6 m and 473.8 m, respectively. It is possible for road networks with different density and spatial distribution to achieve the same potential area for skidding operations (Rogers, 2005). Where forest access is needed, it would be necessary to determine a minimum required road network with maximum potential area for skidding operations (Sessions & Boston, 2006; Xie & Levinson, 2006).

In surface distribution algorithm the mean distances of geometrical centers to road in district I was 199.45 m, whereas in district II the mean distances was 149.31 m. Therefore forest road network distribution in district II was better than that of district I. Indeed, the skidding distances and consequently the skidding cost in district II was lower than that of district I. For achieving a high-quality and efficient management of forest, it is necessary to provide an optimal distributed network of forest road (Pentek *et al.*, 2010). Nowadays, different skidding distance models were programmed to help the road planners judge the efficiency of existing or planned road networks while considering the real skidding distance and the proposed technology of skidding (Heinimann, 1998; Tucek & Pacola, 1999). Each algorithm has its own limitations and advantages that will be further investigated in our ongoing research.

It has been recommended by many authors that depending on terrain and forest conditions the optimum density of forest road network should be between 17 and 30 m ha^{-1} for the needs of timber harvesting (Pentek *et al.*, 2005; Sessions, 2007). In agreement with the mentioned report, road density in forest district I and II were 17.68 and 21.1 m ha^{-1}, respectively. Abdi *et al.* (2009) used geographical information system and multi-criteria evaluation to design a forest road network with the lowest construction cost. This algorithm is useful to evaluate and select optimum road from different alternatives. The value of 250 m has been chosen by Pentek *et al.* (2005) for skidding distance on slope less than 55%. In addition, they emphasized that on slope more than 55% another system of forest opening should be selected instead of ground-based skidding. Skidding distances in districts I and II were 283 and 237 m, respectively. In evaluating the forest openness by road systems which has been accepted by Iranian foresters till 60% is insufficient openness, from 60 to 70% is hardly good openness, from 70 to 85% is very good openness and over 85% the overlap increases and so it cannot be suitable (Lotfalian & Parsakhoo, 2012). Values of forest openness are above 70% in district I (Table 2) and II (Table 4) in Shast Kalate forest and these values have been given the adequate road density and distribution in our study area. Numeric data of forest openness by road says the actual state of forest road infrastructure and defines the need for future opening of forests.

Conclusions

The analysis of forest road network using corrected Backmund algorithm is necessary to design new routes in unopened areas. In addition, to identify the unopened area, PASO estimating under constraints and obstacle conditions may also be provided by corrected Backmund. It was concluded that uncorrected Backmund was not precise enough to assess forest road network, while corrected Backmund could exhibit a real PASO by removing skidding constraints. In recent study we found that forest openness in two districts were in range of 69.5- 70.3% and Therefore, according to standards of foresters the studied road networks can be classified as very good openness. Beside, SDA can investigate the surface distribution of road network in area. It was detected that forest road network distribution in district II was better than that of district I. Totally, forest road network performance in district II was better than district I. Integrating these two algorithms help us to reach optimum PASO with lowest overlaps. By constructing roads and skid trails with suitable density the costs can be reduced.

Acknowledgements

The authors wish to thank the helpful comments from anonymous reviewers, whose remarks and indications significantly improved the original manuscript. Thanks also to Dr. Negar Rasouli for editing and improving the English quality of manuscript.

References

Anderson AE, Nelson J, 2004. Projecting vector-based road networks with a shortest path algorithm. Canadian J For Res 34: 1444-1457. http://dx.doi.org/10.1139/x04-030.

Aruga K, Sessions J, Akay AE, 2005. Heuristic planning techniques applied to forest road profiles. J For Res 10: 83-92. http://dx.doi.org/10.1007/s10310-004-0100-4.

Aruga K, Tasaka T, Sessions J, Miyata ES, 2006. Tabu search optimization of forest road alignments combined with shortest paths and cubic splines. Croatian J For Eng 27: 37-47.

Abdi E, Majnounian B, Darvishsefat A, Mashayekhi Z, Sessions J, 2009. A GIS-MCE based model for forest road planning. Journal of Forest Science 55(4): 171–176.

Backmund F, 1968. Indices for the degree of accessibility of forest districts via roads. Schweizerishe, Zeitschrift Fuer Forstwesen (SZF) 119: 445-452.

Contreras M, Chung W, 2007. A computer approach to finding an optimal log landing location and analyzing influencing factors for ground-based timber harvesting. Canadian J For Res 37: 276-292. http://dx.doi.org/10.1139/x06-219.

Dean JD, 1997. Finding optimal routes for networks of harvest diet access roads using GIS-based techniques. Canadian J For Res 27: 11-22. http://dx.doi.org/10.1139/x96-144.

Eastaugh CS, Molina DM, 2012. Forest road and fuelbreak siting with respect to reference fire intensities. Forest Systems 21: 153-161. http://dx.doi.org/10.5424/fs/2112211-12232.

Eshaghi Rad J, Seyyedi N, 2012. Effects of strip and single-tree selection methods on woody species diversity in Caspian forests (Iran). Int J For Soil Erosion 2: 171-174.

Ghaffariyan MR, Sobhani H, 2008. Optimum road spacing of ground-based skidding operations in Nowshahr, Iran. Caspian J Env Sci 6: 105-112.

Gumus S, Acar HH, Toksoy D, 2008. Functional forest road network planning by consideration of environmental impact assessment for wood harvesting. Environ Monit Assess 142: 109–116. http://dx.doi.org/10.1007/s10661-007-9912-y.

Hayati E, Majnounian B, Abdi E, 2012. Qualitative evaluation and optimization of forest road network to minimize total costs and environmental impacts. Italian Soc Silv For Ecol 5: 121-125. http://dx.doi.org/10.3832/ifor0610-009.

Heinimann HR, 1998. A computer model to differentiate skidder and cable-yarder based road network concepts on steep slopes. J Forest Res 3: 1-9. http://dx.doi.org/10.1007/BF02760286.

Heralt L, 2002. Using the ROADENG system to design an optimum forest road variant aimed at the minimization of negative impacts on the natural environment. J For Sci 48: 361-365.

Ismail MH, 2009. Planning of access road using satellite technology and best path modeling. Modern Applied Science 3: 83-95. http://dx.doi.org/10.5539/mas.v3n3p83.

Jaafari A, Najafi A, Rezaeian J, Sattarian A, Ghajar I, 2015. Planning road networks in landslide-prone areas: A case study from the northern forests of Iran. Land Use Policy 47: 198-208. http://dx.doi.org/10.1016/j.landusepol.2015.04.010.

Klassen A, 2006. Planning, location, survey, construction and maintenance for low-impact forest roads. Fourth Technical Procedures Manual, pp: 63-74.

Lotfalian M, Parsakhoo A, 2012. Forest roads network planning. Ayiizh Press, Tehran, Iran. 155 pp.

Mohammadi J, Shataei Sh, Namiranian M, 2014. Comparison of quantitative and qualitative characteristics of forests structure and composition in natural and managed forest stands (Case study: Shast Kalate forests of Gorgan). J Wood & For Sci Tech 21: 65-83. (In persian).

Musliman IA, Abdul Rahman A, Coors V, 2008. Implementing 3D network analysis in 3D-GIS. The International Archives of the Photogrammetry, Remote Sensing and Spatial Information Sciences. Vol. XXXVII. Part B2. Beijing, pp: 913-918.

Najafi A, Sobhani H, Saeed A, Makhdom M, Marvi Mohajer M, 2008. Planning and assessment of alternative forest road and skidding networks. Croatian J For Eng 29: 63-73.

Naghdi R, Mohammadi Limaei S, 2009. Optimal forest road density based on skidding and road construction costs in Iranian Caspian Forests. Caspian J Env Sci 7: 79-86.

Naghdi R, Babapour R, 2009. Planning and evaluating of forest roads network with respect to environmental aspects via GIS application (Case study: Shafaroud forest, northern Iran). Proceeding of Second International Conference on Environmental and Computer Science, Dubai, UAE, pp: 424-427.

Pentek T, Pičman D, Potočnik I, Dvorščak P, Nevečerel H, 2005. Analysis of an existing forest road network. Croatian J For Eng 26: 39-50.

Pentek T, Nevečerel H, Dasović T, Poršinsky M, Šušnjar I, Potočnik I, 2010. Analysis of secondary relative openness in hilly area as a basis for selection of winch rope length. Šumarski list 134: 241-248.

Puya K, Majnounian B, Feghhi J, Lotfalian M, Abdi E, 2009. The efficiency of Backmund method for evaluation of forest road networks with regard to capabilities of wheeled skidders in ground skidding method. Iranian J For 1: 35-42.

Rogers LW, 2005. Automating contour-based route projection for preliminary forest road designs using GIS. A thesis submitted in partial fulfillment of the requirements for the degree of Master of Science. University of Washington, Management and Engineering Division, College of Forest Resources, pp: 78.

Sessions J, Boston K, 2006. Optimization of road spacing for log length shovels logging on gentle terrain. Int J Forest Eng 17: 53-60.

Sessions J, 2007. Forest road operations in the Tropics. ISBN-10 3-540-46392-5 Springer-Verlag Berlin Heidelberg New York. pp: 170.

Tucek J, Pacola E, 1999. Algorithms for skidding distance modelling on a raster digital terrain model. Int J Forest Eng 10: 67-79.

Wang Z, Crowcroft J, 1992. Analysis of shortest-path routing algorithms in a dynamic network environment. ACM Computer Communication Review 22: 63-71. http://dx.doi.org/10.1145/141800.141805.

Xie F, Levinson DM, 2006. Measuring the structure of road networks. Geographical Analysis 39: 336-356. http://dx.doi.org/10.1111/j.1538-4632.2007.00707.x

2

Incorporating stand level risk management options into forest decision support systems

Kyle Eyvindson[1], Rami Saad[2] and Ljusk Ola Eriksson[2]

[1]*University of Jyvaskyla, Dept. of Biol. Environ. Sci., P.O. Box 35, 40014 University of Jyvaskyla, Finland.* [2]*Swedish University of Agricultural Sciences, Dept. of Forest Resour. Manage., Skogsmarksgränd, SE-901 83, Umeå, Sweden.*

Abstract

Aim of study: To examine methods of incorporating risk and uncertainty to stand level forest decisions.

Area of study: A case study examines a small forest holding from Jönköping, Sweden.

Material and methods: We incorporate empirically estimated uncertainty into the simulation through a Monte Carlo approach when simulating the forest stands for the next 100 years. For the iterations of the Monte Carlo approach, errors were incorporated into the input data which was simulated according to the Heureka decision support system. Both the Value at Risk and the Conditional Value at Risk of the net present value are evaluated for each simulated stand.

Main results: Visual representation of the errors can be used to highlight which decision would be most beneficial dependent on the decision maker's opinion of the forest inventory results. At a stand level, risk preferences can be rather easily incorporated into the current forest decision support software.

Research highlights: Forest management operates under uncertainty and risk. Methods are available to describe this risk in an understandable fashion for the decision maker.

Additional keywords: risk preferences; forest management; inventory error; value at risk; conditional value at risk.

Abbreviations used: BA (basal area); CVaR (conditional value at risk); DA (data assimilation); DM (decision maker); DSS (decision support system); E(NPV) (expected net present value); LP (linear programming); VaR (value at risk).

Authors′ contributions: Writing and editing manuscript; data processing: KE, RS, LOE. Method development; interpretation of data: KE, RS. Supervision: LOE.

Funding: The authors received no specific funding for this work.

Competing interests: The authors have declared that no competing interests exist.

Correspondence should be addressed to Kyle Eyvindson: kyle.j.eyvindson@jyu.fi

Introduction

Forest information serves as the basis for the forest planning process. The information allows the use of growth and yield models to predict future states given different management options. However, forest information is not free from errors, and errors in forest information have the potential to lead to suboptimal decisions. To manage these errors, one option is to update the information by conducting a new inventory (Ståhl *et al.*, 1994). Another option is to use the information you have, including the information with the errors inherent in the data. Empirical studies have been conducted which can be used to provide estimates for the quality of the forest information (see Haara & Leskinen, 2009; Mäkinen *et al.*, 2010). To ease the management of updating the information, a recent proposal has suggested the use of the data assimilation (DA) procedure (Ehlers *et al.*, 2013; a part of the data acquisition research). Very simply, the idea is to assimilate new information with the existing information, resulting in better quality information than either could provide separately. The use of this procedure has two main advantages; the first is to reduce uncertainty in itself and secondly to provide an estimate of uncertainty regarding the forest information.

When making decisions for forest management, the use of stand level forest information is essential when making decisions at either a stand or forest scale. Forest scale problems are characterized by the need to simultaneously consider the actions in several stands. This is especially the case when optimizing the even-flow of timber over time. To many forest owners the sophisticated and computationally demanding methods needed for planning at a forest scale are not always motivated. At the forest scale, some studies have evaluated the cost or benefit of improving the data quality with respect to management or data acquisition policies (Duvemo & Lämås, 2006; Kangas, 2010). When considering issues of risk at the forest scale, Eyvindson & Cheng (2016) have compared the issue of maximizing net present value while managing the risk of achieving a specific periodic flow of income. For decisions to be made at the stand scale, the quality of the information available is a major concern. Forest owners may be more interested in making decisions for individual stands. As more than 90% of all forest holdings in Europe encompass 10 or fewer hectares (Hirsch *et al.*, 2007), forest owners may not consider the need to conduct forest scale management. As a result, research has been invested in solving stand level risk and the problems of uncertainty (Pasalodos *et al.*, 2013).

Akin to multiple criteria problems in general, problems which include risk and uncertainty also require the involvement of the decision maker's (DM) preferences. As most forest owners may not be aware of the implications of a decision in terms of risks and uncertainties, the planner must be able to provide easy to understand descriptions for managing risk. While risk and uncertainty have been defined where risk has a known probability and uncertainty has an unknown probability (Knight, 1921), we will use the terms in a slightly different fashion. We will use uncertainty (or estimate of uncertainty) to refer to an attribute of the forest data, and risk is a perception of the DM. Comparing the ability of different decision options to provide a specific result can be done through a variety of risk measures. This could be done using a variety of probabilistic framework models – such as Bayesian theory, scenario analysis, sensitivity analysis and other methods (Pasalodos *et al.*, 2013). In general, measures of risk evaluate any deviation from the desired outcome as unwanted. For instance Markowitz (1952) evaluated risk as the variance of a return from specific portfolio of investments. However, this generalization may not hold true for all DMs, as often the downside effects are of much more interest than the potential positive effects of uncertainty. Risk measurement literature refers to these measures as "downside risk measures" and they evaluate the probability of the occurrence of loss (Nawrocki, 1999). Two common risk measures are the value at risk (VaR; Duffie & Pan 1997), which measures the point where a given probability of loss is exceeded and the conditional value at risk (CVaR; Rockafellar & Uryasev, 2000), which measures the average loss exceeding a given probability.

To make use of the information regarding risk and uncertainty forest planners should integrate it into forest decision support systems (DSS). These systems are designed to support the planning process in a general manner. By providing the initial conditions of the forest, forest growth and yield models predict the development of the forest while optimization models support the forest planners in their decisions. Forest DSS are ubiquitous and exist in all countries where forestry is of some importance (Borges *et al.*, 2014). However, as pointed out by Pasalodos *et al.* (2013), unless the DSS is specially designed to account for a certain risk factor, like fire, wind or insect outbreaks, there are very few systems that incorporate risk management tools, as they operate with point estimates of the initial forest condition and only estimate a single point for growth and yield projections. As a result, the plans created for the DM ignore the uncertainty in the forest information.

In their review Pasalodos *et al.* (2013) discuss several likely causes for this situation. In addition to technical and computational reasons they also direct attention to the idea that the DM may only want unambiguous answers. For a DM, assessing or expressing risk preferences can be difficult and they may feel unaccustomed to the notion of risk and uncertainty. Their conclusion is that "If we actually decide to consider some given uncertainty in our DSS, we need to select methods and tools that are suitable for the decision problem in question and that are sufficiently easy to implement and use. Extremely important is to find ways to communicate the uncertainties to the decision makers…"

When approaching the management of risk, researchers have often used a rather myopic view. Either they focus on simply minimizing the risk (Hahn *et al.*, 2014; Robinson *et al.*, 2016) or they compare the difference between the expected value solution with and without considering uncertainty (Forsell & Eriksson, 2014). By avoiding the potential opportunities for trade-offs between risk and the expected solution, the potential benefits to incorporate sources of uncertainty may seem rather small. The approach we examine here is related to earlier work by Eyvindson & Cheng (2016) where at a forest scale they examine the trade-off between maximizing the expected net present value (E(NPV)) and the CVaR associated with the desired periodic even-flow of timber. For this case, at a stand scale, we examine possible methods for portraying the trade-off between maximizing the E(NPV) and the risk of achieving a specific desired NPV. The use of these methods can be linked directly through the appropriate use of data provided by data acquisition methods (see Saad *et al.*, 2017).

The purpose of this study is to examine methods of incorporating risk and uncertainty through a forest DSS at stand level. By incorporating measures of risk to manage the forest, we expect different decisions to be made at a stand level for risk neutral and risk avoiding DMs. To maximize NPV, we expect that risk neutral DMs should conduct treatments earlier than those DMs that have a risk avoiding preference. The methods are illustrated through a case study by taking uncertainty stemming from measurement errors of the initial state. Two approaches to portray the impact of managing risk are highlighted in addition to the case of evaluating the expected value approach.

Material and methods

Generating data with the DSS

The input data for the DSS consists of a core set of information for each stand (a relatively homogenous area of forest). Depending on the forest DSS, this information can be in a variety of formats; depending on the specific application. The forest information can be represented by average values, diameter classes or tree lists depending on the growth projection model.

If we limit the description of forest DSS to those that incorporate multiple periods and long term planning we can discriminate those that answer 'what-if' questions and those which utilize optimizing systems. The former assesses the consequences of a particular set of prescriptions. Some of them also make stochastic information available through Monte Carlo simulation (see Pasalodos *et al.*, 2013). The latter DSSs find an optimal management of a forest given a set of objectives and constraints, *i.e.* the way the stands or strata should be managed is the result, not the input, of the analysis. It is this kind of forest DSS we will use as a platform for introducing risk and uncertainty information.

In the Forest DSS database of the about 60 forest DSS documented, 12 use linear programming (LP) for planning at the forest scale (http://www.forestdss.org/wiki/index.php?title=Category:DSS). The typical forest DSS using LP for long term planning has a structure depicted in Figure 1, going back to the taxonomy established by Johnson & Scheurman (1977) as Model I or Model II (see also Gunn, 2009). In the preparation of the LP model, a number of different management programs are created for each stand where a management program is a sequence of forest management activities stretching over the entire planning horizon. The creation of the wide variety of management programs is done through a decision logic tree (Siitonen, 1993). To create a variety of plans, slight modifications are made, by delaying or bringing forward some actions. The key treatment options used in the development of the management programs was to conduct commercial thinnings, the final harvest and the required silvicultural activities to ensure the re-establishment of the forest stand. The variety of management programs were used as the input data for the optimization problem. This problem is to select the management program which most accurately reflects the preferences of the decision maker. It is the capacity of the LP adapted DSS to compute a large number of management programs that will be used to give the forest owner risk information, not the LP model *per se*.

The description of uncertainty in most forest applications assumes continuous variables. To be able to use this data in the DSS structure, the underlying estimate of uncertainty needs to be approximated through scenarios. Each scenario represents a realization of possible forest attribute values. The probability for each scenario (posterior) is calculated given the estimate of uncertainty (prior). In other words each estimate of uncertainty is discretized and replaced with several point estimates with their associated probabilities. Kangas *et al.* (2014) and Eyvindson & Cheng (2016) are examples of scenario approaches found in forest planning.

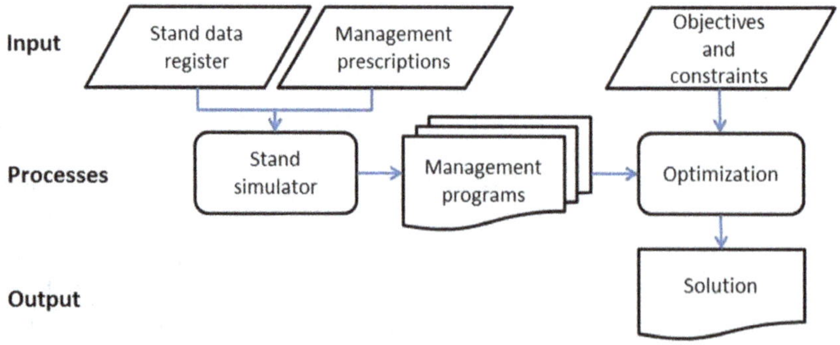

Figure 1. The typical structure of a forest decision support system (DSS) designed according to the Model I or Model II concepts.

Given that the error refers to the initial state, each scenario can be represented as a stand in the DSS. As the data is in the same format as with a single point estimate, this does not require fundamental changes to the structure of the stand management program generating DSS. To generate information about uncertainty with the forest DSS the following steps are followed: (a) the distribution of the stand variable(s) is discretized, where each instance represents a scenario; (b) each scenario is input as a stand (here termed pseudo stand); and (c) a range of management programs over the planning horizon is computed for each pseudo stand.

The key parameter to properly represent the uncertainty is the choice of how many scenarios will be used in representing the true distribution of the uncertainty. In stochastic programming, this has been a key issue of interest, as the tractability of the problems requires as small number of scenarios as possible, while still providing a suitable solution quality (King & Wallace, 2012). For the VaR and the CVaR the number of scenarios must be high enough to allow for the calculations at the specific probability desired.

Expected net present value - E(NPV)

Probably the most proliferous measure involving risk and uncertainty is the expected value. In this case, the maximum expected NPV, defined as

$$\overline{\overline{\mathrm{E}(\mathrm{NPV}_J)}} = \max_{i\epsilon J}[\mathrm{E}(\mathrm{NPV}_i)] \quad (1)$$

where J is the set of management programs and $\mathrm{E}(\mathrm{NPV}_i)$ is the expected NPV of management program i, which in turn is defined as

$$\mathrm{E}(\mathrm{NPV}_i) = \sum_{n\epsilon N} p_n \cdot \mathrm{NPV}_{in} \quad (2)$$

where N is the set of scenarios, p_n the probability of scenario n, and NPV_{in} the NPV of program i under scenario n. The maximum expected NPV can of course be complemented by information of variance as an assessment of its uncertainty.

CVAR / VAR

With each scenario having the same probability, both VaR and CVaR are relatively easy to calculate. For instance, at a 5 % level VaR is equal to the 5th lowest NPV of the set of 100 scenarios, while CVaR is the average of the results lower than the VaR at the specified percentage. In this case, the lowest gains at a specific probability. Put otherwise, there is 1-α probability that the gain will be at least at the level of VaR or that the expected gain will be at least at the level of CVaR.

For this concept, the VaR can be formulated as finding a specific quantile for each management program. Assume that the NPV values for management program j are sorted in ascending order. The VaR for program j at quantile α is then found by

$$VaR(\alpha)_j = \mathrm{NPV}_{jn}\ where\ n = inf_{k\in N}\left\{\sum_{i=1}^{k} p_i \geq \alpha\right\} \quad (3)$$

i.e. $VaR(\alpha)_j$ is the NPV value corresponding to the nth scenario where scenarios 1 to n will sum up to probability α. CVaR is then the average of those scenarios, *i.e.*

$$CVaR(\alpha)_j = \sum_{n\epsilon N} p_n \cdot \mathrm{NPV}_{jn} \quad (4)$$

In the case each scenario is given the same probability n will simply be α*N.

Based on these definitions the optimal management program j* that best balances the risk neutral *E(NPV)* and the risk component (expressed as either VaR or CVaR) is found as

$$j^* = argmax_{j\epsilon J}\left[\lambda \cdot \overline{\overline{\mathrm{E}(\mathrm{NPV_J})}} + (1-\lambda)\cdot R(\alpha)_j\right] \quad (5)$$

where $\lambda \in [0,1]$ represents the tradeoff between maximizing the E(NPV) and ensuring the highest value of the particular risk measure $R(j,\alpha)$ where R_j is expressed either as VaR or CVaR at a specified confidence limit α. At one extreme λ=1 results in the management program with maximum E(NPV); at the other extreme λ=0 gives the management program associated with the maximum NPV or maximum expected NPV at confidence level α according to VaR or CVaR, respectively.

The choice of management program thus depends on two subjective parameters, both related to risk awareness. One possibility to ease the task of the DM would be to calculate the optimal programs for ranges of λ at a given α.

Program choice based on DM's personal view on the true basal area (BA)

Rather than relying on numerical analysis, another approach could be used to inform the forest owner would be to see what programs are efficient from the perspective of the uncertain variable or variables. For each scenario the best management program, in this case the program yielding the maximum NPV, would be presented. Put otherwise, the method presents a Pareto front of management programs linked on the property of the uncertain parameter. One benefit would be that it gives the forest owner a visual illustration of the relation between erroneous variable and optimal management program.

Results. Case study

To highlight how estimates of uncertainty can be incorporated into a forest DSS, we provide a small case study with eight stands from Jönköping, Sweden. These stands were inventoried in 2011, using a plot sampling inventory method. The original stands' variables consist of mean values (corresponding to stand register mean values). The selection of the stands was done subjectively, with an aim of having different stands to cover the variety in stand age, different species and site index. The characteristics of the stands can be found in Table 1.

In this study, errors were simulated for basal area value of each stand using the DSS tool Heureka (Wikström *et al.*, 2011). Heureka uses a large set of models to predict the potential future development of the range of forest conditions found in the Swedish forest. The models are based on the single tree approach, while the input data is at a stratum level, tree lists are generated based on estimations of the distribution of trees. Costs related to silvicultural activities and the stumpage values of the trees were based on the recent data. Heureka/Planwise is a forest DSS which generates a number of management programs as input to a LP or mixed integer programming problem. The simulated errors were replicated with 100 scenarios of a normal distribution with mean corresponding to the point estimate and variance 30 % of the corresponding stand's basal area value. This error is on the higher end of the spectrum, especially considering modern inventory techniques and DA. Only a single source of error has been used in this case for illustration purposes; however multiple sources of errors can be considered using a rather similar approach. For each stand, the 100 different pseudo stands were imported to the Heureka DSS and a maximum of 100 management programs were created for each scenario. Only those management programs which were created for all scenarios were considered. This means that for each stand, all scenarios can be managed using the same range of management programs. This was done to properly calculate the expected value, CVaR and VaR of a stand, for these calculations, a 3% discount rate was used. For stands 1–8, the number of management programs was respectively 59, 45, 84, 22, 71, 60, 74, and 87.

The management program with the maximum NPV based on the point estimate (using only scenario 50 – representing the initial stand level data) and the maximum expected NPV program are presented in Table 2. The harvest decisions are different in only one stand (no. 6) out of the eight study stands. However, the differences between the harvesting decisions are not substantial. The suggested management program obtained from the point estimate is to conduct the same actions one period earlier than the management program suggested by the maximum expected NPV program.

Table 1. The key characteristics of the eight stands used in the study

Stand	Basal area (m^2)	Site index (m)	Age	Dominant species
1	30	27	37	Pine
2	20	24	53	Pine
3	16	28	73	Pine
4	27	24	36	Spruce
5	18	31	50	Spruce
6	27	24	53	Spruce
7	43	35	58	Spruce
8	24	22	87	Spruce

Both the CVaR and VaR methods suggest the use of rather similar management programs depending on the risk preferences of the DM (Tables 3 and 4). Nearly all of the efficient management plans suggested are the same, with the CVaR method suggesting two additional plans for stands 2 and 5. Both measures show a tendency to favor longer rotations with increased risk aversion. This shift to longer rotations allows for the stand to grow, reducing the possibility that a stand will be harvested with a relatively low BA, while increasing the possibility that a stand will be harvested while heavily stocked.

Figure 2 presents the management programs yielding maximum NPV for different scenarios, *i.e.* different BA values. For each stand, the 100 BA scenarios were divided into broader intervals where in each interval the dominant management plan is presented. The BA

Table 2. The optimal management plan (F= final felling; T= thinning; number = 5-year period) when using a point estimate and when using a distribution of alternatives. This highlights that similar management programs can be expected when only net present value (NPV) is of importance.

Stand	Point estimate program to maximize NPV	Scenario approach to maximize expected NPV program
1	T3-T5-F8	T3-T5-F8
2	T2-F6	T2-F6
3	F3	F3
4	T2-T4-F7	T2-T4-F7
5	F4	F4
6	T2-F5	T1-F4
7	F1	F1
8	F2	F2

Table 3. The trade-offs between the conditional value at risk (CVaR) with α=0.05 of the net present value (NPV) and the expected NPV for stands 1-8 (F= final felling; T= thinning; number = 5-year period). Only a selection of management plans is Pareto efficient.

Stand	Lambda	Management plan	CVaR (0.05)	E(NPV)
1	[0,0.57)	T3-T5-F9	24887	47781
	[0.57,1]	T3-T5-F8	24726	47905
2	[0,0.08)	T1-F7	13244	32714
	[0.08,0.61)	T1-F6	13187	33405
	[0.61,1]	T1-F5	12870	33608
3	[0,1]	F3	16036	35749
4	[0,0.32)	T3-F6	35079	67471
	[0.32,1]	T2-T4-F7	33958	69904
5	[0,0.18)	F3	32853	66380
	[0.18]	F4	32714	67033
6	[0,0.37)	T2-F6	30509	71383
	[0.37,0.71)	T2-F5	30166	71971
	[0.71,1]	T1-F4	29046	72433
7	[0,0.28)	F2	87314	175797
	[0.28,1]	F1	86054	179094
8	[0,0.07)	F5	16377	38325
	[0.07,0.34)	F4	16298	39462
	[0.34,0.84)	F3	15934	40195
	[0.84,1]	F2	15170	40346

intervals were then selected subjectively based on the list of best programs over the scenarios. Additionally, from Figure 2 it can be noted that for almost all stands the same management programs are suggested by the VaR/CVaR measure. The results from the expected NPV method and the results from the CVaR and VaR can be seen to provide similar management suggestions. The tendency that increased risk aversion implies longer rotations observed for VaR/CVaR is here associated with the expectation of smaller BAs.

Discussion

The results of this study highlight the potential of incorporating estimates of uncertainty into the forest planning process. To be of practical use, tools for managing uncertainty should be integrated into the application of forest DSSs. This allows the DMs the opportunity to consider their risk preferences, such as risk-neutral or risk-averse, and adjust forest management decisions to reflect these attitudes. For instance, if we examine the decisions taken for the second stand, the E(NPV) and the point estimate decision suggests the conduct of a thinning in period two and a final felling in period six. If the DM is risk averse, the option to conduct the final felling in period 7 or 8 may be preferable. Even though the E(NPV) is lower, during that delay the BA will increase and the probability of harvesting the forest with a low NPV is reduced.

The use of most current DSS assumes the analysis is deterministic and the models are run with a deterministic framework. In the example used in

Table 4. The trade-offs between the value at risk (VaR) with α=0.05 of the net present value (NPV) and the expected NPV for stands 1-8 (F= final felling; T= thinning; number = 5-year period). Only a selection of management plans is Pareto efficient.

Stand	Lambda	Management plan	VaR (0.05)	E(NPV)
1	[0,0.13)	T3-T5-F9	28980	47781
	[0.13,1]	T3-T5-F8	28961	47905
2	[0,0.59)	T1-F6	16323	33405
	[0.59,1]	T1-F5	16041	33608
3	[0,1]	F3	19898	35749
4	[0,0.23)	T3-F6	41084	67471
	[0.23,1]	T2-T4-F7	40377	69904
5	[0,1]	F4	38758	67033
6	[0,0.42)	T2-F6	38652	71383
	[0.42,0.71)	T2-F5	38236	71971
	[0.71,1]	T1-F4	37145	72433
7	[0,0.31)	F2	103241	175797
	[0.31,1]	F1	101817	179094
8	[0,0.1)	F5	19727	38325
	[0.1,0.13)	F4	19606	39462
	[0.13,0.86)	F3	19503	40195
	[0.86,1]	F2	18587	40346

this paper, the information about uncertainty added to the DSS through scenarios. Thus, the increase in computational tasks is related to the number of scenarios used to approximate the estimated uncertainty. For this case, only inventory errors have been included. However if the error can be represented in a scenario approach, that error can also be included. For instance, uncertainty regarding the fertility of the specific stand (*i.e.* the site class) could be included through a scenario approach. In earlier research, Eyvindson & Cheng (2016) included inventory, growth model errors and price uncertainty within the simulation process. While all of the forest simulations were conducted in the DSS, the analysis was conducted outside of the DSS. As indicated by Pasalodos *et al.* (2013), this is commonly how risk analysis is conducted with the limited capacity of today's DSS.

To make forest planning under risk and uncertainty more prevalent amongst practitioners analytical tools need to be integrated into the same DSS as other planning tasks. To add the functionality of conducting this type of analysis within the DSS, an applicable software package should be developed. This package should allow for running the same management program for a set of scenarios which incorporates a variety of uncertainty estimates, and should integrate this information for ease of analysis. A scenario approach, as demonstrated here, could be easily implementable in existing DSS. From a programming point of view, the method based on the DM's personal view of the BA is probably the easiest since each scenario, or pseudo stand, can be treated separately, just registering the maximum value program. Probably, the most demanding part is in arranging the visual illustration, where a mechanism for aggregating the programs is needed. The E(NPV) and the CVaR and VaR methods require that a register is held of the simulations over all scenarios for computing the expected value of a sequence of forest activities. Still, the only operation required is to sort the management

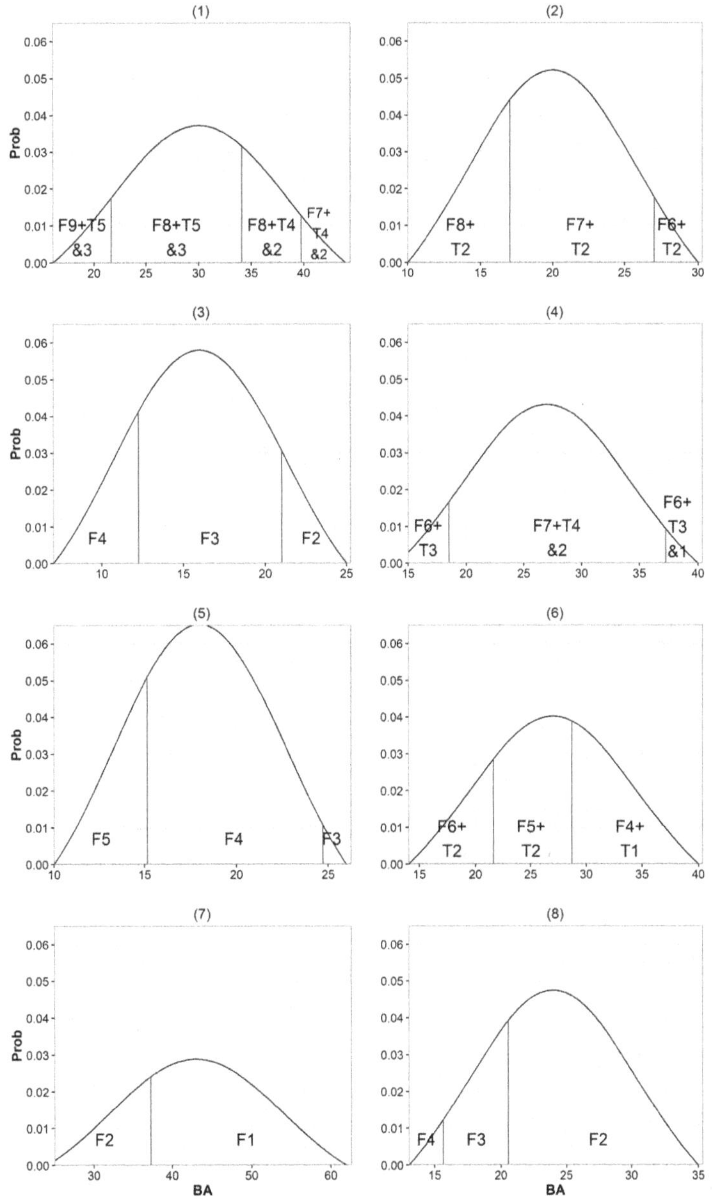

Figure 2. Optimal management plans for different basal areas (BAs) for stand 1-8. F= final felling; T= thinning; number = 5-year period. Risk averse decision makers would select management options which fall below the expected BA, while risk seeking decision makers would select management options which fall above the expected BA.

program according to the criterion of interest (such as NPV).

Risk attitudes and measures can be difficult for forest owners to understand, so it is important that forest planners understand the meaning behind these risk measures. The VaR and CVaR methods provide information regarding the unwanted tail of the distribution. Managing risk is essentially a tradeoff analysis, in this case, between the E(NPV) and the risk of what may happen in the worst cases. Here the trade-off is modeled through the λ parameter, on the one extreme (λ=1) the objective is only focused on the E(NPV), while on the other the objective is focused on the risk measure. All other values of λ represent trade-offs between the E(NPV) and the risk measure. The VaR can be interpreted as at which percentage (α) losses will exceed, while the CVaR is the average of the losses exceeding VaR. For the use of these measures in a multi-criteria context, simplicity in understanding what the measure is may be of utmost importance, and the VaR is simpler to understand than the CVaR (Durbach & Stewart, 2012). However, the VaR is not a coherent risk measure, whereas the CVaR is (Artzner *et al.*, 1999). If risk coherence is of importance to the forest owner, then the use of CVaR could be preferred.

Undeniably, VaR and CVaR can be rather abstract concepts for the typical forest owner. In this respect using the method based on the DM's personal view of the possible errors would seem to offer an advantage. By linking the risk to the forest inventory variable may be easier for forest owners to understand. Most forest oriented DM's can understand measurements of forest properties, however they may find it difficult to understand CVaR or VaR and how it links to forest value. The results displayed in Figure 2 highlight a rather simple way DMs can incorporate elements of risk in the decision making process. This representation allows the DM to select alternative management plans based on their belief on what the 'true' BA for each stand may be.

Both the management proposals selected using the personal view of the DM and those of VaR and CVaR agree with respect to risk (both approaches suggest prolonged rotations with increased risk aversion). The graphics created based on varying the potential personal view of the DM should promote an understanding that there are different management options for managing the forest and they depend on the DMs risk preferences. Thus, the comparative advantage could relate to its pedagogics rather than to its analytical strength. Referring to the latter, it is easy to highlight the shortcomings. One problem is the need of the forest owner to have some idea of the likely state of the stand that is different from the standard density function (if the forest owner is satisfied with the density function risk level can be set by VaR and CVaR). Not all forest owners will be able to consider the inventory state of their forest. Another problem relates to make the illustration convincing when there is more than one source of uncertainty. Evaluating multiple sources of uncertainty would require presenting the figure in multi-dimensional space, or may require some form of interactive process to allow for an evaluation of the risks. A related problem is when moving from stand to forest level planning, as interactions between stands become important. For some of these cases the *E(NPV)*, CVaR or VaR would be more appropriate

One assumption made with the VaR and CVaR methods is that the DM is risk averse. This assumption has not been made with the *E(NPV)* or when providing options based on the personal view of the DM method. Due to this shift in assumption, the *E(NPV)* method suggests management programs a risk seeking DM may select. If we compare the results from stand 1, the same management programs are suggested, if the DM believes that the BA will be lower than the inventoried value. To account for risk seeking preferences, we can adjust the VaR and CVaR method assumption by simply evaluating the opposite side of the tail (this adjustment requires a few minor adjustments to the calculations).

For this case, only economic considerations were evaluated. While this may be a very important aspect for a large proportion of forest owners, the application of risk management can also be applied to ecological considerations As there are a variety of ecological indicators that can be modeled using basic stand level forest information (Mönkkönen *et al.*, 2014; Peura *et al.*, 2016), the risk associated with these values could also be considered. As these indicator values may not be familiar to the DM, special guidance may need to be given towards both understanding the ecological importance of the indicator, and the meaning behind the attached risk measure.

The aim here is to present simple ways of augmenting a common type of DSS with risk management tools, so that such analysis is made more accessible. We have illustrated this with the uncertainty associated with the initial state of the individual stand. There are some rather obvious ways to continue this research. One is to put the methods in actual planning situations. This would give information on how the methods were understood and used. The research could rely on methods in computer science like interactive design (Cooper *et al.*, 2007). Another line of investigation would be to assess for which cases the issues of risk and uncertainty are of utmost importance. Some studies indicate that the point estimate strategies under some conditions fare well compared with risk adapted decisions [climate change (Eriksson *et al.*, 2012), wind throws (Forsell & Eriksson, 2014), and fire (Ferreira *et al.*, 2014)]. This should involve how different risks mix with risk preferences. For instance, prolonging rotation to avoid inventory risk may increase risk exposure to wind throw risk. Yet another area is forest level problems which add another degree of complexity and, subsequently, other demands on the DSS as well as the forest owners' conceptual prowess.

References

Artzner P, Delbaen F, Eber JM, Heath D, 1999. Coherent measures of risk. Math Financ 9 (3): 203-228. https://doi.org/10.1111/1467-9965.00068

Borges JG, Nordstrom E-M, Garcia-Gonzalo J, Hujala T, Trasobares A, 2014. Computer-based tools for supporting forest management. The experience and the expertise world-wide. Report of Cost Action FP 0804 Forest Management Decision Support Systems (FORSYS). Dept. of Forest Resource Management, Swedish Univ. of Agricultural Sciences. Umeå. Sweden.

Cooper A, Reimann R, Cronin D, 2007. About Face 3: The essentials of interaction design. Wiley, Indianapolis, IN, USA. 610 pp.

Duffie D, Pan J, 1997. An overview of value at risk. J Deriv 4 (3): 7-49. https://doi.org/10.3905/jod.1997.407971

Durbach IN, Stewart TJ, 2012. Modeling uncertainty in multi-criteria decision analysis. Eur J Oper Res 223 (1): 1-14. https://doi.org/10.1016/j.ejor.2012.04.038

Duvemo K, Lämås T, 2006. The influence of forest data quality on planning processes in forestry. Scand J Forest Res 21: 327-339. https://doi.org/10.1080/02827580600761645

Ehlers S, Grafström A, Nyström K, Olsson H, Ståhl G, 2013. Data assimilation in stand-level forest inventories. Can J For Res 43(12): 1104-1113. https://doi.org/10.1139/cjfr-2013-0250

Eriksson LO, Backéus S, Garcia F, 2012. Implications of growth uncertainties associated with climate change for stand management. Eur J For Res 131 (4): 1199-1209. https://doi.org/10.1007/s10342-011-0591-4

Eyvindson K, Cheng Z, 2016. Implementing the conditional value at risk approach for even-flow forest management planning. Can J For Res 46 (5): 637-644. https://doi.org/10.1139/cjfr-2015-0270

Ferreira L, Constantino M, Borges JG, 2014. A stochastic approach to optimize maritime pine (Pinus pinaster Ait.) stand management scheduling under fire risk. An application in Portugal. Ann Oper Res 219 (1): 359-377. https://doi.org/10.1007/s10479-011-0845-z

Forsell N, Eriksson LO, 2014. Influence of temporal aggregation on strategic forest management under risk of wind damage. Ann Oper Res 219 (1): 397-414. https://doi.org/10.1007/s10479-011-0966-4

Gunn E, 2009. Some perspectives on strategic forest management models and the forest products supply chain. INFOR 47 (3): 261-272. https://doi.org/10.3138/infor.47.3.261

Haara A, Leskinen P, 2009. The assessment of the uncertainty of updated stand-level inventory data. Silva Fenn 43 (1): 87-112. https://doi.org/10.14214/sf.219

Hahn WA, Härtl F, Irland LC, Kohler C, Moshammer R, Knoke T, 2014. Financially optimized management planning under risk aversion results in even-flow sustained timber yield. Forest Pol Econ 42: 30-41. https://doi.org/10.1016/j.forpol.2014.02.002

Hirsch F, Korotkov A, Wilnhammer M, 2007. Private forest ownership in Europe. Unasylva 228: 23-25.

Johnson KN, Scheurman HL, 1977. Techniques for prescribing optimal timber harvest and investment under different objectives. Discussion and synthesis. Monograph 18, a supplement to Forest Science, vol. 23.

Kangas A, 2010. Value of forest information. Eur J Forest Res129: 863-874. https://doi.org/10.1007/s10342-009-0281-7

Kangas A, Hartikainen M, Miettinen K, 2014. Simultaneous optimization of harvest schedule and measurement strategy. Scand J Forest Res 29: 224-233. https://doi.org/10.1080/02827581.2013.823237

King AJ, Wallace SW, 2012. Modeling with stochastic programming. Springer, NY. https://doi.org/10.1007/978-0-387-87817-1

Knight FH, 1921. Risk, uncertainty and profit. Hart, Schaffner and Marx, NY.

Markowitz H, 1952. Portfolio selection. J Finance 7 (1): 77-91.

Mäkinen A, Kangas A, Mehtätalo L, 2010. Correlations, distributions and trends of inventory errors and their effects on forest planning. Can J For Res 40 (7): 1386-1396. https://doi.org/10.1139/X10-057

Mönkkönen M, Juutinen A, Mazziotta A, Miettinen K, Podkopaev D, Reunanen P, Salminen H, Tikkanen OP, 2014. Spatially dynamic forest management to sustain biodiversity and economic returns. J Environ Manag 134: 80-89. https://doi.org/10.1016/j.jenvman.2013.12.021

Nawrocki DN, 1999. A brief history of downside risk measures. J Invest 8 (3): 9-25. https://doi.org/10.3905/joi.1999.319365

Pasalodos-Tato M, Mäkinen A, Garcia-Gonzalo J, Borges JG, Lämås T, Eriksson LO, 2013. Review. Assessing uncertainty and risk in forest planning and decision support systems: review of classical methods and introduction of new approaches. Forest Syst 22 (2): 282-303. https://doi.org/10.5424/fs/2013222-03063

Peura M, Triviño M, Mazziotta A, Podkopaev D, Juutinen A, Mönkkönen M, 2016. Managing boreal forests for the simultaneous production of collectable goods and timber revenues. Silva Fenn 50 (5): 1-17 . https://doi.org/10.14214/sf.1672

Robinson AP, McLarin M, Moss I, 2016. A simple way to incorporate uncertainty and risk into forest harvest scheduling. For Ecol Manage 359: 11-18.

Rockafellar RT, Uryasev S, 2000. Optimization of conditional value-at-risk. J Risk 2 (3): 21-42. https://doi.org/10.21314/JOR.2000.038

Saad R, Eyvindson K, Gong P, Lämås T, Ståhl G, 2017. Potential of using data assimilation to support forest planning. Can J For Res 47 (5): 690-695. https://doi.org/10.1139/cjfr-2016-0439

Siitonen M, 1993. Experiences in the use of forest management planning models. Silva Fenn 27 (2):167-178. https://doi.org/10.14214/sf.a15670

Ståhl G, Carlsson D, Bondesson L, 1994. A method to determine optimal stand data acquisition policies. For Sci 40: 630-649.

Wikström P, Edenius L, Elfving B, Eriksson LO, Lämås T, Sonesson J, Öhman K, Wallerman J, Waller C, Klintebäck F, 2011. The Heureka forestry decision support system: An overview. Math Comp For Nat Res Sci 3 (2): 87-95.

3

Mapping invasive alien *Acacia dealbata* Link using ASTER multispectral imagery

Filipe Martins[1], Cristina Alegria[2]* and Artur Gil[3]

[1] *Simbiente Açores – Engenharia e Gestão Ambiental, Lda., Azores Parque, 9500-794, Ponta Delgada, Portugal.* [2] *CERNAS – Centro de Estudos de Recursos Naturais, Ambiente e Sociedade, Instituto Politécnico de Castelo Branco, Escola Superior Agrária, Department of Natural Resources and Sustainable Development, Quinta da Senhora de Mércules, Apartado 119, 6001-909 Castelo Branco, Portugal.* [3] *CE3C – Centre for Ecology, Evolution and Environmental Changes, Azorean Biodiversity Group, Faculty of Sciences and Technology, Department of Biology, University of the Azores, 9501-801 Ponta Delgada, Portugal*

Abstract

Aim of the study: *Acacia dealbata* is an alien invasive species that is widely spread in Portugal. The main goal of this study was to produce an accurate and detailed map for this invasive species using ASTER multispectral imagery.

Area of study: The central-eastern zone of Portugal was used as study area. This whole area is represented in an ASTER scene covering about 321.1 x 10^3 ha.

Material and methods: ASTER imagery of two dates (flowering season and dry season) were classified by applying three supervised classifiers (Maximum Likelihood, Support Vector Machine and Artificial Neural Networks) to five different land cover classifications (from most generic to most detailed land cover categories). The spectral separability of the land cover categories was analyzed and the accuracy of the 30 produced maps compared.

Main results: The highest classification accuracy for acacia mapping was obtained using the flowering season imagery, the Maximum Likelihood classifier and the most detailed land cover classification (overall accuracy of 86%; Kappa statistics of 85%; acacia class Kappa statistics of 100%). As a result, the area occupied by acacia was estimated to be approximated 24,770 ha (*i.e.* 8% of the study area).

Research highlights: The methodology explored proved to be a cost-effective solution for acacia mapping in central-eastern of Portugal. The obtained map enables a more accurate and detailed identification of this species' invaded areas due to its spatial resolution (minimum mapping unit of 0.02 ha) providing a substantial improvement comparably to the existent national land cover maps to support monitoring and control activities.

Keywords: remote sensing; invasive alien species; land cover mapping; vegetation mapping.

Funding: The participation of co-author Artur Gil in this study was supported by the Post-Doctoral Research Project # SFRH/BPD/100017/2014 from the "Fundação para a Ciência e Tecnologia" (FCT), funded by the National Budget of the Ministry of Education and Science of Portugal and by the European Social Fund. The participation of co-author Cristina Alegria was supported by CERNAS-IPCB [UID/AMB/00681/2013 funding by FCT].

Competing interests: The authors have declared that no competing interests exist.

Correspondence should be addressed to Cristina Alegria: crisalegria@ipcb.pt

Introduction

Fire and invasive species are becoming two of the most important global problems in natural and anthropogenic ecosystems (Pimentel *et al.*, 2005; Keeley, 2006; Arán *et al.*, 2013). Invasive species constitute a major environmental problem, as they have profound consequences on biodiversity conservation and on ecosystem processes and functioning (Parker *et al.*, 1999; González-Muñoz *et al.*, 2012). In the past few centuries, thousands of woody plant species have been moved out of their natural ranges around the world. As a result, in recent decades many species of trees and shrubs have become naturalized or invasive. Many have spread from planting sites and some are now among the most widespread and damaging of invasive organisms (Richardson & Rejmánek, 2011).

The naturalization/invasion process may be explained by the following key terms: 'introduction' means that the plant (or its propagule) has been transported by humans across a major geographical barrier; 'naturalization' starts when abiotic and biotic barriers to survival are surmounted and when various barriers to regular reproduction are overcome; 'invasion' further requires that introduced plants produce reproductive offspring in areas distant from sites of introduction. Taxa that can cope with the abiotic environment and biota in the general area may invade disturbed, semi-natural communities (Richardson *et al.*, 2000). Asner *et al.* (2008) also consider that some of the life strategies that correlate a plant invasive success are: 1) an ability to grow through the native canopy, or in gaps, and eventually replace it (*e.g.* Yamashita *et al.*, 2000); 2) alteration of fundamental ecosystem processes such as nitrogen (N) cycling (*e.g.* Ehrenfeld, 2003; May & Attiwill, 2003; Hughes & Denslow, 2005); and 3) an ability to alter disturbance regimes such as fire frequency (*e.g.* Hughes *et al.*, 1991; D'Antonio & Vitousek, 1992).

Characterization and distribution of *Acacia dealbata* Link

Australian acacias are a group of leguminous woody plants that include some of the most important plant invaders on a global scale (Richardson & Rejmánek, 2011; Souza-Alonso *et al.*, 2013). Australian acacias have a wide range of impacts on ecosystems that increase with time and disturbance, transform ecosystems and alter and reduce ecosystem service delivery. A shared trait is the accumulation of massive seed banks, which enables them to become dominant after disturbances (Le Maitre *et al.*, 2011). As a result, the invasion of acacias species poses a threat to natural habitats by competition and replacement of native species, decreasing the native biodiversity and homogenizing the community (Lorenzo *et al.*, 2010). In general, invasion takes over sites that have been disturbed by fire, harvesting or other types of anthropogenic disturbance as a result of this species high colonization capacity (Fuentes-Ramírez *et al.*, 2011). Therefore, control and restoration operations should be promoted, particularly active restoration. Despite requiring substantial short to medium term investments these operations can reduce losses of biodiversity and ecosystem services as well as the costs to society in the long term (Le Maitre *et al.*, 2011).

The species *Acacia dealbata* Link (silver wattle) was introduced in Europe during the 19th century and is currently causing huge ecological concerns in Southern Europe (Souza-Alonso *et al.*, 2013; Vazquez-de-la-Cueva, 2014). In Portugal, the species was introduced in 1850 and used for ornamental purpose, for its products (flowers, tannin and timber either for construction or for firewood) and for soil fixation. Despite its invasive behaviour has been reported during the 19th century, restrictive legislation to acacia plantation occurred only in 1937. Nevertheless, its status as an invasive species was only legally established in 1999 (Marchante *et al.*, 2005). At the present, the species acacia can be found all over the country and its high invasive potential comes from the high seed production, dispersal and longevity in the soil, as well as the stimulation of seeds by fire (quite frequent in Mediterranean climates such as Portugal) allowing that recently burnt areas of this species easily recover their stands (Marchante *et al.*, 2005). According to the last National Forest Inventory (AFN, 2010), 35% of the territory (3.2 x 10^6 ha) is covered by forests whereas the afforested area of *Acacia* sp. (stands) in 2010 was 5,351 ha (*i.e.* 0.2%). Furthermore, between 1995, 2005 and 2010 (a 15 year period) an enormous increase of this species afforested area was observed (2,701 ha in 1995, 4,726 ha in 2005 and 5,351 ha in 2010, *i.e.* + 98%). Moreover, when considering acacia's total area as a dominant species its expression is fairly higher (12,278 ha in 1995 that has decreased to 11,803 in 2010) representing around 0.4% of Portuguese forest area (ICNF, 2013). Even so, it must be stressed that in this National Forest Inventory forest areas estimation considered only areas with a minimum of 0.5 ha, a minimum width of 20 m, and a ground cover higher than 10% (ICNF, 2013).

According to Marchante *et al.* (2005) this species grows preferably in fresh valley soils or along water streams banks and it is also very frequently located along road sides. It is also known that the species acacia is widely spread in patches much smaller than 0.5 ha nearby localities and on main road sides (*e.g.* linear patches of width smaller than 20 m). Therefore, it is expected that this species invaded area to be much higher than those from the National Forest Inventory statistics. Acacia areas are characterized as both dense and very short vegetation structures (*e.g.* ground cover around 90%) wherein this species dominates all height classes under 16 m (*e.g.* mean height around 4 m). This species invasion is mainly found in maritime pine (*Pinus pinaster* Aiton), umbrella pine (*Pinus pinea* L.) and eucalyptus (*Eucalyptus* sp.) stands (Godinho-Ferreira *et al.*, 2005).

Application of RS and GIS techniques in mapping biological invasions

The spread of invasive species has generated interest in mapping their present distribution worldwide as the patterns of plant invasions, and the ecological processes which generate these patterns, vary across spatial scales. Thus, consideration of spatial scale may help to illuminate the mechanisms driving biological invasions, and offer insight into potential management strategies (Pauchard & Shea, 2006). Therefore, it becomes essential to have tools to identify and monitor invasive species distributions, in order to obtain reliable and updated information for better management of invaded areas (Joshi *et al.*, 2006; Underwood & Ustin, 2007). This implies being able to delineate the spatial extent and to ascertain the severity or intensity of the invasion, providing therefore a baseline for monitoring future expansion, increasing the effectiveness of control efforts, and assisting in identifying specific targets for control activities such as satellite populations and 'invasion fronts' (Underwood *et al.*, 2003).

Remote sensing (RS) has been an important tool for large-scale ecological studies in the past three decades, but it was not commonly used to study alien invasive plants until the mid 1990s (Huang & Asner, 2009). RS and Geographic Information Systems (GIS) are useful tools for mapping and monitoring invasive species and to predict areas of susceptibility for exotic species invasion (Joshi *et al.*, 2004). RS provides multi-temporal records that can be integrated and used into a GIS in order to support monitoring and control activities of invaded sites (Gil *et al.*, 2013; Gil *et al.*, 2014). Additionally, researchers have sought to exploit unique phenological, spectral, or structural characteristics of invasive species in digital multispectral imagery to distinguish them from the species around them (Underwood *et al.*, 2003; Underwood & Ustin, 2007; Resasco *et al.*, 2007). According to Huang & Asner (2009), moderate resolution satellite imagery (*e.g.* spatial resolution between 10-100 m and spectral resolution with less than twenty spectral bands) such as the ones captured by the satellites LANDSAT (*e.g.* Thematic Mapper and Enhanced Thematic Mapper Plus), SPOT (*Satellite pour l'Observation de la Terre*) and TERRA (particularly, ASTER – Advanced Spaceborne Thermal Emission and Reflection Radiometer) are well suited for mapping at the community level and have been used to map invasive species before (*e.g.* Cobbing, 2007; Lawes & Wallace, 2008; Viana & Aranha, 2010; Gil *et al.*, 2014).

In Portugal the latest official land cover map (COS2007) was produced in 2007 at the scale of 1:25,000 (with a minimum mapping unit of 1 ha, a minimum distance between lines of 20 m, and a minimum polygon width of 20 m). Despite using a five-level classification system of 238 land cover categories, the COS2007 map considers broad classes of invasive species only (*e.g. A. dealbata* and *Ailanthus altissima* Mill.) either as, pure or mixed forest (ground cover higher than 30%) or open forest (ground cover between 10 to 30%) (DGT, 2007). Therefore, the use of moderate resolution satellite imagery to produce detailed presence maps for the species acacia by supervised classification techniques should be explored.

Viana & Aranha (2010) performed a comparison study between ASTER/TERRA and ETM+/LANDSAT 7 imagery for mapping the species acacia, in a study area in the center of Portugal, using two supervised classifiers and a classification system with three categories. These authors obtained overall accuracies of 89% for the ETM+ imagery classification (24 January 2003) and 87% for the ASTER imagery classification (7 October 2003) using the best classifier (*e.g.* maximum likelihood classifier). Despite of the ASTER imagery having a better special resolution (*e.g.* 15 m) than the ETM+ imagery (*e.g.* 30 m) it was the imagery acquisition date that proved to be the most relevant factor on imagery classification accuracy.

Therefore, due to the success of those previous studies and also because of ASTER imagery specific characteristics (low economic cost allied to moderate spatial and spectral resolutions), the aim of this study was to explore ASTER imagery from two acquisition dates (*Acacia* sp. flowering season vs. dry season), five land cover classifications with different degree of generalization and three supervised classifiers, in order to assess the best approach to accurately map this species. The working hypothesis essayed was that ASTER imagery from March (*i.e.* the species flowering season) will allow a better spectral separability and classification/mapping of the species than the imagery from August (dry season) when using the most detailed land cover classification essayed and the maximum likelihood classifier. To test this hypothesis, ASTER imagery from 25 August 2005 and from 24 March 2007 were used. First, both imagery datasets were assessed in terms of spectral separability using the five land cover classifications with different degree of generalization. After, the images were classified using three supervised classifiers (*e.g.* the parametric Maximum Likelihood classifier and the two non-parametric Support Vector Machine and Artificial Neural Networks classifiers). In the end, the accuracy of the produced maps was compared to assess the best approach to accurately map this species. The most accurate map was used to evaluate species' invasion area at a minimum mapping unit of 0.02 ha providing a substantial improvement comparably to the COS2007 map and/or the National Forest Inventory statistics.

Material and methods

Study area and ASTER imagery

The study area consists of an ASTER scene of 321.1 x 10^3 ha located in the central-eastern of Portugal covering 49% of 11 municipalities of the Castelo Branco's district (Fig. 1), located between 40°9′47′′N - 39°34′49′′N and 8°10′36′′W – 7°16′46′′W. The imagery obtained was of two dates: the first, on 25 August 2005 with a cloud cover of 20% and the second, on 24 March 2007 without cloud cover as there were no images available from the same year to performed valuable work (*e.g.* high cloud cover). The months of March and August were chosen, as the first corresponds to acacia flowering season by contrast to the second (dry season).

The multispectral sensor ASTER produces images with moderate spectral and spatial resolutions, namely: three visible and near-infrared bands (VNIR bands 1, 2, 3N and 3D) with a spatial resolution of 15 m; six mid-infrared bands (SWIR bands 4, 5, 6, 7, 8 and 9) with a spatial resolution of 30 m; and five far-infrared bands (TIR bands 10, 11, 12, 13 and 14) with a spatial resolution of 90 m. The imagery used in this study (VNIR and SWIR bands) was acquired as raw data (L1A) that has radiometric correction but no geometric and atmospheric corrections. Therefore, imagery geometric correction was performed using the digital elevation model (DEM) developed from both bands 3N and 3B to ortho-rectify the ASTER images by the resampling method of cubic convolution. Atmospheric correction was accomplished using the FLAASH model (Fast Line-of-sight Atmospheric Analysis of Spectral Hypercubes) through the MODTRAN algorithm (Adler-Golden *et al.*, 1999). To do so, both VNIR and SWIR bands were first aggregate in one file (layer stacking) and the spatial resolution of the SWIR bands expanded to a pixel size of 15 m using the resampling method of cubic convolution once more. Subsequently, the layer stacking was converted to the BIL format so the FLAASH module could be processed (Fig. 2).

Methodology – RS classification scheme

Five land cover classification schemes with different degree of generalization (Table 1 – Level 1 to Level 5) were essayed which allowed studying the spectral separability of acacia class in comparison to the other land cover categories considered (Fig. 2). The classification system of the Portuguese official land cover map for 2007 (COS2007) was used as reference (DGT 2007) and further adapted to the purpose of this study. It should be emphasized that the COS2007 classifica-

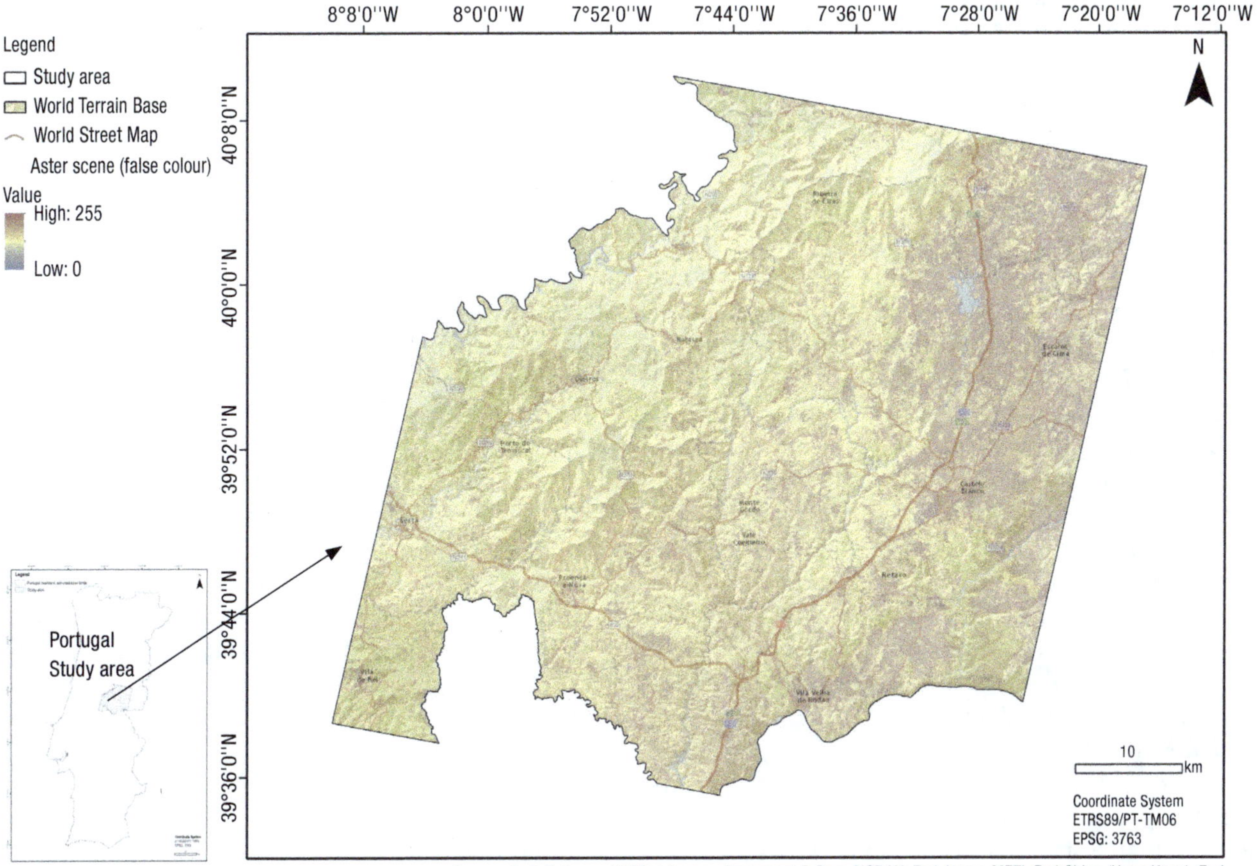

Source: Esri, DeLome, USGS, intermap, increment P Corp, NCRAN, Esri Japan, METI, Esri China (Hong Kong), Esri (Thailand), MapmyIndia, OpenStreetMap contributors, and the GIS User Community; Sources: Esri, USGS, NOAA

Figure 1. Geographic location of the study area – ASTER scene of 3,211 km^2 (false colour composite with elevation, cities names and main roads).

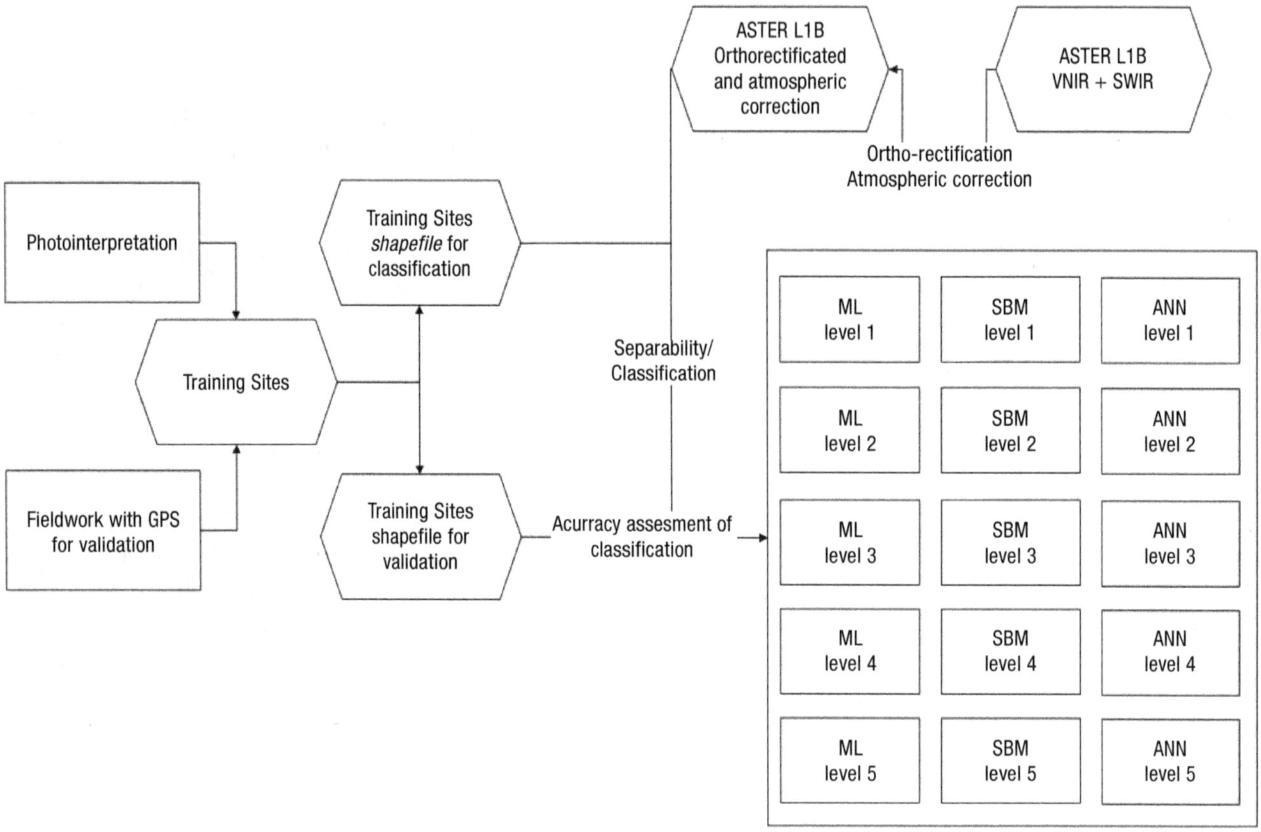

Figure 2. Methodological approach to test acacia mapping using ASTER imagery of two dates (25 August 2005 and 24 March 2007) and three classifiers (ML – Maximum Likelihood; SVM – Support Vector Machine; and ANN – Artificial Neural Networks).

tion system is standardized with the Corine Land Cover (CLC) classification system until the third level of detail (Caetano *et al.*, 2009).

The categories "artificial areas", "agricultural areas", "forests" and "water bodies" used in this study follow the nomenclature referred to above for the broader classification level (Table 1 – Level 1). However, when moving to a more detailed classification (Table 1 – Level 2), the "acacia" category is isolated from the "forests" category. After, the "natural areas" category (*e.g.* scrub and/or herbaceous vegetation associations and open spaces with little or no vegetation) is isolated from the "forests" category (Table 1 – Level 3). Next, "forests" category is sliced into "broadleaves" and "coniferous" categories (Table 1 – Level 4). In the end, "broadleaves" category is divided into "holm oak/cork oak" category (Qr/Qs – *Quercus rotundifolia* Lam. and *Quercus suber* L), "chestnut/other oaks" category (Cs/Q – *Castanea sativa* Mill. and *Quercus* sp.) and "eucalyptus" category (Ec – *Eucalyptus* sp.) (Table 1 – Level 5). In this study area the "coniferous" category comprises essentially the species maritime pine (*Pinus pinaster* Aiton). The goals of sub-dividing the "broadleaves" category (Level 4) into more specific forest categories in Level 5 were: (1) testing spectral separability among more specific and homogeneous forest assemblages; and (2) testing the potential of ASTER imagery for mapping these same more specific and homogeneous forest assemblages, as represented in the Portuguese National Forest Inventory.

A total of 1080 sites (120 for each land cover class, Table 1 – Level 5) were selected over the ASTER imagery. The training sites were obtained by applying a stratified random sampling, in order to cover the whole study area (Congalton, 2001). Ancillary information such as the CLC land cover maps (2000 and 2006), the COS land cover map (1990 and 2007), ortho-rectified aerial photographs (2005) and Google Earth™ 5.2.1 imagery (several dates) were used to cross-validate the sites selection by photo-interpretation and field work with GPS (undertaken between 2011 and 2012). These 1080 field sites identified over both ASTER imagery were randomly divided in two sub-sets as follows: 75% as training sub-set (810 sites: 90 for each land cover class, Table 1 – level 5) and 25% as testing sub-set (270 sites: 30 for each land cover class, Table 1 – level 5). The training sub-set was used to create the spectral signatures, for each land cover category by classification level, needed to support the imagery classification stage. While, the training sub-set was used to assess the accuracy of the 30 imagery classifications produced

Table 1. Land cover categories by classification level (*i.e.* increasing degree of detail)

Land cover categories				
Level 1	**Level 2**	**Level 3**	**Level 4**	**Level 5**
Artificial areas	Artificial areas	Artificial areas	Artificial areas	Artificial areas
Agricultural areas	Agricultural areas	Agricultural areas	Agricultural areas	Agricultural areas
	Acacia	Acacia	Acacia	Acacia
				Qr/Qs
Forests	Forests	Forests	Broadleaves	Cs/Q
				Ec
			Coniferous	Coniferous
		Natural areas	Natural areas	Natural areas
Water bodies	Water bodies	Water bodies	Water bodies	Water bodies

Legend: Broadleaves – Qr/Qs (holm oak/cork oak); Cs/Q (chestnut/other oaks); Ec (eucalyptus); Coniferous (*e.g.* maritime pine).

(*i.e.* two dates imagery x five land cover classifications x three supervised classifiers).

Prior to the ASTER imagery classification stage, the signatures were used to assess the spectral separability between every land cover categories referred to above (Table 1) by calculating the Transformed Divergence (TD). The TD is a measure of the statistical separation between categories response patterns computed for all pairs of categories and presented in the form of a matrix. It is based on the covariance of the samples for each category at the pixel level and estimates a weight which exponentially decreases to enlarge the distances between categories (Richards, 2013). The TD values are normalized to the range of [0 , 2] and as a general rule it can be accepted that good spectral separability occurs for values higher than 1.9, moderate spectral separability for values between 1.7 and 1.9 and poor spectral separability (*i.e.* spectral similarity or spectral confusion) for values below 1.7 (Jensen, 1996).

After, the ASTER imagery for each date was classified at the pixel level wherein three supervised classifiers were applied (Fig. 2). The classifiers essayed were the following: the parametric Maximum Likelihood classifier (ML) and the two non-parametric Support Vector Machine (SVM) and Artificial Neural Networks (ANN) classifiers (Xu *et al.*, 2005; Foody & Mathur, 2004; Filippi & Jensen, 2006).

In the end, the accuracy of the 30 classified images produced (Fig. 2) was assessed by calculating the error matrix (or confusion matrix). This matrix is obtained by comparing the land cover category found in the testing sub-set (ground-truth) to that which was mapped in the image for the same location and it shows the distribution of the percentage of pixels classified correctly and in an erroneous way (Congalton, 1991). The statistical assessments of accuracy derived from the error matrix that were considered are: the producer's accuracy (*P*), the user´s accuracy (*U*), the overall accuracy (*O*) and the Kappa statistics (*K*). Producer's accuracy (*P*) is a measure of omission error that indicates the probability that a sample is correctly classified. User's accuracy (*U*) is a measure of commission and indicates the probability that a classified pixel does represent that category in the field. Overall accuracy (*O*) allows evaluating the overall thematic classification of the map produced (Congalton & Green, 1999). The Kappa statistics measures model accuracy with respect to the accuracy expected with a random assignment of pixels to categories. The Kappa statistics takes into account all elements of the confusion matrix in its evaluation (*i.e.* also includes the elements off the main diagonal, which represent disagreements in classification) as opposed to the overall accuracy which uses only the diagonal elements (real agreement) (Congalton, 1991).

Acacia spatial distribution

The accuracy measures referred to above were used to select the best image classification (from all classifications performed during the previous methodological step) in order to obtain the spatial distribution of the species acacia in the study area. Then, an estimate of the invasion area of this species was obtained with a 15 m spatial resolution (*i.e.* a minimum cartographic unit of 0.02 ha). Additionally, the previous acacia spatial distribution was compared to the one obtained by the overlay of the common areas classified as acacia in all high accuracy (*i.e.* $U \geq 80.00$ and/or $K \geq 0.8$) image classifications for the species, in order to minimize classification uncertainty and errors regarding this same class. After, the estimated areas of acacia spatial distribution were compared to the area of acacia stands over the study area according to the National Forest Inventory statistics in 2005 (AFN, 2010).

Results

Spectral separability of land cover categories

The analysis of the transformed divergence (TD) matrices for each one of the five tested land cover classifications proved that there were more spectral confusion when using the ASTER imagery of August 2005 (dry season) than when using the image of March 2007 (acacia flowering season – bright yellow flowers). In August, spectral separability among categories in the most detailed classifications (Table 2 – Level 5) were mainly moderate to good. Whilst in March, good spectral separability was found between almost all of these same categories of land cover (Table 3 – Level 5). Furthermore, overall separability assessments performed for more generic land cover classifications (*e.g.* Levels 1, 2 and 3; Table 1) for both flowering season and dry season images did not demonstrate to be unequivocally higher than those reached for the most detailed land cover classifications (Levels 4 and 5; Table 1).

It was found spectral confusion between the categories "acacia" and both "forests" (*i.e.* broadleaves – holm oak/cork oak, chestnut/other oaks and eucalyptus – and coniferous – *e.g.* maritime pine) and "natural areas" when using the imagery of August 2005 (Table 2 – Level 5). Whilst, spectral confusion was only found between the categories "acacia" and both "agricultural areas" and "coniferous" (*e.g.* maritime pine) when using the imagery of March 2007. Finally, good spectral separability between the "acacia" category and the remaining categories was observed (Table 3 – Level 5).

Table 2. Land cover categories spectral separability in level 5 (ASTER imagery of 25 August 2005) – Transformed Divergence matrices obtained using the training sub-set

Land cover categories	Transform divergence matrices								
Level 5	Artificial areas	Agricultural areas	Acacia	Qr/Qs	Cs/Q	Ec	Coniferous	Natural areas	Water bodies
Artificial areas	0								
Agricultural areas	1.79	0							
Acacia	1.98	1.78	0						
Qr/Qs	1.99	1.51	1.38	0					
Cs/Q	2.00	1.84	0.97	1.29	0				
Ec	2.00	1.91	1.64	1.54	1.43	0			
Coniferous	2.00	1.99	1.46	1.87	1.48	1.33	0		
Natural areas	1.99	1.88	1.61	1.86	1.67	1.65	1.84	0	
Water bodies	1.90	1.79	1.83	1.75	1.82	1.93	1.97	1.89	0

Legend: Good separability – > 1.9; Moderate separability – 1.7 - 1.9; Poor separability – < 1.7; Qr/Qs – holm oak/cork oak; Cs/Q – chestnut/other oaks; Ec – eucalyptus; Coniferous – *e.g.* maritime pine.

Table 3. Land cover categories spectral separability in level 5 (ASTER imagery of 24 March 2007) – Transformed Divergence matrices obtained using the training sub-set

Land cover categories	Transform divergence matrices								
Level 5	Artificial areas	Agricultural areas	Acacia	Qr/Qs	Cs/Q	Ec	Coniferous	Natural areas	Water bodies
Artificial areas	0								
Agricultural areas	2.00	0							
Acacia	2.00	1.29	0						
Qr/Qs	2.00	2.00	2.00	0					
Cs/Q	2.00	1.90	1.93	2.00	0				
Ec	1.99	2.00	2.00	2.00	2.00	0			
Coniferous	2.00	1.84	1.21	2.00	1.95	2.00	0		
Natural areas	1.52	1.99	1.97	2.00	1.98	1.99	1.99	0	
Water bodies	1.97	1.99	1.98	2.00	1.99	1.99	1.99	1.92	0

Legend: Good separability – > 1.9; Moderate separability – 1.7 - 1.9; Poor separability – < 1.7; Qr/Qs – holm oak/cork oak; Cs/Q – chestnut/other oaks; Ec – eucalyptus; Coniferous – e.g. maritime pine.

Imagery classifications accuracy

The analysis of the evaluated accuracy measures (*P*, *U*, *O* and *K*) for the classified images produced by the three classifiers essayed (ML, ANN, SVM) when using the imagery of August 2005 (dry season) proved that the ML classifier was the one that performed best for all tested land cover classifications. The same result was found when analysing the classified images produced by the three classifiers essayed (ML, ANN, SVM) while using the image of March 2007.

Regarding the classified images overall accuracy (for both *O* and *K*) it was verified that it was higher in all cases when using ASTER imagery of March 2007 (flowering season) when compared to the results obtained by processing the dry season image of August 2005 (Table 4). An explicit trend of increasing or decreasing accuracy associated to the change of land cover classifications (from level 1 to level 5) was not found. The highest overall accuracies ($O \geq 80.0$ and/or $K \geq 0.80$) were always obtained by applying the ML classifier to the flowering season image, which has slightly decreased from the most generic (Table 4 – Level 1: $O = 92.45$ and $K = 0.89$) to the most detailed land cover classification (Table 4 – Level 5: $O = 85.21$ and $K = 0.83$). Conversely, overall accuracy results obtained by processing the dry season image (August 2005) were significantly lower (Table 4 – Level 1: $O = 79.17$ and $K = 0.70$; Table 4 – Level 5: $O = 73.44$ and $K = 0.70$) than those obtained with the flowering season image (March 2007).

To what concerns to the classifiers tested, the ML classifier showed very good accuracies especially when using the flowering season imagery (Table 4 – *K* between 0.83-0.89). As opposed to the lower accuracies observed either with the ANN classifier (Table 4 – *K* between 0.50-0.78) or the SVM classifier (Table 4 – *K* between 0.36-0.54). In fact, the ML classifier obtained the best results in all land cover classifications but the highest overall Kappa was obtained at level 1 (Table 4 – $K = 0.89$). By contrast, the remaining classifiers (ANN and SVM) showed a very low global accuracy (Table 4; ANN always below $K = 0.65$; SVM always below $K = 0.54$).

Acacia spatial distribution

Regarding the "acacia" category, the highest accuracy (Table 5 – Level 5: $U = 100.00$, $P = 69.57$ and $K = 1.0$) was reached by applying the ML classifier to the March 2007 (flowering season) image while using the most detailed land cover classification. Good accuracies were also obtained by applying to this same image the ML classifier to level 4 ($U = 84.21$ and $K = 0.81$) and level 2 ($U = 80.77$ and $K = 0.75$) land cover classifications. Finally, applying the ANN classifier to the flowering season image while using the level 2 classification scheme also allowed to reach a good mapping accuracy for acacia ($U = 84.62$ and $K = 0.80$).

Accuracy results of acacia mapping obtained by processing the dry season image (August 2005) were significantly lower than those obtained with the flowering season image (March 2007), as shown in Table 5. Best results for this image were obtained by applying the ANN classifier to level 5 ($U = 66.67$ and $K = 0.61$) as well as the ML classifier to level 2 land cover classification ($U = 66.67$ and $K = 0.57$), to level 4 ($U = 60.61$ and $K = 0.53$) and to level 5 ($U = 60.00$ and $K = 0.54$).

As a result, to estimate the area of acacia spatial distribution over the study area the best image classification that resulted from the application of ML to the March 2007 image using the level 5 land cover classification (*i.e.* best image classification only; Table 5 – Level 5: $U = 100.00$, $P = 69.57$ and $K = 1.0$) was used (Fig. 3). According to this output, the total area covered by acacia in the study area was estimated approximately in 24,770 hectares.

Table 4. Accuracy assessments – Overall accuracy and Kappa statistics of the 30 imagery classifications *vs.* the testing sub-set: ASTER imagery of two dates (25 August 2005 and 24 March 2007), three supervised classifiers (Maximum Likelihood, Support Vector Machine and Artificial Neural Networks) and five levels of land cover classification with different degree of generalization

Land cover categories	ASTER imagery classification of 25 August 2005 *vs.* testing sub-set						ASTER imagery classification of 24 March 20075 *vs.* testing sub-set					
	ML		SVM		ANN		ML		SVM		ANN	
	O	*K*	*O*	*K*	*O*	*K*	*O*	*K*	*O*	*K*	*O*	*K*
Level 1	79.17	0.70	60.42	0.41	79.17	0.70	92.45	0.89	65.09	0.79	84.90	0.78
Level 2	83.96	0.79	39.58	0.24	63.54	0.54	88.54	0.86	63.20	0.54	72.64	0.65
Level 3	77.42	0.73	42.98	0.31	55.26	0.46	87.72	0.85	53.26	0.43	66.13	0.59
Level 4	72.53	0.68	44.70	0.35	49.24	0.41	87.12	0.85	47.18	0.38	55.63	0.48
Level 5	73.44	0.70	40.21	0.32	47.93	0.42	85.21	0.83	43.50	0.36	55.93	0.50

Legend: ML – Maximum Likelihood; SVM – Support Vector Machine; ANN – Artificial Neural Network; *O* – overall accuracy; and *K* – *Kappa* statistics.

Table 5. Accuracy assessments – Confusion matrices of imagery classifications *vs.* the testing sub-set: ASTER imagery of two dates (25 August 2005 and 24 March 2007), three supervised classifiers (Maximum Likelihood, Support Vector Machine and Artificial Neural Networks) and the most detailed level of land cover classification (Level 5)

Land cover categories Level 5	ASTER imagery classification of 25 August 2005 *vs.* testing sub-set								
	ML			**SVM**			**ANN**		
	P	***U***	***K***	***P***	***U***	***K***	***P***	***U***	***K***
Artificial areas	77.78	100.00	1.00	44.44	42.11	0.35	72.22	41.94	0.35
Agricultural areas	81.82	85.71	0.84	50.00	75.00	0.72	55.56	90.91	0.90
Acacia	78.26	60.00	0.54	69.57	32.65	0.22	8.70	66.67	0.61
Qr/Qs	35.00	100.00	1.00	56.52	30.95	0.20	39.13	47.37	0.39
Cs/Q	77.27	51.52	0.45	26.67	50.00	0.45	40.00	50.00	0.45
Ec	88.89	80.00	0.78	11.11	20.00	0.10	16.67	30.00	0.22
Coniferous	66.67	66.67	0.63	0.00	0.00	0.00	72.22	27.08	0.18
Natural areas	61.11	64.71	0.61	44.44	57.14	0.52	66.67	75.00	0.72
Water bodies	94.44	100.00	1.00	44.44	53.33	0.48	72.22	68.42	0.65
O		73.44			40.24			47.93	
K		0.70			0.32			0.42	

Land cover categories Level 5	ASTER imagery classification of 24 March 20075 *vs.* testing sub-set								
	ML			**SVM**			**ANN**		
	P	***U***	***K***	***P***	***U***	***K***	***P***	***U***	***K***
Artificial areas	83.33	93.75	0.93	88.89	69.57	0.66	77.78	70.00	0.67
Agricultural areas	88.89	84.21	0.82	81.82	58.06	0.52	77.27	77.27	0.74
Acacia	69.57	100.00	1.00	56.52	26.00	0.15	56.52	59.09	0.53
Qr/Qs	86.96	90.91	0.89	55.00	32.35	0.24	45.00	60.00	0.55
Cs/Q	86.67	92.86	0.92	4.55	33.33	0.24	13.64	100.00	1.00
Ec	83.33	68.18	0.64	22.22	66.67	0.63	44.44	47.06	0.41
Coniferous	94.44	80.95	0.79	0.00	0.00	0.00	66.67	46.15	0.40
Natural areas	94.44	77.27	0.75	5.56	100.00	1.00	44.44	25.00	0.17
Water bodies	83.33	88.24	0.87	72.22	44.83	0.39	83.33	75.00	0.72
O		85.21			43.5			55.93	
K		0.83			0.36			0.50	

Legend: ML – Maximum Likelihood; SVM – Support Vector Machine; ANN – Artificial Neural Network; *P* – producer's accuracy; *U* – user's accuracy; *O* – overall accuracy; *K* – *Kappa* statistics; Qr/Qs – holm oak/cork oak; Cs/Q – chestnut/other oaks; Ec – eucalyptus; Coniferous – *e.g.* maritime pine.

Additionally, the acacia spatial distribution obtained by the overlay of the common areas classified as acacia in all high accuracy (*i.e.* $U \geq 80.00$ and/or $K \geq 0.8$) image classifications for the species was produced (Fig. 4). Therefore, the aforementioned imagery classifications meeting those conditions were the following: (1) March 2007 image, ML classifier and Level 5 legend ($U = 100.00$ and $K = 1.0$); (2) March 2007 image, ANN classifier and Level 2 legend ($U = 84.62$ and $K = 0.80$); (3) March 2007 image, ML classifier and Level 4 legend ($U = 84.21$ and $K = 0.81$); and (4) March 2007 image, ML classifier and Level 2 legend ($U = 80.77$ and $K = 0.75$). According to this output, the total area covered by acacia in the study area was estimated approximately in 12,178 hectares.

By analysing both acacia spatial distribution maps (Figs. 3 and 4) it can be stated that although acacia vegetation patches occur all over the study area, they are especially dense in the western and northern zones.

Discussion

Spectral separability of land cover categories

The analysis of the spectral separability of the land cover categories used in each of the five tested land cover classifications showed that the flowering season ASTER imagery provided higher spectral separability between almost all categories of land cover considered (*e.g.* good spectral separability) than the dry season ASTER imagery (*e.g.* moderate to good spectral separability). This result highlights how detection by RS can be facilitated by the timing of image acquisition to correspond with particular phenological periods of a target

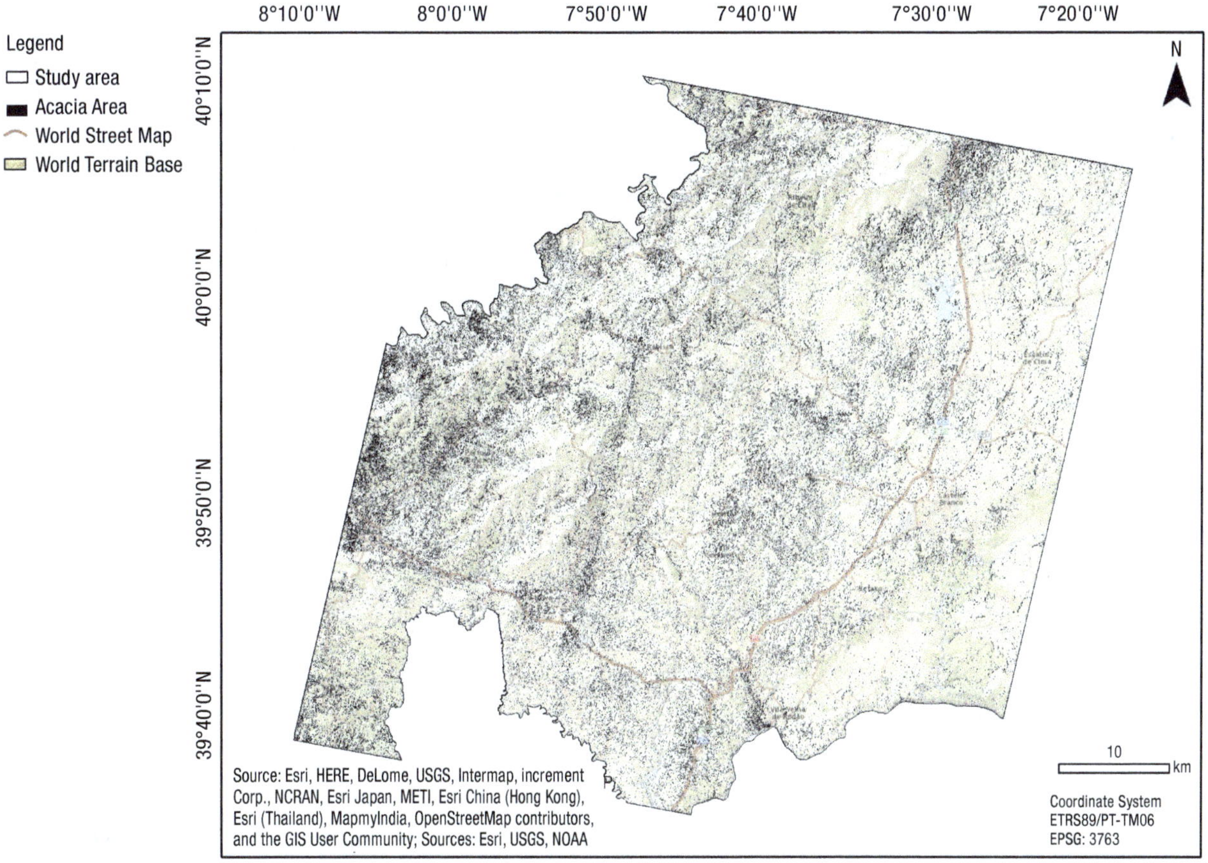

Figure 3. Spatial distribution of acacia over the study area (with elevation, cities names and main roads) according to the best accuracy classification for acacia category (*i.e.* ML classification using the level 5 land cover classification and the ASTER imagery of March 2007).

invasive species (*e.g.* maximizing phenological differences between the invasive plant species and the native species) (see Underwood & Ustin, 2007; Pauchard & Maheu-Giroux, 2007). In this case, the species acacia provides a clear and intense yellow pattern during its flowering season (*e.g.* between January and March) (*e.g.* Marchante *et al.*, 2005; Pauchard & Maheu-Giroux, 2007) that can be distinguished by RS techniques, providing a tremendous potential for monitoring the spread of invasion of this species. Therefore, the analysis of the spectral separability between land cover categories allows to identify where spectral confusion exists which is useful to better define optimal imagery acquisition date for each species. The findings in this study showed that in March some spectral confusion between "acacia" and "coniferous" category was observed thus suggesting that maybe a late winter imagery acquisition date (*e.g.* January-February) would be more suitable.

Imagery classifications accuracy

The analysis of the accuracy of the 30 classification maps proved that the highest classification accuracy was found using both the imagery of flowering season and the Maximum Likelihood classifier ($K \geq 0.80$). Moreover, acacia mapping accuracy obtained excellent results when using the most detailed land cover classification ($K = 1.00$) which indicate the high potentiality of ASTER imagery for mapping this species distribution when the most appropriate imagery date is used. As a result, mapping the spatial distribution of the species acacia across central-eastern Portugal was achieved with good accuracy by applying both the ML classifier and the most detailed land cover classification to the ASTER image of March (flowering season).

These findings are in accordance to the study of Viana & Aranha (2010) in which the imagery acquisition date was found to be the most relevant factor on imagery classification accuracy. The ML classifier was the one that performed the best as well. However, the accuracy of ASTER imagery classification, having an acquisition date of 7 October 2003 and using only three land cover categories, was slightly lower (O=86.69 and K=0.81) when compared to the ones obtained in the present study using the flowering season imagery, the ML classifier and five land cover categories (Level 2: O=88.54 and K=0.86). Despite producer´s accuracy

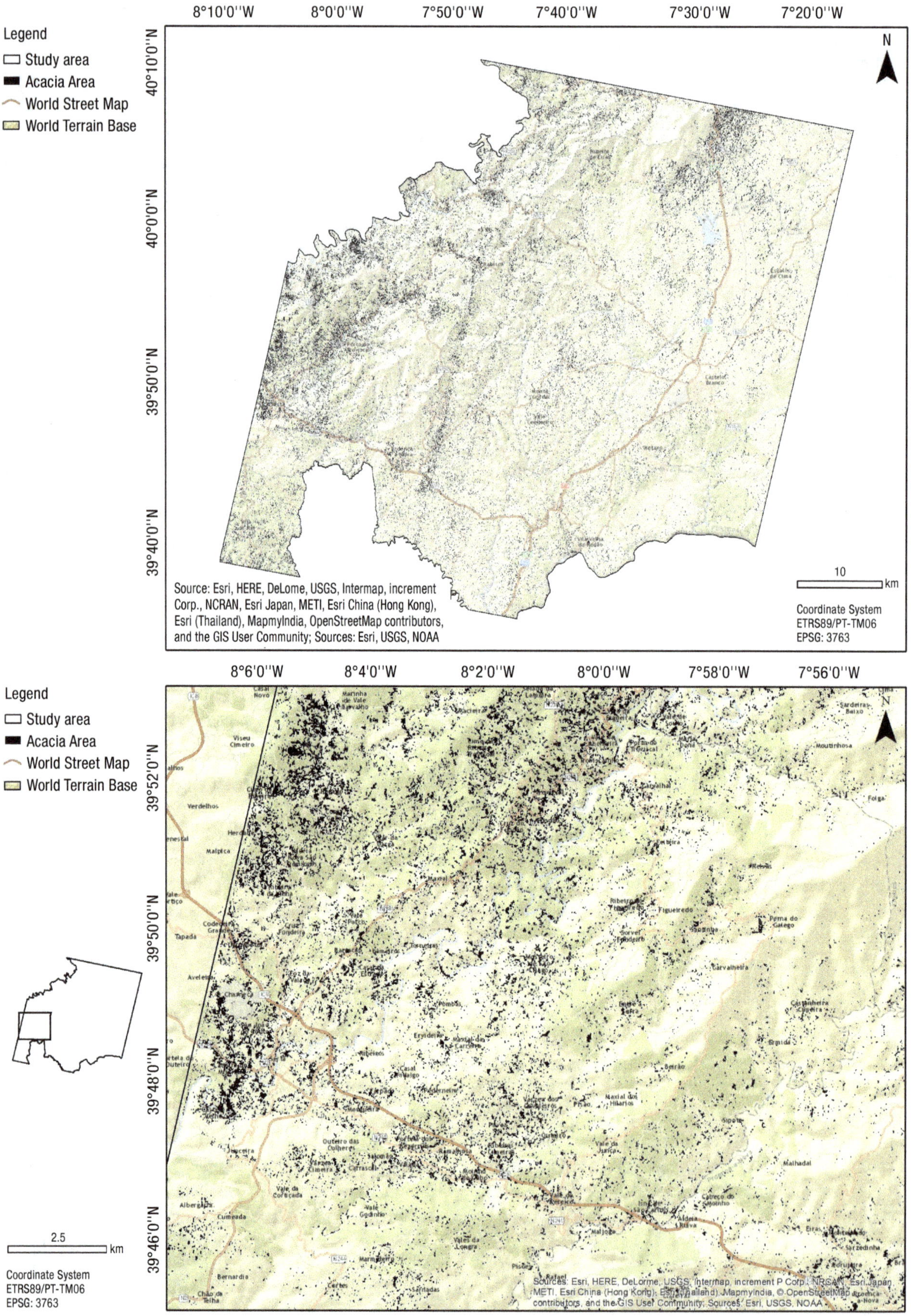

Figure 4. a) Spatial distribution of common areas classified as acacia in all high accuracy (*i.e.* $U \geq 80.00$ and/or $K \geq 0.8$) image classifications obtained for the species over the study area (with elevation, cities names and main roads); **b)** Zoom over an area located at the western zone of the study area having a vast incidence of acacia invasion (*e.g.* nearby localities and along roads).

being the same in both studies (*P*=100), a paramount improvement in user's accuracy was attained in the present study (Level 2: *U*=80.77) when compared to the one of those authors (*U*=11.10).

Acacia spatial distribution

The approximated total area occupied by this invasive woody species was estimated in 24,770 hectares (8% of the study area) according to the most accurate image classification for the species (i.e. applying both the ML classifier and the most detailed land cover classification to the ASTER image of March). An estimate of approximately 12,178 hectares was obtained when considering only the common areas classified as acacia in all high accuracy (*i.e.* $U \geq 80.00$ and/or $K \geq 0.8$) image classifications for the species.

By contrast, according to the National Forest Inventory in 2005 (AFN, 2010), the surface covered by acacia stands in the study area was estimated in 225 hectares only (0.1% of the study area). This can be explained by the standards used in this forest inventory (*e.g.* area with a minimum of 0.5 hectares, a minimum width of 20 meters and a ground cover higher than 10%). Therefore, due to this species fragmented distribution pattern it is argued that the estimated area in this study (24,770 hectares) is a more realistic approach of acacia occupation and geographic distribution as a spatial resolution of 15 m was used (*i.e.* minimum cartographic area of 0.02 ha) and no generalization process was applied.

Despite of the results obtained in this study, it is expected that acacia mapping accuracy may be slightly increased if both the number of training sites are increased and ancillary GIS-based information (*e.g.* roads, distance to water lines, altitude, slope, soil type, lithology, etc.) are included in the classification process (Van der Wouw *et al.*, 2011; Gil *et al.*, 2014) to overcome the poor spectral separability (TD < 1.7) observed between some of the land cover categories.

The spatial distribution map of acacia obtained allows decision-makers, stakeholders and landowners to both setup and implement more realistic, adequate and cost-effective invasive woodland management in central-eastern Portugal, by clearly identifying the location, the dimension and the logistics constraints (distance to roads, distance to water streams, slope, etc.) associated to the sites to be intervened and managed. Additionally, this new and more detailed data on acacia spatial distribution also allows more accurate ecological modeling studies of this invasive alien species in this same geographic area (*e.g.* Bradley & Mustard, 2006; Andrew & Ustin, 2009).

Directions for future research – Satellite Remote Sensing and acacia mapping

At the present the numbers of satellites for earth observations are constantly changing as new are being launched while others are getting inoperative (*e.g.* Richards, 2013). Available commercial imagery is not always cheap or has the desired resolutions (spatial, spectral, radiometric and temporal) for the study to be developed (*e.g.* ASTER imagery offer only four bands and a spatial resolution of 15 m; Landsat (E)TM imagery offer 6 or 11 bands but a spatial resolution of 30 m). However, the ASTER imagery proved to constitute a cost-effective solution for acacia mapping in central-eastern of Portugal. Furthermore, the use of flowering season date ASTER imagery with the application of the ML classifier when using levels 1, 2 and 3 land cover classifications also provided an effective and reliable solution to perform generic land cover mapping for both this species and type of landscape (rural territories in central inland of Portugal).

In sum, the use of ASTER imagery obtained during the species flowering season may therefore be recommended for detailed land cover/vegetation and especially acacia mapping purposes and management decision-support on this type of southern Europe's ecosystems. Despite of the ASTER SWIR detectors no longer function since April 2008 (only VNIR and TIR sensors are still operational), the archived and currently available ASTER multispectral data may be used for multi sensor-based acacia temporal studies (*e.g.* Lawes & Wallace, 2008), in order to assess its invasive distribution trends and to enable a more effective control of these populations.

Finally, further efforts on improving acacia mapping are needed to overcome the poor spectral separability observed between "acacia" category and both "agricultural areas" and "coniferous" categories when using the flowering season date imagery. This situation is tightly related to this specific rural landscape under study where maritime pine forest dominates (flowering season March-April; golden yellow masculine flowers) and the distribution pattern of this invasive species consists mostly of both very small patches nearby agricultural areas and very narrow strips along main road sides.

Although all the positive results achieved in this study and their potential application for improving decision-making on acacia management in Portugal, spatial and spectral information provided by moderate spatial and spectral resolution satellite images as ASTER is still insufficient to decipher the complexity of natural environment and further fully delineate the distribution of alien plants, as stated by Huang & Asner

(2009) and Joshi *et al.* (2004). Nevertheless, due to its innovative technical features, applying a multi-method processing approach to the new ESA Copernicus Sentinel-2 sensor data might be able to cost-effectively address this relevant information gap at regional level, by providing accurate, detailed and periodic thematic cartography and change detection assessment (*e.g.* Immitzer *et al.*, 2016; Gil *et al.*, 2012). Furthermore, observing alien plants as acacia requires data collected from sensors pushing the limits of at least one type of resolution (spatial, temporal or spectral resolution) since the profiles of these species may be quite similar to those of native plants, from a RS perspective (Asner, 2008). For instance, the use of very high spatial resolution in several studies on mapping invasive species has pointed out very encouraging results (*e.g.* Gil *et al.*, 2013). On the other hand, hyperspectral images are currently the most heavily used imaging source for studies of alien plants because detailed spectral profiles can be developed for native and non-native plants, allowing the analysis of specific spectral regions that are most sensitive to the abundance of the species of interest (Underwood *et al.*, 2003). Therefore, as stated by Gillespie *et al.* (2008) and Bradley (2014), future research on satellite RS of alien invasive plants as the species acacia should focus on the collection and dissemination of high-quality field data coupled with the incorporation and integration of available data acquired by existing (especially the most recent) spaceborne multispectral (*e.g.* WorldView-3, Landsat-8 OLI, Sentinel-2), hyperspectral (*e.g.* Chris-Proba, Hyperion) and synthetic aperture radar (*e.g.* Sentinel-1) sensors. Finally, this improved data (in quality and quantity) on the spatial distribution of this alien invasive plant species may also allow more accurate ecological modelling studies (*e.g.* Gutierres *et al.*, 2011; Costa *et al.*, 2015, 2013; Pereira & Figueiredo, 2015).

References

Adler-Golden SM, Matthew MW, Bernstein LS, Levine RY, Berk A, Richtsmeier SC, Acharya PK, Anderson GP, Felde G, Gardner J *et al.*, 1999. Atmospheric correction for shortwave spectral imagery based on MODTRAN4. Proc. of SPIE's International Symposion on Optical Science, Engineering and Instrumentation, International Society for Optics and Photonics. pp: 61-69.

AFN, 2010. FloreStat. Aplicação para consulta dos resultados do 5° Inventário Florestal Nacional. AFN – Autoridade Florestal Nacional, Ministério da Agricultura, do Desenvolvimento Rural e das Pescas. http://www.icnf.pt/portal/florestas/ifn/ifn5/.

Andrew ME, Ustin SL, 2009. Habitat suitability modelling of an invasive plant with advanced remote sensing data. Divers Distrib 15: 627-640. http://dx.doi.org/10.1111/j.1472-4642.2009.00568.x.

Arán D, García-Duro J, Reyes O, Casal M, 2013. Fire and invasive species: Modifications in the germination potential of *Acacia melanoxylon, Conyza canadensis* and Eucalyptus globulus. For Ecol Manage 302: 7-13.

Asner GP, 2008. Hyperspectral remote sensing of canopy chemistry, physiology and diversity in tropical rainforests. In: Hyperspectral Remote Sensing of Tropical and Subtropical Forests; Kalacska M, Sanchez-Azofeifa GA (eds.). Taylor and Francis, Oxon, UK. pp: 261-288. http://dx.doi.org/10.1201/9781420053432.ch12.

Asner GP, Jones MO, Martin RE, Knapp DE, Hughes RF, 2008. Remote sensing of native and invasive species in Hawaiian forests. Remote Sens Environ 112: 1912-1926. http://dx.doi.org/10.1016/j.rse.2007.02.043.

Bradley BA, 2014. Remote detection of invasive plants: a review of spectral, textural and phenological approaches. Biol Invasions 16: 1411-1425. http://dx.doi.org/10.1007/s10530-013-0578-9.

Bradley BA, Mustard JF, 2006. Characterizing the landscape dynamics of an invasive plant and risk of invasion using remote sensing. Ecol Appl 16: 1132-1147. http://dx.doi.org/10.1890/1051-0761(2006)016[1132:CTLDOA]2.0.CO;2.

Caetano M, Nunes V, Nunes A, 2009. CORINE Land Cover 2006 for Continental Portugal, Technical Report, Lisbon, Portugal. Instituto Geografico Português. 81 pp.

Cobbing BL, 2007. The use of Landsat ETM imagery as a suitable data capture source for alien Acacía species for the WFW programme. Master thesis. Rhodes University, Rhodes, Greece.

Congalton RG, 1991. A review of assessing the accuracy of classifications of remotely sensed data. Remote Sens Environ 49: 1671-1678. http://dx.doi.org/10.1016/0034-4257(91)90048-b.

Congalton RG, 2001. Accuracy assessment and validation of remotely sensed and other spatial information. Int J Wildl Fire 10(4): 321-328. http://dx.doi.org/10.1071/WF01031.

Congalton RG, Green K, 1999. Assessing the accuracy of remotely sensed data: principles and practices. New York, USA. Lewis Publishers. 200 pp.

Costa H, Bettencourt MJ, Silva CMN, Teodósio J, Gil A, Silva L, 2013. Invasive alien plants in the Azorean protected areas: invasion status and mitigation actions. In: Plant Invasions in Protected Areas; Foxcroft LC, Richardson DM, Pyšek P, Genovesi P (eds.). Springer, The Netherlands. pp. 375-394. http://dx.doi.org/10.1007/978-94-007-7750-7_17.

Costa H, Ponte NB, Azevedo EB, Gil A, 2015. Fuzzy set theory for predicting the potential distribution and cost-effective monitoring of invasive species. Ecol Model 316: 122-132. http://dx.doi.org/10.1016/j.ecolmodel.2015.07.034.

D'Antonio CM, Vitousek PM, 1992. Biological invasions by exotic grasses, the grass/fire cycle, and global change. Ann Rev Ecol Syst 23: 63–87. http://dx.doi.org/10.1146/annurev.es.23.110192.000431.

DGT, 2007. COS 2007 - Carta de Ocupação do Solo 2007 Nível 2 para Portugal Continental. Lisboa, Portugal. DGT - Direção-Geral do Território. http://www.dgterritorio.pt/cartografia_e_geodesia/cartografia/cartografia_tematica/carta_de_ocupacao_do_solo__cos_/cos__2007/.

Ehrenfeld JG, 2003. Effects of exotic plant invasions on soil nutrient cycling processes. Ecosystems 6: 503-523. http://dx.doi.org/10.1007/s10021-002-0151-3.

Filippi AM, Jensen JR, 2006. Fuzzy learning vector quantization for hyperspectral coastal vegetation classification. Remote Sens Environ 100: 512-30. http://dx.doi.org/10.1016/j.rse.2005.11.007.

Foody GM, Mathur A, 2004. A relative evaluation of multiclass image classification by support vector machines. IEEE T Geosci Remote 42: 1336-1343. http://dx.doi.org/10.1109/TGRS.2004.827257.

Fuentes-Ramirez A, Pauchard A, Cavieres LA, García RA, 2011. Survival and growth of *Acacia dealbata* vs. native trees across an invasion front in south-central Chile. For Ecol Manage 261: 1003-1009.

Gil A, Fonseca C, Lobo A, Calado H, 2012. Linking GMES Space Component to the development of land policies in Outermost Regions - the Azores (Portugal) case-study. Eur J Remote Sens 45: 263-281. http://dx.doi.org/10.5721/EuJRS20124524.

Gil A, Lobo A, Abadi M, Silva L, Calado H, 2013. Mapping invasive woody plants in Azores Protected Areas by using very high-resolution multispectral imagery. Eur J Remote Sens 46: 289-304. http://dx.doi.org/10.5721/EuJRS20134616.

Gil A, Yu Q, Abadi M, Calado H, 2014. Using ASTER multispectral imagery for mapping woody invasive species in Pico da Vara Natural Reserve (Azores Islands, Portugal). Rev Árvore 38: 391-401. http://dx.doi.org/10.1590/S0100-67622014000300001.

Gillespie TW, Foody GM, Rocchini D, Giorgi AP, Saatchi S, 2008. Measuring and modelling biodiversity from space. Prog Phys Geo 32: 203-221. http://dx.doi.org/10.1177/0309133308093606.

Godinho-Ferreira P, Azevedo A, Rego F, 2005. Carta da tipologia florestal de Portugal Continental. Silva Lusit 13(1): 1-34.

González-Muñoz N, Costa-Tenorio M, Espigares T, 2012. Invasion of alien *Acacia dealbata* on Spanish *Quercus robur* forests: impact on soils and vegetation. For Ecol Manage 269: 214-221.

Gutierres F, Gil A, Reis E, Lobo A, Neto C, Calado H, Costa JC, 2011. *Acacia saligna* (Labill.) H. Wendl in the Sesimbra Council: Invaded habitats and potential distribution modeling. J Coastal Res SI 64: 403-407.

Huang C, Asner G, 2009. Applications of remote sensing to alien invasive plant studies. Sensors 9: 4869-4889. http://dx.doi.org/10.3390/s90604869.

Hughes FR, Denslow JS, 2005. Invasion by a N2-fixing tree alters function and structure in wet lowland forests of Hawaii. Ecol Appl 15: 1615-1628. http://dx.doi.org/10.1890/04-0874.

Hughes FR, Vitousek PM, Tunison T, 1991. Alien grass invasion and fire in the seasonal submontane zone of Hawai'i. Ecology 72: 743-746. http://dx.doi.org/10.2307/2937215.

ICNF, 2013. IFN6 - Áreas dos usos do solo e das espécies florestais de Portugal continental. Resultados preliminares. Lisboa, Portugal. ICNF - Instituto da Conservação da Natureza e das Florestas. http://www.icnf.pt/portal/florestas/ifn/ifn6.

Immitzer M, Vuolo F, Atzberger C, 2016. First Experience with Sentinel-2 Data for Crop and Tree Species Classifications in Central Europe. Remote Sens 8(3): 166. http://dx.doi.org/10.3390/rs8030166.

Jensen J, 1996. Introductory Digital Image Processing: a Remote Sensing Perspective. Englewood Cliffs, New Jersey, USA. Prentice-Hall. 316 pp.

Joshi C, De Leeuw J, Van Andel J, Skidmore AK, Lekhak HD, Van Duren IC, Norbu N, 2006. Indirect remote sensing of a cryptic forest understorey invasive species. For Ecol Manage 225: 245-256.

Joshi C, Leeuw JD, Van Duren IC, 2004. Remote sensing and GIS applications for mapping and spatial modelling of invasive species. Proc. ISPRS, Istanbul, July 12-23. pp: 669-677.

Keeley JE. 2006. Fire management impacts on invasive plants in the Western United States. Conserv Biol 20: 375-384. http://dx.doi.org/10.1111/j.1523-1739.2006.00339.x.

Lawes RA, Wallace JF, 2008. Monitoring an invasive perennial at the landscape scale with remote sensing. Ecol Manage Restor 9: 53-59. http://dx.doi.org/10.1111/j.1442-8903.2008.00387.x.

Le Maitre DC, Gaertner M, Marchante E, Ens E, Holmes PM, Pauchard A, O'Farrell PJ, Rogers AM, Blanchard R, Blignaut J *et al.*, 2011. Impacts of invasive Australian acacias: Implications for management and restoration. Divers Distrib 17: 1015-1029. http://dx.doi.org/10.1111/j.1472-4642.2011.00816.x.

Lorenzo P, Gonzalez L, Reigosa M, 2010. The genus Acacia as invader: the characteristic case of Acacia dealbata Link in Europe. Ann For Sci 67: 101-112. http://dx.doi.org/10.1051/forest/2009082.

Marchante H, E Marchante, H Freitas. 2005. Invasive plant species in Portugal: an overview. Proc. of the International Workshop on Invasive Plants in Mediterranean Type Regions of the World. Montpellier, France. Council of Europe Publishing. pp: 99-103.

May BM, Attiwill PM, 2003. Nitrogen-fixation by Acacia dealbata and changes in soil properties 5 years after mechanical disturbance or slash-burning following timber harvest. For Ecol Manage 181: 339-355.

Parker IM, D Simberloff, WM Lonsdale *et al.* 1999. Impact: toward a framework for understanding the ecological effect of invaders. Biol Invasions 1: 3-19. http://dx.doi.org/10.1023/A:1010034312781.

Pauchard A, Maheu-Giroux M, 2007. Case study 11.1: *Acacia dealbata* invasion across multiple scales: Conspicuous flowering species can help us study invasion pattern and processes. In: Sourcebook on Remote Sensing and Biodiversity Indicators; Strand H, Höft R, Strittholt J, Miles L, Horning N, Fosnight E (eds.). pp: 166-167. CBD Technical Series 32. Montreal.

Pauchard A, Shea K, 2006. Integrating the study of non-native plant invasions across spatial scales. Biol Invasions 8: 399-413. http://dx.doi.org/10.1007/s10530-005-6419-8.

Pereira J, Figueiredo A, 2015. Assessing suitable area for *Acacia dealbata* Mill. in the Ceira River Basin (Central Portugal) based on maximum entropy modelling approach. GOT - Revista de Geografia e Ordenamento do Território 8: 171-190.

Pimentel D, Zuniga R, Morrison D, 2005. Update on the environmental and economic costs associated with alien-invasive species in the United States. Ecol Econ 52: 273-288. http://dx.doi.org/10.1016/j.ecolecon.2004.10.002.

Resasco J, Hale AN, Henry MC, Gorchov DL, 2007. Detecting an invasive shrub in a deciduous forest understory using late fall Landsat sensor imagery. Int J Remote Sens 28: 3739-3745. http://dx.doi.org/10.1080/01431160701373721.

Richards JA, 2013. Remote Sensing Digital Image Analysis. Berlin, Germany. Springer. 496 pp. http://dx.doi.org/10.1007/978-3-642-30062-2.

Richardson DM, Pysek P, Rejmánek M, Barbour MG, Panetta FD, West CJ, 2000. Naturalization and invasion of alien plants: concepts and definitions. Divers Distrib 6: 93-107. http://dx.doi.org/10.1046/j.1472-4642.2000.00083.x.

Richardson DM, Rejmánek M, 2011. Trees and shrubs as invasive alien species - a global review. Divers Distrib 17: 788-809. http://dx.doi.org/10.1111/j.1472-4642.2011.00782.x.

Souza-Alonso P, Lorenzo P, Rubido-Bará M, González L, 2013. Effectiveness of management strategies in *Acacia dealbata* Link invasion, native vegetation and soil microbial community responses. For Ecol Manage 304: 464-472.

Underwood E, Ustin S, 2007. Chapter 11. Trends in Invasive Alien Species. In: Sourcebook on Remote Sensing and Biodiversity Indicators; Strand H, Höft R, Strittholt J, Miles L, Horning N, Fosnight E (eds.). pp. 161-177. CBD Technical Series 32. Montreal.

Underwood E, Ustin S, DiPietro D, 2003. Mapping non-native plants using hyperspectral imagery. Remote Sens Environ 86: 150-161. http://dx.doi.org/10.1016/S0034-4257(03)00096-8.

Van de Wouw P, Echeverría C, Rey-Benayas JM, Holmgren M, 2011. Persistent acacia savannas replace Mediterranean sclerophyllous forests in South America. For Ecol Manage 262: 1100-1108.

Vazquez-de-la-Cueva A, 2014. Case studies of the expansion of *Acacia dealbata* in the valley of the river Miño (Galicia, Spain). Forest Syst 23: 3-14. http://dx.doi.org/10.5424/fs/2014231-02531.

Viana H, Aranha J, 2010. Mapping invasive species (*Acacia dealbata* Link) using ASTER / TERRA and LANDSAT 7 ETM + imagery. Proc. IUFRO Landscape Ecology Working Group International Conference. Bragança. pp: 443-448.

Xu M, Watanachaturaporn P, Varshney PK, Arora MK, 2005. Decision tree regression for soft classification of remote sensing data. Remote Sens Environ 97: 322-36. http://dx.doi.org/10.1016/j.rse.2005.05.008.

Yamashita N, Ishida A, Kushima H, Tanaka N, 2000. Acclimation to sudden increase in light favoring an invasive over native trees in subtropical islands, Japan. Oecologia 125: 412-419. http://dx.doi.org/10.1007/s004420000475.

4

Using the SUBER model for assessing the impact of cork debarking rotation on equivalent annual annuity in Portuguese stands

Joana A. Paulo and Margarida Tomé

Centro de Estudos Florestais, Instituto Superior de Agronomia, Universidade de Lisboa, Tapada da Ajuda, 1349-017 Lisboa, Portugal

Abstract

Aim of study: Use the SUBER model to evaluate the influence of the cork debarking rotation period (CDR) on equivalent annual annuity (EAA) value.

Area of study: Nine simulated stands, varying in site index (14.4, 15.6, 17.1) and cork quality characteristics (high, medium, low).

Material and methods: EAA values were computed considering CDR periods varying from 9 to 14 years, two contrasting structures of cork prices (high and low cork price scenarios), and three discount rate values (0.5%, 2% and 5%).

Main results: For discount rates of 0.5% and 2% the impact of different CDR on the EAA is similar. In stands characterized by high to average site index values or high to medium cork quality characteristics, CDR of 9 and 11 years are associated with similar values of EAA. The variation of the CDR in stands characterized by low site index values and/or low cork quality characteristics did not have a relevant effect on the variation of EAA. For the simulations carried out with a discount rate of 5% the EAA decreases with the increase of CDR, indicating that the minimum legal value of 9 years for CDR should be applied.

Research highlights: In stands characterized by high to average site index values or high to medium cork quality characteristics, a delay in the debarking may result in a significant increase of cork thickness and, as a result, of cork price. Detailed knowledge of cork and stand characteristics and updated information on cork prices structure and values are essential for the best usage of management tools such as the SUBER model, which can contribute to the decision-making process concerning the debarking operation.

Keywords: adaptive forest management; cork production; cork price; cork quality; growth and yield model; site index.

Funding: This work was supported by Fundação para a Ciência e a Tecnologia (Portugal) under contracts SFRH/BD/23855/2005 and SFRH/BPD/96475/2013. Support was also given by EU projects StarTree (Grant Agreement 311919) and AgForward (Grant Agreement 613520), financed under the Seventh Framework Program for Research and Technological Development.

Competing interests: The authors have declared that no competing interests exist.

Correspondence should be addressed to Joana Amaral Paulo: joanaap@isa.ulisboa.pt

Introduction

Because of the importance of cork oak trees in *montado* and *dehesas* ecosystems where cork oak (*Quercus suber* L.) is present, several management options are regulated by national legislation in some countries along its distribution area. One example is the interval between two consecutive cork extractions occurring in the same tree, here designated as cork debarking rotation (CDR), which, according to Portuguese legislation, should be of at least 9 years. Other countries have established different values for this management option. For example in Spain, as in Portugal, the debarking can take place after a minimum period of 9 years, but in Italy and Tunisia this minimum period is extended to 12 years.

The cork industry transforms this raw material into different products, but the economic feasibility of the whole sector is dependent on the production of stoppers of natural cork for use in the bottling of wines (Pereira, 2007; Corticeira Amorim, 2013). Therefore, the suitability of the raw material for this product establishes its commercial value. Two features are important when defining the value of the cork for the industry of cork stoppers: the thickness of the cork plank and the cork quality. Cork thickness can be measured either before or after boiling, but the latter is the reference for the industry and will be considered throughout this research (see details of this industrial process in Pereira, 2007). Cork quality is visually evaluated by an operator who, considering the cork porosity and the presence of cork defects, classifies the cork into 7 classes: from quality class 1 (best quality) to quality class 7 or refuse (worst quality). Due to the great variation of cork thickness (e.g. Almeida *et al.,* 2010; Paulo *et al.,* 2016) and cork porosity (e.g. Oliveira *et al.*, 2012) between trees in the same stand, and since different combinations of these two variables result in different prices, average cork price is not a suitable variable for characterizing the cork produced in a particular stand. Instead, we suggest the usage of a two -dimensional structure of cork prices that includes the proportions of both cork thickness and cork quality classes that characterise the stand for determining revenues associated with cork extraction (Almeida & Tomé, 2010).

In practice three aspects mainly influence the decision of the forest manager concerning the debarking: evolution of the structure of cork prices, labour costs and weather conditions during the cork growth period. For instance, an increase in the prices of thin corks (< 22.6 mm) may lead to the anticipation of cork extraction, while an increase in the price of thicker corks (> 40.6 mm) will make landowners postpone cork debarking one or two years, in the expectation of increasing the percentage of thick cork. In the years between 2000 and 2010, cork price structure experienced considerable fluctuations, generally showing a significant decrease in the value for most of the cork assortments. This was more evident in cork classified as poor and medium quality classes (APCOR, 2014). Even if in recent years the prices for some cork quality classes have increased and have been more stable, there is an increasing number of forest managers extending the CDR in their stands in an attempt to increase the value of the extracted product and therefore increase the resulting income.

In the matter of cork oak stands, there are three available models that can be used for this task: the SUBER model (Tomé, 2004; Paulo, 2011; Tomé & Faias, 2014) and the CORKFITS model (Ribeiro & Surový, 2011) for Portuguese cork oak stands, and the ALCORNOQUE model (Sánchez-González *et al.*, 2007; Tomé & Faias, 2014) for Spanish cork oak stands. The SUBER model has been selected because:

- the model is included in the sIMfLOR platform (Faias *et al.*, 2012) accessible for download at www.isa.ulisboa.pt/cef/fctools
- the model simulates annual cork growth based on tree level cork growth index values
- the model considers tree cork quality at tree level
- the model includes an economic module that uses cork prices structure

The objectives of the present work are: (i) to analyse the impact of site index, cork quality and thickness distribution and cork price structure on the equivalent annual annuity (EAA); (ii) to analyze the effectiveness of the extension of the CDR on the increase of EAA values for stands with different characteristics.

Although the simulation unit of the SUBER model is the tree, for a time step of one year, its outputs are produced at the stand level. Since forest management is made at this scale, this approach is considered adequate for the purpose of the present work. A multi-stand approach addressing cork oak stands in Portugal has been carried out by Borges *et al.* (1997) using previous simpler versions of the SUBER model that did not allow the simulation of different CDR. Palma *et al.* (2015), using the present version of the SUBER model, focused on an approach for the establishment of adaptive management plans to climate change on a regional approach, but the solutions found for the optimal CDR calendar are, in practice, difficult to implement in the current land management scenario. These studies are relevant at a strategical level, e.g. definition of forest policies or regional prediction of cork availability, but are not appropriate to support the landowner's decisions.

Revenues and costs resulting from pastoral or agricultural activities are not accounted for either, since the tree and cork growth modules included in the SUBER model do not take into account the influence of these practices on tree and cork growth. These have been considered by Campos *et al.* (2007a, 2007b, 2008), Costa *et al.* (2010) and Fragoso *et al.* (2011) using alternative simulation tools that do not account for the increase of cork thickness with increasing cork age, the existence of cork prices structures or the differences in cork growth and quality between trees in the same stand.

Material and methods

Tree data

Tree data was collected on three existing permanent plots established in three young cork oak plantations. The stands present similar tree density and management operations up to the moment of the measurement (Table 1). A characterization of the stands at the year of the measurement is presented in Table 1, including site index value for a base age of 80 years estimated with the model developed by Sánchez-González *et al.* (2005).

The three selected plots are part of a larger database that includes measurements made in 42 permanent plots established in young plantations in Portugal, which present site index values ranging from 11.5 to 18.4 m (reference age of 80 years), with an average value of 16.6 m. Fig. 1 presents the measurements of dominant

Table 1. Stand variables for each one of the stands used in the simulations.

Stand	N_t	t	Gu	dug	hdom	FAO Soil type[d]	Rock type
Good site	273	15 [a]	3.49	13.2	7.54	Luvisol	Sedimentar gravel
	270	20 [bc]	5.77	16.6	8.17		
	68	80 [b]	19.56	60.4	17.1 [SI]		
Average site	225	13 [a]	0.24	8.3	4.09	Cambisol	Sedimentar sandy
	215	34 [bc]	4.95	17.1	10.06		
	105	80 [b]	14.96	42.6	15.6 [SI]		
Poor site	222	14 [a]	0.02	5.5	3.14	Cambisol	Metamorphic granite
	205	40 [bc]	4.95	17.5	9.51		
	128	80 [b]	14.43	38.0	14.4 [SI]		

Nt - number of trees per hectare at the age t; t – stand age (years); Gu – basal area under cork at age t (m^2); dug – quadratic mean diameter under cork (cm) at the age t; hdom – dominant height at the age t (m); SI - site index (m) estimated using the Sánchez-González *et al.* (2005) model for a base age of 80 years. [a] Measured values; [b] Simulated values obtained with the SUBER model; [c] Age when the first debarking occurs in the stand, corresponding to the year when dg is equal to or greater than 22.3 cm (70 cm of perimeter at breast height over cork). [d] According to IUSS Working Group WRB (2006). Variables referring to values under cork are estimated by the SUBER model using the Paulo & Tomé (2014) model for the relation between diameter over (d) and under cork (du).

height taken in these plots, and five curves representing the Sánchez-González *et al.* (2005) dominant height curves for 10m, 12m, 14m, 16 and 18m classes, the range of site indices observed in Portugal. The selection of the three stands was made in order to include stands with different site index values, characterizing different growth conditions of new cork oak plantations. The three stands present the following values of site index: 14.4 m, 15.6 m and 17.1 m. Also, in order to homogenize the data from the different stands and make the equivalent annual annuity analysis comparable, stands were selected in order to present similar age and number of trees per hectare in the year of the measurement.

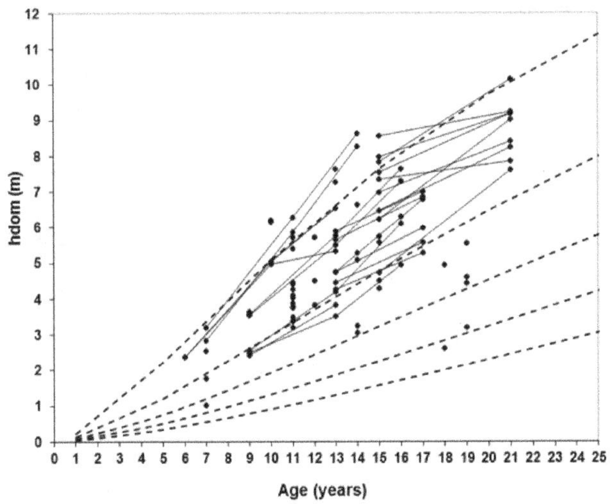

Figure 1. Dominant height value measured in permanent plots installed in young cork oak plantations in Portugal. Dots refer to permanent plots only measured once; black lines refer to plots already measured twice; hatched lines are the site index curves for cork oak, according to Sánchez-González *et al.* (2005) model for a base age of 80 years. The five curves represent the range of site indices observed in Portugal 10 m, 12 m, 14 m, 16 m and 18 m.

Cork data

The SUBER model needs the input of the empirical distribution values for the cork thickness and for the cork quality variables. These two distributions may be applied to the model from two alternative sources: i) at the tree level: information collected during a forest inventory and cork sampling procedures; ii) at the stand level: information derived from forest inventory or other source but not only available at the stand level (see details in Paulo, 2011). In the latter option the model uses the distribution values to assign cork growth index and cork quality class values to each tree in the stand, using Monte Carlo simulation (e.g. Gilks *et al.*, 1996). In the present research the information provided refers to frequency distributions.

As no relationship has been found between cork quality and cork growth index (Pereira, 2007) the distributions considered as model inputs were selected from a large data set including 187 cork inventories made in different stands for estimation of the cork value before the extraction. Each inventory included more than 30 cork samples. The average cork price for each of the 187 cork inventories was computed using the real cork prices structures from 2009 (Associação dos Produtores Florestais de Coruche, personal communication). The ones resulting in a higher, medium and lowest cork price value were used for the simulations and assumed

Table 2. Cork quality and cork growth index distributions used to characterise stands with corks of different value for the industry

	Classes	High	Medium	Low
Cork quality	1	0.08	0.1	0
	2	0.08	0.1	0
	3	0.08	0.1	0.03
	4	0.35	0.32	0.13
	5	0.21	0.14	0.4
	6	0.12	0.04	0.35
	refuse	0.08	0.2	0.09
Cork growth index (mm)	< 18	0	0.14	0.1
	18-23	0	0.26	0.25
	23-27	0.08	0.28	0.2
	27-41	0.7	0.32	0.38
	>41	0.22	0	0.07

High, medium and low classes were defined as the ones presenting higher, medium and lowest cork price values computed using a real cork prices structure.

to characterize stands with high (H), medium (M) and low (L) cork characteristics from an industrial point of view (Table 2).

It is important to notice that passing from high to low cork prices generally represents a decrease in the fraction of good quality corks (1, 2 and 3 cork quality classes) and/or thick corks (>41 mm cork growth index class), and simultaneously an increase in the fraction of low quality corks (6 and refuse cork quality classes) and/or thinner corks (< 23 mm cork growth index class). In detail the variation of cork price may be more complex, since it results from the simultaneous combination of the different classes of thickness and quality classes.

The SUBER model includes the Almeida *et al.* (2010) system of equations for predicting the evolution of individual tree mature cork caliper over time, and the Paulo & Tomé (2010) system of equations for estimating cork biomass production depending on cork age. Cork quality class of each tree is assumed to be constant along the years of the simulations and regardless of the cork age at the debarking by the model, as no relation between cork quality and age has been demonstrated in the literature. This assumption is reasonable if cork defects, for instance due to insect presence (e.g. *Coroebus undatus*, *Crematogaster scutellaris*), are discarded. These assumptions allow the model to determine the cork value, assuming different cork debarking rotation periods on each simulation.

Forest management options and silvicultural operations

The main aspects regarding the management of cork oak stands in Portugal are regulated by national legislation (Decree-Law n.º 169/2001 and Decree-Law n.º 155/2004). These determine:

- The minimum size of the tree perimeter at breast height over cork (p) after which the first debarking is allowed: 70 cm.
- The minimum interval between two consecutive cork extractions on the same tree, designated by the cork debarking rotation (CDR): 9 years.
- The maximum values for debarking coefficient (ratio between debarking height and tree perimeter over cork): 2 for the first debarking, 2.5 for the second debarking and 3.0 for the following debarking operations.

Regarding the forest management operations defined for the simulations, the assumptions take into account both national legislation and common practices for high natural and cultural value systems management in Portugal. Details on the forest management operations are presented in Table 3.

The maximum value of 14 years, considered for the interval between two cork extractions, is based on the fact that this is the longest period recorded in

Table 3. Forest management operations defined for the simulations.

Operation	1st event	Periodicity	Details
Debarking operation	$dg \geq 22.3$ cm	9 to 14 years	At tree level debarking is only made when the $p \geq 70$ cm Debarking coefficients*: 2 for the 1st debarking, 2.5 for the following debarking operations
Formation pruning	year of the 1st debarking		
Mechanical control of the understory	15 years old	5 year	Carried out with mounted knifes or chains
Sanitary thinning		annually	Only dead trees are thinned
Thinning operations	year of the 1st debarking	Debarking years	To reach crown cover percentage of 58%

dg – stand quadratic mean diameter, measured over cork, in cm; p – tree perimeter at breast height, measured over cork, in cm;
* debarking coefficient = hd/p; hd – debarking height in cm (defined according to Decree-Law n.º 169/2001 and Decree-Law n.º 155/2004);

the database of the Forchange research group, that was used by Almeida *et al.* (2010) and Paulo & Tomé (2010) models included in the SUBER. Intervals longer than 12 years between debarking operations are only encountered in very particular situations and in a reduced number of stands.

Economic data

Two real cork prices structures were considered for characterizing two distinct moments in the cork market in recent decades: the highest and lowest average cork prices, which occurred in the years 2000 and 2009, respectively. These distinct cork prices structures are the basis for the definition of two distinct scenarios: the first characterizing a favourable management scenario and the second an unfavourable management scenario. An average price scenario was not considered since this is difficult to define when considering a cork price structure instead of one single average cork price. For instance, the definition of a cork price structure characterized by an average cork price for all the cork thickness and quality classes would be completely unrealistic and not useful for fulfilling the objectives of the present study. These market fluctuations have originated from a combination of drivers, such as the evolution of the world wine market, the development of new cork applications and the improvement of methods of conservation of the raw material in the industry in order to cope with the fluctuations of material supply.

Although the classification of the cork is made by the industry considering five thickness classes and seven quality classes, cork prices are not differentiated for the resulting 35 combinations. For example, cork quality classes 1, 2 and 3 are typically attributed the same price, as the use given to this cork material is generally the same (e.g. Oliveira *et al.*, 2012). This is reflected in Tables 4 and 5. The values presented in these tables already include the costs from the debarking operation. Virgin cork price is equal to the price attributed to refuse cork.

When comparing the cork price values from the two tables, although referring to two different years, the differences regarding the two cork price structures and values are clear. Poor and medium cork quality classes were strongly affected by the decrease in price in 2009, but good cork even revealed some increase in the prices from 2000 to 2009. Another important aspect derived from these tables is the decrease in the number of cork price categories, going from 9 to 6 between 2000 (Table 4) and 2009 (Table 5).

In order to compare the simulations, the equivalent annual annuity (see for example Davis *et al.*, 2001) is computed by the SUBER model as

$$EAA = (r\ NPV) / [\ 1 - (1+r)^{-n}\ ]$$

Table 4. Cork price structure (€ @$^{-1}$) characterizing a 'high cork price scenario' (Cork prices for the year of 2000). 1@ = 15 kg. Source: Associação dos produtores florestais de Coruche (APFC)

<table>
<tr><th rowspan="2">Cork thickness class (mm)</th><th colspan="7">Quality class</th></tr>
<tr><th>1</th><th>2</th><th>3</th><th>4</th><th>5</th><th>6</th><th>7</th></tr>
<tr><td>< 18</td><td colspan="3" rowspan="2">36.39</td><td colspan="3">11.91</td><td></td></tr>
<tr><td>18 – 23</td><td colspan="2">21.43</td><td>12.45</td><td></td></tr>
<tr><td>23 – 27</td><td colspan="5">44.62</td><td>15.44</td><td></td></tr>
<tr><td>27 – 41</td><td colspan="5">64.82</td><td>21.43</td><td></td></tr>
<tr><td>> 41</td><td colspan="6">32.65</td><td></td></tr>
</table>

Prices already include the costs from the debarking operation (see Table 6).

Table 5. Cork price structure (€ @$^{-1}$) characterizing a 'low cork price scenario' (Cork prices for the year of 2009). 1@ = 15 kg. Source: Associação dos produtores florestais de Coruche (APFC).

<table>
<tr><th rowspan="2">Cork thickness class (mm)</th><th colspan="7">Quality class</th></tr>
<tr><th>1</th><th>2</th><th>3</th><th>4</th><th>5</th><th>6</th><th>7</th></tr>
<tr><td>< 18</td><td colspan="3" rowspan="3">11.75</td><td colspan="3">0.70</td><td></td></tr>
<tr><td>18 – 23</td><td rowspan="2">11.65</td><td colspan="2" rowspan="4">6.50</td><td></td></tr>
<tr><td>23 – 27</td><td></td></tr>
<tr><td>27 – 41</td><td colspan="3" rowspan="2">88.25</td><td rowspan="2">51.50</td><td></td></tr>
<tr><td>> 41</td><td></td></tr>
</table>

Prices already include the costs from the debarking operation (see Table 6).

where EAA is the equivalent annual annuity (€ ha^{-1} $year^{-1}$), NPV is the net present value (€ ha^{-1}), r is the discount rate assumed as 0.5%, 2% or 5% depending on the simulation, and n is the number of years of the simulation (80 years for all the simulations).

NPV (see for example Davis *et al.*, 2001) was calculated as

$$NPV = \sum^{n} [(R_t - C_t) / (1 + r)^t]$$

where R_t is the revenue for year t, C_t are the costs for year t, r and n are as previously defined.

NPV value includes the revenues resulting only from cork production. No revenues resulting from wood selling, cattle raising, or any additional activity (e.g. subsidies) were considered.

Costs from all silvicultural operations previously defined in section 2.3 are included in the C_t value. Costs are considered constant for the period of the simulation and were obtained in the CAOF (Comissão de Acompanhamento das Operações Florestais) reference matrices for 2013/2014, available in http://www.anefa.pt/. Since the objective of the paper is the evaluation of the impact of site productivity, cork growth index and quality distributions and cork prices structure on cork debarking rotation for long-term management of new cork oak plantations, the stands were assumed to be similar regarding site conditions, namely slope. Table 6 presents the value for each considered operation.

Costs and revenues associated to sanitary thinning operations are both considered equal to zero. Although some revenue may come from the selling of woody material with larger diameter, this is often consumed in domestic use. Other biomass residuals (small branches and leaves) are frequently discarded in the field, sometimes incorporated in the soil.

Simulations

Through the combination of tree data collected in three young existing stands (undebarked and for this reason without information on cork characteristics), with real cork quality and cork growth index frequencies distribution, nine simulated stands were defined for simulations. Growth and cork production of each stand was simulated until the stand reaches 80 years of age, considering two contrasting cork price structures and cork debarking rotations (CDR) ranging from 9 to 14 years. Simulations were carried out using the SUBER 5.0 model (Paulo, 2011; Tomé & Faias, 2014). A total of 324 simulations were carried out (3 stands x 3 cork distributions x 2 cork price structure x 6 CDR alternatives x 3 alternative discount rate).

Results

Figures 2, 3 and 4 present the results from the simulations carried out considering discount rate values of 0.5%, 2% and 5%, respectively. They present the values of the equivalent annual annuity (EAA) obtained for each simulated stand and each cork price structure considered, as a function of the cork debarking rotation (CDR). The analysis of the effects of different site indices, cork quality and thickness distributions and cork price structures on the CDR that maximizes equivalent annual annuity (EAA) are presented in the discussion section.

Discussion

The impact of different site index of the stand on the EAA

The impact of site index on EAA can be observed by comparing the three curves presented on each of the graphics in Figures 2, 3 and 4 separately. As expected the EAA values are smaller in sites with lower site indices. This reduction results from lower tree growth rates in sites with lower site index value, according to Tomé *et al.* (2006) diameter growth model that is used by the SUBER model. Since cork production (cork weight) is related to tree perimeter at breast height (Paulo & Tomé 2010), stands with lower site index are characterized, for the same age, by trees with lower perimeter and consequently lower values of cork produced for a same CDR. These results point out the importance of the

Table 6. Cost of the silvicultural operations considered along the simulations (see Table 3). Includes all valid legal taxes. Source: Comissão de Acompanhamento para as Operações Florestais (CAOF) reference matrices for 2013/2014, available in http://www.anefa.pt/

Operation	Unit	Cost
Mechanical control of the understory vegetation with mounted knifes or chains	€ ha^{-1}	195.57
Formation pruning	€ $tree^{-1}$	0.45
Cork debarking *	€ $@^{-1}$	3.50

*Value already deducted in the cork prices presented in Tables 4 and 5

location of cork oak stands, since the relationship of local edafoclimatic variables and tree productivity has been demonstrated on several occasions (Hidalgo *et al.*, 2008; Lacambra *et al.*, 2010; Paulo *et al.*, 2015).

When looking at the relationship between the curves, the ones denoting high and medium site indices are nearly parallel in all the graphics, indicating that for stands located in good sites the site index does not have a strong impact on the definition of the CDR that maximizes EAA. For scenarios using discount rates of 0.5% and 2% (Figures 2 and 3) CDRs of 9 or 11 years provide similar results. For scenarios using the discount rate of 5% (Fig. 4) a decreasing tendency of the EAA with CDR is observed, showing that CDR of 9 years should be implemented if EAA maximization is the main objective of forest management.

The simulations of the low site index stands show that, despite the discount rate considered, EAA values were low and not very sensitive to the variation of the CDR. For these stands the importance of the multifunctional management of the *montado* is clear, since the income resulting from other products such as game, agriculture production, wild mushrooms picking etc, not accounted for in the SUBER model, will have an important input in the final economic return for landowners (Campos *et al.*, 2007a). Subsidies derived from the common agricultural policy are also expected to impact economic return (Fragoso *et al.*, 2011). Without this economic support some *montado* areas may be driven to land use conversion, depletion of oak trees due to tree aging and lack of tree renewal, or even land abandonment (Campos *et al.*, 2007b; Campos *et al.*, 2008).

The impact of cork growth index and cork quality distributions

The effect of the cork quality and cork growth index distributions can be assessed by comparing the same type of curve along each of the three graphics named with prefix 'H', 'M' or 'L', separately in the upper and lower lines of graphics presented in figures 2, 3 and 4. As expected a decrease in the fractions of thicker and good quality cork, observed when moving from graphic H to M and then to L, results in a significant decrease of EAA values. This result is related to the fact that these two variables are the ones determining cork price, and consequently the income resulting from the cork extraction.

The impact of the cork growth index and cork quality distributions on the CDR that maximizes the EAA was more evident in the simulations carried out with discount rates of 0.5% and 2% (Figures 2 and 3). For high and medium average cork an increase of the CDR to 11 years may result in very similar or even higher EAA values. This suggests that postponing the extraction of the cork for one year is not compensated by an increase of the cork thickness, since this does not create a shift in the price to a higher cork price class. This is only obtained when the debarking is delayed by two years, increasing the CDR to 11 years.

Stands characterized by low and medium cork are

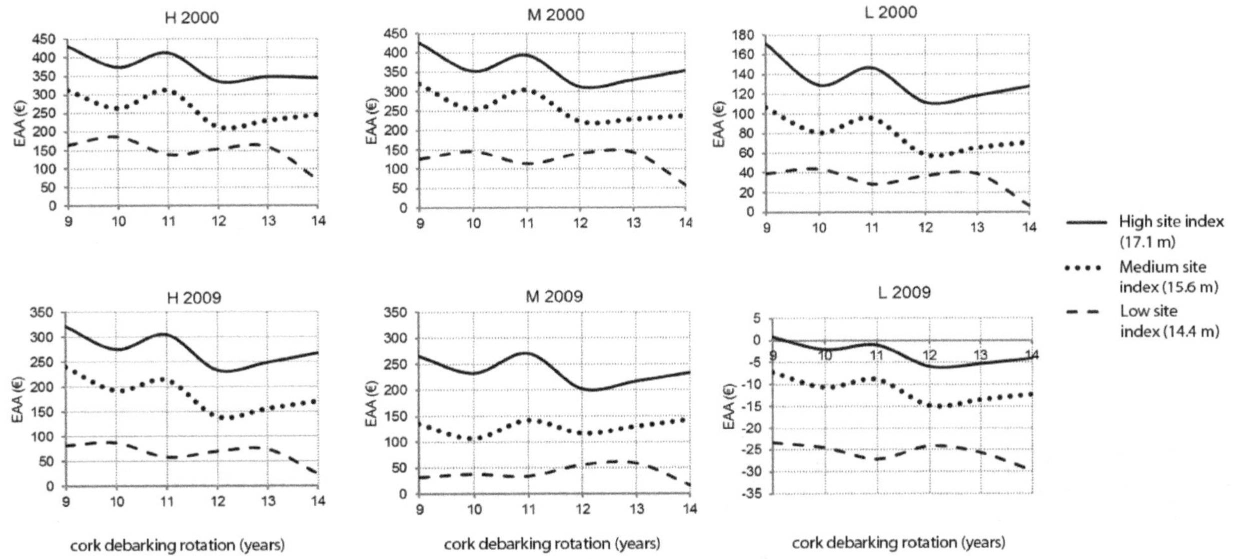

Figure 2. Equivalent annual annuity (EAA) for discount rate of 0.5% as a function of cork debarking rotation for stands with different site index (14.4 m, 15.6 m or 17.1 m). Cork prices structure from the year 2000 (high price scenario) is presented in the above part of the figure, and cork prices structure from the year 2009 (low price scenario) is presented below. H, M and L initials identify stands with high, medium or low fraction of good quality and thick corks, respectively. Vertical axis limits are different in upper and lower graphics in order to improve the analysis.

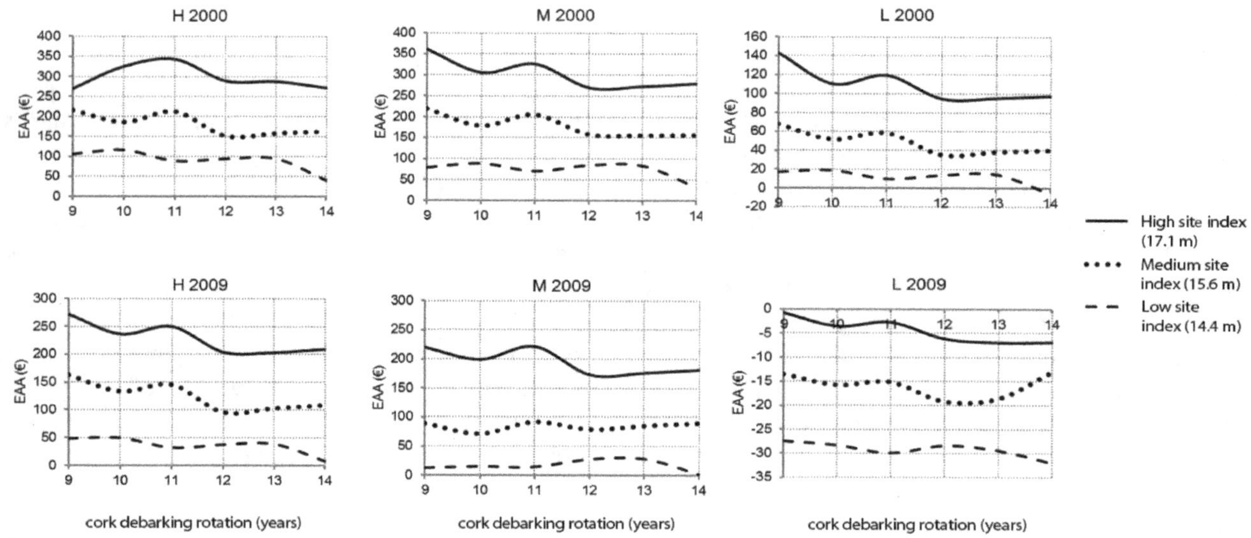

Figure 3. Equivalent annual annuity (EAA) for discount rate of 2% as a function of cork debarking rotation for stands with different site index (14.4 m, 15.6 m or 17.1 m). Cork prices structure from the year 2000 (high price scenario) is presented in the above part of the figure, and cork prices structure from the year 2009 (low price scenario) is presented below. H, M and L initials identify stands with high, medium or low fraction of good quality and thick corks, respectively. Vertical axis limits are different in upper and lower graphics in order to improve the analysis.

characterized by a reduction of the EAA value for increasing CDR. Although the increase of the CDR gives rise to an increase in cork thickness, low quality cork characteristics are related to lower cork price values, in particular cork quality classes 5, 6 and refuse (see tables 4 and 5), the ones more represented in the simulated stands presented in graphics 'L'.

For a discount rate of 5% (Fig. 4) the impact of extending CDR to 11 or more years is reduced since a decrease or stabilization of the EAA value is observed in all curves, suggesting that carrying out debarking operations are best practised in periods of 9 years.

Results demonstrate the importance of the cork growth index and quality distributions on the definition of CDR. In order to provide accurate data to the SUBER model and improve the accuracy of the simulated outputs, *prior* forest inventory and cork sampling should be carried out (cork sampling). This is already a standard operation in some of the major cork production regions of Portugal, carried on with specific sampling designs that optimise the information about cork value (Almeida & Tomé, 2010).

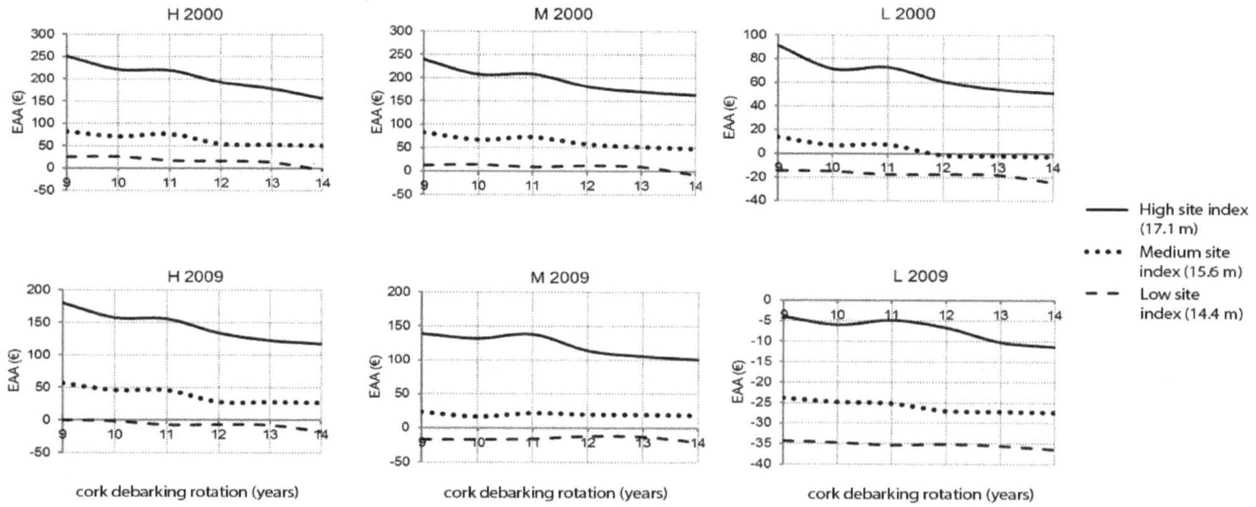

Figure 4. Equivalent annual annuity (EAA) for discount rate of 5% as a function of cork debarking rotation for stands with different site index (14.4 m, 15.6 m or 17.1 m). Cork prices structure from the year 2000 (high price scenario) is presented in the above part of the figure, and cork prices structure from the year 2009 (low price scenario) is presented below. H, M and L initials identify stands with high, medium or low fraction of good quality and thick corks, respectively. Vertical axis limits are different in upper and lower graphics in order to improve the analysis.

Impact of different cork price structures

The effect of the cork price structures on EAA can be observed by comparing the same type of curve (full, broken and dot lines) of the upper and lower graphics (suffix '2000' and '2009') in each of the Figures 2, 3 and 4.

As expected, and regardless of the discount rate taken for the simulation, the values of the EAA are strongly dependent on the cork price structure, with a clear reduction in the resulting values when moving from cork prices structure for the year 2000 to the one for the year 2009 (high and low cork price scenarios respectively). It is also observed that the same curves presented in the upper (2000) and lower (2009) graphics are nearly parallel. This shows that the CDR values that maximize the EAA are not related to the cork price structure.

Decision on cork debarking rotation

Increasing income resulting from cork extraction may be achieved by: i) reduction of labour costs; ii) increasing cork price; iii) increasing the amount of cork extracted by increasing debarking coefficients and consequently debarking intensity; iv) increasing cork thickness and/or the percentage of good quality cork. The evolution of the first two is market driven, and forest managers are limited in their influence. Still, due to the impact of cork price structures on the final income, annual assessments considering the historical records of cork price structures should be carried out when cork age is 7 or more years, in order to provide additional information relevant to the decision regarding CDR. This can be achieved using the SUBER model, which allows the realization of simulations considering cork thickness and cork quality data at tree level, different CDR and different cork price structures.

In the matter of the amount of cork extracted, it is common practice not to debark trees up to the maximum value allowed by law. Although the relationship between debarking intensity and tree growth and resilience is still an area in need of research (Oliveira & Costa, 2012), landowners are usually conservative concerning debarking intensity, and even adapt it to the tree growing and sanitary conditions. Decision on CDR should consider not only the variables considered in this research, but also growth conditions, taking into account the importance of cork harvesting and climate for stem radial growth of *Quercus suber* L. (Oliveira *et al.*, 2002; Leal *et al.*, 2008). It is frequent, when the price of thinner cork is low, for landowners to decrease the debarking height since cork thickness is known to decrease along the stem height (Montero, 1992) and simultaneously cork extraction costs are known to increase when debarking at upper tree heights and along tree main branches.

Since the relationship between management practices and cork quality is still not well established, it is difficult to influence cork quality. Although cork quality is known to be largely related to genetic characteristics (e.g. Teixeira *et al.*, 2014), further research on the development of reproduction material is needed. However, cork thickness is a result of cork age (determined by the applied CDR) and influenced by climatic conditions during the growing period, such as precipitation and temperature (e.g. Caritat *et al.*, 2000; Paulo *et al.*, 2016).

The large range of variables influencing cork production implies that any decision regarding CDR is a complex task. Decision support tools such as the SUBER model can provide additional insight during this process. It is suggested that this should be made in a two-stage procedure:

1. Initial assessment of the evolution of the EAA for extending CDR periods from 9 to 14 years at tree level: this should be made using the SUBER model, following a similar approach to the one presented here. This long-term simulation will give a first definition of the best CDR according to the stand characteristics. This analysis can also be incorporated in forest management plans.
2. Around 7 to 9 years of cork age, using updated data on cork quality and thickness distributions and recent cork prices structure information, simulation of cork thickness evolution and expected EAA value, for CDR values ranging from 9 to 14 years. This can also be made using the SUBER model. It is expected that this simulation will contribute to the decision on whether to debark or not. If the decision is made to postpone the debarking, new simulations can be done the following year if up-to-date data on cork prices structure is available.

Conclusions

The three considered drivers (site index, cork quality and thickness distribution, and cork price structure) influence the EAA value.

The two most influencial ones, which allow maximizing the EAA, are site index and the cork quality and cork growth index distributions. The variation of EAA due to the CDR extension was evidenced in scenarios considering discount rates of 0.5% or 2% of. For these scenarios similar values of EAA were obtained by postponing the debarking for CDR values to 11 years for stands characterized by high or medium site index and high or medium cork characteristics (quality and growth index). Stands characterized by low site index or low cork characteristics presented lower, nearly constant, and in some cases negative EAA values, irrespective of the CDR considered in the simulations.

For 5% discount rates the extension of the CDR period systematically originates a reduction of the EAA value, suggesting an increase in the number of cork extractions instead of an increase in cork thickness as a result of the CDR postponement.

Growth and yield models such as the SUBER model can be helpful tools for decision makers in the decision to undertake or postpone a debarking operation. The need for accurate and updated data on cork stand characteristics and cork prices is crucial for the accomplishment of similar studies.

Acknowledgements

Authors thank the APFC, ACHAR and AFLOSOR forest owners associations for their collaboration and the facilitation of data regarding cork inventories.

References

Almeida A, Tomé J, Tomé M, 2010. Development of a system to predict the evolution of individual tree mature cork caliper over time. Forest Ecol Manag 260: 1303 – 1314. https://doi.org/10.1016/j.foreco.2010.07.017

Almeida A, Tomé M, 2010. Field sampling of cork value before extraction in Portuguese 'montados'. Agr Syst 79: 419–430. https://doi.org/10.1007/s10457-009-9260-8

APCOR, 2014. Anuário da Cortiça 2014. 78 pp. Available on line:http://www.apcor.pt/userfiles/File/Publicacoes/AnuarioAPCOR2014.pdf.

Borges JG, Oliveira AC, Costa MA, 1997. A quantitative approach to cork oak forest management. Forest Ecol Manag 97: 223–229. https://doi.org/10.1016/S0378-1127(97)00064-9

Campos P, Bonnieux F, Caparrós A, 2007a. Measuring total sustainable incomes from multifunctional management of Corsican Maritime Pine and Andalusian Cork oak Mediterranean forests. J Environ Plan Manag 50(1): 65-85. https://doi.org/10.1080/09640560601048424

Campos P, Daly-Hassen H, Ovando P, 2007b. Cork Oak Forest Management in Spain and Tunisia: Two Case Studies of Conflicts between Sustainability and Private Income. Internat Forestry Rev 9 (2): 610-626. https://doi.org/10.1505/ifor.9.2.610

Campos P, Ovando P, Montero G, 2008. Does private income support sustainable agroforestry in Spanish dehesa? Land Use Policy 25(4): 510–522. https://doi.org/10.1016/j.landusepol.2007.11.005

Caritat A, Gutiérrez E, Molinas M, 2000. Influence of weather on cork-ring with. Tree Physiol 20: 893-900. https://doi.org/10.1093/treephys/20.13.893

Corticeira Amorim, 2013. Relatório e Contas 2013 – Corticeira Amorim, S.G.P.S., S.A. Mozelos. 142 pp. Available on: http://www.amorim.com/xms/files/Investidores/5_Relatorio_e_Contas/PT_R_C_Corticeira_Amorim2013.pdf.

Costa A, Oliveira A C, Vidas F, Borges JG, 2010. An approach to cork oak forest management planning: a case study in southwestern Portugal. Eur J Forest Res 129(2): 233-241. https://doi.org/10.1007/s10342-009-0326-y

Davis LS, Johnson KN, Bettinger PS, Howard TE, 2001. Forest Management: To sustain ecological, economic and social values. Fourth edition. MacGraw-Hill series in forest resources. Boston. 804 pp.

Faias SP, Palma JHN, Barreiro SM, Paulo JA, Tomé M, 2012. Resource communication. sIMfLOR – platform for the Portuguese forest simulators. Forest Syst 21(3): 543 – 548. https://doi.org/10.5424/fs/2012213-02951

Fragoso R, Marques C, Lucas MR, Martins MB, Jorge R, 2011. The economic effects of common agricultural policy on Mediterranean montado/dehesa ecosystem. J Policy Model 33(2): 311–327. https://doi.org/10.1016/j.jpolmod.2010.12.007

Gilks WR, Richardson S, Spiegelhalter D, 1996. Markov Chain Monte Carlo in Practice. Chapman and Hall. CRC Interdisciplinary Statistics. 481 pp.

Hidalgo PJ, Marín JM, Quijada J, Moreira JM, 2008. A spatial distribution model of cork oak (Quercus suber) in Southern Spain: A suitable tool for reforestation. Forest Ecol Manag 255: 25-34. https://doi.org/10.1016/j.foreco.2007.07.012

IUSS Working Group WRB, 2006. World reference base for soil resources 2006. 2nd edition. World Soil Resources Reports No. 103, FAO, Rome, 133 pp.

Lacambra LCJ, Andray AB, Francés FS, 2010. Influence of soil water holding capacity on the potential distribution of forest species. A case study: the potential distribution of cork oak (Quercus suber L.) in central-western Spain. Eur J Forest Res 129: 111-117. https://doi.org/10.1007/s10342-008-0251-5

Leal S, Nunes E, Pereira H, 2008. Cork oak (Quercus suber L.) wood growth and vessel characteristics variation in relation to climate and cork harvesting. Eur J Forest Res 127: 33-41. https://doi.org/10.1007/s10342-007-0180-8

Montero G, 1992. Variacion del calibre del corcho medido a distintas alturas. Inv Agrar. Sis. Recur For 1(2): 181-188.

Oliveira G, Costa A, 2012. How resilient id Quercus suber L. to cork harvesting? A review and identification of knowledge gaps. Forest Ecol Manag 270: 257-272. https://doi.org/10.1016/j.foreco.2012.01.025

Oliveira G, Martins-Loução MA, Correia O, 2002. The relative importance of cork harvesting and climate for stem radial growth of Quercus suber L. Ann Forest Sci 59: 439 – 443. https://doi.org/10.1051/forest:2002018

Oliveira V, Knapic S, Pereira H, 2012. Natural variability of surface porosity of wine cork stoppers of different commercial classes. J Int Sci Vigne Vin 46(4): 331-340. https://doi.org/10.20870/oeno-one.2012.46.4.1523

Palma JHN, Paulo JA, Faias SP, Garcia-Gonzalo J, Borges JG, Tomé M, 2015. Adaptive management and debarking schedule optimization of Quercus suber L. stands under climate change: case study in Chamusca, Portugal. Reg Environ Change 15(8): pp 1569-1580. https://doi.org/10.1007/s10113-015-0818-x

Paulo JA, 2011. Desenvolvimento de um sistema para apoio à gestão sustentável de montados de sobro. PhD Thesis. Universidade Técnica de Lisboa, Instituto Superior de Agronomia, Lisboa, Portugal, 188 pp. http://hdl.handle.net/10400.5/3850

Paulo JA, Palma JHN, Gomes AA, Faias SP, Tomé J, Tomé M, 2015. Predicting site index from climate and soil variables for cork oak (Quercus suber L.) stands in Portugal. New Forests 46(2): 293-307. https://doi.org/10.1007/s11056-014-9462-4

Paulo JA, Pereira H, Tomé M, 2016. Analysis of variables influencing tree cork caliper in two consecutive cork extractions using cork growth index modelling. Agr Syst 90: 1-17. https://doi.org/10.1007/s10457-016-9922-2

Paulo JA, Tomé M, 2010. Predicting mature cork biomass with t years of growth based in one measurement taken at any other age. Forest Ecol Manag 259: 1993 – 2005. https://doi.org/10.1016/j.foreco.2010.02.010

Pereira H, 2007. Cork: biology, production and uses. Elsevier, Lisbon, 336 pp.

Ribeiro NA, Surový P, 2011. Growth modeling in complex forest systems: CORKFITS a tree spatial growth model for cork oak woodlands. Formath 10: 263 - 278 https://doi.org/10.15684/formath.10.263

Sánchez-González M, Tomé M, Montero G, 2005. Modelling height and diameter growth of dominant cork oak trees in Spain. Ann Forest Sci 62: 633 - 643. https://doi.org/10.1051/forest:2005065

Sánchez-González M, Calama R, Cañellas I, Montero G, 2007. Management oriented growth models for multifunctional Mediterranean forests: the case of the cork oak (Quercus suber L.) In: Scientific tools and research needs for multifunctional Mediterranean forest ecosystem managemen (Palahí M., Birot Y., Rois M., eds). EFI proceedings 56: 71-84.

Teixeira RT, Fortes AM, Pinheiro C, Pereira H, 2014. Comparison of good- and bad-quality cork: application of high-throughput sequencing of phellogenic tissue. J Exp Bot 65(17): 4887-4905 https://doi.org/10.1093/jxb/eru252

Tomé M, 2004. Modelo de crescimento e produção para a gestão do montado de sobro em Portugal. Projecto POCTI/AGR/35172/99. Relatório Final – Relatório de Execução Material (Volume I). Publicações GIMREF RFP 1/2004. Universidade Técnica de Lisboa. Instituto Superior Agronomia. Centro de Estudos Florestais. Lisboa. 89 pp. http://hdl.handle.net/10400.5/2355

Tomé M, Barreiro S, Paulo JA, Tomé J, 2006. Modelling tree and stand growth with growth functions formulated as age independent difference equations. Canadian J Forest Res 36(7): 1621-1630. https://doi.org/10.1139/x06-065

Tomé M, Faias SP, 2014. State of the art, review of silviculture, models and decision support tools for multipurpose trees (MPT) and non-wood forest products (NWFP). Deliverable 2.1 of the StarTree project. 176 pp. http://star-tree.eu/images/deliverables/WP2/Deliverable2_1.pdf

5

Biomass conversion and expansion factors in Douglas-fir stands of different planting density: variation according to individual growth and prediction equations

Pasquale A. Marziliano, Giuliano Menguzzato, Angelo Scuderi, Clemente Scalise and Vittoria Coletta

Agraria Department, University "Mediterranea" of Reggio Calabria. Loc. Feo Di Vito - 89060 Reggio Calabria, Italy

Abstract

Aim of study: We built biomass expansion factors (BCEFs) from Douglas-fir felled trees planted with different planting densities to evaluate the differences according tree size and planting density.

Area of study: The Douglas-fir plantation under study is located on the northern coastal chain of Calabria (Tyrrhenian side) south Italy.

Materials and methods: We derived tree level BCEFs, relative to crown ($BCEF_c$), to stem ($BCEF_{st}$ = basic density, BD) and total above-ground ($BCEF_t$) from destructive measurements carried out in a Douglas-fir plantation where four study plots were selected according to different planting densities (from 833 to 2500 trees per hectare). The measured BCEFs were regressed against diameter at breast height and total height, planting density, site productivity (SP) and their interactions to test the variation of BCEFs. Analysis of variance (ANOVA) and the post hoc Tukey comparison test were used to test differences in $BCEF_t$, $BCEF_c$ and in BD between plots with different planting density.

Main results: BCEFs decreased with increasing total height and DBH, but large dispersion measures were obtained for any of the compartments in the analysis. An increasing trend with planting density was found for all the analyzed BCEFs, but together with planting density, BCEFs also resulted dependent upon site productivity. $BCEF_t$ average values ranged between 1.40 Mg m^{-3} in planting density with 833 trees/ha (PD833) to 2.09 Mg m^{-3} in planting density with 2500 trees/ha (PD2500), which are in the range of IPCC prescribed values for Douglas-fir trees.

Research highlights: Our results showed that the application of BCEF to estimate forest biomass in stands with different planting densities should explicitly account for the effect of planting density and site productivity.

Keywords: biomass; biomass expansion factor; planting density; *Pseudotsuga menziesii* (Mirb.) Franco.

Funding: authors received no specific funding for this work.

Competing interests: The authors have declared that no competing interests exist.

Correspondence should be addressed to Pasquale Antonio Marziliano: pasquale.marziliano@unirc.it

Introduction

International agreements like the United Nations Framework Convention on Climate Change (UNFCCC) and its Kyoto Protocol (1997) have recognized the potential of forests to mitigate the anthropogenic increase of atmospheric CO_2 concentrations, leading countries to give major attention to the quantification of the amount of carbon stored in the biomass. Under these agreements, countries are requested to estimate and report CO_2 emissions and removals of forest, and the credited sinks may be used as emission reductions (Somogyi *et al.*, 2006).

Biomass may be calculated from measured tree attributes using allometric equations (Zianis *et al.*, 2005; Coletta *et al.*, 2016), or calculated indirectly by multiplying the volume estimates by biomass factors that convert timber volumes to dry weight (density factor) and thereafter to whole tree biomass (expansion factor) (i.e. BCEF, biomass conversion expansion factor, Somogyi *et al.*, 2006; IPCC, 2006; Teobaldelli *et al.*, 2009). These factors are calculated as the ratio between the biomass of the compartment under consideration (e.g. aboveground; aboveground and roots) and the stem volume (s) of tree (s). The Good practice guidance for land use, land-use change and forestry (LULUCF, Penman *et al.*, 2003) developed under the Kyoto protocol, lists BCEF approach as the preferred method for some of the tiers. Therefore national and regional aboveground biomass (AGB) estimates are generally calculated based on estimates of standing stem volume from forest inventories and from default biomass conversion and expansion factors (BCEFs). However, BCEF values can vary according to vegetation type, precipitation regime, mean annual temperature and tree age and size (e.g. Lehtonen *et al.*, 2004, Tobin & Nieuwenhuis, 2007, Petersson *et al.*, 2012), thus, use of default values for national- or regional-scale estimates might result in unreliable assessments of biomass, and carbon (Magalhães & Seifert, 2015). This is recognized by LULUCF guidance, since the higher tier methods call for greater specificity, such as country-level factors and factors specific to species.

In the present study we derived tree level BCEFs from destructive measurements carried out in a Douglas-fir plantation arranged in 4 different planting densities (from 833 to 2500 trees per hectare). The study allowed the evaluation of the planting density effects on the main stand dendrometric parameters through analysis of variation of the ratio of above-ground biomass to stem volume. The choice of the proper planting density is an extremely important factor, especially in wood plantations. Significant changes in planting density affect single tree and whole stand growth, which in turn affect yield potential and economic return. In Italy a BCEF value of 1.4 Mg m^{-3} for exotic conifer plantations C-reporting is suggested by the INFC (National Forest and Carbon Inventory, 2005). It was shown though that BCEF changes with age in Douglas-fir plantations in southern Italy (Marziliano *et al.*, 2015a), therefore using the constant value as reported by INFC leads to an underestimation of biomass for the younger stands and an overestimation of biomass for the older stands (Marziliano *et al.*, 2015b). In specific studies, prediction models for BCEFs have been proposed to better reflect stand characteristics comparatively to the use of a constant and unique (average value) for the species (e.g. Sanquetta *et al.*, 2011; Soares & Tomé, 2012; Enes Duque & Fonseca Fidalgo, 2014). The advantages of using locally derived biomass expansion factors are that while they compensate for regional environmental conditions (Schoene, 2002), they will also redress contemporary differences in growth patterns.

The objective of this research was to analyze the variation of total above-ground, crown and stem BCEFs in pure stands of Douglas-fir according to the planting density. We hypothesized that (i) the biomass expansion factors vary with stands characteristics and (ii) planting density could affect the value of the factors. BCEFs were calculated for every above-ground tree compartment (crown, stem and total) and different BCEFs for each planting density have been obtained. A regression analysis with the two major dendrometric parameters (diameter at breast height and total height) highlighted that dependencies of expansion factor on dendrometric characteristics are highly variable.

Materials and Methods

Study area

The Douglas-fir (*Pseudotsuga Menziesii*, (Mirb.) Franco) plantation under study is located on the northern

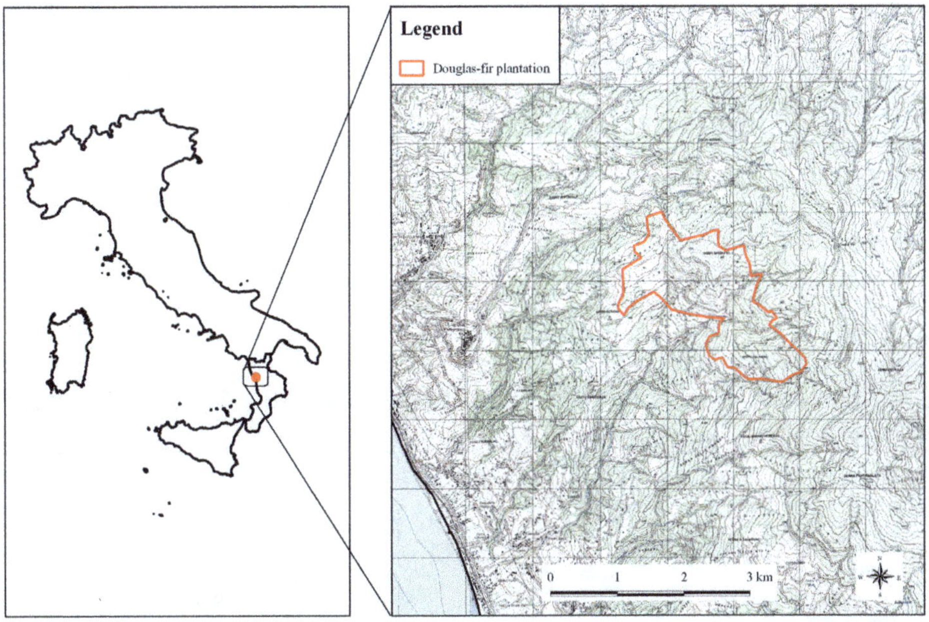

Figure 1. The map shows the location of the plots

coastal chain of Calabria, Tyrrhenian side of south Italy. The locality is called Serra Salinaro (39°, 25' N, 16°, 2' E), average altitude 900 m above sea level (Fig. 1). Regional Service of State Forest afforested the area in 1967 mainly with Douglas-fir. At the time of plantation (1967), Douglas-fir was planted with different planting densities (from 833 to 2500 trees per hectare). No silvicultural interventions were performed thereafter, and mortality is solely attributed to natural mortality. The climate of the area is typical Mediterranean. The annual rainfall is 1233 mm, with minimum precipitation in summer (88 mm) and a maximum during the winter (501 mm). Average annual temperature is 11.5 °C at an altitude of 950 m a.s.l. Average temperature of the coldest month is 3 °C, while the warmest has an average temperature of 20.8 °C. The problem of summer aridity is absent, owing to abundant precipitation, frequent fogs and exposure to moist air from the Tyrrhenian sea (Cantore & Iovino, 1989). Geologically the site is underlain by green and purple Paleozoic schist. According to FAO soils classification (FAO, 1998) soils are brown Podzols, with a plinthic horizon starting within 50 cm from the soil surface, deriving from basic metamorphic rocks, with highly decomposed organic material in more arid sites.

Data collection

The planting densities (PD) were: (i) PD2500 (2500 trees ha^{-1}); (ii) PD1667 (1667 trees ha^{-1}); (iii) PD1200 (1200 trees ha^{-1}); (iv) PD833 (833 trees ha^{-1}). When plantation was 15 years old, a randomized block design was used to test the effects of planting density. Totally, 12 permanent plots were established (4 treatments x 3 replicates). Plots had a square shape, with a surface area of 900 m^2. In order to avoid a "border effect" a 10 m buffer was left between contiguous plots. From this year and later on at age 25, 32 and 40, two orthogonal diameters at breast height (*DBH*) of all trees and the total height (*Ht*) of a representative sample (about 50%) of trees were measured in each plot (surveys at ages 32 and 40 have been used for other research studies).

When the plantation was 25 years old, the total height of 9 largest (diameter) trees (the 100 largest trees per hectare, according to Hägglund (1981)) was used to calculate the site productivity (SP) for each plot (Table 1). At the same age (25 years old), 35 trees were randomly selected for each planting density, according to the randomized block design (4 treatments x 3 replicates), but outside plots marked for dendrometric measurements (to avoid affecting tree-density). On the whole 420 selected trees were felled (stumps were cut at 20 cm height from soil) and divided into the following components: (1) stem; (2) branches; (3) needles; (4) crown (2 + 3); and (5) whole above-ground tree (1 + 4). Felled trees were scaled up to a 2.5 cm top diameter. The stem was defined as the length of the trunk from the stump to the height that corresponded to 2.5 cm diameter. The crown was divided into two sub-components: branches and foliage. Primary branches, originating from the stem, were classified in two categories: primary branches with diameters at the insertion point on the stem ≥ 2.5 cm were classified as large branches, and those with diameter < 2.5 cm were classified as fine branches.

The trees were cut from the stump and their heights were measured from breast height up to the height where diameter becomes 5 cm. Afterwards the stem was divided into sections, the first with 1.1 m length, the second with 1.7 m, and the remaining with 3 m, except the last, which length depended on the length of the stem. The volume of the highest part of each tree (where diameter was < 5 cm) was calculated using the cone's volume equation. The volumes of the stem were obtained by summing up the volumes of remaining sections calculated using Smalian's formula and the volume of the top. All the sections were weighted using a digital scale. Discs were removed at the bottom and top of the first section, and on the top of the remaining sections. The discs were dipped in drums filled with water, until constant weight (about 3 months), for its saturation and subsequent determination of the basic density, obtained by dividing the oven dry weight of the discs by the relevant saturated wood volume (calculated according to Archimedes' principle). Oven drying of all samples (of all the components) was done at 105°C to constant weight. Dry weights of each stem section were obtained by multiplying respective basic densities by relevant stem section volumes.

For what concerns crown, each sub-component (needles, large and fine branches) was weighted and

Table 1. Dendrometric characteristics and site productivity (SP). G = Basal area; Dg = Mean DBH; Ht = Total height; V= Volume; SP = Site productivity. Age of the trees = 25 years old

PD	N° trees ha^{-1}	Mortality	G (m^2 ha^{-1})	Dg (cm)	Ht (m)	V (m^3 ha^{-1})	SP (m)
2500	1634	35%	49.83	19.7	19.9	579.18	22.7
1667	1409	16%	51.53	21.6	21.4	628.94	23.4
1200	1056	16%	44.63	23.2	19.9	492.30	21.9
833	689	17%	41.24	27.6	19	410.99	20.7

from each a sample was taken to the laboratory for oven drying. Dry weight (biomass) of all the components of crown was estimated as the product of the fresh weight of a component and the ratio between the dry weight and the fresh weight of the sample.

Calculation of BCEF

The BCEF is a multiplier with dimension (Mg m^{-3}) that is used to convert timber volumes to dry weight (density factor) and thereafter to whole tree biomass (expansion factor) to estimate the biomass of whole tree. In this paper V refers to the stem volume with bark (m^3) and B_i to the biomass (dry weight, Mg) of the i compartment (total above-ground, stem and crown).

$$BCEF = \frac{B_i}{V} \quad \text{Eq.1}$$

BCEF relative to stem (corresponding to basic density, BD), to crown ($BCEF_c$) and to total above-ground biomass ($BCEF_t$) were calculated from sample trees belonging to different planting density. They were then regressed against DBH and height.

Stem volume and biomass of tree components were evaluated as described before.

Statistical analysis

Data from felled trees were used to analyze changes in BCEF varying with stand variables and to develop models to estimate BCEFs (of total above-ground, crown and stem). The Shapiro-Wilk test was performed to evaluate the normal distribution of data. Analysis of variance (ANOVA) and the post hoc Tukey comparison test were used to test differences in BCEFs between plots with different planting density. A 5% significance level was used throughout.

The effects influencing BCEFs accounted for the statistical analysis were DBH, height, planting density (PD), stand density at the age of 25 years old (SD_{25}), site productivity (SP) and their interactions.

As a preliminary analysis, these variables were scattered against each other to gain a general understanding of their behaviour. Afterwards, to estimate BCEFi, the variables were combined as in the following model:

$$Ln(BCEF_i) = b_0 + b_1 \cdot DBH + b_2 \cdot H + b_3 \cdot PD + b_4 \cdot SD_{25} + b_5 \cdot SP + b_n \cdot \text{interaction terms}_n \quad \text{Eq. 2}$$

The equation was built on the basis of the models proposed by Curtis (1967) and Clutter *et al.*, (1983) and already adopted in a modified form by Marziliano *et al.*

Table 2. Descriptive statistics of the attributes measured on the sampled trees (age = 25 years old); the Shapiro-Wilk test shows that all attributes have a normal distribution. DBH Diameter at breast height, Ht total height, $BCEF_t$ (Total BCEF), $BCEF_c$ (Crown BCEF), BD (basic density of stem), SD standard deviation, SWT Shapiro-Wilk test (n=420)

PD (tree/ha)	Variable	Mean	SD	Skewness	Kurtosis	SWT Sig.
833	DBH (cm)	23.30	3.961	-0.338	0.129	0.080
	Ht (m)	18.50	1.037	0.521	-0.023	0.348
	$BCEF_t$ (Mg m^3)	1.40	0.191	0.596	0.303	0.367
	$BCEF_c$ (Mg m^3)	0.36	0.098	0.880	0.387	0.068
	BD (Mg m^3)	0.66	0.094	0.409	-0.325	0.570
1200	DBH (cm)	21.40	4.376	0.324	-0.789	0.392
	Ht (m)	19.36	1.802	-0.025	-0.267	0.733
	$BCEF_t$ (Mg m^3)	1.88	0.491	0.176	-0.586	0.784
	$BCEF_c$ (Mg m^3)	0.58	0.214	0.478	-0.006	0.535
	BD (Mg m^3)	0.71	0.123	0.913	0.059	0.051
1667	DBH (cm)	19.81	3.982	0.833	0.232	0.052
	Ht (m)	20.85	1.722	-0.197	-0.980	0.341
	$BCEF_t$ (Mg m^3)	1.55	0.340	0.710	-0.155	0.140
	$BCEF_c$ (Mg m^3)	0.48	0.173	0.723	-0.079	0.082
	BD (Mg m^3)	0.57	0.099	0.120	-0.170	0.887
2500	DBH (cm)	18.21	2.285	0.199	0.489	0.793
	Ht (m)	19.66	1.499	0.639	0.455	0.310
	$BCEF_t$ (Mg m^3)	2.09	0.535	0.660	-0.149	0.166
	$BCEF_c$ (Mg m^3)	0.63	0.224	0.944	0.633	0.073
	BD (Mg m^3)	0.81	0.184	0.073	-1.106	0.265

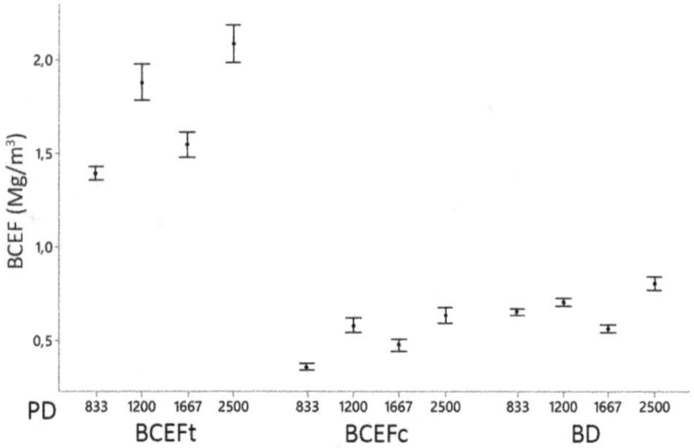

Figure 2. Mean +/- 2 standard errors of total above-ground ($BCEF_t$), stem (BD) and crown BEF ($BCEF_c$) (Mg m^{-3}) in the plots with different planting densities. Bars are 2 standard errors from mean.

(2013) and Marziliano *et al.* (2015c). The logarithmic transformation of the dependent variable produces a curve which, with the increase of the diameter, tends to be parallel to *x* axis, which is fully compatible to the biological growth of trees.

Candidate models were developed and the final model was selected on the basis of statistics criteria, considering the coefficient of determination (R^2) and the root mean square error (RMSE). The process involved a regression analysis with a stepwise procedure. Durbin Watson statistic was used to test the independence of residuals. VIF (variance inflation factor) was analyzed to check for multicollinearity.

Statistical analyses were made with R statistical software 2.3.0 (R development Core Team, 2008).

Results

The main descriptive statistics of the measured variables on the sampled trees (age = 25 years old), all with normal distribution, are summarized in Table 2. $BCEF_t$ average values ranged between 1.40 Mg m^{-3} (PD833) to 2.09 Mg m^{-3} (PD2500); $BCEF_c$ ranged from 0.36 Mg m^{-3} (PD833) to 0.63 Mg m^{-3} (PD2500) and BD varied from 0.57 Mg m^{-3} (PD1667) to 0.81 Mg m^{-3} (PD2500). The ANOVA showed significant differences between planting density for all the BCEFs ($BCEF_t$: $F_{(3/414)}$ =60.936; p=0.000; $BCEF_c$: $F_{(3/414)}$=46.063; p=0.000; BD: $F_{(3/414)}$=64.188; p=0.000), while no significant differences were found between replicates ($BCEF_t$: $F_{(2/414)}$ =0.508; p=0.602; $BCEF_c$: $F_{(2/414)}$=1.156; p=0.316; BD: $F_{(2/414)}$=0.083; p=0.920). Tukey test showed that $BCEF_t$ was significantly different between

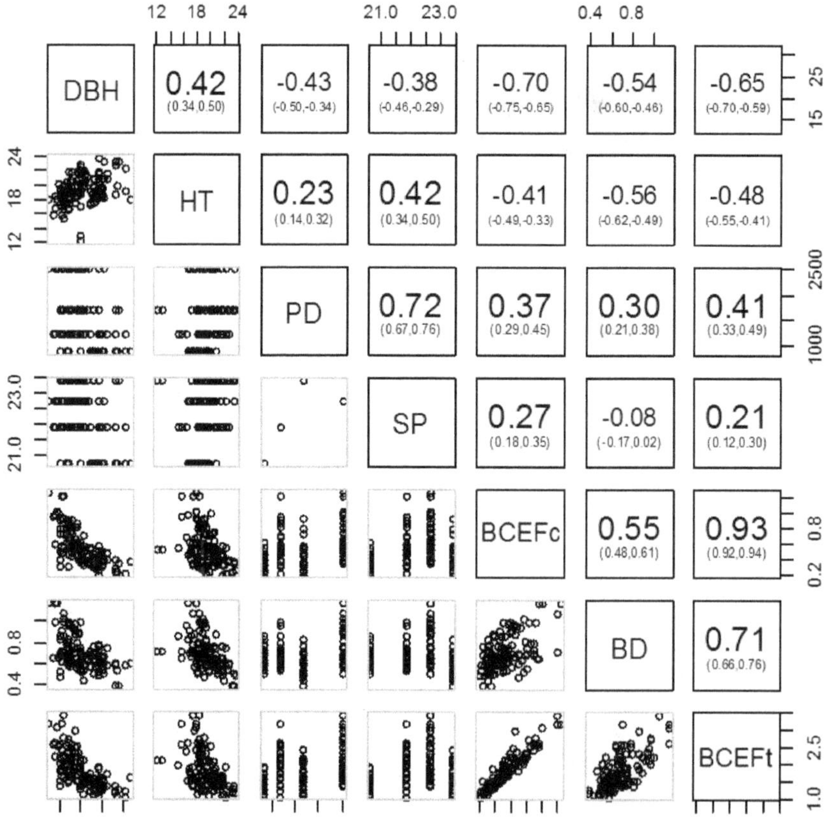

Figure 3. Scatterplot matrix (DBH = diameter at breast height; HT = Total height; PD = planting density; SP = site productivity; $BCEF_c$ = crown biomass expansion factor; BD = stem biomass expansion factor; $BCEF_t$ = total biomass expansion factor) with Pearson correlations and confidence interval, n=420.

all the PD. $BCEF_c$ had not significant differences between PD1200 and PD2500 but differences were significant between PD1667 and PD833. PD2500 and PD1667 had BD values statistically different from all the other PD (Fig. 2).

DBH, Ht, and PD were correlated with all BCEFs, instead SP was correlated only with $BCEF_t$ and $BCEF_c$; no correlation was found between SP and BD (Fig. 3). Large dispersion measures were obtained for any of the compartments in the analysis, especially for smaller trees (Fig. 3).

Starting from Eq. (2), after comparing different combinations and transformations of the variables through the stepwise procedure, the best fitting model was found to be the following:

$$Ln(BCEF_i) = b_0 + b_1 \cdot \frac{1}{DBH \cdot H} + b_2 \cdot PD + b_3 \cdot SP \quad \text{Eq. 3}$$

In no case the stand density at the age of 25 years old (SD_{25}) was significant, whereas planting density (PD) was significant for all the BCEFs. In the $BCEF_c$ equations, the SP did not appear to be significant ($P > 0.05$). As a result of this finding this variable was not included in the equation of $BCEF_c$. In general, all models provided a reasonable predictive power with R^2 values ranging from 0.57 (BD) to 0.69 ($BCEF_t$) (Table 3).

The strongest linear regression (R^2adj=69%; RMSE=0.158) was found for $BCEF_t$, with all variables as predictors. The variable formed by the combination of DBH and H could explain 66% of the total variance (Table 3), however, the PD and SP provided a further significant contribution in the estimate of $BCEF_t$. $BCEF_c$ was explained by the combination of DBH and H, and PD with a fitting of 65% (RMSE=0.253), with the first combined variable able to explain 64% of the total variance. BD was explained by all variables as predictor, with a fitting of 57% (RMSE=0.145). In the BCEFs estimates, values of Durbin Watson statistic ranged from 1.3 to 1.7. The variance inflation factor was always under 2 (for all the variance predictors used in the models), therefore absence of multicollinearity was corroborated (Zuur *et al.*, 2010).

Table 4 shows the final models with the estimated parameters and significance values for each model of BCEFs (p-values always < 0.001). The modeled relationship between all the BCEFs and DBH were plotted for all the planting densities (Fig. 4). BCEFs values decreased with increasing DBH, and were higher for higher planting density, with lower values for PD1667. For smaller DBH differences were higher and tended to disappear for larger DBH, for basic density this trend was less evident.

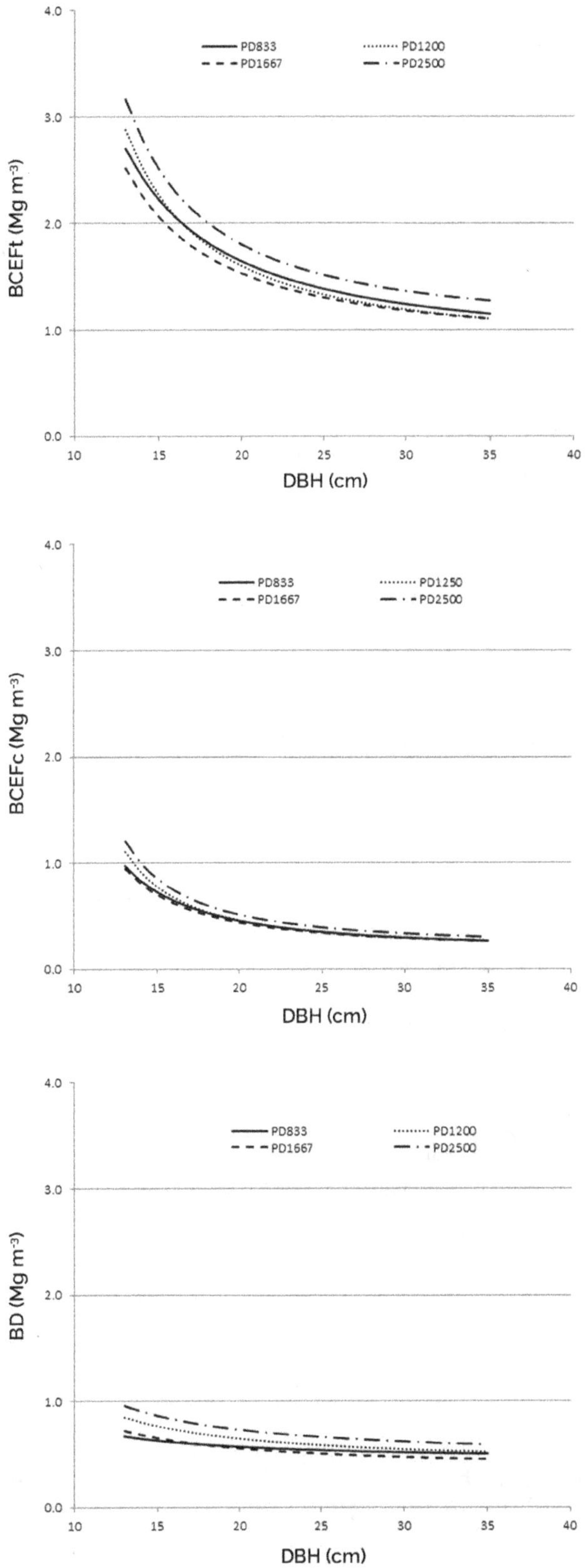

Figure 4. Total, crown and stem BCEFs (biomass expansion factor) models against DBH for the 4 planting densities (833 trees ha^{-1}; 1667 trees ha^{-1}; 1200 trees ha^{-1}; 2500 trees ha^{-1}).

Table 3. Model summary of stepwise regression in the estimate of BCEFs, derived by Eq. 2

	Step	Variable	R^2	R^2 Adj.	RMSE	Durbin-Watson
$BCEF_t$	1	1/DxH	0.657	0.656	0.166	
	2	PD	0.672	0.670	0.163	
	3	SP	0.690	0.687	0.158	1.307
$BCEF_c$	1	1/DxH	0.642	0.641	0.257	
	2	PD	0.654	0.652	0.253	1.391
BD	1	1/DxH	0.390	0.389	0.174	
	2	PD	0.439	0.436	0.167	
	3	SP	0.576	0.573	0.145	1.669

Discussion

Variation of BEFs with tree size

Total above-ground biomass expansion factor ranged from 1.4 to 2.1 Mg m^{-3} which are in the range of IPCC prescribed values for Douglas-fir trees (IPCC, Chapter IV, 2006). BCEFs values decreased with increasing DBH and tree height. Similar results have been observed in specific forest types, for example, aboveground BCEF for tropical forest (Brown *et al.*, 1989) and *Picea sitchensis* forest (Tobin & Nieuwenhuis, 2007) and total BCEF for *Chamaecyparis obtusa* forest and *Cryptomeria japonica* forest (Fukuda *et al.*, 2003).This trend is due to the changes in the biomass allocation as stands grow. While stem volume and stem biomass proportion in total tree biomass increase as the tree becomes larger, the other tree components decrease proportionally (Landsberg & Sands, 2011). Thus, larger and older trees have proportionally less foliage biomass as compared to smaller and younger ones. A decreasing trend in crown biomass across a range of tree ages (e.g. Kauppi *et al.*, 1995; Brown, 2002; Lehtonen *et al.*, 2004) may be explained by the physiological and competition attitudes of trees, since in the earlier growth phases they require greater photosynthetic biomass to win competition, whereas later, after canopy closure, shading of the tree crown, limitation of root expansion and natural pruning lead to a lower allocation into crown (Sanquetta *et al.*, 2011). The reverse dependence of BCEF on tree size is a result of an inverse relationship between wood density and tree size (Ducta *et al.,* 2010), since wood density slowly decreases as the trees grow older, even if it may increase again in older stands when the growth rate declines (Pajtík *et al.*, 2008).

Correlation between BCEF, DBH, H, SP and PD

The correlation matrix (Fig. 3) indicates that DBH is the variable more closely associated with crown and total BCEF, followed by height. This is in contrast with other authors findings (Levy *et al.*, 2010; Brown & Schroeder 1999; Lehtonen *et al.*, 2004) where it was shown that tree height had the greatest explanatory power in estimating BCEF. In our case only BD showed an higher dependence on H. As stated by Sanquetta *et al.* (2011) DBH can be chosen as explanatory variable, rather than height, because it is easier to measure, less time consuming and higher precision.

Table 4. Model equations for total ($BCEF_t$), crown (BCEFc) and stem (BD) with equation parameters, standard error of parameters (S.E.) and significance values (t and P)

Model equation		b_0	b_1	b_2	b_3
Total BCEF	Parameter	0.354	265.281	0.00015	-0.35
$Ln(BCEF_t) = b_0 + b_1 \cdot \frac{1}{DBH \cdot H} + b_2 \cdot PD + b_3 \cdot SP$	S.E.	0.229	10.950	0.00002	0.011
	t	3.350	27.043	6.581	-4.920
	Sig.	0.001	<0.0001	<0.0001	<0.0001
Crown BCEF	Parameter	-2.043	406.774	0.00015	
$Ln(BCEF_c) = b_0 + b_1 \cdot \frac{1}{DBH \cdot H} + b_2 \cdot PD$	S.E.	0.056	19.866	0.000	
	T	-36.662	20.476	7.125	
	Sig.	<0.0001	<0.0001	<0.0001	
Stem BD	Parameter	1.737	196.694	0.00019	-0.133
$Ln(BD) = b_0 + b_1 \cdot \frac{1}{DBH \cdot H} + b_2 \cdot PD + b_3 \cdot SP$	S.E.	0.210	11.074	0.00002	0.010
	T	8.272	17.761	11.620	-13.124
	Sig.	<0.0001	<0.0001	<0.0001	<0.0001

The lowest crown BEF was found in the less dense plot (PD833). An opposite trend was shown by Correia *et al.,* (2010). Our results seem in accordance with some authors findings (e.g. Ilomaki *et al.*, 2003), who refer that trees in conditions of poor competition tend to invest more in diameter growth instead of height growth. In PD1667 site productivity was the highest and this can explain the lower total and crown BCEFs than PD1200 and the absolute lowest BD, since better growing conditions led to a more rapid growth with larger tree rings width and then a lower basic density of wood. Our results are however in accordance with Kerr (2003), who found production of higher quality and quantity stem volume in higher planting densities of a plantation of *Fraxinus excelsior*, and Pfister *et al.*, (2007) who found larger branch diameter of log in high density *Picea abies* stands.

Even if planting density alone had a low explanatory power in estimating BCEFs, its effect was significant and thus it ameliorated the variance explanation in total BCEF modelling. The same happened for site productivity. Conversely Antonio *et al.*, (2007) found no effect of planting density, site index or climate on tree allometry, whereas Black *et al.*, (2004) suggested that DBH, tree height and planting density should be used as inputs for biomass and BCEF models to reduce the error of estimate. Also Lehtonen *et al.,* (2004) showed that BCEF in Finland are affected by planting density, which can vary mainly due to varying forest management intensity. $BCEF_c$ and BD showed less the effect of PD and SP, or however other factors may be influencing allometry in the component together with these, and this effect is disguised when considering total BCEF. However when we modeled BCEFs considering the effect of planting density and site productivity we found significant differences among planting spacing due to density and DBH, with larger differences among close spacings and smaller differences among wide spacings (Fig. 4) in accordance with Harrington *et al.* (2009).

PD1667 showed the lowest BCEFs probably due to the highest site productivity, as also showed in studies carried out in spruce pine- and broadleaved-dominated stands in Norway, which gave lower BCEFs when the productivity in terms of site index increased (Viken, 2012).

Considering the variability of BCEFs with stand characteristics, if a constant BCEF is applied across all site productivity classes, even though the BCEFs are divided into age classes (Lehtonen *et al.*, 2004) or planting densities (Marziliano *et al.*, 2015a), the forest biomass of forest with low productivity can be underestimated, while the biomass of high productivity can be overestimated.

Conclusions

Regression of BCEFs against DBH and H showed a greater dispersion for smaller trees. Single default values are often used (UN-ECE/FAO, 2000; FAO, 2000; IPCC, 2006), but it is evident how these factors may vary, especially in young stands, depending on the species to be planted, growth phase, and site productivity. Therefore, calculations of BCEF under specific conditions shall be preferred.

Results from the regression analysis pointed out that the application of BCEF factors to estimate forest biomass in stands with different planting densities should explicitly account for the effect of planting density. Although this effect is expected to reduce as stand grows (Marziliano *et al.*, 2015a), during a period of non-disturbance, better estimates of biomass can be obtained when planting density is accounted for.

References

António N, Tomé M, Tomé J, Soares P, Fontes L, 2007. Effect of tree, stand, and site variables on the allometry of Eucalyptus globulus tree biomass. Can J Forest Res 37: 895-906. https://doi.org/10.1139/X06-276

Black K, Tobin B, Siaz G, Byrne KA, Osborne B, 2004. Allometric regressions for an improved estimate of biomass expansion factors for Ireland based on a Sitka spruce chronosequence. Irish Forestry 61(1): 50–65.

Brown S, 2002. Measuring carbon in forests: current status and future challenges, Environ Pollut 116: 363–372. https://doi.org/10.1016/S0269-7491(01)00212-3

Brown S, Gillespie AJR, Lugo AE, 1989. Biomass estimation methods for tropical forests with applications to forest inventory data. Forest Sci 35(4): 381–902.

Brown SL, Schroeder PE, 1999. Spatial patterns of aboveground production and mortality of woody biomass for Eastern US forests. Ecol Appl 9: 968-980.

Cantore V, Iovino F, 1989. Effetti dei diradamenti sull'umidità del suolo in popolamenti di douglasia della Catena Costiera (Calabria). Annali dell'Istituto Sperimentale per la Selvicoltura 20: 13-39.

Clutter JL, Fortson JC, Pienaar LV, Brister GH, Bailey RL, 1983. Timber Management: A Quantitative Approach. John Wiley & Sons, New York, USA.

Coletta V, Menguzzato G, Pellicone G, Veltri A, Marziliano PA, 2016. Effect of thinning on above-ground biomass accumulation in a Douglas-fir plantation in southern Italy. J For Res 27(6): 1313–1320. https://doi.org/10.1007/s11676-016-0247-9

Correia AC, Tomé M, Pacheco CA, Faias S, Dias AC, Freire J, Carvalho PO, Pereira JS, 2010. Biomass allometry and carbon factors for a Mediterranean pine (Pinus pinea L,) in Portugal. Forest Systems 19 (3): 418-433. https://doi.org/10.5424/fs/2010193-9082

Curtis RO, 1967. Height–diameter and height–diameter–age equation for secondgrowth Douglas-fir. Forest Sci 13 (4): 365–375.

Dutca I, Abrudan IV, Stancioiu PT, Blujdea V, 2010. Biomass conversion and expansion factors for young Norway spruce (Picea abies (L,) Karst,) trees planted on non-forest lands in Eastern Carpathians. Not Bot Hort Agrobot Cluj 38(3): 286–92.

Enes Duque T, Fonseca Fidalgo T, 2014. Biomass conversion and expansion factors are affected by thinning. Forest Systems 23(3): 438-447. https://doi.org/10.5424/fs/2014233-05128

Food and Agriculture Organization of the United Nations, 1998. World Reference Base for Soil Resources, Rome: FAO, ISRIC, ISSS [Soil Resource Report n. 84].

Food and Agriculture Organization of the United Nations, 2000. Global Forest Resource Assessment, FAO Forestry Paper 140, FAO. Rome, Italy.

Fukuda M, Iehara T, Matsumoto M, 2003. Carbon stock estimates for sugi and hinoki forests in Japan. Forest Ecol Manag 184: 1–16. https://doi.org/10.1016/S0378-1127(03)00146-4

Hägglund B, 1981. Evaluation of forest site productivity. Forestry Abstracts 42: 515–527.

Harrington TB, Harrington CA, DeBell DS, 2009. Effects of planting spacing and site quality on 25-year growth and mortality relationships of Douglas-fir (Pseudotsuga menziesii var, menziesii). Forest Ecol Manag 258: 18–25. https://doi.org/10.1016/j.foreco.2009.03.039

Ilomaki S, Nikinmaa E, Makela A, 2003. Crown rise due to competition drives biomass allocation in silver birch. Can J Forest Res 33: 2395-2404. https://doi.org/10.1139/x03-164

IPCC, 2006. Intergovernmental Panel on Climate Change, Guidelines for National Greenhouse Gas Inventories [http://www.ipcc.ch/].

Kauppi P, Tomppo E, Ferm A, 1995. C and N storage in living trees within Finland since 1950s, Plant soil 168-169: 633-638. https://doi.org/10.1007/BF00029377

Kerr G, 2003. Effects of spacing on the early growth of planted Fraxinus excelcior L. Can J Forest Res 33: 1196¬1207.

Landsberg JJ, Sands PJ, 2011. Physiological Ecology of Forest Production: Principles, Processes and Modelling. Oxford, UK: Elsevier Inc.

Lehtonen A, Mäkipää R, Heikkinen J, Sievänen R, Liski J, 2004. Biomass expansion factors (BEF) for Scots pine, Norway spruce and birch according to stand age for boreal forests. Forest Ecol Manag 188: 211–224. https://doi.org/10.1016/j.foreco.2003.07.008

Levy PE, Hale SE, Nicoli BC, 2010. Biomass expansion factors and root: shoot ratios for coniferous tree species in Great Britain. Forestry 77(5): 421-430. https://doi.org/10.1093/forestry/77.5.421

Magalhães TM, Seifert T, 2015. Estimation of tree biomass, carbon stocks, and error propagation in mecrusse woodlands. Open J Forestry 5: 471-488. https://doi.org/10.4236/ojf.2015.54041

Marziliano PA, Lafortezza R, Colangelo G, Davies C, Sanesi G, 2013. Structural diversity and height growth models in urban forest plantations: a case-study in northern Italy. Urban Forestry & Urban Greening, 12(2): 246–254. https://doi.org/10.1016/j.ufug.2013.01.006

Marziliano PA, Coletta V, Menguzzato G, Nicolaci A, Pellicone G, Veltri A, 2015a. Effects of planting density on the distribution of biomass in a Douglas-fir plantation in southern Italy. iForest 8: 368-376 [online 2014-09-09] URL: http://www.sisef.it/iforest/contents/?id=ifor1078-007.

Marziliano PA, Veltri A, Menguzzato G, Pellicone G, Coletta V, 2015b. A comparative study between "default method" and "stock change method" of Good Practice Guidance for Land Use, Land-Use Change and Forestry (IPCC, 2003) to evaluate carbon stock changes in forest. In: Atti del II Congresso Internazionale di Selvicoltura, Progettare il futuro per il settore forestale, Firenze, 26-29 novembre 2014, Accademia Italiana di Scienze Forestali, Firenze, Italy. Vol. 1, pp. 551-557.

Marziliano PA, Lafortezza R, Medicamento U, Lorusso L, Giannico V, Colangelo G, Sanesi G, 2015c. Estimating belowground biomass and root/shoot ratio of Phillyrea latifolia L. in the Mediterranean forest landscapes. Ann Forest Sci 72:585–593. https://doi.org/10.1007/s13595-015-0486-5

Pajtík J, Konopka B, Lukac M, 2008. Biomass functions and expansion factors in young Norway spruce (Picea abies [L.] Karst.) trees. Forest Ecol Manag 256: 1096–1103. https://doi.org/10.1016/j.foreco.2008.06.013

Penman J, Gytarsky M, Hiraishi T, Krug T, Kruger D, Pipatti R, Buendia L, Miwa K, Ngara T, Tanabe K, and Wagner F, 2003. Definitions and Methodological Options to Inventory Emissions from Direct Human-induced Degradation of Forests and Devegetation of Other Vegetation Types. The Institute for Global Environmental Strategies for the IPCC and The Intergovernmental Panel on Climate Change, Hayama, Kanagawa, Japan.

Petersson H, Holm S, Ståhl G, Alger D, Fridman J, Lethonen A, Lundstrøm A, Mäkipää R, 2012. Individual tree biomass functions or biomass expansion factors for assessment of carbon stock changes in living biomass – A comparative study. Forest Ecol Manag 270: 78–84. https://doi.org/10.1016/j.foreco.2012.01.004

Pfister O, Wallentin C, Nilsson U, Ekö PM, 2007. Effects of wide spacing and thinning strategies on wood quality in Norway spruce (Picea abies) stands in southern Sweden. Scand J For Res 22: 333– 343. https://doi.org/10.1080/02827580701504951

R Development Core Team, 2008. R: a language and environment for statistical computing. R Foundation for Statistical Computing, Vienna, Austria. https://www.r-project.org/.

Sanquetta CR, Dalla Corte AP, Silva FS, 2011. Biomass expansion factor and root-to-shoot ratio for Pinus in Brazil. Carbon Balance Manag 6(6): 1–8. https://doi.org/10.1186/1750-0680-6-6

Schoene D, 2002. Terminology in assessing and reporting forest carbon change. In: Second expert meeting on harmonizing forest-related definitions for use by various stakeholders, FAO, Rome, Italy.

Soares P, Tome M, 2012. Biomass expansion factors for Eucalyptus globules stands in Portugal. Forest Systems 21(1): 141-152. https://doi.org/10.5424/fs/2112211-12086

Somogyi Z, Cienciala E, Mäkipää R, Muukkonen P, Lehtonen A, Weiss P, 2006. Indirect methods of large-scale forest biomass estimation. Eur J Forest Res 126 (2): 197-207. https://doi.org/10.1007/s10342-006-0125-7

Teobaldelli M, Somogyi Z, Migliavacca M, Usoltsev VA, 2009. Generalized functions of biomass expansion factors for conifers and broadleaved by stand age, growing stock and site index. Forest Ecol Manag 257(3): 1004–1013. https://doi.org/10.1016/j.foreco.2008.11.002

Tobin B, Nieuwenhuis M, 2007. Biomass expansion factors for Sitka spruce (Picea sitchensis (Bong) Carr,) in Ireland. Eur J Forest Res 126: 189-196. https://doi.org/10.1007/s10342-005-0105-3

UN-ECE/FAO, 2000. Forest Resources of Europe, CIS, North America, Australia, Japan and New Zealand (industrialized temperate/boreal countries). UNECE/ FAO Contribution to the Global Forest Resources Assessment 2000 Main Report United Nations, New York, Geneva.

United Nations Framework Convention on Climate Change – UNFCCC, 1997. Kyoto Protocol to the United Nations Framework Convention on Climate Change. FCCC/CP/L7/Add.1, 10 December 1997, UN. New York, USA.

Viken KO, 2012. Biomass equations and biomass expansion factors (BEFs) for pine (Pinus spp.), spruce (Picea spp.) and broadleaved dominated stands in Norway. Mastergradsoppgave ved Institutt for naturforvaltning, Universitetet for milijø- og biovitenskap 43 s. + vedlegg.

Zianis D, Muukkonen P, Ma¨kipa¨a¨ R, Mencuccini M. 2005. Biomass and stem volume equations for tree species in Europe. Silva Fennica Monographs N. 4, pp. 1–2, 5–63

Zuur AF, Ieno EN, Elphick CS, 2010. A protocol for data exploration to avoid common statistical problems. Methods Ecol Evol 1: 3–14. https://doi.org/10.1111/j.2041-210X.2009.00001.x

6

Assessment of different remote sensing data for forest structural attributes estimation in the Hyrcanian forests

Nooreddin Noorian, Shaban Shataee-Jouibary* and Jahangir Mohammadi

Forestry Department, Gorgan University of Agricultural Sciences, and Natural Resources, 386, Gorgan, Iran

Abstract

Aim of the study: The objective of the study was the comparative assessment of various spatial resolutions of optical satellite imagery including Landsat-TM, ASTER, and Quickbird data to estimate the forest structure attributes of Hyrcanian forests, Golestan province, northern Iran.

Material and methods: The 112 square plots with area of 0.09 ha were measured using a random cluster sampling method and then stand volume, basal area, and tree stem density were computed using measured data. After geometric and atmospheric corrections of images, the spectral attributes from original and different synthetic bands were extracted for modelling. The statistical modelling was performed using CART algorithm. Performance assessment of models was examined using the unused validation plots by RMSE and bias measures.

Main Results: The results showed that model of Quickbird data for stand volume, basal area, and tree stem density had a better performance compared to ASTER and TM data. However, estimations by ASTER and TM imagery had slightly similar results for all three parameters.

Research highlights: This study exposed that the high-resolution satellite data are more useful for forest structure attributes estimation in the Hyrcanian broadleaves forests compared with medium resolution images without consideration of images costs. However, regarding to be free of the most medium resolution data such as ASTER and TM, ETM+ or OLI images, these data can be used with slightly similar results.

Keywords: Forest structure attributes; quickbird; ASTER; TM; CART algorithm; Hyrcanian forests.

Funding: The author(s) received no specific funding for this work.

Competing interests: The authors have declared that no competing interests exist.

Correspondence should be addressed to Shaban Shataee Jouibary: shataee@yahoo.com

Introduction

The forest has a significant role in human life. It not only has to provide goods for consumption but it also has an ecological, environmental, and aesthetic role in human life. Forest structural attributes such as volume, basal area, and number of trees per unit area are important data needed for effective forest management (Gebreslasie *et al.*, 2010). Hyrcanian forests comprise a diverse vegetation cover in the north of Iran and are increasingly fragmented, degraded and converted to other forms of land use (Mohammadi *et al.*, 2010). The forest's structural attributes are traditionally gathered by ground-based field measurements using hand-held equipment. These measurements are generally expensive, time-consuming and labour intensive, as well as difficult to perform, especially in mountainous and dense forests (Mohammadi *et al.*, 2011). Satellite remote sensing is an alternative source and new tool for forest management and surveying, particularly in large areas. Rapid improvements in remote sensing technology have led to various types of sensors, such as multispectral, hyper spectral, ultraviolet, thermal sensors, light detection and ranging (Lidar), radio detection and ranging (radar), and other sensors. Each type of sensor has been designed for specialized purposes, tasks and different applications (Kalbi, 2011). These new potential sources have been shown to be appropriate tools to assess and monitor forest attributes with reasonable accuracy levels (Hyyppa *et al.*, 2000). Satellite sensor data have recently been used in a multisource forest inventory for estimating forest characteristics due to their advantages including large geographic coverage and large spectral range (Tuominen & Haakana, 2005). In the past two decades, many researchers (see table 1) have focused on the extraction and retrieving of forest stand parameters such as stand volume, basal area, DBH, and tree stem density using medium-to-high resolution optical sensor data.

This research has used different remote sensing sources from aerial photos to satellite based images, with various spatial, radiometric, or spectral resolutions. In some occasions, the results of previous studies were not satisfactory for managers according to forest condition such as forest composition, forest structure, and topography complexity. The comparative studies for investigation into the capability of various image sources based on spatial or spectral characteristics can help to choose suitable images for the extraction of the desired information. Few research studies have compared the effect of different spatial resolutions from different image sources including very high-spatial resolution (airborne- or space-borne systems) to medium spatial resolution imagery in a same forest. For instance, Hyyppa *et al.*, (2000) used various image sources, such as aerial photographs, SPOT Pan, SPOT XS, and Landsat-TM, and compared the accuracy of retrieving the following forest stand variables: stem volume, mean tree height and basal area. They found that using high-resolution image types had better estimates compared to medium resolution images (i.e., SE% of 46, 50, 56 in stem volume and 38, 42, 47 for basal area estimation, respectively for aerial photos, SPOT-XS and Landsat-TM).

It has been confirmed that using high-spatial resolution imagery may lead to estimates of high-precision results, especially for the extraction of forest attributes, due to its ability for finer detection and recognition of spectral reflections of the canopy crown and lower mixed pixels. However, medium spatial resolution imagery has been the most popularly used data in stand and plot level estimations. Comparative studies in mixed and multi-layered forests such as Hyrcanian forests could be useful to determine suitable image sources to extract forest structure attributes.

In recent years, non-parametric algorithms such as decision tree-based algorithms (Breiman *et al.,* 1984) and their variants like CART (classification and regression tree) have been widely used in different studies due to their simple interpretation, high precision, and ability to characterize complex interactions among variables (Cutler *et al.,* 2007). Non-parametric algorithms have obvious advantages over parametric-based algorithms for multisource predictive forest mapping. One major drawback of parametric-based algorithms is that they assume a particular statistical distribution in the dataset, which is usually not compatible with multisource data. Nevertheless, non-parametric-based

Table 1. The previous studies performed by different sources and algorithms

Study	Variables	Data sources	Algorithm	RMSE (%)	R^2
Trotter *et al.*, (1997)	Stand volume	TM	Linear	-	0.3
Kilpelainen and Tokola (1999)	Stand volume	TM	Knn RSP	-	0.60
Muukkonen and Heiskanen (2005)	Stand volume	ASTER	Nonlinear ANN	44.7 41	
Hall *et al.*, (2006)	Stand volume	ETM+	Linear	-	0.3
Sivanpillai *et al.*, (2006)	Tree density	ETM+	linear	-	0.60
Huiyan *et al.*, (2006)	Stand volume	TM	Knn	44.2	-
Wolter *et al.*, (2009)	Stand volume, Basal area	SPOT-HRG	OLS	-	0.53, 0.74
Mohammadi *et al.*, (2010)	Stand volume, Tree density	ETM+	CART Linear	43, 73 67, 80	0.43, 0.73
Gebreslasie *et al.*, (2010)	Stand volume, Basal area, Tree density	ASTER	Linear	-	0.51, 0.67, 0.65
Kalbi *et al.*, (2014)	Stand volume, basal area	ASTER, SPOT-HRG	CART	18.2, 15.8	0.59
Shataee *et al.*, (2012)	Stand volume, Basal area, Tree density	ASTER	Knn SVM, RF	26.86,20.20, 21.53 28.54, 19.35, 22.09 25.86, 18.39, 20.64	- - -

Abbreviations: RMSE, Root mean square error; Knn, K nearest neighbor; RSP, Reference sample plot; ANN, Artificial neural network; OLS, Ordinary least squares; CART, Classification and regression tree; SVM, Support vector machine; RF, Random forest.

algorithms make no assumption on data distribution, and therefore avoid the significant error source. This means that they are free from assumptions of any given probability distribution, and observations are assumed independent of each other (Sironen *et al.,* 2010). Many studies have shown that non-parametric methods provide better estimation results. In some studies, such as Sarunas (1997), it has been demonstrated that even with small training samples, non-parametric estimation algorithms provide better results than parametric ones. Among non-parametric algorithms, tree based algorithms are more famous and more commonly used for both forest attribute estimation and classification. In this paper, the capability of Quickbird, ASTER, and TM data were compared to estimate forest structural attributes using classification and regression trees algorithm (CART) as one of the non-parametric algorithms.

In the past several years, classification and regression tree-based analyses have been implemented in several software programs, and used in many remote-sensing applications (Huang & Jensen, 1997; Lawrence & Wright, 2001; Cooke & Jacobs, 2005). The use of non-parametric methods for land cover classification has increased in the past decade. Aertsen *et al.*, (2010) investigated the performance of non-parametric techniques such as CART compared to parametric techniques for the prediction of site index in Mediterranean mountainous forests. Moisen & Frescino (2002) and Wang *et al.,* (2005) evaluated these techniques for the prediction of several species-independent forest characteristics in the interior western United States and for the spatial prediction of site index of Lodge pole pines in Canada. These studies concluded that non-parametric approaches were more effective compared to parametric ones.

The objective of research was comparison of various medium (TM and ASTER) and high (Quickbird) spatial resolution satellite images by CART algorithm for estimation the quantitative forest attributes of the Shastkalateh's forest as a part of the Hyrcanian forests with mixed and multi-layered hardwood stands, in the Golestan province, north of Iran.

Materials

Study area

The Hyrcanian vegetation zone is a green belt stretching over the northern slopes of Alborz Mountain and covers the southern coasts of the Caspian Sea. This zone is rich in point of species diversity and includes 80 woody species (trees and shrubs) dominated with hardwoods. The Iranian Hyrcanian forests extend into the three provinces of Gilan, Mazandaran, and Golestan. This research was performed in the Shastkalateh forest as a small part of the eastern Hyrcanian forest located in the Golestan province with a 1,714 hectares area. The study area is positioned between 36° 43′ to 36°45′N latitudes and 54°21′ to 54°24′E longitudes (Fig. 1) and elevations range from 210 to 1010 m above mean sea level, with slopes of 5% to 45%. The main slope aspects of the site are west and southwest aspects. The main and dominant tree species are beech (*Fagus orientalis*), hornbeam (*Carpinus betulus*), Persian parotia (*Parrotia persica*), chestnut-leaved oak (*Quercus castaneafolia*), coliseum maple (*Acer cappadocicum*), velvet maple (*Acer velutinum*), Caucasian alder (*Alnus subcordata*) and date palm (*Diospyros lotus*). The Shastkalateh's forest is under a forestry management plan with a selective cutting treatment method.

Data

Field data

The in-situ data were gathered in the summer of 2010 (in line with season of images) and on 23 clusters with 5 plots and 112 plots in the natural stands (95 plots) and planted stands (17 plots) with homogenous unities in terms of slopes, aspects, and forest types. The distance between plots in each cluster was 75 m and the plots were square with an area of 30×30 (0.09 ha) m (Fig. 1). The coordinates of the centre of each plot were accurately registered using a DGPS device in a post-processing kinematics (PPK) method. In each plot, the species name, stand height (measured to the nearest tree) and diameter at breast height (DBH) of trees with a diameter greater than 7.5 cm at breast height were taken. Singletree volume in plots was calculated using a local volume table, containing diameter at breast height ($d_{1.3}$) and height, to estimate the volume of different species in the plots. Plot level volume was computed throughout, adding total singletree volumes. Finally, the volume per hectare (m^3/ha) was estimated using the total volume of all trees in each plot. The measured DBH of trees was used to determine basal area. In addition, the tree density was computed through counting of measured trees in each plot.

Remote sensing data

Data used for this study included Quickbird images from 7 October 2007, ASTER images from 3 July 2006,

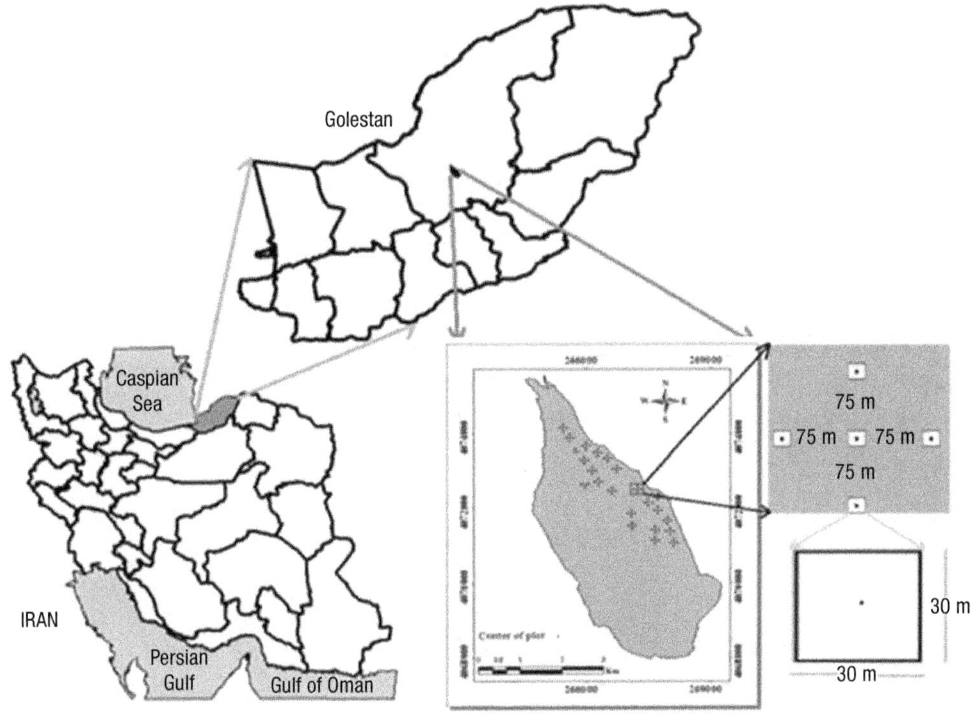

Figure 1. Location of study area in the Golestan province, northern Iran, and spatial distribution of field plots.

and Landsat-5 TM images from 17 September 2010. The dates of images are ranging during summer season so that trees do not have phenology differences due to completing leaf growing and there were not considerable cover changes in the period of five years. The projection system was UTM zone 40N, Datum WGS 1984. Table 2 summarizes the acquisition of the data in detail.

Methods

Pre-processing of satellite data

In this study, the image datasets were georeferenced to the UTM coordinate system using ground control points (GCPs) and second-order polynomial equations. The Quickbird images were georeferenced and orthorectified using 24 GCPs collected from DGPS and a digital elevation model (DEM) derived from a 1:25,000 topography map (Yazdani, 2011). The ASTER and TM images were rectified with Quickbird images using 25 and 30 GCPs, respectively. The total root mean square errors (RMSE) of images were obtained approximately of 0.36 and 0.25 for ASTER and TM images, respectively. The geometric precision of images was also verified using road and river vectors.

In this study, due to a lack of some information required for accurate atmospheric models, the cosine estimation of atmospheric transmittance (COST) absolute radiometric correction model (Chavez & Pat, 1996) was applied on each image. This model consists of a modification of the dark-object subtraction (DOS) method by including a simple multiplicative correction for the effect of atmospheric transmittance.

Image processing

After geometric rectification and atmospheric corrections, some suitable processing analyses including tasselled cap transformation (greenness, brightness, and wetness components), standardized principal components (PCA), Pansharpning (for Quickbird images), vegetation indices (Table 3) and texture analyses were performed to produce useful and correlated artificial for quantifying and enhancing biophysical characteristics. The texture analysis is done on VNIR band of ASTER and Quickbird images with grey level co-occurrence matrix (GLCM) indices (Mean, Variance, Homogeneity, Contrast, Dissimilarity, Entropy, Angular second moment, Correlation, GLDV angular second moment, GLDV entropy, GLDV mean, GLDV contrast, Inverse difference). These were included in the statistical analysis via main bands for prediction of forest structural attributes in the regression modelling.

Table 2. Information of satellite imagery used in this study

Source	Day and month	Year	Bands	Number of bands	Pixel size (m)	Quantization level
Quickbird	October 7	2007	Multispectral	4	2.5	11-bit
			Panchromatic	1	0.68	
ASTER	July 3	2006	VNIR	3	15	8-bit
			SWIR	6	30	
Landsat5-TM	September 17	2010	Multispectral	6	30	8-bit

Table 3. Most popular used spectral vegetation indices related to forest structure attributes

Index	Equation	References
Normalized difference vegetation index (NDVI)	$(NIR - RED)/(NIR + RED)$	37
Green Normalized Difference Vegetation Index (GNDVI)	$(NIR - G)/(NIR + G)$	37
Normalized ratio (NR)	$RED - NIR$	38
Simple ratio (SR)	NIR/RED	38
Near Infrared Ratio (NIR)	$TM4/TM3$	39
Difference vegetation index (DVI)	$NIR - RED$	40
Modified soil adjusted vegetation index (MSAVI2)	$(0.5) * \{2 * (NIR + 1) - \text{Sqrt}[(2 * NIR + 1)^2 - 8 * (NIR - R)\}$	41
Ratio Vegetation Index (RVI)	$NIR/(RED + GREEN)$	40
Transformed Vegetation Index (TVI)	$(TM5 - TM3)/ (TM5 + TM3)$	39
Contrast Reflectance in Visible and Near Infrared (VNIR1)	$(TM4 - TM1)/ (TM4 + TM1)$	39
Middle Infrared Vegetation Index (MIRV2)	$(MIR - RED)/(MIR + RED)$	42

NIR, Near infra red; G, Green; MIR, Middle infra red.

Statistical analyses

In this study, statistical analyses were performed by R-2.14.2 (2011-12-22) software (R core Team, 2012). The spectral values of the main and artificial images of the three sensors were extracted by the Zonal statistics function in ArcGIS software corresponding to ground plots. Using 85% of plots (95 plots), the CART algorithm was conducted to evaluate relationships between forest structural attributes of volume, basal area, and number of stem trees per hectare as dependent and spectral data extracted from main and artificial images as independent variables. The CART algorithm is a nonparametric modelling approach that can explain responses of a dependent from a set of independent continuous or categorical variables. The CART models recursively partitioned the data to find increasingly homogeneous subsets based on independent variable splitting criteria using variance-minimizing algorithms. This model produces outcomes that are unaffected by monotone transformations and differing scales of measurement among predictors. Regression trees are insensitive to outliers and can accommodate missing data in predictor variables using surrogates. The dependent data are partitioned into a series of descending left and right child nodes derived from parent nodes. Once the partitioning has ceased, the child nodes are designated as terminal nodes. Pruning of trees is often necessary to avoid over-fitting of data, often accomplished by setting aside a portion of the training data for pruning. A cross validation methodology is applied for pruning to outcome prediction. Regression trees are insensitive to outliers, and can accommodate missing data in predictor variables using surrogates. For more details on the CART model, refer to Breiman *et al.,* (1984).

Model evaluation

Of the plots used, 85% (95 plots) were used for modelling and the remaining 15% (17 plots) were used to evaluate the model outputs. The reliability of estimates was measured by adjusted coefficient of determination (R^2adj), root mean square error (RMSE) [1], percentage RMSE [2], Bias [3], and percentage bias [4] (Makela & Pekkarinen, 2004).

$$\text{RMSE} = \sqrt{\frac{1}{n}\sum_{i=1}^{n}\left(y_i - \hat{y}_i\right)^2} \qquad \text{[Eqs. 1]}$$

$$\text{RMSE \%} = \frac{\sqrt{\frac{1}{n}\sum_{i=1}^{n}\left(y_i - \hat{y}_i\right)^2}}{\hat{y}} \times 100 \qquad \text{[Eqs. 2]}$$

$$\text{Bias} = \frac{1}{n}\sum_{i=1}^{n}\left(y_i - \hat{y}_i\right) \quad \text{[Eqs. 3]}$$

$$\text{Bias}\,\% = \frac{\frac{1}{n}\sum_{i=1}^{n}\left(y_i - \hat{y}_i\right)}{\hat{y}} * 100 \quad \text{[Eqs. 4]}$$

Where, $\hat{y}_i$ is predicted value, y_i is the observed value, $\bar{y}$ is the mean of the observed values, and n is the number of observations in the test dataset.

Results and Discussion

Descriptive statistics of field data

The preliminary descriptive statistics results showed that stand volume, basal area and tree density ranged from 22.84 to 647.82 m^3/ha, 3.29 to 52.57 (m^2/ha) and 111.10 to 966.57 (n/ha), respectively. The mean stand volume was 294.64 m^3/ha with standard deviation of 141.21 m^3/ha, the mean basal area was 25.40 m^2/ha with standard deviation of 10.09 m^2/ha and the mean tree density was 366.98 n/ha with a standard deviation of 194.55. Table 4 represents a full range of stand structure attributes in the study area.

Estimating forest characteristics

Estimating stand volume

The results of CART model implementation to create the relationship between spectral values of used images and stand volume are shown in the table 5 as the best predictors of stand volume, with the adjusted R^2 of 0.61, 0.76 and 0.71 and RMSE of 50.96, 120.92, 102.39 m^3/ha, for Quickbird, ASTER and TM images, respectively (Table 6).

The obtained R^2 values of this study were higher than obtained by others (Trotter *et al.*, 1997; Hall *et al.*, 2006; Mohammadi *et al.*, 2010; Gebreslasie *et al.*, 2010; Wolter *et al.*, 2009; Kalbi *et al.*, 2014). In contrast, the obtained RMSE values were lower than RMSE of different studies (Tokola & Heikkila, 1997; Fazakas *et al.*, 1999; Hyyppa *et al.*, 2000, Hyvonen, 2002; Makela & Pekkarine, 2004; Huiyan *et al.*, 2006; Muukkonen & Heiskanen, 2005).

The regression tree model for stand volume estimation using Quickbird images as the best image, which could produce better estimations, is shown in fig. 2 (for ASTER and TM images please see Figures S1 and S2 [supplementary]). The models indicate that blue, mean (VNIR1) and mean of blue bands were the most important variables to model the stand volume, and could be explained by three image dates (Table 5). The performance results of using different data sources by CART for stand volume estimation showed that Quickbird images could produce estimations with lower absolute and percentage RMSE and bias compared to using ASTER and TM data. However, estimations produced using ASTER and TM data were almost similar and had very slight differences. For volume estimation, the results obtained from TM data were better than those obtained from the ASTER data were. Comparative results of implementations are given in Table 6.

Estimating basal area

The results of CART model implementation to create the relationship between spectral values of used images and basal area are shown in the table 5 as the best predictors of basal area with the adjusted R^2 of 0.50, 0.73, and 0.70 and RMSE of 2.44, 10.03, and 9.08 $m^2\ ha^{-1}$, for Quick bird, ASTER and TM images, respectively (Table 6).

The R^2 values obtained in this study to predict basal area using Quickbird image were lower than ones were obtained by other researches (Gebreslasie *et al.*, 2010; Wolter *et al.*, 2009; Kalbi *et al.*, 2014). In addition, the RMSE values were lower than those values obtained

Table 4. Descriptive statistics of the field inventory data

	Stand volume ($m^3\ ha^{-1}$)		Basal area (m^2/ha)		Tree density (n/ha)	
	Model	Validation	Model	Validation	Model	Validation
Number	95	17	95	17	95	17
Mean	294.64	300.61	25.40	25.96	366.98	373.81
Standard Deviation	141.21	119.62	10.09	9.49	194.55	208.66
Range	624.98	467.93	49.28	34.46	855.47	644.38
Minimum	22.84	65.93	3.29	8.03	111.10	177.76
Maximum	647.82	533.81	52.57	42.49	966.57	822.14

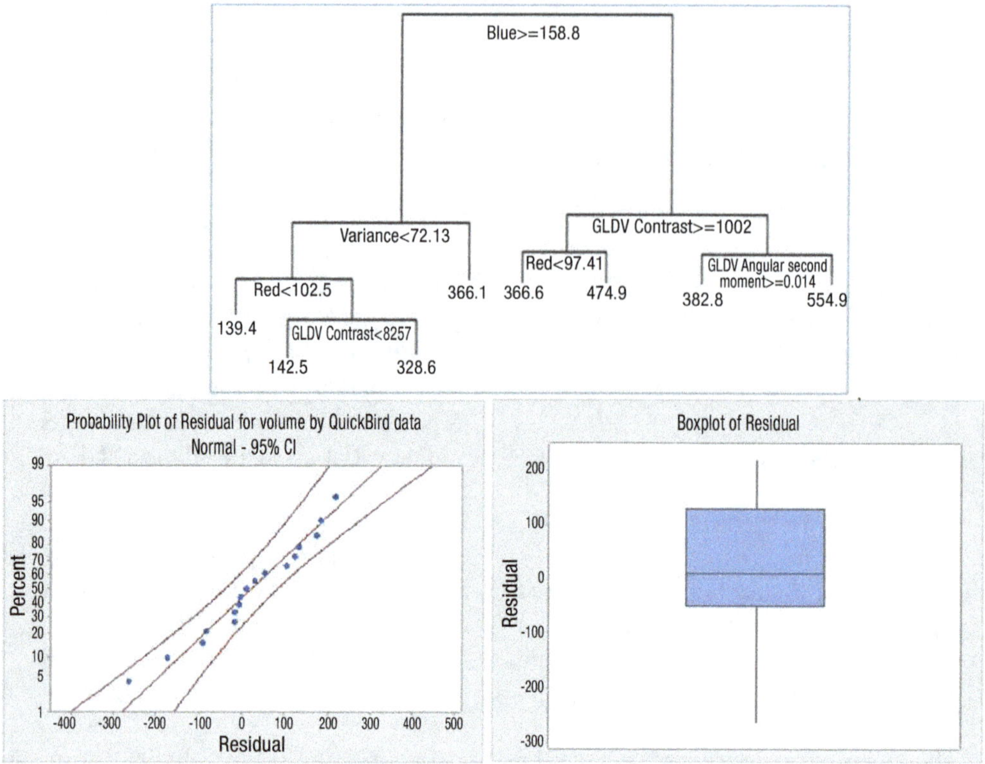

Figure 2. Binary regression tree (top) and probability (left), and box plot (right) of residual CART model for stand volume using Quickbird data.

Table 5. The variables selected by the best models developed for each of three image sources

Attribute	Source	Selected variables by model
Stand volume	Quickbird	band1, band 3, Variance, GLDV Contras and GLDV Angular second moment
	ASTER	Mean (VNIR1), PCA2, Homogeneity (VNIR1), Contrast (VNIR2) and Correlation (VNIR3)
	TM	band 1, band 4 (NIR), greenness, and GNDVI, Ratio and NR
Basal area	Quickbird	GLDV Entropy, Band1, ARVI index, Brightness and GLDV Contrast
	ASTER	PCA1, PCA2, Mean (VNIR1), Variance (VNIR1), Homogeneity (VNIR3) and Correlation (VNIR3)
	TM	Greenness, band1, PCA3, GNDVI, Ratio and ARVI
Tree density	Quickbird	Entropy, Variance, Mean and GLDV Contrast,
	ASTER	greenness, brightness, Msavi2, GLDV Angular second, Dissimilarity and Homogeneity
	TM	PCA3, band1, band2, band3 and Msavi2

by direct estimation to predict basal area (Gebreslasie *et al.,* 2010; Kalbi *et al.,* 2014).

The R^2 values obtained in this study using ASTER images were higher than those that obtained through direct estimation to predict basal area (Gebreslasie *et al.,* 2010; Wolter *et al.,* 2009; Kalbi *et al.,* 2014).

The R^2 values obtained in this study using TM images were higher than those that obtained through direct estimation to predict basal area (Hyyppa *et al.,* 2000; Gebreslasie *et al.,* 2010). In addition, the obtained RMSE values were higher than those obtained by direct estimation and used to predict basal area (Gebreslasie *et al.*, 2010; Wolter *et al.*, 2009; Kalbi *et al.,* 2014).

The results of regression tree model for basal area estimation using Quickbird images as the best image, which could produce better estimations, is shown in Fig. 3 (for ASTER and TM images please see Figures S3 and S4 [supplementary]). The results indicated that GLDV entropy, PCA1, and greenness were the most important variable determining basal area, could be explained by three image dates above, respectively (Table 5). The results of using different data sources in CART performances for estimate basal area/ha showed that Quickbird images could produce estimations with lower absolute and percentage RMSE and bias compared to using ASTER and TM data. However, estimation using ASTER and TM data produced slightly

Table 6. Results of the best CART models performance to estimate the variables using data sources

Attribute	Source	Validation metrics				
		Adjusted R^2	RMSE (m^3 ha^{-1})	RMSE (%)	Bias (m^3 ha^{-1})	Bias (%)
Stand volume (m^3 ha^{-1})	Quickbird	0.61	50.98	20.35	3.71	1.48
	ASTER	0.76	120.92	40.22	52.71	17.5
	TM	0.71	102.39	34.06	0.57	0.19
Basal area (m^2 ha^{-1})	Quickbird	0.50	2.44	12.10	0.71	3.52
	ASTER	0.73	10.03	38.67	2.08	8
	TM	0.70	9.08	35	1.47	5.68
Tree density/ha (nha^{-1})	Quickbird	0.59	125	30.36	2.98	0.71
	ASTER	0.80	219.4	58.68	10.18	2.72
	TM	0.67	210.64	56.34	9.71	2.62

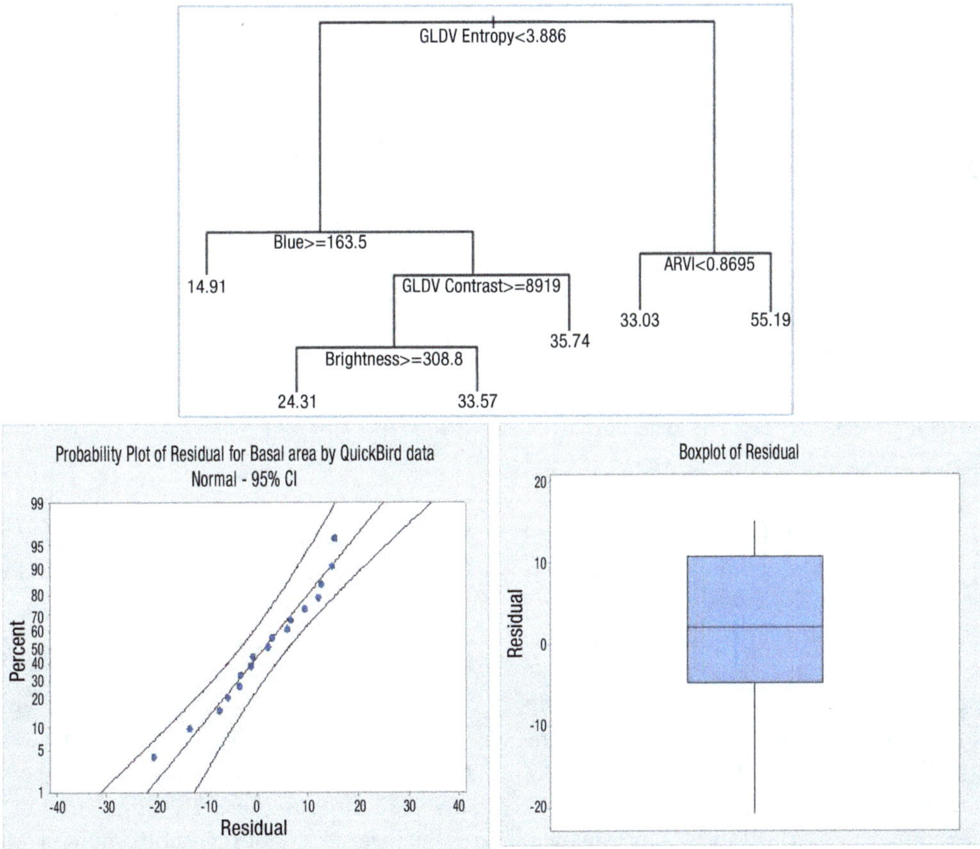

Figure 3. Binary regression tree (top) and probability (left), and box plot (right) of residual CART model for basal area/ha using Quickbird data.

similar results. Comparative results of implementations are given in Table 6.

Estimating tree density/ha

The results of CART model implementation to create the relationship between spectral values of used images and basal area are shown in the table 5 as the best predictors of basal area with the adjusted R^2 of 0.59, 0.80 and 0.67; and RMSE of 125, 219.4 and 210.64 n/ha for Quick bird, ASTER and TM images, respectively (Table 6). The obtained R^2 values using Quickbird were lower than ones that obtained through direct estimation to predict tree density by Sivanpillai *et al.*, (2006), Gebreslasie *et al.*, (2010), Mohammadi *et al.*, (2010) and Kalbi *et al.*, (2014). In addition, RMSE values obtained in this study were lower than values obtained when direct estimation was used to predict tree density by Sivanpillai *et al.*, (2006).

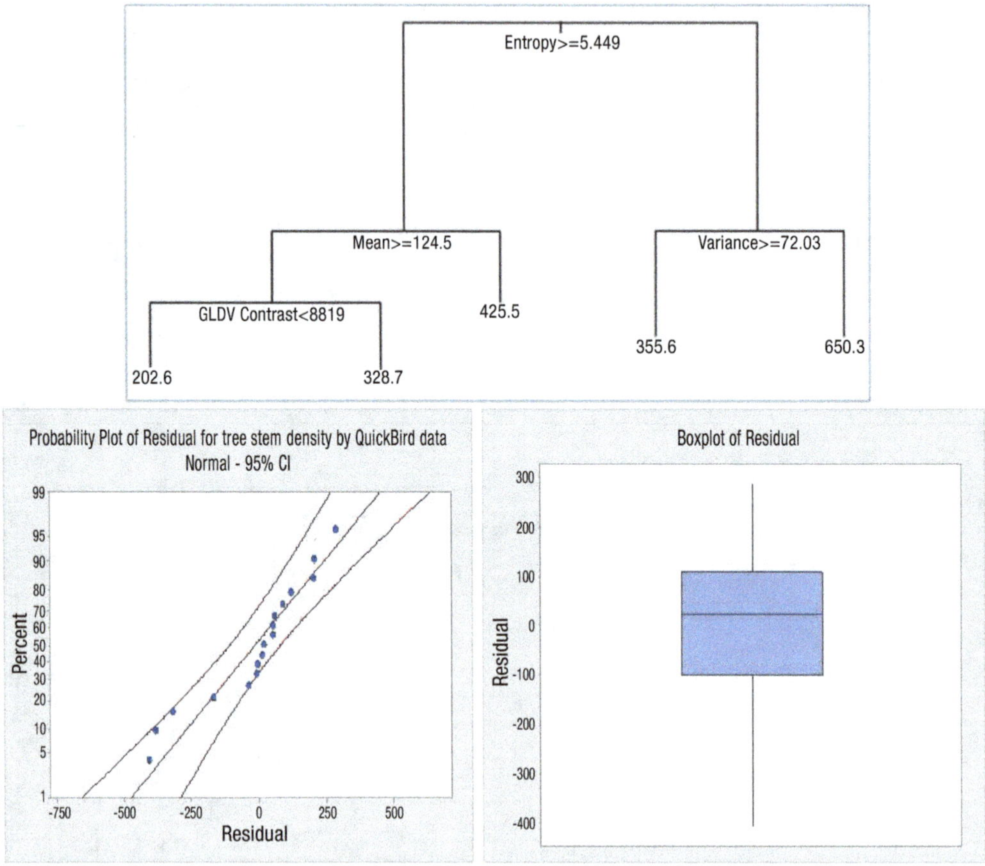

Figure 4. Binary regression tree (top) and probability (left), and box plot (right) of residual CART model for tree density/ha using Quickbird data.

The R^2 values obtained in this study using ASTER images were higher than obtained through direct estimation to predict tree density by Sivanpillai *et al.,* (2006) and Gebreslasie *et al.,* (2010). In addition, the RMSE values obtained in this study were higher than results obtained the studies that were used direct estimation to predict stand volume (Kalbi *et al.,* 2014 and Shataee *et al.,* 2012). The RMSE and R^2 values of tree density values were favourably compared to those obtained by Sivanpillai *et al.,* (2006); Gebreslasie *et al.,* (2010); and Freitas *et al.,* (2005).

The regression tree model for tree density estimation using Quickbird images as the best image, which could produce better estimations, is shown in Fig. 4 (for ASTER and TM images please see Figures S5 and S6 [supplementary]). The results indicated that entropy, greenness, and PCA3 were the most important variables in determining tree density/ha, which could be explained by the three-mentioned images, respectively (Table 5). The results of CART performances by different data sources to estimate tree density showed that Quickbird images could produce estimations with lower absolute and relative RMSE and bias compared to ASTER and TM data. However, estimations of both data were slightly similar (Table 6). Fig. 5 show maps of estimates by CART algorithm using Quickbird data in the study area.

Conclusions

The relationships between reflectance data recorded by the spectral data and forest structural attributes were analysed through the CART algorithm in this study.

Performance assessment of models was examined using RMSE and bias on the unused plots. The results showed that Quickbird satellite data for each three attributes (stand volume, basal area, and tree stem density) have better results than ASTER and TM satellites. However, estimations of ASTER and TM images showed slightly similar results. It seems that the results obtained from TM data were better than results obtained from ASTER data. This priority could be due to using texture analysis on the ASTER image, and retrieval quantities variables using texture analysis ASTER image cannot be very precise.

Although results of this study are valuable and important for extracting and retrieving forest quantity information, and it could provide valuable information

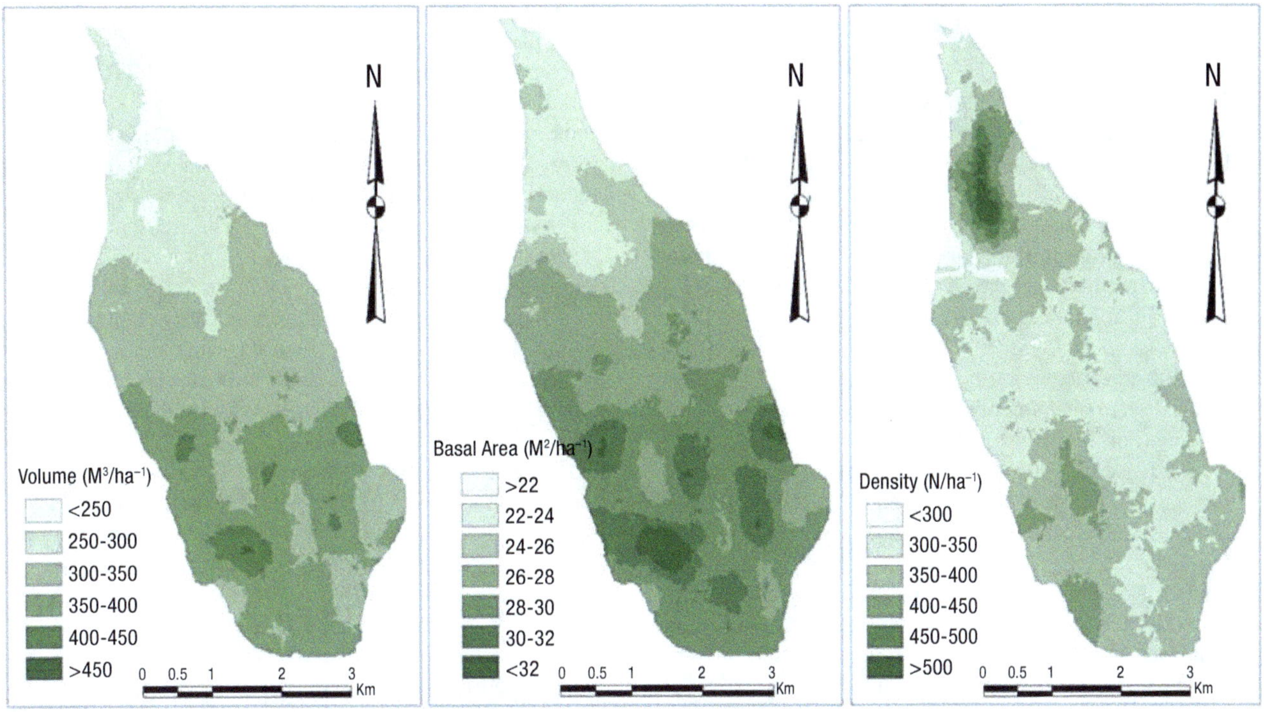

Figure 5. The maps of variables estimated by CART algorithm using Quickbird data.

about changes in stand structure and help forest resource managers to devise suitable management plans. However, the outcomes of this study should be again tested in similar forests elsewhere and/or be adopted in other types of forests, which have same composition and structure. We suggest that the procedure adopted in this study is tested in other areas while investigating the effects of other satellite data.

The results of modelling showed that in spite of priority of Quickbird data compared to ASTER and TM data to estimate the forest attributes, none of these estimation is not enough for accurate variable mapping for executive planning. The causes of do not be adequate the estimations refer to some things. First, these images are coming from optical remote sensing sources, which are producing the 2D information from reflections of canopy surfaces; however, for estimation of stand volume using the 3D information such as using Lidar data could be improve the estimations. Second, complexity of forest stands in points of multi-layers and species composition can be effect on the inaccurate estimations. Using of combination the optical with Lidar data can be useful for improving the estimations.

In conclusion, the results of this study demonstrated that the reflectance values recorded by satellite data are related to spatial resolution. Updating information periodically through satellite remote sensing technology could provide valuable information about changes in stand structure and help forest resource managers to devise suitable management plans.

References

Aertsen W, Kint V, Orshoven JV, Ozakan KA, Wuysa B, 2010. Comparison and ranking of different modelling techniques for prediction of site index in Mediterranean mountain forests, Ecol Model 221: 1119-1130. http://dx.doi.org/10.1016/j.ecolmodel.2010.01.007.

Breiman L, Friedman JH, Olshen RA, Stone CJ, 1984. Classification and Regression Trees. Wadsworth & Brooks/Cole Advanced Books & Software. Monterey, C, 358.

Chavez JR, Pat S, 1996. Image-based atmospheric corrections-revisited and improved. Photogramm Eng Rem S 62: 1025-1036.

Cohen WB, Spies TA, 1992. Estimation structural attributes of Douglas fir/western hemlock Forests stand from Landsat and Spot imagery, Rem Sens Envi 41: 1-17. http://dx.doi.org/10.1016/0034-4257(92)90056-P.

Cooke WH, Jacobs DM, 2005. Rapid classification of Landsat TM imagery for phase stratification using the automated NDVI threshold supervised classification (ANTSC) methodology. In: Proceedings of the fourth annual forest inventory and analysis symposium, 2002 November 19-21, New Orleans, LA. Gen. Tech. Rep. NC-252. St. Paul, MN: US Department of Agriculture, Forest Service, North Central Research Station, 81-86.

Cutler DR, Edwards TC, Beard KH, Cutler A, Hess KT, Gibson J, Lawler JJ, 2007. Random forests for classification in ecology. Ecology 88: 2783-2792. http://dx.doi.org/10.1890/07-0539.1.

Fazakas Z, Nilsson M, Olsson H, 1999. Regional forest biomass and wood volume estimation using satellite data and ancillary data. Agri For Met 98: 417-425. http://dx.doi.org/10.1016/S0168-1923(99)00112-4.

Franco-Lopez H, Ek AR, Bauer ME, 2001. Estimation and mapping of forest stand density, volume, and cover type using the k-nearest neighbour's method. Rem Sen Env 77: 251-274. http://dx.doi.org/10.1016/S0034-4257(01)00209-7.

Freitas SR, Mello MCS, Cruz CBM, 2005. Relationships between forest structure and vegetation indices in Atlantic Rainforest. For Ecol Manage 218: 353-362.

Gebreslasie MT, Ahmed FB, Jan AN, Adrdt V, 2010. Predicting forest structural attributes using ancillary data and ASTER satellite data. Int J App Ear Obs Geoin 125: 523-526. http://dx.doi.org/10.1016/j.jag.2009.11.006.

Hall RJ, Skakun RS, Arsenault EJ, Case BS, 2006. Modelling forest stand structure attributes using Landsat ETM+ data: application to mapping of aboveground biomass and stand volume. For Ecol Manage 225: 375-390.

Huang X, Jensen J, 1997. A Machine- Learning Approach to Automated Knowledge-Base Building for Remote sensing Image Analysis with GIS Data. Photogram Eng Rem Sen 63: 1185-1194.

Huiyan G, Limin D, Gang W, Dong X, Shunzhong W, Hui W, 2006. Estimation of forest volumes by integrating Landsat TM imagery and forest inventory data. Science in China Series E. Tec Sci 49: 54-62.

Hyvonen P, 2002. Kuvioittaisten puustotunnsteja toimenpide-ehdotusten estimointi k-kuviotason tukianeiston Avulla. Metsat Aikaka 3: 363-379.

Hyyppa J, Hyyppa H, Inkinen M, Engdahi M, Linko S, Zhu YH, 2000. Accuracy comparison of various remote sensing data sources in the retrieval of stand attributes. For Ecol Manage 128: 109-120.

Kalbi S, 2011. Estimation of Forest Structural Attributes Using ASTER and SPOT-HRG Data (Case study: Darabkola forest). M. Sc thesis, Department of Forestry, Sari University of Agricultural Sciences & Natural Resources, 107 pp.

Kalbi S, Fallah A, Shataee Sh, 2014. Estimation of forest attributes in the Hyrcanian forests, comparison of advanced space-borne thermal emission and reflection radiometer and satellite poure I'observation de la terre-high resolution grounding data by multiple linear, and classification and regression tree regression models. J App Rem S 8: 2-18. http://dx.doi.org/10.1117/1.jrs.8.083632.

Kilpelainen P, and Tokola T, 1999. Gain to be achieved from stand delineation in Landsat TM image-based estimation of stand volume. For Ecol Manage 124: 105-111.

Lawrence L, Wright A, 2001. Rule-Based Classification and Regression Tree (CART) Analysis. Photogramm Eng Rem Sen 67: 1137-1142.

Makela H, Pekkarine A, 2004. Estimation of forest stands volumes by Landsat TM imagery and stand- Level field-inventory data. For Manage 196: 245-255.

Mohammadi J, Shataee Sh, Babanezhad M, 2011. Estimation of forest stand volume, tree density and biodiversity using Landsat-ETM+ Data, comparison of linear and regression tree analyses. First Conference on Spatial Statistics. Proced Env Sci 7: 299-304.

Mohammadi J, Shataee Sh, Yaghmaee F, Mahiny AS, 2010. Modelling Forest Stand volume and Tree Density Using Landsat ETM+ Data. Int J Remote Sens 31: 2959-2975. http://dx.doi.org/10.1080/01431160903140811.

Moisen GG, Frescino TS, 2002. Comparing five modelling techniques for predicting forest characteristics. Ecol Mod 157: 209-225. http://dx.doi.org/10.1016/S0304-3800(02)00197-7.

Muukkonen P, Heiskanen J, 2005. Estimation biomass for boreal forests using ASTER satellite data combined with stand wise forest inventory data. Remote Sen Env 99: 434-447. http://dx.doi.org/10.1016/j.rse.2005.09.011.

R Core Team, 2012. R: A language and environment for statistical computing R Foundation for Statistical Computing, Vienna, Austria: Available: http://www.R-project.org/.

Sarunas, R. 1997. On dimensionality, sample size, and classification error of nonparametric linear classification algorithms. IEEE Trans. Pattern Anal Mach Intel 19: 667-671. http://dx.doi.org/10.1109/34.601254.

Shataee Sh, Kalbi S, Fallah A, Pelz DR, 2012. Forest attributes imputation using machine-learning methods and ASTER data: comparison of k-NN, SVR, and random forest regression algorithms. Int J Rem Sen 33: 6254-6280. http://dx.doi.org/10.1080/01431161.2012.682661.

Sironen S, Kangas A, Maltamo M, 2010. Comparison of different non-parametric growth imputation methods in the presence of correlated observations. Forestry 83(1): 39-51. http://dx.doi.org/10.1093/forestry/cpp030.

Sivanpillai R, Smith CT, Srinivasan R, Messina MG, BenWU X, 2006. Estimation of managed loblolly pine stands age and density with Landsat ETM+ data. For Manage 223: 247-254.

Tokola T, Heikikkila J, 1997. Improving Satellite image based forest inventory by using a priori site quality information. Silva Fen 31: 67-78. http://dx.doi.org/10.14214/sf.a8511.

Trotter CM, Dymond JR, Goulding CJ, 1997. Estimation of timber volume in a coniferous plantation forest using Landsat TM. Int J Rem Sens 18: 2209-2223. http://dx.doi.org/10.1080/014311697217846.

Tuominen S, Haakana M, 2005. Landsat TM Imagery and High Altitude Aerial Photographs in Estimation of Forest Characteristics, Silva Fen 39: 573-584. http://dx.doi.org/10.14214/sf.367.

Wang YH, Raulier F, Ung CH, 2005. Evaluation of spatial predictions of site index obtained by parametric and non-parametric methods a case study of Lodgepole Pine productivity. For Ecol Manage 214: 201-211.

Wolter TP, Townsend PA, Sturtevant BR, 2009. Estimation of forest structural parameters using 5 and 10 meter SPOT-5 satellite data. Rem Sens 113: 2019-2036. http://dx.doi.org/10.1016/j.rse.2009.05.009.

Yazdani S, 2011. Estimation of Forest Structural Attributes Using Quickbird Data. M. Sc thesis, Department of Forestry, Gorgan University of Agricultural Sciences & Natural Resources, 156.

7

Assessment of crown fire initiation and spread models in Mediterranean conifer forests by using data from field and laboratory experiments

Francisco Rodríguez y Silva[1], Mercedes Guijarro[2,3], Javier Madrigal[2,3], Enrique Jiménez[4], Juan R. Molina[1], Carmen Hernando[2,3], Ricardo Vélez[5], and José A. Vega[4]

[1] Univesity of Córdoba,Dept. Forest Engineering, Edificio Leonardo da Vinci-Campus de Rabanales, 14071 Córdoba, Spain. [2] INIA, Forest Research Center, Dept. Silviculture and Forest Management, Forest Fire Laboratory, Ctra de La Couña km 7,5, 28040 Madrid, Spain. [3] iuFOR, Sustainable Forest Management Institute UVa-INIA, Avda. Madrid 44, 34071 Palencia, Spain. [4] Xunta de Galicia, Lourizán Forest Management Research institute, PO Box 127, 36080 Pontevedra, Spain. [5] Spanish Society of Forest Sciences, PO Box 127, 36080 Pontevedra, Spain.

Abstract

Aim of study: To conduct the first full-scale crown fire experiment carried out in a Mediterranean conifer stand in Spain; to use different data sources to assess crown fire initiation and spread models, and to evaluate the role of convection in crown fire initiation.

Area of study: The Sierra Morena mountains (Coordinates ETRS89 30N: X: 284793-285038; Y: 4218650-4218766), southern Spain, and the outdoor facilities of the Lourizán Forest Research Centre, northwestern Spain.

Material and methods: The full-scale crown fire experiment was conducted in a young *Pinus pinea* stand. Field data were compared with data predicted using the most used crown fire spread models. A small-scale experiment was developed with *Pinus pinaster* trees to evaluate the role of convection in crown fire initiation. Mass loss calorimeter tests were conducted with *P. pinea* needles to estimate residence time of the flame, which was used to validate the crown fire spread model.

Main results: The commonly used crown fire models underestimated the crown fire spread rate observed in the full-scale experiment, but the proposed new integrated approach yielded better fits. Without wind-forced convection, tree crowns did not ignite until flames from an intense surface fire contacted tree foliage. Bench-scale tests based on radiation heat flux therefore offer a limited insight to full-scale phenomena.

Research highlights: Existing crown fire behaviour models may underestimate the rate of spread of crown fires in many Mediterranean ecosystems. New bench-scale methods based on flame buoyancy and more crown field experiments allowing detailed measurements of fire behaviour are needed.

Additional key words: wildland fire; fire behaviour; *Pinus pinea*; small-scale experiment; bench-scale experiment; convection

Abbreviations used: CBD (Canopy Bulk Density); CBH (Crown Base Height); CFIS (Crown Fire Initiation and Spread); CFL (Crown Fuel Load); FMC (Fuel Moisture Content); ICFME (International Crown Fire Modelling Experiment); NFFL (Northern Forest Fire Laboratory); ROS (Rate of Spread); TFO (Time to Flame Out); TTI (Time to Ignition); t_r (Residence Time of the Flame)

Authors´ contributions: FRS conceived the field experimental work, obtained field data and co-wrote the manuscript. MG participated in data collection and paper writing and supervised the manuscript. JM designed the laboratory experimental work, obtained data, run data analysis and co-wrote de manuscript. EJ and JRM participated in data collection and analysis, and paper writing. CH supervised the work and is coordinating the GEPRIF Project. RV contributed to the conception of field experimental work and discussions with its expertise in forest fires fighting. JAV designed the small-scale experiment, supervised the work and coordinated the INFOCOPAS Project.

Funding: panish Ministry of Science and Innovation (Projects INFOCOPAS, RTA2009-00153-C03 and GEPRIF, RTA2014-00011-C06), cofunded by FEDER funds, INIA and European Social Fund (grant to EJ).

Competing interests: The authors have declared that no competing interests exist.

Correspondence should be addressed to Mercedes Guijarro: guijarro@inia.es

Introduction

Crown fires display extreme fire behaviour (Werth *et al.*, 2016) as the fire spreads through both the surface and tree canopy fuel layers, with the surface and crown fire phases more or less linked (Alexander & Cruz, 2016). These fires are generally of high intensity, are difficult to suppress and can have severe ecological and socioeconomic impacts (Moreira *et al.*, 2011). However, despite the social, economic and ecological significance of crown fires, modelling of crown fire behaviour has not been adequately addressed from scientific and technical points of view, and it is far from being solved (Cruz & Alexander, 2010). There are two main reasons for the difficulties in modelling crown fires: (i) the extreme complexity of the physical phenomenon (Rothermel, 1991), due to the heterogeneous fuel involved and the multiple influencing factors such as wind, slope, relative humidity, fuel moisture, atmospheric stability and surface fire intensity; and (ii) the complexity of conducting experimental crown fires (Stocks *et al.*, 2004a). These difficulties are common throughout the world, including the most advanced countries where crown fire experiments have been carried out in a limited extension (for a review, see Alexander & Cruz, 2016).

Although current physical models (Linn *et al.*, 2005; Mell *et al.*, 2009) represent substantial advances in crown fire modelling, from an operational point of view, two main empirical approaches in predicting crown fire spread continue being the base of the commonly used crown fire prediction systems. The first approach is that used in the models proposed by Rothermel (1991) and Finney (1998), based on the relationship between the predicted rate of fire spread at the surface and the fire rate of spread observed in crown fires in the Rocky Mountains (North America). The second approach, used for active crown fires, is based on wind speed, canopy bulk density and the estimated fine dead fuel moisture content (Cruz *et al.*, 2005). Crown fires are serious and dangerous events for forest management, as fire suppression efforts are much more complex than for surface fires, due to high fire-line intensity, spread rate, smoke production, spotting and turbulent fire spread (Alexander & Cruz, 2016). In Spain, technical advances, available allometric equations and previous experience in tackling large fire in Andalusia have demonstrated the need for development of new approaches (Molina, 2015).

It must be taken into account that the most widely used crown fire initiation and spread models (Van Wagner, 1977; Rothermel, 1991; Finney, 1998; Cruz *et al.*, 2005) have an empirical basis developed and tested in boreal forests (Alexander & Cruz, 2016). The models of Van Wagner (1977) and Rothermel (1991) are integrated in prediction systems (www.fire.org) as well as in the widely applied FARSITE fire area simulation tool (https://www.firelab.org/project/farsite). Other fire area simulation systems (*e.g.* VISUAL-CARDIN® and WILDFIRE ANALYST®) also include these models. These tools are frequently used by forest managers to estimate fire risk (*e.g.* Bradstock *et al.*, 2012), to design fuel management treatments (*e.g.* Ager *et al.*, 2012) and fuelbreak networks (*e.g.* Oliveira *et al.*, 2016), and to prioritize firefighting strategies (*e.g.* Butler & Cohen, 1998) in a wide range of forest ecosystems. Crown fire behaviour modelling is also used within the framework of wildfire research to assess the efficacy of fuel treatments (*e.g.* Arca *et al.*, 2007; Duguy *et al.*, 2007; González-Olabarría *et al.*, 2012; Jiménez *et al.*, 2016; Oliveira *et al.*, 2016; Madrigal *et al.*, 2017). Nevertheless, crown fire models have not been tested by experimental burns in Mediterranean conditions and the predictions may therefore be misleading (Cruz & Alexander, 2010; Benali *et al.*, 2016), leading to the application of inappropriate forest and fire management activities.

Laboratory experiments offer limited insight into forest fire behaviour (Fernandes & Cruz, 2012; Sullivan & Cruz, 2015); however, bench-scale and small-scale experiments can be designed in a partially controlled environment to help explain some physical and chemical aspects with a complex interpretation at full-scale (Finney *et al.*, 2015). In this sense, full-scale crown fire experiments can benefit from observations and data obtained in field experiments at smaller scale.

The aims of this paper are as follows: (1) to present the main results of the first full-scale crown fire experiment conducted in a Mediterranean conifer stand in Spain; (2) to compare the results obtained for crown fire initiation and propagation in Mediterranean conifer forests (*Pinus pinea* L. and *Pinus pinaster* Ait.) with current crown fire models by using data from field and laboratory experiments; (3) to obtain information on pine crown ignition in the absence of wind-forced convection; (4) to assess the flame residence time during crown fires in Mediterranean conditions. This work is part of a more comprehensive research project on the initiation, propagation and socio-economic impacts of crown fires in pine stands.

Material and methods

Full-scale crown fire experiment:"Las Traviesas" (Córdoba, southern Spain)

A young *P. pinea* (stone pine) stand (<40 years old) located in the Sierra Morena mountains in southern Spain (coordinates ETRS89 30N: X: 284793-285038; Y: 4218650-4218766) was selected as the study site for full-scale experiments (Fig. 1). *P. pinea* is one of the most important tree species in the Mediterranean region. It is patchily distributed all

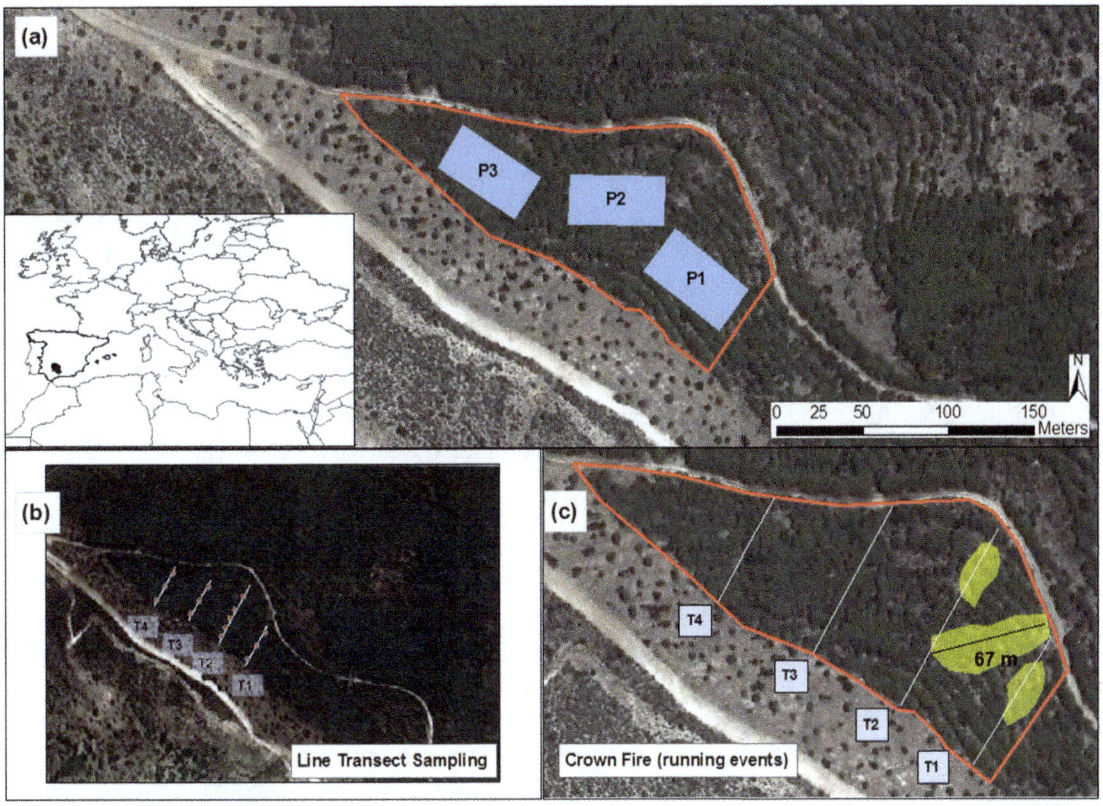

Figure 1. Full-scale crown fire experiment: (a) map showing location of the study area and photograph of the area (red line) and the experimental plots (blue rectangles), (b) transects for monitoring the fuels, and positions of the thermocouple stations (red points), and (c) positions of the crown fire events (indicated in yellow).

around the northern and eastern Mediterranean, from Portugal to Syria as well as along some coastal areas of the Black Sea, from sea level up to 500-600 m in the northern Mediterranean and up to 800-1400 m in the eastern Mediterranean (Fady *et al.*, 2004). In Spain, *P. pinea* occupies an area of approximately 480,000 ha in both natural and afforested stands (Bravo-Oviedo & Montero, 2008). For the present study, this species of pine was chosen for an opportunity reason. The area of burning is included in the priority areas of fuel treatment for the reduction of fire hazard planned by the Provincial Forest Fire Service. The selection was based on the annual fuel treatment planning and counted with the support of suppression teams and equipment for a fire escape contingency. In addition, the structure of the young stone pine stands is similar to that of other species of Mediterranean pines, and therefore it is possible to extrapolate the results to other species of pine associated with the model 7 of the Northern Forest Fire Laboratory (NFFL) System (Anderson, 1982). The climate in the selected area is classified as continental Mediterranean (mean annual precipitation, 492.92 mm; mean annual temperature, 15.47 °C; and mean temperature of coldest and hottest months, respectively, 6.62 and 26.31 °C). Altitude ranges from 664 and 716 m, with a mean slope of 34.25 %. The soils are siliceous with acid metasedimentary rocks.

Three rectangular plots (60 m × 30 m) were established in the experimental site. Vegetation in Plots 1 and 2 (Fig. 1a) corresponds to the HPM5 surface fuel model (Rodríguez y Silva & Molina, 2012) and is characterized by a combination of litter, grasses and

Table 1. Characteristics of tree stand and canopy (*Pinus pinea*, fuel moisture content of needles = 140%) in the full-scale experimental crown fire in the "Las Traviesas" plots. Values are shown as means ± standard deviation.

Density (N)	525 (± 293) trees/ha
Diameter at breast height (DBH)	17.20 (± 4.49) cm
Mean height (H)	6.85 (± 1.77) m
Live crown base (LCB)	1.61 (± 0.49) m
Crown diameter (D)	4.08 (± 0.60) m
Crown fuel load (CFL)	1.239 (± 0.18) kg/m^2
Crown bulk density (CBD)	0.236 (±0.04) kg/m^3

shrubs under canopy. Plot 3 corresponds to model HR, litter under canopy, which is similar to model 9 in the NFFL System.

Litter, duff (humus) and understory fuel (vegetation under canopy) variables (average understory height and cover per species, litter and duff depth and cover) were characterized by continuous measurements made on two perpendicular transects across the plots. The fuel loads were determined by using destructive sampling in adjacent areas (30 quadrats of 0.4 m x 0.4 m for litter and duff and 28 samples of 1 m × 1 m for understory). Litter and understory fuel of thickness less than 6 mm was considered available fuel. The inventories were repeated in the plots after carrying out the experiment to estimate the fuel load consumed during burning.

Canopy and tree variables (diameter at breast height, height, crown base height, crown diameter) were measured in two circular plots (radius 10 m) in each rectangular plot (Table 1). Three representative trees were felled in an adjacent area. The fractions available for burning (branches and needles sorted by diameter size class <6 mm and 6-25 mm) were weighed and dried (100 ºC, 24 h) to enable calculation of the Crown Fuel Load (CFL, kg/m^2). The load over depth method (Van Wagner, 1977) was used to calculate Canopy Bulk Density (CBD, kg/m^3) (Table 1). This inventory was repeated after burning to estimate consumed fuel load by harvesting six trees.

In order to estimate the wind speed correction factor, micrometeorological parameters were measured by three weather stations placed under the canopy (at a height of 1.30 m) for 6 days before the experiment. An additional weather station was sited (at a height of 10 m) in an adjacent area as control (Tables S1, S2, S3 and S4 [suppl]). These automatic weather stations were used to monitor relative humidity, temperature, wind speed and wind direction throughout the experiment.

Samples of fine fuels (dead and live) were collected before the experiment and stored in hermetically sealed plastic bags for subsequent estimation of Fuel Moisture Content (FMC) in the laboratory within 24 hours.

In order to monitor the crown fire Rate of Spread (ROS) and temperatures reached during the experiment, 56 K type thermocouples (1 mm of diameter) were located at four different positions across four transects (Fig. 1b): (1) two thermocouples were placed at the Crown Base Height (CBH) level (roughly 2 m) in each tree: one in a windward (and upslope) position and the other one at an angle of 90º from the latter, and (2) two thermocouples were placed in the middle of the length of the crown (at a height of 4 m, in the same positions as before), *i.e* in the layer with highest canopy bulk density. The flame residence time t_r was estimated in (a) the passive crown fire events as the residence time of flame temperature (above 300 ºC), measured by thermocouples positioned in the same tree (2 and 4 m height), and in (b) the active crown fire events as the residence time of flame temperature (above 300 ºC), measured by thermocouples located in the crown (height 4 m) in two consecutively burned trees.

The fire was initiated with a dip torch. A fire line was established along the border of the experimental area and an upslope fire front was established. Crown fire events were monitored visually by three independent observers and automatically by thermocouples (see above). Passive crown fire events were observed along the fire line, and automatically-collected data were completed with visual observations of flame residence time. All field data were used as inputs to compare with the theoretical predictions from the most common crown fire propagation models.

In previous studies carried out in the Region of Andalusia (southern Spain), validation of crown fire models has relied on wildfires technical reports, cases study of large fires occurred and the experimental fires undertaken in the region. In these studies, comparisons between rates of fire spread observed in the field and their predictions have transmitted a certain degree of uncertainty, as the predictions did not yield similar behaviour parameters to those observed (Molina, 2015). A new approach to crown fire spread was developed on the basis of the models proposed by Rothermel (1991), Finney (1998) and Cruz *et al.* (2005), but with an integration parameter included in order to adapt the simulations to Mediterranean conditions. It was thus necessary to incorporate another two variables (wind speed and canopy cover) to explain real events (Eq. [1]). Information about the uncertainty (Ʊ) was important in testing and validating the new model for Mediterranean forest conditions. In this sense, the uncertainty measurement was lower for the new integrated equation (Ʊ = 43.19) than for approaches used by Rothermel and Finney (Ʊ = 74.70) and by Cruz *et al.* (Ʊ =58.49) according to some crown fire events (Molina *et al.*, 2013).

$$V_{UCO} = \frac{16.6 * V_{RF} * V_C}{(V_{RF} + 0.85 * V_C)^2 * V_{10m}{}^{0.298}} + 0.855 * (0.65 * V_{RF} + 0.35 * V_C) * CC^{0.35} \quad [1]$$

where "V_{UCO}" is the crown rate of spread according to the proposed new mathematical integration (m/min), "V_{RF}" is the crown rate of spread rate based on the approach used by Rothermel and Finney (m/min), "V_C" is the crown rate of spread according to the approach used by Cruz *et al.* (m/min), "V_{10m}" is the 10 m height wind speed (km/h) and "CC" is the canopy cover (expressed as parts per unit). This equation was also used in the present study to compare field data with modelling predictions.

Small-scale experiment: observing crown ignition without wind-driven convection in *Pinus pinaster*

This experiment, conducted in the field facilities of the Lourizán Forest Research Centre (Pontevedra, NW Spain), aimed to obtain direct information on *P. pinaster* crown ignition caused by the radiation from a high intensity surface fire front, spreading towards a clump of pine trees, simulating conditions in absence of wind-forced convection. A flat terrain and meteorological conditions with very low wind velocity were selected to minimize the effect of air of slope-forced convection on energy transport from the surface fire to tree crown. A plot of length 12 m and width 10 m was installed and surrounded by a fire break. A surface fuel layer made from a mixture of plants of the shrub species *Pterospartum tridentatum* (L.) Willk. and *Erica* spp., previously collected in the field and air-dried for several weeks, was constructed and used to carry a surface fire. This added material was weighted and homogenously distributed across the plot to obtain 9.4 kg/m^2 (oven-dry basis) of fuel load, with relatively low moisture content to enhance a high combustion rate and the development of a sustained high vertical flame and linear intensity. Twenty measurements of the shrub layer fuel height were systematically made. Eight *P. pinaster* trees (diameter at breast height from 8.3 to 11.8 cm, total height from 8.2 to 10.5 m, and height to live crown from 1.8 to 4.2 m) were felled in a nearby area and immediately placed in two rows at one end of the plot. The trees were inserted vertically, with the help of a small crane, in previously excavated cylindrical holes, which were filled with water to minimize foliar moisture loss. Eight samples of the added shrub fuel (1 m × 1 m) were collected and classified by size classes (1 h, 10 h and 100 h) and oven-dried in the laboratory for 24 h (100ºC) to determine pre-fire surface fuel load.

Wind direction and speed (2 m, 4 m and 6 m height), relative air humidity and air temperatures during the experiment were measured and recorded every second in a meteorological station installed in the nearby in the upwind position. Immediately before ignition, samples of shrub, trees needles and branches were collected to determine the fuel moisture content after oven-dried for 24 h (80ºC).

The fire was ignited along the edge of the plot opposite the trees rows. Fire behaviour was monitored during the experiment. Eight fine thermocouples were regularly located inside the shrub fuel area to determine fire rate of spread, and four thermocouples were installed in each tree crown at heights of 2 m, 4 m, 6 m and 8 m, on needles fascicles. The thermocouples were inconel-sheathed, type K (chromel-alumel), of 1 mm of diameter, with bared wires of 0.13 mm diameter with an exposed junction. Temperatures were recorded at 1-s intervals during the fire by dataloggers. Images of fire development and flame characteristics were recorded, at ground level, with one infrared thermographic camera and one video camera, and allowed estimate flames length. After flaming, a smouldering phase took place making it inadequate post-fire sampling to calculate fuel consumption during the flaming phase. Hence, linear intensity of the surface fire was calculated following two approaches. First, from the relationship between this parameter and flame length (Thomas, 1963). Second, according to Byram's equation (Byram, 1959), considering total 1 h material and half 10 h material as consumed fuel load during flame phase. A value of 19,500 kJ/kg was used for the heat content of the fuel complex.

Flammability bench-scale experiments: estimating flame residence time

Stocks *et al.* (2004b) and Taylor *et al.* (2004) reported that a physically-based model proposed by Nelson & Adkins (1988) yielded better cross-validations than the models developed by Rothermel (1991) and Finney (1998) during the International Crown Fire Modelling Experiment (ICFME) carried out in Canada (Stocks *et al.*, 2014a). This model (Eq. [2]) requires the value of residence time of the flame t_r to predict ROS:

$$ROS = \frac{0.39W^{0.25}U^{1.51}}{t_r} \qquad [2]$$

where W is the fuel consumed and U is the wind speed. Due to the difficulties in measuring t_r in the field (visually and automatically observed, see above) a laboratory experiment was designed to enable comparison of full-scale and bench-scale tests and to determine the t_r value for its use in the model proposed by Nelson & Adkins (1988).

Sub-samples of material (500 g) were obtained from the crown of six *P. pinea* trees during collection of samples to estimate the FMC in the "Las Traviesas" experimental fire (see above). The samples were transported to the laboratory in hermetically sealed plastics bags and stored in a refrigerator chamber (4ºC). A series of mass loss calorimeter tests (N=10) was carried out during the course of one week by applying the method proposed by Madrigal *et al.* (2013) to produce at least 5 replicates that comply with repeatability criteria

Table 2. Surface fuel load and fuel consumed (mean ± standard deviation), for class size and type of fuel (dead and live) in the full-scale crown fire experiment. Live fuel comprised *Rosmarinus officinalis* (FMC[1]=193%), *Ulex parviflorus* (FMC=80%), *Pistacia lentiscus* (FMC=101%), *Quercus ilex* (FMC=98%) (total cover 86.79±9.83%, mean height 1.22±0.03 m).

	Litter and humus layer	Dead fuel			Live fuel	Total
		<0.6 cm	0.6-2.5 cm	>2.5 cm		
Fuel load (kg/m²)	0.64 (±0.009)	0.21 (±0.02)	0.46 (±0.08)	0.07 (±0.01)	0.986 (±0.13)	2.37 (±0.22)
Fuel consumed (%)	83.12 (±1.17)	95.89 (±19.82)	54.37 (±16.44)	9.27 (±4.08)	70.48 (±22.68)	83.87 (±19.81)

[1]FMC = fuel mtoisture content

Table 3. Canopy fuel load and fuel consumed during the full-scale experimental crown fire (mean±standard deviation). FMC=fuel moisture content

	Needles (FMC=140%)	Branches		Total
		<0.6 cm	0.6-2.5 cm	
Fuel load (kg/m²)	0.56 (±0.16)	0.255 (±0.05)	0.424 (±0.05)	1.239
Fuel consumed (%)	100	51.5 (±19.51)	36.82 (±16.35)	64.85 (±15.04)
Fuel consumed (kg/m²)	0.56 (±0.16)	0.131 (±0.03)	0.156 (±0.02)	0.847 (±0.17)

Table 4. Validation of crown fire rate of spread models for the main crown fire events (A, B, C) in the full-scale crown fire experiment.

Event	Slope (%)	DFMC (%)	U (km/h)	CDB (kg/m³)	ROS[1] (m/min)	ROS (m/min)[2]		
						Rothermel-Finney	Cruz *et al.*	UCO
A	30	11	5.9	0.21	8	7.5	3.5	9.3
B	30	11	5.9	0.11	10	7.5	4	11
C	30	11	14.04	0.22	21	13.5	8.5	17.3

DFMC=dead fuel moisture content; U=wind speed; CBD=crown bulk density; ROS= rate of spread; [1] observed; [2] estimated with Rothermel and Finney, Cruz *et al.* and University of Córdoba (UCO) models.

Table 5. Fuel moisture values of fuel samples immediately before ignition in the small-scale experiment at the Lourizán Forest Research Centre.

Vegetation	Sample	Fuel moisture content (%)
Trees	Current needles	219
	One-year-old needles	164
	Two-year-old needles	157
	Branches (<0.6 cm)	122
Shrub	1 h (<0.6 cm)	17
	10 h (0.6-2.5 cm)	17

Figure 2. Images of crown fire events during burning in the fullscale crown fire experiment: (a,b) general view of study area, (c,d,e) crown fire events, (f) detail of a crown fire.

(Madrigal *et al.*, 2009). Samples were tested at an average of 130% of FMC (similar to the field-estimated FMC of 140%). A heat flux of 50 kw/m^2 was selected to simulate the crown fire combustion during the field experiments (Cruz *et al.*, 2006). A type K thermocouple (1 mm diameter) was sited at fuel level to measure the increase in temperature during the tests for comparison with field data (see previous section). Difference between Time to Flame Out (TFO) and Time to Ignition (TTI) of the samples was used as an estimate of t_r. The results of this series of tests were compared with the time dependent curve of temperatures during crown fire events determined in the full-scale experiment (see previous section). Different t_r values obtained in bench scale and full scale experiments were used to validate the model proposed by Nelson & Adkins (1988), considering the average values of W and U determined in the "Las Traviesas" experiment.

Results

Full-scale crown fire experiment: "Las Traviesas" (Córdoba, southern Spain)

Crown fire activity was only observed in Plot 1 and in the lower part of Plot 2 (Fig. 1). Surface fire spread was observed in Plots 1, 2 and 3. The average total surface fuel load consumed was 2.05 kg/m^2 (Table

2) and the estimated canopy fuel load consumed was 0.85 kg/m^2 (Table 3). Three crown fire events (A, B and C) were monitored in the "Las Traviesas" experiment (Fig. 1c, Fig. 2). Although wind speed was different in event C than in events A and B, the differences in the ROS between "events A and C" and "event B" seem to be due to differences in CBD (Table 4). Wind speed was moderate (5.9 km/h), but a mass fire (Finney & McAllister, 2016) was observed, generating an increase in local wind speed (14 km/h) detected by the meteorological station closest to the crown fire event C (Table 4). The signal recorded by the thermocouples located in the tree crowns (Fig. 3) indicated that ROS ranged between 8 and 21 m/min (Table 4). ROS of event C (21 m/min) differed from that of events A and B (8-10 m/min). Field ROS were compared with the theoretical predictions from the Rothermel-Finney approach, the Cruz *et al.* approach and the new mathematical integrated approach (Eq. [1]).

The most commonly used crown fire models (Rothermel, 1991; Finney, 1998; Cruz *et al.*, 2005) underestimated ROS relative to observed values (particularly the Cruz *et al.* approach) (Table 4). Although the best-fit for event A was yielded by the Rothermel and Finney model, the best-fit for events B and C was achieved with the proposed new integrated model (Table 4).

Small-scale experiment: observing crown ignition without wind-driven convection in *Pinus pinaster*

Shrub fuel load was 5.6 kg/m^2 for 1 h material (<0.6 cm of thickness) and 3.8 kg/m^2 for 10 h material

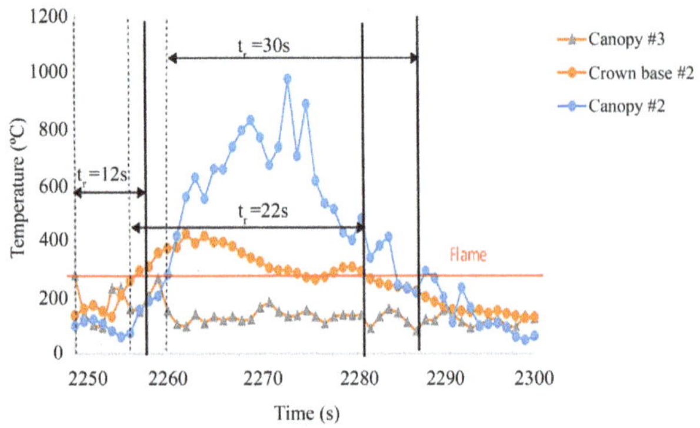

Figure 3. Full-scale crown fire experiment. Time-dependent temperature curves during the burning. Horizontal red line indicates assumed crown ignition threshold. Vertical lines indicate the residence time of flame (t_r). Minimum, average and maximum t_r are shown.

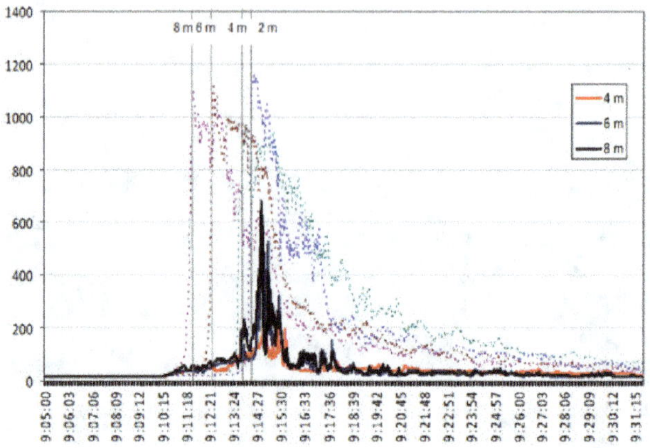

Figure 4. Small-scale experiment. Temporal variation (x-axis, s) in crown temperature (y-axis, ºC) at different heights (solid lines) and temporal variation in temperatures in surface fuel at different distances from the first trees line (dotted lines).

(between 0.6 and 2.5 cm). Shrub layer bulk density was 7.3 kg/m^3. FMC immediately before ignition of fuel samples is shown in Table 5. Air temperature was 20ºC and air humidity was 75%. Mean wind speed during the experimental fire was less than 1 km/h.

Despite low FMC, surface fire ROS was low (2.2 m/min), likely due to low wind velocity and high fuel compactness (more than twice the usual value for similar shrub species; Arellano *et al.*, 2016). Flames reached 7-12 m and were vertical during the fire run. This was likely due to high combustion rate (the heat generated was sufficient to partially melt some 20 mm diameter steel rods placed in the area to indicate the flame height), low wind velocity and fire front indraft, resulting in a high fireline intensity (9,500 kW/m following Thomas (1963) approach, and 5,363 kW/m following Byram (1959) approach). The maximum temperatures reached in the shrub layer ranged from 954 to 1,168 ºC, whereas in the tree crowns maximum temperatures ranged from 82 to 1,224 ºC.

Analysis of the information recorded by thermocouples (Fig. 4) and the images of the infrared thermography camera (Fig. 5) revealed that crowns did not start to ignite (temperatures higher than 300 ºC) until the distance between the fire and trees was less than 2 m. The average of t_r was 22 s (range between 12 s and 31 s). A crown fire simulation using the Crown Fire Initiation and Spread (CFIS) model (Alexander *et al.*, 2006) using the meteorological and fuel load information of the experimental fire predicted a probability of crown fire initiation of 61%, consistent with that observed in the experiment.

Table 6. Validation of Nelson & Adkins (1988) model considering different criteria for fuel consumed (W) and flame residence time (t_r) in the full-scale crown fire experiment. The best estimate of the observed rate of spread (ROS) is shown in bold.

Event	SFC (kg/m^2)	CFC (kg/m^2)	W (kg/m^2)	U (km/h)	ROS (m/min)	Nelson-Adkins (m/min)			
						t_r = 44 s	t_r = 30 s	t_r = 20 s	t_r = 12 s
A	2.05	0.847	SFC+CFC	5.9	8	1.46	2.14	3.21	5.35
B	2.05	0.423	SFC+CFC	5.9	10	1.40	2.06	3.09	5.15
C	2.05	0.860	SFC+CFC	14.04	21	5.42	7.95	11.93	**19.88**
A	2.05	0.847	CFC	5.9	8	1.07	1.57	2.36	3.94
B	2.05	0.423	CFC	5.9	10	0.90	1.32	1.98	3.31
C	2.05	0.860	CFC	14.04	21	3.99	5.86	8.79	14.66

SFC=surface fuel consumption; CFC=crown fuel consumption; U=wind speed

Flammability bench-scale experiments: estimating flame residence time

The Mass Loss Calorimeter experiments (Fig. 6) yielded a difference between TFO and TTI values of 44±10 s (average±standard error) as an estimate of t_r (range between 19 s and 71 s). This is consistent with visually observed values for passive crown fire during the "Las Traviesas" experiment (range, 12-30 s, and mean value, 22 s), small-scale experiments (range, 12-31 s, and mean value, 21 s) and is within the range obtained by Taylor *et al.* (2004) during the ICFME (range 29-37 s, and mean value, 34 s). Madrigal *et al.* (2013) reported an average t_r value of 30 s obtained during an experimental series of crown fire initiation tests in an outdoor wind tunnel. The prediction yielded by the Nelson & Adkins (1988) model using t_r = 30 s (Table 6) was similar to the rate of spread estimated by Cruz *et al.* (2005) model and was a poorer estimate than provided by the Rothermel-Finney model (Table 4).

Analysis of time-dependent temperature curve in the crowns during the full-scale experiment (Fig. 3) indicated a t_r of between 12 s and 30 s. The use of average values of t_r = 20 s observed during crown events in the full-scale and small scale experiments (t_r = 22s) and minimum values during bench-scale experiment (t_r = 19 s) yielded a better overall estimation of ROS but greatly underestimated the rate of spread (Table 6). The estimates closest to the observed ROS corresponding to t_r = 12 s for event C (considering the sum of consumed surface and crown fuel load as the fuel consumed *W* in the Nelson and Adkins equation) (Table 6).

Discussion

The fit for the new integrated model was validated in the "Las Traviesas" experiment. Variations in fire spread predicted by the models proposed by Rothermel-Finney and by Cruz *et al.* and actual field behaviour can be observed (Table 4). The new integrated equation (Eq. [1]) yielded better results for the mean absolute error and standard deviation (14.62±4.06) than obtained by the approaches used by Rothermel (1991) and Finney (1998) (23.11±16) and by Cruz *et al.* (2005) (58.59±2.04) and also provided a reasonable fit for observed values of ROS in the three events analysed during the full-scale "Las Traviesas" experiment (Table 4). Although Molina (2015) obtained overestimates for wind speeds, the most commonly used models (Rothermel 1991, Finney 1998, Cruz *et al.*, 2005) underestimated the ROS. This may be due to the low wind speed during the experiment (Table 4). Stocks *et al.* (2004b) obtained underestimations of Rothermel (1991) and Finney (1998) models but a reasonable fit from Cruz *et al.* (2002). The model proposed by Cruz *et al.* (2002) was developed in boreal forests characterized by low surface fuel load and lower CBD than in the *P. pinea* stands under study here (at least for events A and C, Table 4). This crown biomass structure is typical in the Mediterranean area (Molina *et al.*, 2013) and could be extrapolated to other young conifer stands arising from afforestation or post-fire natural regeneration in Spain (Montero *et al.*, 2005; Ruiz-Peinado *et al.*, 2011). Taking into account that meteorological conditions during the full-scale experiment were moderate relative to summer conditions in southern Spain, underestimation of the spread rate is expected if these models are applied during suppression activities in the study area.

Although crown fire is a very frequent phenomenon in pine wildfire, detailed records of temperature in crown flaming during experimental fires are extremely scarce. Our data are a first contribution, referred to the Mediterranean conifers, and show temperature peak values and transient responses fairly similar to those

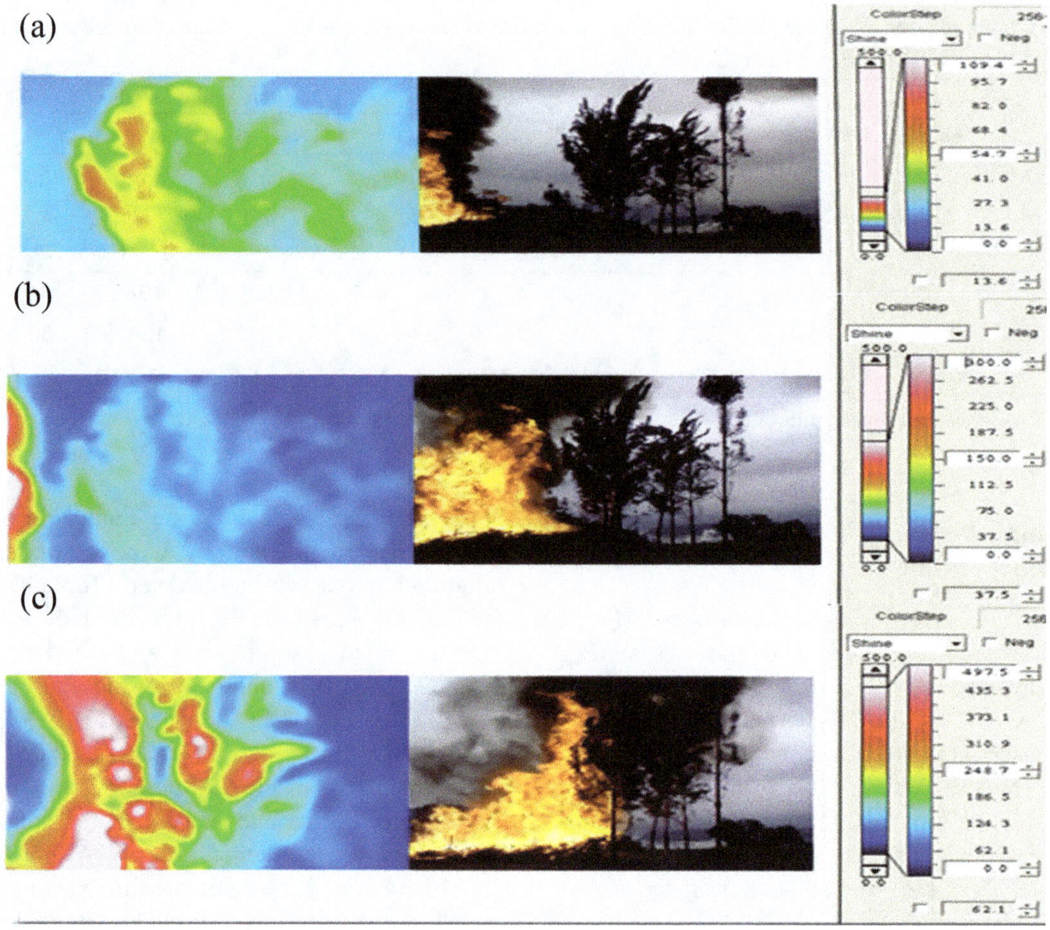

Figure 5. Small-scale experiment. Simultaneous images taken withan infrared thermography recording camera (left) and with a video camera (right), showing the flameat a distance of (a) 6 m and (b) 2 m from the tree crowns, and (c) in contact with the crowns and just when the ignition of them is starting.

observed in boreal forests (Taylor *et al.*, 2004; Butler *et al.*, 2004b; Butler, 2010). Although the limitations of the thermocouples to reflect the true values of temperature in the flaming area are well known (*e.g.* Pitts *et al.*, 2002; Shannon & Butler, 2003; Bova & Dickinson, 2008), the errors when fine thermocouples are used can be considered as to relatively be minor for peaks temperatures (Butler *et al.*, 2004b). Still, in our study, temperature records were primary used to determine fire spread in the full-scale experiment and the distance between fire front and crown to initiate a passive crown fire in the small-scale experiment. Moreover, both in the field and bench experiments, thermocouples records supplied values for flaming residence time in different conditions of heat transfer and assuming a common ignition threshold (573 ºK). The measured time were quite similar among them, consistent with data referred by Despain *et al.* (1996), obtained from video records in the Yellowstone fire (1988) and those measured in the above referred crown fire experiments.

Despite the high intensity of the fire developed during the small-scale experiment and the substantial radiant surface from a well-developed "flames wall" of the surface fire, crown torching did no occurred until flames were very close to the trees crowns (Fig. 5). The images recorded do not suggest a piloted ignition of pyrolisates from radiation, but by buoyancy with strong turbulent convection in concordance with observations in the full-scale experiment and data from other crown fires (Clark *et al.*, 1999; Coen *et al.*, 2004; Taylor *et al.*, 2004). Therefore, it would seems to question the ability of just radiation to explain crown fire (Butler *et al.*, 2004a) and support the need to consider the role played by convection (Finney *et al.*, 2013, 2015). Measurements in scanty crown fire experiments (Frankman *et al.*, 2012) show that relative contribution of convective and radiative energy transfer can be highly variable from a fire to another. That suggests that likely models considering a more flexible contribution of convective and radiative components could be more realistic (Finney *et al.*, 2015). Both the full and small-scale experiments, with light-moderate wind and absence of wind respectively, showed the apparent need for contact between the flame and the crowns to generate crown fire.

Stocks *et al.* (2004b) and Taylor *et al.* (2004), from observations during the ICFME carried out in a boreal forest dominated by *Pinus banksiana* Lamb. with a minor *Picea mariana* Mill. component, suggested the need of a mean flame duration of at least 30 s for crown ignition. That time was assessed in the bench-scale experiments (Fig. 6) and through observations of the occurrence of passive crown fire events in the border of the field plots in the full-scale experiment. The values measured with thermocouples placed in the crowns (Fig. 3) and the validation of the Nelson & Adkins (1988) model (Table 6) indicate t_r required to generate a crown fire much shorter (12 s) in young stone pine stands than in boreal stands. The results also show that bench-scale tests, based on radiation heat flux, offer limited insight into what actually occurs in full-scale phenomena (Fernandes & Cruz, 2012; Sullivan & Cruz, 2015), thus highlighting the urgent need to develop new bench-scale methods based on flame buoyancy (Finney *et al.*, 2015).

Our initial results obtained at different scales suggest that flame buoyancy and likely air-forced convection are determinant for the ignition and fire propagation process in crown fires (Finney *et al.*, 2015). Furthermore, the reasonable estimation of ROS using the total fuel (surface and crown) consumed in the fire run C relative to runs A and B (better fit using only crown fuel), during fuel-scale experiment (Table 6) suggests interactions between both fuel layers and between fuel consumed and wind speed. This therefore suggests the need to develop specific fuel vegetation models that include the effect of fuel vertical structure and the simultaneous description of surface and crown biomass (Arellano *et al.*, 2016). This initial work calls for developing additional experiments with measurements of convective and radiative heat transfer components to confirm these first observations and help to shed some light on those questions.

Conclusions

The results of the first crown fire full-scale experiment in a Mediterranean conifer stand in Spain indicate that the current crown fire behaviour models

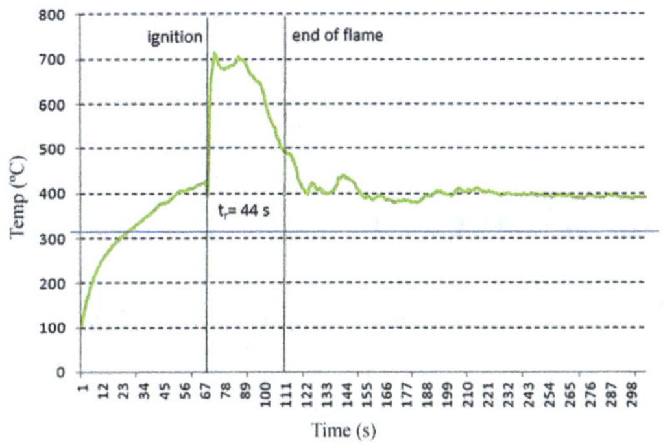

Figure 6. Flammability bench-scale experiment. Mean temperature curve (green) for pine needles (FMC 130%) during the Mass Loss Calorimeter experiments (N=10). The blue line represents the baseline temperature from the radiant heater (Heat flux 50 kW/m^2, piloted ignition). The grey lines represent the time to ignition and time to flame out; difference between both times is used as an estimation of flame residence time (t_r)

may underestimate the rate of spread of crown fires in many Mediterranean ecosystems. In spite of the particular crown shape of *P. pinea*, we think that the results of the current study could be likely extrapolated to other Mediterranean conifers in their young stage. In this stand stage, fuel complex structure is very similar, in terms of understory and canopy fuel layers, to other pine species, and arguably, a similar response in relation to the transition from surface fire to crown fire and the crown fire spread can be expected, although that requires complementary research.

The findings also suggest that empirical crown fire models developed in boreal ecosystems must be calibrated in Mediterranean conditions in order to minimize the observed discrepancies. Evaluation of crown fire behaviour in Mediterranean forests is urgently required and more experimental fires must be conducted to enable development of new models or to calibrate available models under Mediterranean conditions. The new integrated model proposed in this study may be useful for optimizing prevention activities and for addressing difficulties in fire suppression. Nonetheless, that new model needs to be improved as new data become available. The new model programming carried out with Visual Behave software (Rodríguez y Silva *et al.*, 2010) is a simple tool for predicting crown fire behaviour in Mediterranean environments, which is urgently needed by fire managers to help in preventing and extinguishing forest fires.

These initial results confirm the complexity of crown fire behaviour (Alexander & Cruz, 2016) and reinforce the idea of Finney *et al.* (2015) regarding the critical importance of convection and contact between flame and the crowns for crown fire initiation.

Acknowledgements

The authors are grateful to Antonio Arellano for his special technical assistance during field and laboratory experiments. "Las Traviesas" crown fire experiment was carried out with the technical and legal support of Andalusia Autonomous Region (Plan INFOCA, Consejería de Medio Ambiente de la Junta de Andalucía). The assistance of Jose R. Gonzalez, Jesus Pardo, Emilia Puga y Marina Peleteiro is also acknowledged. The authors also thanks the anonymous reviewers for their suggestions to improve the manuscript.

References

Ager AA, Vaillant NM, Owens DE, Brittain S, Hamann J, 2012. Overview and example application of the Landscape Treatment Designer. USDA Forest Service, General Technical Report PNW-GTR-859, Portland, OR, USA, 11 pp.

Alexander ME, Cruz MG, 2016. Crown fire dynamics in conifer forests. In: Synthesis of knowledge of extreme fire behaviour: Volume 2 for fire behaviour specialists, researchers, and meteorologists; Werth PA et al., USDA Forest Service, General Technical Report PNW-GTR-891, Portland, OR, USA, pp: 163-258.

Alexander ME, Cruz MG, Lopes AMG, 2006. CFIS: a software tool for simulating crown fire initiation and spread. Proc 5th Int Conf on Forest Fire Research, Figueira da Foz (Portugal), November 27-30, Amsterdam, The Netherlands: Elsevier B.V. pp. 1-13. https://doi.org/10.1016/j.foreco.2006.08.174

Anderson HE, 1982. Aids to determining fuel models for estimating fire behaviour. USDA Forest Service, Research Papert INT-122, Ogden, UT, USA, 20 pp. https://doi.org/10.2737/INT-GTR-122

Arca B, Duce P, Laconi M, Pellizzaro G, Salis M, Spano D, 2007. Evaluation of FARSITE simulator in Mediterranean maquis. Int J Wildland Fire 16: 563-572. https://doi.org/10.1071/WF06070

Arellano S, Vega JA, Ruiz AD, Arellano A, Álvarez JG, Vega DJ, Pérez E, 2016. Foto-guía de combustibles forestales de Galicia, Versión I. Andavira Editora SL, Santiago de Compostela, Spain, 224 pp.

Benali A, Ervilha AR, Sá ACL, Fernandes PM, Pinto RMS, Trigo RM, Pereira JMC, 2016. Deciphering the impact of uncertainty on the accuracy of large wildfire spread simulations. Sci Total Environ 569-570: 73-85. https://doi.org/10.1016/j.scitotenv.2016.06.112

Bova AS, Dickinson MB, 2008. Beyond "fire temperatures": calibrating thermocouple probes and modeling their response to surface fires in hardwood fuels. Can J For Res 38 (5): 1008-1020. https://doi.org/10.1139/X07-204

Bradstock RA, Cary GJ, Davies I, Lindenmayer DB, Price OF, Williams RJ, 2012. Wildfires, fuel treatment and risk mitigation in Australian eucalypt forests: insights from landscape-scale simulation. J Environ Manage 105: 66-75. https://doi.org/10.1016/j.jenvman.2012.03.050

Bravo-Oviedo A, Montero G, 2008. Descripción de los caracteres culturales de las principales especies forestales de España. In: Compendio de Selvicultura Aplicada en España; Serrada R, Montero G, Reque JA (eds). INIA-Ministerio de Educación y Ciencia, Madrid. 1178 pp + CD.

Butler BW, 2010. Characterization of convective heating in full-scale wildland fires. Proc 6th Int Conf on Forest Fire Research, Coimbra (Portugal), Nov 15-18, DX Viegas (Ed), 9 pp.

Butler BW, Cohen JD, 1998. Firefighter safety zones: a theoretical model based on radiative heating. Int J Wildland Fire 8: 73-77. https://doi.org/10.1071/WF9980073

Butler BW, Finney MA, Andrews PL, Albini FA, 2004a. A radiation-driven model for crown fire spread. Can J For Res 34 (8): 1588-1599. https://doi.org/10.1139/x04-074

Butler BW, Cohen MJ, Latham RD, Schuette P, Sopko K, Shannon S, Jiménez D, Bradshaw LS, 2004b. Measurements of radiant emissive power and temperatures in crown fires. Can J For Res 34 (8): 1577-1587. https://doi.org/10.1139/x04-060

Byram GM, 1959. Combustion of forest fuels. In: Forest fire: control and use; KP Davis (ed), pp: 61-80. McGraw-Hill, NY.

Clark TL, Radke R, Coen J, Middleton D. 1999. Analysis of small-scale convective dynamics in a crown fire using infrared video camera imagery. J Appl Meteor 38: 1401-1420. https://doi.org/10.1175/1520-0450(1999)038<1401:AOSSCD>2.0.CO;2

Coen J, Mahalingam S, Daily J, 2004. Infrared imagery of crown-fire dynamics during FROSTFIRE. J Appl Meteor 43:1241-1259. https://doi.org/10.1175/1520-0450(2004)043<1241:IIOCDD>2.0.CO;2

Cruz MG, Alexander ME, 2010. Assessing crown fire potential in coniferous forests of western North America: a critique of current approaches and recent simulation studies. Int J Wildland Fire 19: 377-398. https://doi.org/10.1071/WF08132

Cruz MG, Alexander ME, Wakimoto RH, 2002. Predicting crown fire behavior to support forest fire management decision-making. In: Forest fire research and wildland fire safety [CD-ROM]; Viegas DX (Ed). Millpress Sci Publ, Rotterdam, Netherlands, 11 pp.

Cruz MG, Alexander ME, Wakimoto RH, 2005. Development and testing of models for predicting crown fire rate of spread in conifer forest stands. Can J For Res 35: 1626-1639. https://doi.org/10.1139/x05-085

Cruz MG, Butler BW, Alexander ME, 2006. Predicting the ignition of crown fuels above a spreading surface fire. Part II: model evaluation. Int J Wildland Fire 15 (1): 61-72 https://doi.org/10.1071/WF05045

Despain DG, Clark DL, Reardon JJ, 1996. Simulation of crown fire effects on canopy seed bank in Lodgepole Pine. Int J Wildland Fire 6: 45-49. https://doi.org/10.1071/WF9960045

Duguy B, Alloza JA, Röder A, Vallejo R, Pastor F, 2007. Modelling the effects of landscape fuel treatments on fire growth and behaviour in a Mediterranean landscape (eastern Spain). Int J Wildland Fire 16: 619-632. https://doi.org/10.1071/WF06101

Fady B, Fineschi S, Vendramin GG, 2004. EUFORGEN Technical Guidelines for genetic conservation and use for Italian stone pine (*Pinus pinea*). IPGRI, Rome, 6 pp. ISBN 92-9043-663-8.

Fernandes P, Cruz MG, 2012. Plant flammability experiments offer limited insight into vegetation fire dynamics interactions. New Phytol 194 (3): 606-609. https://doi.org/10.1111/j.1469-8137.2012.04065.x

Finney MA, 1998. FARSITE: Fire area simulator-Model development and evaluation. USDA Forest Service, Research Paper RMRS-RP-4, Ogden, UT, USA, 47 pp.

Finney MA, McAllister SS, 2016. Fire interactions and mass fires. In: Synthesis of knowledge of extreme fire behaviour, Vol 2; Werth PA et al. (eds), USDA Forest Service, General Technical Report PNW-GTR-891, Portland, OR, USA, pp: 83-104.

Finney MA, Cohen JD, McAllister SA, Jolly WM, 2013. On the need for a theory of wildland fire spread. Int J Wildland Fire 22: 25-36. https://doi.org/10.1071/WF11117

Finney MA, Cohen JD, Forthofer JM, McAllister SS, Gollner MJ, Gorham DJ, Saito K, Akafuah NK, Adam BA, English JD, 2015. Role of buoyant flame dynamics in wildfire spread. Proc Natl Acad Sci USA 112 (32): 9833-9838. https://doi.org/10.1073/pnas.1504498112

Frankman D, Webb BW, Butler BW, Jimenez D, Forthofer JM, Sopko P, Shannon KS, Hiers JK, Ottmar RD, 2012. Measurements of convective and radiative heating in wildland fires. Int J Wildland Fire 22: 157-167. https://doi.org/10.1071/WF11097

González-Olabarría JR, Rodríguez F, Fernández-Landa A, Mola-Yudego B, 2012. Mapping fire risk in the Model Forest of Urbión (Spain) based on airborne LiDAR measurements. For Ecol Manage 282: 149-156.

Jiménez E, Vega-Nieva D, Rey E, Fernández C, Vega JA, 2016. Midterm fuel structure recovery and potential fire behaviour in a *Pinus pinaster* Ait. forest in northern central Spain after thinning and mastication. Eur J Forest Res 135 (4): 675-686. https://doi.org/10.1007/s10342-016-0963-x

Linn R, Winterkamp J, Colman JJ, Edminster C, Bailey JD, 2005. Modeling interactions between fire and atmosphere in discrete element fuel beds. Int J Wildland Fire 14: 37-48. https://doi.org/10.1071/WF04043

Madrigal J, Hernando C, Guijarro M, Diez C, Marino E, de Castro AJ, 2009. Evaluation of forest fuel flammability and combustion properties with an adapted mass loss calorimeter device. J For Sci 27 (4): 323-342. https://doi.org/10.1177/0734904109102030

Madrigal J, Hernando C, Guijarro M, 2013. A new bench-scale methodology for evaluating the flammability of live forest fuels. J Fire Sci 31 (2): 131-142. https://doi.org/10.1177/0734904112458244

Madrigal J, Fernández-Migueláñez I, Hernando C, Guijarro M, Vega-Nieva DJ, Tolosana E, 2017. Does forest biomass harvesting for energy reduce fire hazard in the Mediterranean Basin? A case study in the Caroig Massif (Eastern Spain). Eur J Forest Res 136: 13-26. https://doi.org/10.1007/s10342-016-1004-5

Mell W, Maranghides A, McDermott, R, Manzello SL, 2009. Numerical simulation and experiments of burning Douglas fir trees. Combust Flame 156: 2023-2041. https://doi.org/10.1016/j.combustflame.2009.06.015

Molina JR, 2015. Validación y ajustes de los modelos de propagación de fuego de copa en la ordenación del paisaje forestal. In: Lecciones aprendidas en los incendios forestales; Rodríguez y Silva F (Ed). SECF-Universidad de Córdoba-MAGRAMA-Junta de Andalucía. Córdoba, Spain, pp: 146-166. ISBN: 978-84-608-3864-7.

Molina JR, Rodríguez y Silva F, Mérida E, Herrera MA, 2013. Evaluando las propagaciones de copa en incendios acontecidos en Andalucía con vistas a su modelización. Una aproximación empírica al inicio y propagación del fuego de copas en masas de pinares. 6º Congreso Forestal Español, SECF, Vitoria-Gasteiz, Jun 10-14. http://www.secforestales.org/

Montero G, Ruiz-Peinado R, Muñoz M, 2005. Producción de biomasa y fijación de CO2 por los bosques españoles. Monogr INIA: Ser For Nº 13. INIA, Madrid. 270 pp.

Moreira F, Viedma O, Arianoutsou M, Curt T, Koutsias N, Rigolot E, Barbati E, Corona P, Vaz P, Xanthopoulos G, et al., 2011. Landscape–wildfire interactions in southern Europe: implications for landscape management. J Environ Manage 92 (10): 2389-2402. https://doi.org/10.1016/j.jenvman.2011.06.028

Nelson RM, Adkins CW, 1988. A dimensionless correlation for the spread of wind-driven fires. Can J For Res 18: 391-397. https://doi.org/10.1139/x88-058

Oliveira TM, Barros AM, Ager AA, Fernandes PM, 2016. Assessing the effect of a fuel break network to reduce burnt area and wildfire risk transmission. Int J Wildland Fire 25: 619-632. https://doi.org/10.1071/WF15146

Pitts WM, Brown E, Peacock RD, Mitler HE, Johnsson EL, Reneke PA, Blevis LG, 2002. Temperatures uncertainties for bare-bead and aspirated thermocouples measurements in fire environments. In: Thermal measurements: The foundation of Fire Standards, ASTM STP 1427, LA Grizo & NJ Alvares, Eds., ASTM International, West Conshohocken, PA, USA.

Rodríguez y Silva F, Molina JR, 2012. Modeling Mediterranean forest fuels by integrating field data and mapping tools. Eur J Forest Res 131: 571-582. https://doi.org/10.1007/s10342-011-0532-2

Rodríguez y Silva F, Molina JR, Martínez JF, 2010. Manual técnico de aplicaciones informáticas para la defensa contra incendios forestales. MANPAI XXI. Córdoba. 117 pp.

Rothermel RC, 1991. Predicting behaviour and size of crown fires in the Northern Rocky Mountains. USDA Forest Service, Research Paper INT-438, Ogden, UT, USA, 46 pp.

Ruiz-Peinado R, del Río M, Montero G, 2011. New models for estimating the carbon sink capacity of Spanish softwood species. Forest Syst 20 (1): 176-188. https://doi.org/10.5424/fs/2011201-11643

Shannon KS, Butler BW, 2003. A review of errors associated with thermocouple temperature measurement in fire environments. Proc 2nd Fire Ecol Congr, Orlando, FL, USA, Nov 16-20

Stocks BJ, Alexander ME, Lanoville RA, 2004a. Overview of the international crown fire modelling experiment (ICFME). Can J For Res 34: 1543-1547. https://doi.org/10.1139/x04-905

Stocks BJ, Alexander ME, Wotton BM, Stefner CN, Flannigan MD, Taylor SW, Lavoie N, Mason JA, Hartley GR, Maffey ME, et al., 2004b. Crown fire behaviour in a northern jack pine – black spruce forest. Can J For Res 34: 1548-1560. https://doi.org/10.1139/x04-054

Sullivan AL, Cruz MG, 2015. Small-scale flame dynamics provide limited insight into wildfire behaviour. Proc Natl Acad Sci USA 112 (31): E4165. https://doi.org/10.1073/pnas.1506877112

Taylor SW, Wotton BM, Alexander ME, Dalrymple GN, 2004. Variation in wind and crown fire behaviour in a northern jack pine – black spruce forest. Can J For Res 34: 1561-1576. https://doi.org/10.1139/x04-116

Thomas PH, 1963. The size of flames from natural fires. Int Symp on Combustion 9, pp: 844-859. Elsevier. https://doi.org/10.1016/S0082-0784(63)80091-0

Van Wagner CE, 1977. Conditions for the start and spread of crown fire. Can J For Res 7: 23-34. https://doi.org/10.1139/x77-004

Werth PA, Potter BE, Alexander ME, Cruz MG, Clements CB, Finney MA, Forthofer JM, Goodrick SL, Hoffman C, Jolly WM, et al., 2016. Synthesis of knowledge of extreme fire behaviour, Vol 2. USDA Forest Service, General Technical Report PNW-GTR-891, Portland, OR, USA, 258 pp.

8

Effects of basal area on survival and growth of longleaf pine when practicing selection silviculture

Ferhat Kara[1*], Edward F. Loewenstein[2] and Dale G. Brockway[3]

[1*]*Faculty of Forestry, Kastamonu University, Kastamonu, Turkey, 37100.* [2]*School of Forestry and Wildlife Sciences, 602 Duncan Drive, Auburn University, Auburn, AL, 36849, USA.* [3]*Southern Research Station, USDA Forest Service, 521 Devall Drive, Auburn, AL, 36849, USA.*

Abstract

Aim of study: Uneven-aged (UEA) management systems can achieve multiple-use objectives, however, use of UEA techniques to manage longleaf pine (*Pinus palustris* Mill.) forests are still open to question, because of the species' intolerance of competition. It was our aim to examine the influence of different levels (9.2, 13.8 and 18.4 m^2 ha^{-1}) of residual basal area (RBA) on longleaf pine seedling survival and growth following three growing seasons.

Area of study: This study was conducted at the Escambia Experimental Forest, located on the Southern Coastal Plain of Alabama, in the southeastern United States.

Material and Methods: Selection silviculture was implemented with the Proportional-Basal Area (Pro-B) method. Prescribed burning was conducted before seed dispersal and in the second year after germination. Photosynthetically active radiation (PAR) was measured under the canopy in the study plots. Survival and growth of longleaf pine seedlings were observed for three growing seasons.

Main results: An inverse relationship was found between the number of germinants and RBA, but the mortality of germinants and planted seedlings was not affected by RBA. At age three, an inverse relationship was observed between root-collar diameter (RCD) growth of the germinants and RBA, but RCD growth of planted seedlings was not affected by RBA. Most of the study plots contained more than the projected number of seedlings needed to sustain the target diameter structure.

Research highlights: Long-term continuous monitoring of seedling development and recruitment into canopy is required to determine the efficacy of UEA management. However, current data suggest that UEA methods may be a viable alternative to the use of even-aged (EA) methods in longleaf ecosystems.

Keywords: Prescribed burning; Recruitment; Single-tree selection; Sustainability; Uneven-aged management.

Abbreviations used: EA: even-aged, PAR: photosynthetically active radiation, Pro-B: Proportional-Basal Area method, RBA: residual basal area, RCD: root collar diameter, S-N: The Stoddard-Neel approach, UEA: uneven-aged, VGDL: volume guiding diameter limit.

Funding: The author(s) received no specific funding for this work.

Competing interests: The authors have declared that no competing interests exist.

Correspondence should be addressed to Ferhat Kara: fkara@kastamonu.edu.tr

Introduction

Longleaf pine (*Pinus palustris* Mill.) ecosystems exhibit a rich species diversity containing more than 40 vascular plant species in 1 m^2 (Walker & Peet, 1983), produce wild game and forage grasses (Franklin, 1997), and provide high quality wildlife habitat (Brockway *et al.*, 2006). As a result of the historic decline (by 97%) in longleaf pine acreage (Boyer, 1990), concern about restoration and management of longleaf pine ecosystems has increased in recent years (Brockway & Outcalt, 2000). Achieving restoration and management goals for longleaf pine forests will require application of practical silvicultural methods that lead to sustainable production of goods and services. Even-aged (EA) methods, particularly shelterwood regeneration methods, have been successfully and effectively used to regenerate existing longleaf pine stands (Croker & Boyer, 1975). However, ecosystem services have been episodic due to complete overstory removal at the end of each rotation (Brockway *et al.*, 2006) and the intensity of logging during shelterwood harvesting appears to result in a sharp decline for native species such as wiregrass (*Aristida beyrichiana* Trin. & Rupr.) and silverthread goldaster (*Pityopsis graminifolia* (Michx) Nutt.) in the understory

(Brockway & Outcalt, 2015). Ecosystem services are the benefits people obtain from ecosystems, and they include provisioning (i.e. food, fresh water and wood), regulating (i.e. climate regulation, wildlife habitat and pollination) and cultural services (i.e. recreation, aesthetic and education) (Millennium Ecosystem Assessment, 2005). There is a need for an approach that ensures continuous ecosystem services while producing high quality timber.

Recently, more attention has been given to multiple-use objectives such as aesthetics, wildlife management, water quality, and recreation (Guldin, 2006). It has been suggested that uneven-aged (UEA) management techniques can meet multiple-use objectives, and achieve restoration of longleaf pine forests (Guldin, 1996) because of the continuous forest cover and multi-cohort structure provided by UEA systems (Moser *et al.*, 2002). Field trials to demonstrate and assess the utility of UEA silviculture in these forests have been attempted to a limited degree. The Stoddard-Neel (S-N) approach to single-tree selection is one example of UEA silviculture in longleaf pine forests (Moser *et al.*, 2002; Mitchell *et al.*, 2006; Neel *et al.*, 2010). However, the S-N approach is not quantitative, and it cannot be systematically evaluated because there is no defined target residual structure. It can only be applied by experienced practitioners and training someone to properly apply this approach is time consuming. The S-N approach suggests that UEA systems work for longleaf pine; however, the implementation of the approach, and the required residual basal area (RBA) and number of seedlings are not explicit. Volume guiding diameter limit (VGDL) and BDq methods (Farrar, 1996) are more quantifiable than the S-N approach, but they still require a high degree of professional skill to properly apply in the field. It is also difficult to mark the stand in one pass when using these approaches. In this study, we used an easy-to-apply and scientifically-based tree-marking approach, the Proportional-Basal Area (Pro-B) method, to implement UEA management via single-tree selection in longleaf pine forests (Brockway *et al.*, 2014). Pro-B is based on structural control, allows stand marking in one pass and does not require extensive field experience (Brockway *et al.*, 2014).

The natural gap-phase regeneration pattern of longleaf pine can be closely achieved using group selection (Farrar & Boyer, 1991; Brockway & Outcalt, 1998; McGuire *et al.*, 2001) since longleaf pine can regenerate by filling openings regardless of the gap size (Schwarz, 1907). Distribution of large longleaf pine seeds by wind may be a problem in larger canopy gaps (Boyer, 1963). An additional consideration is that small canopy gaps can become closed before regeneration is established (Guldin, 2006), and the spatial control of many small gaps may be difficult in the long-term (Roach, 1974). Some scientists have suggested larger gaps (Palik *et al.*, 2002; Brockway *et al.*, 2006) while others recommend that smaller gaps such as those created by single-tree selection may be sufficient for securing adequate reproduction (McGuire *et al.*, 2001; Jack *et al.*, 2006). However, there has not yet been enough long-term research in the use of UEA management to verify that selection silviculture can sustain these forests.

The degree of tolerance of longleaf pine to shade has been questioned (Samuelson & Stokes, 2012). Longleaf pine has been classified as shade intolerant species (Boyer, 1990), but its seedlings may survive under shaded conditions for a prolonged time during the grass stage (Croker & Boyer, 1975). During the grass stage, longleaf seedlings concentrate growth in their root system, and this stored energy in the taproot facilitates recovery after fire (Chapman, 1932). It has been suggested that longleaf pine is moderately tolerant to shade when young, and then becomes more intolerant of shade with increasing age (Bhuta *et al.*, 2008). The RBA level at which seedling survival and development is acceptable under UEA silviculture is still open to question. This study aims to address such questions by assessing the influence of differing RBA levels on longleaf pine seedling survival and growth. Because regeneration of longleaf pine may be hindered by its intolerance to shade and competition (Croker & Boyer, 1975), canopy openness as measured by RBA seems to be an important factor in seedling establishment. Thus, it seems reasonable to test a range of RBAs (9.2, 13.8, and 18.4 m^2 ha^{-1}) to assess their effects on longleaf pine survival and growth. These RBA levels were selected to contrast this study with previous studies that examined rates of longleaf pine regeneration under varying stand densities (Croker & Boyer, 1975; Boyer, 1979; Brockway *et al.*, 2014). The RBA suggested for shelterwood methods ranges from 6.9 to 9.2 m^2 ha^{-1}, because seed production usually peaks within this range of RBA (Croker & Boyer, 1975) and lower stand densities may, because of insufficient needle fall, inhibit the effectiveness of prescribed fire. Therefore, we selected an RBA of 9.2 m^2 ha^{-1}, the upper limit suggested for shelterwood, as our lowest RBA. Brockway *et al.* (2014) examined development of longleaf pine seedlings under an RBA of 11.5 $m^2 ha^{-1}$ and noted encouraging levels of longleaf pine regeneration and stand development under that RBA. We selected an RBA of 13.8 m^2 ha^{-1} as our mid-level of RBA, somewhat higher than that observed by Brockway *et al.* (2014). Finally, an RBA of 18.4 m^2 ha^{-1} was chosen as the high level of RBA, to determine whether a dense stand of longleaf pine can be managed to obtain adequate numbers of well-developed seedlings through selection silviculture.

Longleaf pine is adapted to survive in ecosystems that are subjected to frequent surface fires (Landers, 1991). Prescribed fire prepares a seedbed for longleaf pine seedlings, facilitates germination by exposing the mineral soil (Boyer & White, 1990), decreases the competition for longleaf pine seedlings from other species (Heyward, 1939) and reduces the risk of brown-spot needle blight disease (Chapman, 1932). Periodic fire is also necessary for restoring and maintaining native groundcover plants and wildlife communities (Brockway & Outcalt, 2000). Even though prescribed fire is an essential silvicultural tool in longleaf pine forests, it may be partly responsible for seedling mortality, especially when seedlings are newly germinated (Boyer, 1963). For this reason, we included an examination of fire-inducted mortality during the establishment phase to quantify its magnitude.

In examining the efficacy of selection silviculture with the Pro-B method in longleaf pine ecosystems, our primary objectives were to (1) assess the effects of RBA on longleaf pine seedlings survival and growth following three growing seasons, (2) estimate future seedling abundance across varying levels of RBA, and (3) evaluate the influence of RBA on the interaction between a dormant season prescribed fire and survival of planted longleaf pine seedlings.

Materials and methods

Study Site

This study was conducted on the Escambia Experimental Forest, which is located in the Gulf Coastal Plain of the southeastern United States (USDA, 2014) (Fig. 1a). The Escambia Experimental Forest was established in 1947 to study the ecology and management of longleaf pine forests. About 80% of the forest is dominated by longleaf pine and the remainder consists of slash pine (*Pinus elliottii* Engelm.) and mixed hardwoods, occurring mostly along small streams. Average site index for longleaf pine is 22 m (base age 50). Soils are coarse to fine, loamy, siliceous thermic Paleudults (Mattox, 1975). Annual precipitation is about 1520 mm and average temperatures range from 5 to 33 °C. Topography is flat to rolling with slopes mostly less than 10%. At this experimental forest, various age classes are present, ranging from new germinants to trees up to 160 years old. Prior to harvesting in our study area, average tree diameter ranged from 19 to 29 cm, while average density ranged from 200 to 680 tree ha^{-1} across all plots. Basal area ranged from 11.5 to 30 m^2 ha^{-1}. Stand structures varied across all plots. In general, plots contained trees of all diameter classes representing UEA conditions; however, they did not exhibit an explicit reverse j-shaped distribution typical of balanced UEA stands.

Study Design

The study was installed as a completely randomized design. In the winter of 2010, nine 2-ha square plots were established, and randomly assigned to one of three RBA levels: 9.2, 13.8, and 18.4 m^2 ha^{-1} (40, 60, and 80 ft^2 ac^{-1}, respectively). Plots were designated H_1, H_2 and H_3 for high-RBA; M_1, M_2 and M_3 for mid-RBA; and L_1, L_2 and L_3 for low-RBA treatments, with each RBA treatment replicated three times. Assigned treatments were applied to the entire plot (the experimental unit); treatment response was estimated by measurements conducted on subplots. Each study plot contained six 100-m^2 square overstory measurement subplots (Fig. 1b), and each overstory measurement subplots contained three 10-m^2 circular regeneration subplots (Fig. 1b). Overstory and understory subplots were systematically located within each plot (Fig. 1b).

Harvest operations were completed during May 2011, after stands were marked to the defined treatment RBA using single tree-selection implemented with the Pro-B method (Brockway *et al.*, 2014). A standard 'target structure' defined by a q-value of 1.3 (for 5-cm diameter class) and a largest diameter tree (LDT of 45 cm) was used. This structure apportions RBA among three product classes (<15 cm; 15-30 cm; >30 cm) in a ratio of approximately 1:2:3 (Loewenstein, 2005; Brockway *et al.*, 2014). Given a properly selected RBA, this distribution has been shown to allocate sufficient growing space for the recruitment of new cohorts in studies with longleaf pine (Dyson *et al.*, 2009) and with various hardwood species (Loewenstein, 2005).

Longleaf pine seedlings were planted in the event that seed production was poor or failed during 2011-2012, since longleaf pine exhibits a high degree of annual variation in seed production. Three longleaf pine seedlings were hand-planted, in early December 2011, in each regeneration subplot (486 total) following a growing-season prescribed fire. Seedlings were 15-cm deep-plug containerized planting stock with a rooting volume of 100 cm^3. Planting was performed following a rainy period. Seedlings were planted approximately 1-m away from plot center and equidistant from each other to minimize intraspecific competition. Seedlings were tagged and numbered after planting in order to monitor their survival and growth.

A dormant-season fire was applied in January-February 2014 to reduce competition for longleaf pine seedlings from hardwoods, and note the influence of

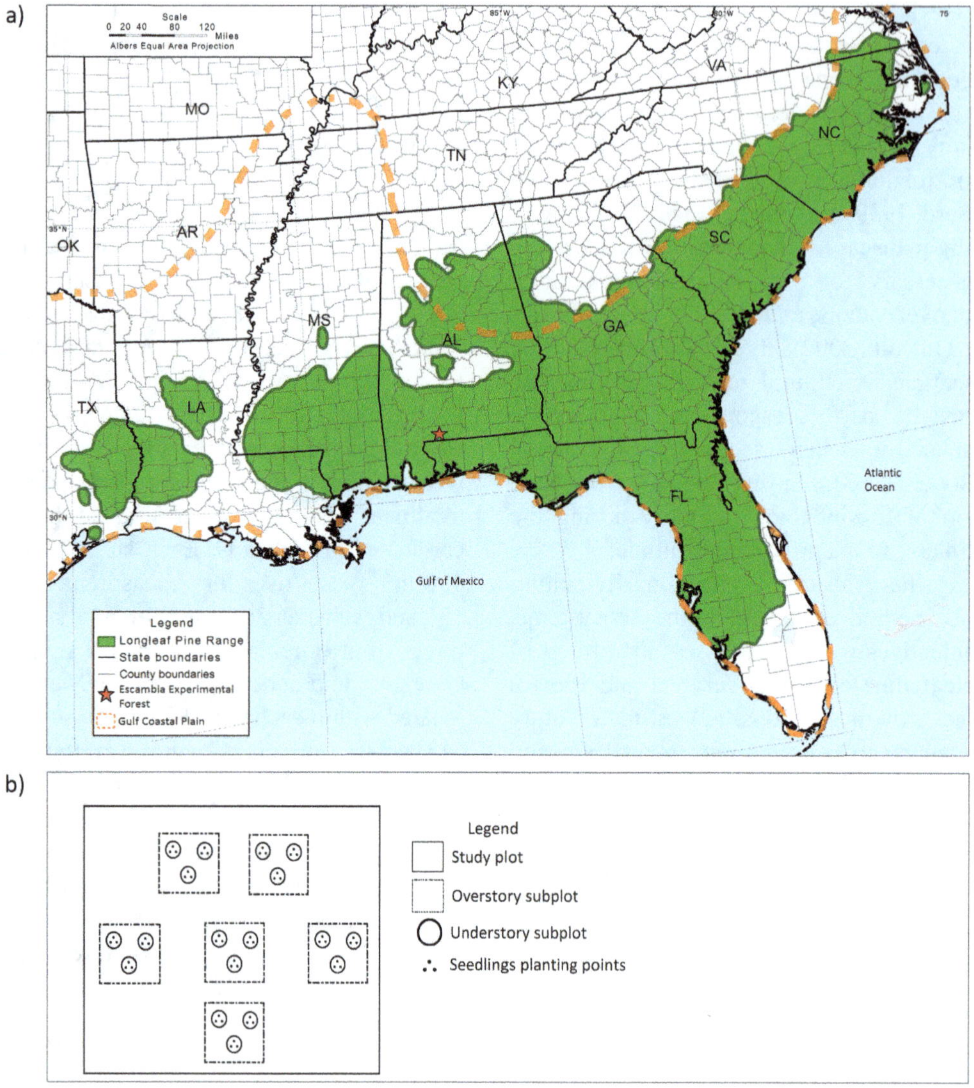

Figure 1. (a) Escambia Experimental Forest within the native range of longleaf pine in the southeastern United States (USDA, 2014). (b) Study plot design.

RBA on the interaction between burning and survival of planted seedlings. During the dormant-season fire, the air temperature ranged from 11 to 23 0C, while relative humidity ranged from 23 to 47 percent. Within the study plots, most of the understory vegetation was consumed and resulted in more than 93% of topkill for hardwoods during the growing-season and dormant-season prescribed fires. Hardwood seedlings larger than 5 cm in their root-collar diameters (RCD) mostly survived.

Data collection and analysis

Following the dormant-season prescribed fire at age 2 (January-February 2014), three germinants among those that survived were randomly selected in each regeneration subplot, and their RCDs were measured to the nearest millimeter to record their growth at the end of the third growing season (August 2014). In addition, the number of germinants was also counted at the end of the third growing season in each regeneration subplot.

The RCD growth of the planted seedlings were recorded in the second (July 2013) and third (August 2014) growing seasons to calculate the growth of planted seedlings. Mortality of planted seedlings was recorded at the end of each growing season (i.e. July 2013 and August 2014). Survival of the planted seedlings from the dormant-season fire was also observed two months after the fire, in the first week of April (2014) when the new needles begin to emerge.

Photosynthetically active radiation (PAR) was measured under the canopy in each regeneration subplot, on nearly cloudless days between 11:00 and 14:00 hours, during each measurement period, using an AccuPar Linear PAR/LAI Model PAR-80 ceptometer (Decagon

Devices, Inc., Pullman, WA). PAR measurements from regeneration subplots were averaged in each treatment plot. A HOBO weather station PAR sensor (Onset Computer Corporation, 2009) was also installed in a treeless area of about 8 ha. Intercepted PAR (IPAR) was calculated using the following formula;

$$IPAR = 1 - (\frac{PAR\ under\ canopy}{PAR\ in\ open}) \times 100$$

Since PAR measurement is affected by factors such as topography IPAR=y, elevation and solar zenith angle, readings were corrected for zenith angles and topography elevations using the correction formula suggested by Liang *et al.* (2012). The change in PAR values, following the correction, was negligible.

Rather than using ANOVA to test difference among treatments, because of deviation from the target RBA levels (Kara & Loewenstein, 2015a) we decided that simple linear regression (α-level=0.05) would be preferable for testing the relationships between RBA and the (1) number of germinants, (2) growth of germinants, (3) growth of planted seedlings, (4) survival of planted seedlings in the absence of fire, and (5) survival of planted seedlings following fire. Relationships between IPAR and the same variables were also examined using simple linear regression (α-level=0.05). In order to more appropriately model the relationship between RBA and number of germinants, Poisson regression was used. For count data, such as the number of germinants, a Poisson regression model is usually recommended (Rodriguez, 2007). Seedling survival percentages were averaged for each plot and these percentages were arcsine transformed before regression analysis (Davis *et al.*, 1999). Growth data were log-transformed to improve residual homogeneity and normality (McDonald, 2014). The models used plot level data defined by the mean value of all subplots within a given 2-ha experimental unit. R-Statistical software (R-Project, 2008) was used for the analyses. It should be noted that a wildfire occurred during May 2012 on one of the mid-level BA plots, M_2 (13.8 m^2 ha^{-1}), and all new germinants were consumed. Therefore, data from this plot were not included in the analysis.

Reliable Recruitment

When a seedling bank is created across the observed range of RBA, the concern at this point becomes whether the stand conditions are suitable for providing reliable recruitment into the overstory. Three-year observations and data from the literature provide some evidence as to whether the current stand conditions are adequate for contributing to the sustainable management of these stands when implementing selection silviculture. Although views concerning the stated assumptions might vary (in particular the q-value of our defined target diameter structure), our predictions are presented to provide a rough idea concerning the number of seedlings that may be expected to recruit into the overstory.

Longleaf pine seedlings typically bolt from the grass stage (i.e. begin rapid height growth) and recruit into the overstory, after they reach an RCD of approximately 25 mm (Boyer, 1990). The age required to reach the RCD of 25 mm for each level of RBA was projected using the 3-year growth rates, and the growth rates reported by Boyer (1963). The number of seedlings that reach the RCD of 25 mm was determined using the mortality rates measured: (1) following the prescribed fire, (2) in the absence of fire, and (3) following the logging operation.

Longleaf pine seedlings become more resistant to fire when they reach an RCD of 13 mm (Boyer, 1974). For this reason, average mortality rates observed during the prescribed fire at age 2 were assumed for subsequent fires until the seedlings reach the RCD of 13 mm. In addition, during the prescribed fire at age 2, planted seedlings were already larger than the RCD of 13 mm, thus, for the fires after seedlings reach the RCD of 13 mm, we used the mortality rates of planted seedlings observed during the prescribed fire at age 2. Moreover, average mortality rates observed in the absence of fire were assumed for subsequent years when no burning is conducted. An average mortality of 50 % during logging operations at age 10 was assumed as suggested by Maple (1977) and Boyer (1990).

Results

IPAR

IPARs ranged from 77 to 88% across all study plots before the harvest operations. Average IPARs following the harvesting ranged from 57 to 78% across the study plots. PAR levels ranged from 192 to 536 μmol m-2 s-1 across all plots before harvesting, and higher PAR levels were measured following the harvesting ranging from 557 to 1087 μmol m^{-2} s^{-1}. There was a statistically significant inverse relationship between RBA and IPAR following the harvest operations (p = 0.0025).

Seedling survival and growth

At the end of the second growing season (July 2013), there was no significant relationship between RBA and the mortality of germinants (p=0.85). Nor did IPAR affect the survival of germinants in the second growing season

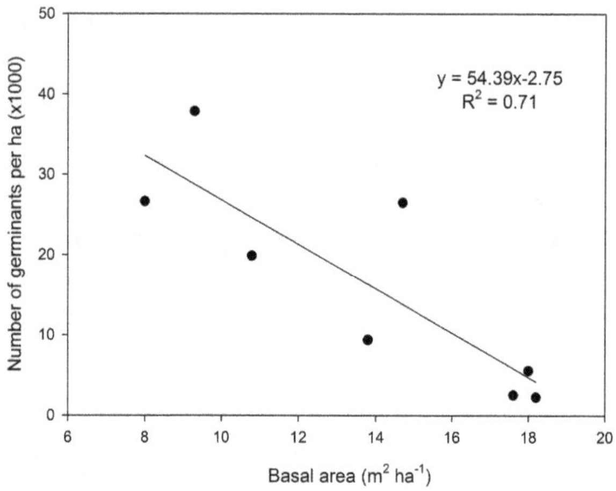

Figure 2. Relationship between the number of germinants and RBA during the third growing season.

(p=0.79). Survival rate of germinants was high across all plots in the absence of fire (i.e. in the second growing season), ranging from 95 to 98%. At the end of the third growing season (August 2014), following the dormant-season fire at age 2, the number of germinants ranged from 2,270 to 37,870 per hectare across all study plots. There was a statistically significant inverse relationship between RBA and the number of germinants in the third growing season (p<0.0001) (Fig. 2) as well as a significant relationship between IPAR and the number of germinants (p<0.0001). Higher numbers of germinants were observed from the lower RBA plots. Although stand density is known to impact cone production in longleaf pine forests (Croker & Boyer, 1975) and consequently may influence the number of germinants, there was no statistically significant effects of initial basal area on the number of germinants (p=0.94) nor was there an effect from the interaction of initial basal area with RBA on the number of germinants (p=0.93).

Mortality among planted seedlings was not affected by RBA or IPAR during the second year following planting (p=0.87 and p=0.76, respectively) (Fig.3a). Survival was high during the first two growing seasons, ranging from 96 to 100 % across all plots. In addition, a statistically significant negative relationship was noted between RBA and the survival of planted seedlings following the dormant-season prescribed fire at age 2 (p=0.04) (Fig.3b). The survival rate ranged from 39 to 85% across all plots and increased with decreasing RBA following the dormant-season fire (Fig.3b).

At the end of the third growing season (August 2014), following the dormant-season prescribed fire, there was a statistically significant relationship between RCD size of germinants and RBA (p=0.006) (Fig. 4a), and a significant relationship between IPAR and RCD size of germinants (p<0.0001). Average RCD of germinants at the end of the third growing season ranged from 4.11 to 5.90 mm across all plots (Fig. 4a). It should be noted that the RCD of the germinants was measured only in the third growing season, thus, the growth of germinants was cumulative growth during the three-year period.

As for the planted seedlings, it should be noted that seedling growth in the second and third growing seasons (i.e. July 2013 and August 2014, respectively) was cumulative growth during the two-year and three-year periods, respectively. There was no statistically significant relationship between RBA and the RCD of planted seedlings in the second growing season (p=0.41) (Fig. 4b). Nor did we observe a significant relationship between IPAR and RCD (p=0.54). Average RCD at the end of the second growing season ranged from 2.48 to 4.37 mm across all plots (Fig. 4b). RBA did not significantly affect the RCD of planted seedlings at the end of the third growing season either (p=0.13) (Fig. 4c). However, the relationship between RBA and RCD growth in the third year itself (i.e. RCD growth between the second and the third growing seasons) was significant (p=0.0009) (Fig. 4d). Average RCD ranged

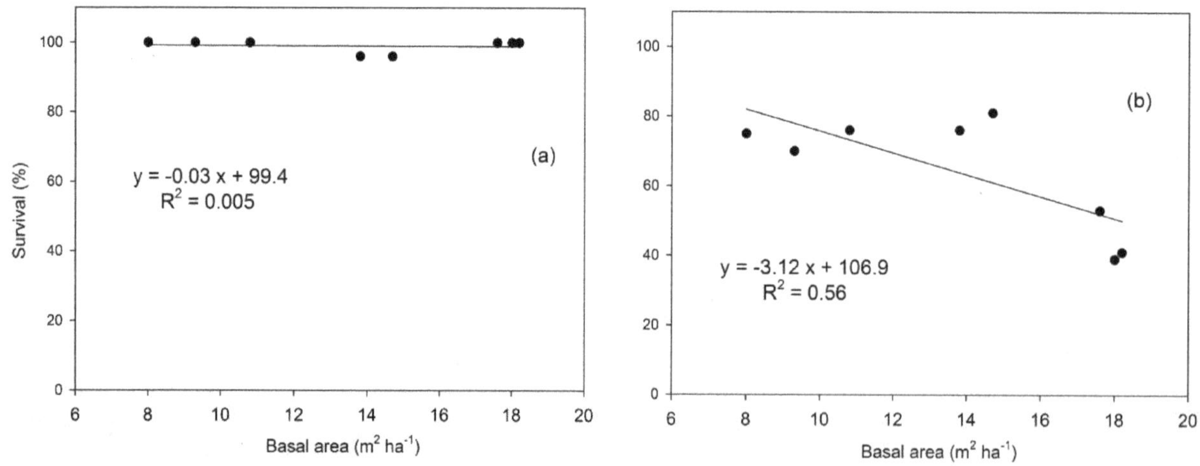

Figure 3. Relationships between RBA and survival (%) of planted seedlings in the (a) second growing season, and (b) following dormant-season fire.

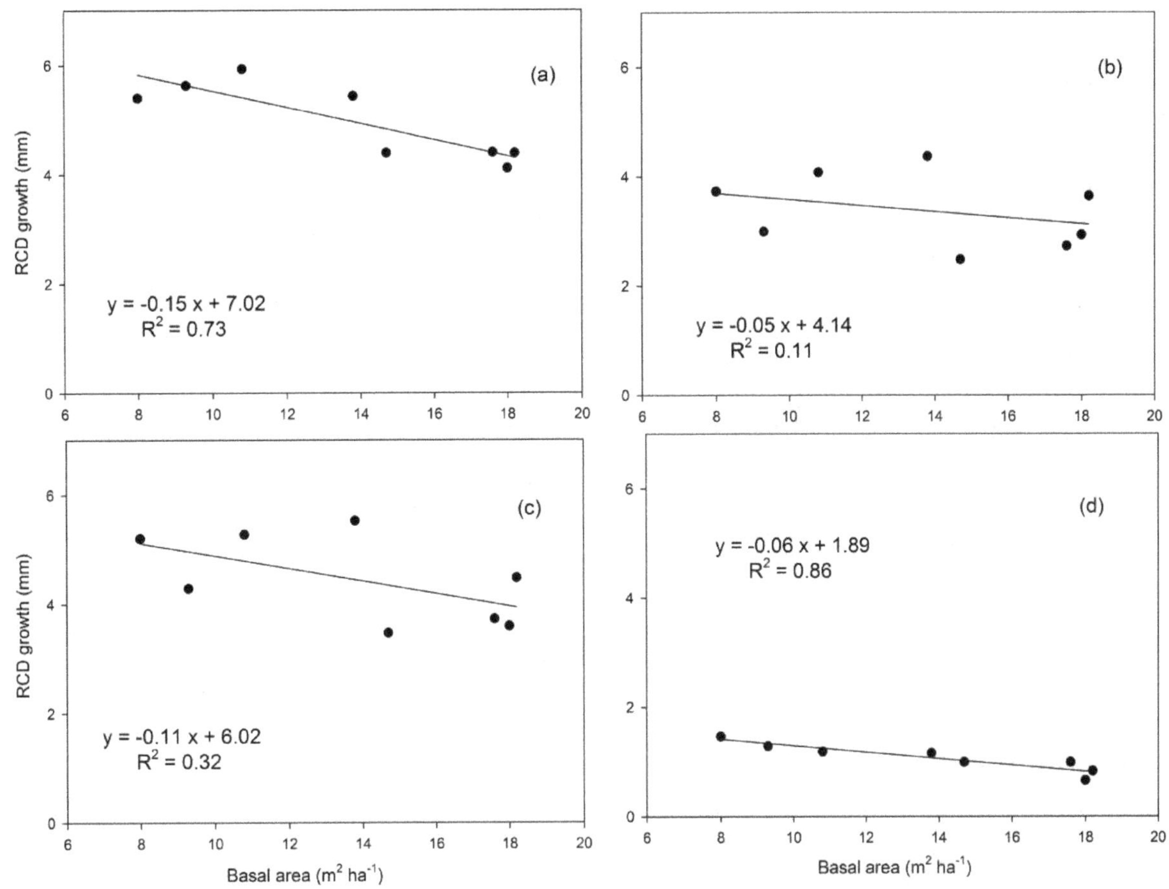

Figure 4. Relationship between RBA and RCD growth of germinants at end of the third growing season (a). Relationship between RBA and RCD growth of planted seedlings in the (b) second growing season and (c) third growing season. Growth in the second and third growing seasons was cumulative growth during the two-year and three-year periods, respectively. Relationship between RBA and RCD growth of planted seedlings in the third growing season itself (d).

from 3.47 to 5.27 mm across all plots during three-year period (Fig. 4d).

Reliable Recruitment

Average mortality rates of germinants observed during the prescribed fire at age 2 were 20, 44, and 72 % in the low, mid and high-RBA plots, respectively (Kara & Loewenstein, 2015b). It should be noted that these rates were assumed for subsequent fires until the seedlings reach the RCD of 13 mm. For the fires after seedlings reach the RCD of 13 mm, the mortality rates were 20, 20, and 46 % in the low, mid and high-RBA plots, respectively (Fig. 3b). Average mortality in the absence of fire was 3, 2, and 3 % in the low, mid and high-RBA plots, respectively (Fig. 3a).

In low-RBA plots (9.2 m^2 ha^{-1}), our average RCD growth was 1.88 mm per year. If the seedlings under this RBA follow a similar trend, seedlings would reach a RCD of 25 mm and bolt to breast height by age 14 or sooner, thereafter eventually recruiting into the forest canopy. In low-RBA plots, the average number of germinants in the third growing season was 28,000 per hectare. Until age 14, these plots would be burned every two years (at ages 4, 6, 8, 10 and 12). RBA is once again adjusted to its target (i.e., 9.2 m^2 ha^{-1}) during the next cutting cycle, by removing trees across a broad range of diameter-classes in year 10. It should be noted that this 10-year cutting cycle may be modified due to operational constraints. If the growth of the residual stand is not sufficient to produce a merchantable harvest, the cutting cycle may be extended. If we assume a mortality of 20 % from each prescribed fire before seedlings reach the RCD of 13 mm (at age 8), 20 % from each prescribed fire after they reach the RCD of 13 mm, 3 % in the absence of fire, and 50 % during logging operations at age 10, then there would be about 3900 seedlings per hectare when seedlings reach the RCD of 25 mm at age 14.

With an average RCD growth of 1.50 mm per year under the mid-RBA plots (13.8 m^2 ha^{-1}), seedlings would reach a RCD of 25 mm at age 17. In mid-RBA plots, the average number of germinants in the third growing season was 18000 per hectare. With the 44 % mortality during fires before they reach a RCD of 13 mm (at age 9), 20 % mortality during fires after they reach a RCD of 13 mm, 2 % in the absence of fire, and 50 % during logging operations at age 10, there would be about 1000 seedlings per hectare when seedlings reach the RCD of 25 mm at age 17.

Under our high-RBA (18.4 m^2 ha^{-1}), the average RCD growth was 1.43 mm per year. If the seedlings under this RBA follow a similar trend, an RCD of 25 mm would be reached at age 18. In high-RBA plots, the average number of germinants in the third growing season was 3500 per hectare. Assuming the 72 % mortality during fires before they reach a RCD of 13 mm (at age 9), 46 % mortality during fires after they reach a RCD of 13 mm, 3 % in the absence of fire, and 50 % during logging operations at age 10, there would be about 3 seedlings per hectare when seedlings reach the RCD of 25 mm at age 18.

Discussion

IPAR

Few light measurements have been reported for longleaf pine forests (Battaglia *et al.*, 2003), and scientists came to different conclusions using different approaches (Palik *et al.*, 1997; Brockway & Outcalt, 1998; McGuire *et al.*, 2001). This inconsistency has been explained by highly variable light availability temporally and spatially in longleaf pine ecosystems (McGuire *et al.*, 2001). Other reasons for these differences in light transmittance are the differences in forest canopy structures (Battaglia *et al.*, 2003) and differences in the densities of surrounding stands in longleaf pine forests (Gagnon *et al.*, 2003). Palik *et al.* (1997) stated that the relationship between stand density and light availability was curvilinear, while Brockway & Outcalt (1998) suggested that the distribution of solar radiation was relatively uniform across canopy gaps. Using a similar approach, Brockway & Outcalt (1998) investigated gap-phase regeneration in an UEA longleaf pine forest, and measured relatively lower amounts of light transmittance ranging from 500 to 900 μmol m^{-2} s^{-1} depending on the four cardinal directions that the measurements were taken. Following the harvest operations, our PAR measurements ranged from 557 to 1087 μmol m^{-2} s^{-1} under varying levels of RBA during midday and seemed to be consistent with that measured by Brockway & Outcalt (1998).

Seedling survival and growth

The higher number of longleaf pine germinants observed in stands of lower RBA was a result of higher levels of light penetration to the ground with decreasing RBA across the plots, rather than the amount of tree removal or soil disturbance caused by logging and skidding operations in the plots. The level of soil disturbance was proportional to the amount of timber skidded from the stand because the higher amount of removal usually requires more skid trails and causes more disturbances (Whitford & Mellican, 2011). Although a greater amount of timber was usually removed from the mid-RBA and high-RBA plots (Table 1), fewer germinants were observed in these stands suggesting that light penetration had more influence than soil disturbance on germination.

Survival rates of the planted seedlings in this study were consistent with reports from previous studies. South *et al.* (2005) observed 87% survival two years after planting on a nearby cutover site in Escambia County, Alabama. Palik *et al.* (1997) monitored the effects of canopy structure on longleaf pine seedling survival and found a 100% survival soon after planting, and an average of 97% survival 12 months after planting. Franklin (2008) suggests that early planting, as early as October with adequate soil moisture, usually results in a better developed root system, more drought tolerance and improved competitiveness with other vegetation in spring and summer. Therefore, our planting following a rainy period in early December favored seedling survival during their first two years. In addition, the higher survival was also related to the fact that containerized longleaf pine seedlings have much lower mortality rates in comparison to bare-root seedlings, as observed by previous studies (Barnett *et al.*, 1996; Rodriguez-Trejo *et al.*, 2003). Several studies support our observations that seedling survival is not influenced by overstory density in the absence of prescribed fire (Boyer, 1993; Palik *et al.*, 1997; McGuire *et al.*, 2001). However, higher litter accumulation occurs in denser stands, and it results in fires of higher intensity, and thus, higher mortality rates among understory seedlings occurring within these plots during fire (Boyer, 1963;

Table 1. Residual basal areas for pre-harvest, target and post-harvest conditions in each study plot.

Plot	Pre-harvest (m^2 ha^{-1})	Target (m^2 ha^{-1})	Post-harvest (m^2 ha^{-1})
L_1	11.5	9.2	8.0
L_2	13.7	9.2	9.3
L_3	15.9	9.2	10.8
M_1	24.9	13.8	14.7
M_2	29.6	13.8	14.5
M_3	19.2	13.8	13.8
H_1	30	18.4	17.6
H_2	29.8	18.4	18.2
H_3	21.8	18.4	18.0

Platt *et al.*, 1988; Grace & Platt, 1995; Brockway & Outcalt 1998). Our findings during dormant-season prescribed fire substantiate this expectation as well. Another consideration for explaining fire-inducted mortality is the relationship between RCD of seedlings and RBA observed in the third growing season. Seedling size positively affects the survival rate of longleaf pine seedlings following burning (Croker & Boyer, 1975; Brockway *et al.*, 2006), thus, seedlings under the lower stand densities reached relatively larger size, had better root development, and were better able to tolerate fire.

Average RCD growth rates of the seedlings in this study were consistent with several studies in the literature (Gagnon *et al.*, 2003; Dyson, 2010; Hu *et al.*, 2012; Knapp *et al.*, 2013). Longleaf pine seedling growth usually decreases as overstory stocking increases (Pecot *et al.* 2007). Although the early growth of longleaf seedlings is quite slow, even under low levels of overstory RBA (Boyer, 1993), it is usually expected that higher rates of RCD growth would be observed under low RBA (Palik *et al.*, 1997; Palik *et al.*, 2003). Larger RCD is usually measured near the gap centers, while smaller RCD is noted along the gap edges (Gagnon *et al.*, 2003). The significant relationship between RBA and the RCD of germinants in this study substantiates this expectation. Although the relationship between RBA and RCD growth for planted seedlings during the three-year period was insignificant, the statistically significant relationship between RBA and RCD growth during the third year itself (i.e. between the second and third year) can also be attributed to the influence of RBA on RCD growth of longleaf pine seedlings.

Reliable Recruitment

Boyer (1963) stated that longleaf seedlings reached an average RCD of 12.2 mm at age 7, under the same low-RBA (9.2 m^2 ha^{-1}) as in our study. This growth rate is consistent with our average RCD growth of 1.88 mm per year, and with other studies (Gagnon *et al.*, 2003; Dyson, 2010). In addition, Wahlenberg (1946) noted that longleaf pine seedlings typically reach breast height within 13 years, depending on stand density. In our projection, the age required to reach the RCD of 25 mm under the low-RBA (i.e., age 14) was also consistent with that stated by Wahlenberg (1946). The projected number of seedlings under the low-RBA (i.e., 3900 seedlings ha^{-1}) is substantially higher than the recommended target stand structure of ~70 seedlings per hectare for the smallest diameter class (0-5 cm in DBH) (Fig. 5).

The growth rates observed under the mid-RBA (13.8 m^2 ha^{-1}) and high-RBA (18.4 m^2 ha^{-1}) were also consistent with those reported by Boyer (1963). Under the same mid-RBA as in our study, Boyer (1963) found that seedlings reached an average RCD of 10.2 mm at age 7, while it was 9.6 mm under the same high-RBA as in our study. The projected number of seedlings under the mid-RBA (i.e., 1000 seedlings ha^{-1}) also exceeds the recommended target stand structure of 110 seedlings per hectare (Fig. 5). For the high-RBA plots, the required number of seedlings per hectare in the smallest diameter class (0-5 cm in DBH) is 145 seedlings per hectare (Fig. 5). Thus, the number of seedlings projected (i.e., 3 seedlings ha^{-1}) to recruit into the smallest diameter class is substantially lower than required to maintain the target diameter structure of longleaf pine forests under the high-RBA.

In summary, our seedling projection suggests that an RBA of 9.2 m^2 ha^{-1} is lower than necessary, because the predicted number of seedlings far exceeds the recommended target stand structure (Fig. 5). At the other end of the spectrum, the 18.4 m^2 ha^{-1} is too high to ensure reliable recruitment into the canopy of longleaf pine forests when implementing selection silviculture. Therefore, the mid-RBA (13.8 m^2 ha^{-1}), providing 1000 seedlings ha^{-1}, is closer to an optimal target for reliably recruiting enough seedlings into the canopy of longleaf pine forests, thus contributing to sustainability of the composition, structure and function of these ecosystems.

Conclusions

Fire is one of the primary factors in longleaf pine growth and survival. Modeling the fire effects is considered difficult because the influence of fire on growth and survival is usually associated with several

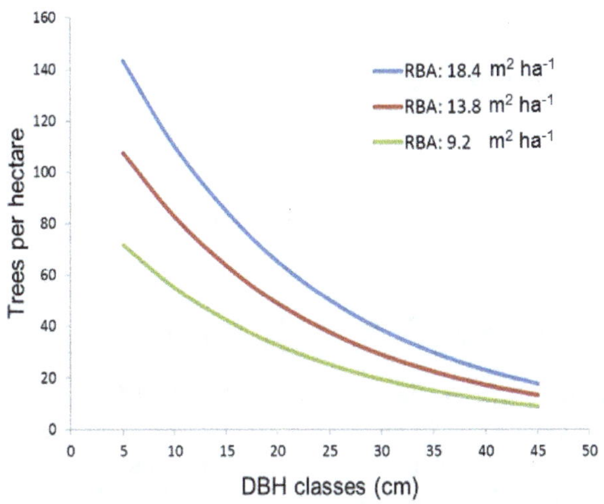

Figure 5. Target stand structure, based on RBA, calculated via the Pro-B method using a LDT = 45 cm and a q-value = 1.3 (5 cm).

factors such as the timing of burn, weather conditions, fuel conditions, stand conditions, methods of burning, fire behavior, etc. We usually have no control over what a burning will do or how it will behave. Thus, our approach of seedling projection may have some shortcomings in predicting the mortality and growth of longleaf pine seedlings. However, our assumptions are based on the data available, and our observations confirm those prior publications. The assumption that these mortality rates should persist going forward seems reasonable. Despite its shortcomings, the predictions provide a rough idea concerning the number of seedlings that may be expected to recruit into the overstory.

Successful development and implementation of UEA management systems requires an understanding of the linkage between overstory density and its influence on seedling responses. As research demonstrates successful management of longleaf pine forests with selection silviculture, UEA systems may become an increasingly used alternative relative to EA methods, especially under circumstance where UEA management better fulfills a land owner's objectives. Additional study is required to improve our understanding of the applicability and efficacy of single-tree selection in longleaf pine forests. Current data are yet insufficient to fully address the matter of reliable recruitment into the overstory. However, nothing in these data suggests that this approach cannot be successful in establishing a new cohort of trees in an UEA longleaf pine stand. Three-year results show that this approach may be a viable alternative to traditional EA methods in longleaf pine forests, if the appropriate RBA and disturbance regime are considered. Current data also suggest that understory planting of longleaf pine may be an option for converting from an EA to UEA structure in longleaf pine forests, and for species conversion.

Acknowledgements

We thank David Dyson and Ronald Tucker for assistance in the field and USDA Forest Service, Southern Research Station for research support.

References

Barnett JP, Haywood JD, Sword MA, 1996. Artificial regeneration of longleaf pine; reevaluation technique. In: Proceedings of the first longleaf alliance conference; Kush JS (ed), September 17-19, Mobile, AL, USA. pp. 53–55.

Battaglia MA, Mitchell RJ, Mou PP, Pecot SD, 2003. Light transmittance estimates in a longleaf pine woodland. For Sci 49: 752-762.

Bhuta AA, Kenneday RLM, Copenheaver CA, Sheridan PM, Campbell JB, 2008. Boundary-line patterns to determine disturbance history of remnant longleaf pine (Pinus palustris Mill.) in mixed forests of southeastern Virginia. J Torrey Bot Soc 135: 516–529. https://doi.org/10.3159/08-RA-024R.1

Boyer WD, 1963. Development of longleaf pine seedlings under parent trees. Res. Pap. SO-4. New Orleans, LA, USA: USDA Forest Service publication.

Boyer WD, 1974. Impact of prescribed fire on mortality of released and unreleased longleaf pine seedlings. Res. Note SO-182. New Orleans, LA, USA: USDA Forest Service publication.

Boyer WD, 1979. Regenerating the Natural Longleaf Pine Forest. J For 77: 572-575.

Boyer WD, 1990. Pinus palustris Mill. Longleaf pine. In: Silvics of North America; Burns RM, Honkala BH (eds), 1st ed. Washington, DC, USA: pp. 405-412.

Boyer WD, 1993. Regenerating longleaf pine with natural seeding. In: Proceeding of the 18th Tall Timbers Fire Ecology Conference; Hermann SM (eds), 30 May-2 June 1991; Tallahassee, FL, USA, pp. 299-303.

Boyer WD, White JB, 1990. Natural regeneration of longleaf pine. In: Proceedings of the Symposium on the Management of longleaf pine; Farrar RM (eds), 4-6 April 1989; Long Beach, Mississippi, USA, pp. 94-113.

Brockway DG, Loewenstein EF, Outcalt KW, 2014. Proportional basal area method for implementing selection silviculture systems in longleaf pine forests. Can J For Res 44: 977-985. https://doi.org/10.1139/cjfr-2013-0510

Brockway DG, Outcalt KW, 1998. Gap-phase regeneration in longleaf pine wiregrass ecosystems. For Ecol Manage 106: 125-139.

Brockway DG, Outcalt KW, 2000. Restoring longleaf pine wiregrass ecosystems: Hexazinone application enhances effects of prescribed fire. For Ecol Manage 137: 121-128.

Brockway DG, Outcalt KW, 2015. Influence of selection systems and shelterwood methods on understory plant communities of longleaf pine forests in flatwoods and uplands. For Ecol Manage 357: 138-150.

Brockway DG, Outcalt KW, Boyer WD, 2006. Longleaf pine regeneration ecology and methods. In: The longleaf pine ecosystem: ecology, silviculture, and restoration; Jose S, Jokela EJ, Miller DL (eds), New York, USA: Springer, pp. 95-133. https://doi.org/10.1007/978-0-387-30687-2_4

Chapman HH, 1932. Is the longleaf type a climax? Ecology 13: 328-334. https://doi.org/10.2307/1932309

Croker TC, Boyer WD, 1975. Regenerating longleaf pine naturally. Res. Pap. SO-105. New Orleans, La, USA: USDA Forest Service publication.

Davis MA, Wrage KJ, Reich PB, Tjoelker MG, Schaeffer T, Muermann C, 1999. Survival, growth and photosynthesis of tree seedlings competing with herbaceous vegetation along a water-light-nitrogen gradient. Plant Ecol 145: 341-350. https://doi.org/10.1023/A:1009802211896

Decagon Devices, Inc., Pullman, WA. AccuPAR PAR/LAI ceptometer manual. http://manuals.decagon.com/Manuals/10242_Accupar%20LP80_Web.pdf. [2 May 2011].

Dyson DS, 2010. Influence of competition on longleaf pine seedling recruitment in selection silviculture. Master thesis, Auburn University, Auburn, AL, USA.

Dyson DS, Loewenstein EF, Jack SB, Brockway DG, 2009. Influence of light and moisture on longleaf pine seedling growth in selection silviculture. USDA For. Ser. e-Gen. Tech. Rep. SRS-156. Asheville, NC, USA. 109-116.

Farrar RM, 1996. Fundamentals of uneven-aged management in southern pine. Tall Timbers Research Station Miscellaneous Publication No. 9. Tallahassee, FL. 68 pp.

Farrar RM, Boyer WD, 1991. Managing longleaf pine under the selection system – promises and problems. In: Proceedings of the 6th biennial southern silvicultural research conference; Coleman SS, Neary DG (eds), USDA Gen. Tech. Rep. SE-70. Asheville, NC, pp. 357-368.

Franklin RM, 1997. Stewardship of longleaf pine forests: a guide for landowners. Longleaf Alliance Rep. 2. Andalusia, AL: Solon Dixon Forestry Education Center. 44 pp.

Franklin RM, 2008. Stewardship of longleaf pine forests: a guide for landowners. The Longleaf Alliance Report No. 2 (rev.), 58 p. Andalusia, AL, USA.

Gagnon JL, Jokela EJ, Moser WK, Huber DA, 2003. Dynamics of artificial regeneration in gaps within a longleaf pine flatwoods ecosystem. For Ecol Manage 172: 133-144.

Grace SL, Platt WJ, 1995. Effects of adult tree density and fire on the demography of pregrass stage juvenile longleaf pine (Pinus palustris Mill.). J Ecol 83: 75-86. https://doi.org/10.2307/2261152

Guldin JM, 1996. The role of uneven-aged silviculture in the context of ecosystem management. West J Appl For 11: 4-12.

Guldin JM, 2006. Longleaf pine regeneration ecology and methods. In: The longleaf pine ecosystem: ecology, silviculture, and restoration; Jose S, Jokela EJ, Miller D L (eds), New York, NY, USA: Springer, pp. 217-251. https://doi.org/10.1007/978-0-387-30687-2_7

Heyward, F. 1939. The relation of fire to stand composition of longleaf pine forests. Ecology, 20: 287-304. https://doi.org/10.2307/1930747

Hu H, Wang GG, Walker JL, Knapp BO, 2012. Silvicultural treatments for converting loblolly pine to longleaf pine dominance: effects on planted longleaf pine seedlings. For Ecol Manage 276: 209-216.

Jack SB, Mitchell RJ, Pecot SD, 2006. Silvicultural alternatives in a longleaf pine/wiregrass woodland in southwest Georgia: understory hardwood response to harvest-created gaps. In: Proceedings of the 13th biennial southern silvicultural research conference; Connor KF (eds), USDA Southern Research Station Gen. Tech. Rep. SRS-92, Asheville, NC, pp. 85-89.

Kara F, Loewenstein EF, 2015a. Influence of residual basal area on longleaf pine (Pinus palustris Mill.) first year germination and establishment under selection silviculture. Open Journal of Forestry 5: 10-20. https://doi.org/10.4236/ojf.2015.51002

Kara F, Loewenstein EF, 2015b. Impacts of logging and prescribed burning in longleaf pine forests managed under uneven-aged silviculture. Turk J Agric For 39: 99-106. https://doi.org/10.3906/tar-1405-32

Knapp BO, Wang GG, Walker JL, 2013. Effects of canopy structure and cultural treatments on the survival and growth of Pinus palustris Mill. Seedlings underplanted in Pinus taeda L. stands. Ecological Engineering 57: 46-56. https://doi.org/10.1016/j.ecoleng.2013.04.014

Landers JL, 1991. Disturbance influences on pine traits in the southeastern United States. In: Proceedings of 17th Tall Timbers fire ecology conference, Tallahassee, FL, USA. 17: 61–98.

Liang S, Li X, Wang J, 2012. Advanced remote sensing: Terrestrial information extraction and applications, 1stedn. Academic Press. ISBN: 0123859549.

Loewenstein EF, 2005. Conversion of uniform broadleaved stands to an uneven-aged structure. For Ecol Manage 215: 103-112.

Maple WR, 1977. Planning longleaf pine regeneration cuttings for best seedling survival and growth. J For 75:25-27.

Mattox MG, 1975. Soil Survey of Escambia County, Alabama. Natural Resource Conservation Service. US Department of Agriculture, Washington, DC.

McDonald JH, 2014. Handbook of Biological Statistics, 3rd edition. Sparky House Publishing. Baltimore, Maryland, USA.

McGuire JP, Mitchell RJ, Moser EB, Pecot SD, Gjerstad DH, Hedman CW, 2001. Gaps in a gappy forest: plant resources, longleaf pine regeneration, and understory response to tree removal in longleaf pine savannas. Can J For Res 31: 765-778. https://doi.org/10.1139/x01-003

Millennium Ecosystem Assessment, 2005. Ecosystems and Human Well-Being. Island Press, Washington, DC.

Mitchell RJ, Hiers JK, O'Brien JJ, Jack SB, Engstrom RT, 2006. Silviculture that sustains: the nexus between silviculture, frequent prescribed fire, and conservation of biodiversity in longleaf pine forests of the southeastern United States. Can J For Res 36: 2724-2736. https://doi.org/10.1139/x06-100

Moser WK, Jackson SM, Podrazsky V, Larsen DR, 2002. Examination of stand structure on quail plantations in the Red Hills region of Georgia and Florida managed by the Stoddard-Neel system: an example for forest managers. Forestry 75: 443-449. https://doi.org/10.1093/forestry/75.4.443

Neel L, Sutter PS, Way AG, 2010. The art of managing longleaf: a personal history of the Stoddard-Neel Approach. The University of Georgia Press, Athens, GA, USA.

Onset Computer Corporation (2009) HOBO® weather station user's guide. http://www.onsetcomp.com/files/manual_pdfs/6106-I-MAN-H21.pdf. [May 2011].

Palik BJ, Mitchell RJ, Hiers JK, 2002. Modeling silviculture after natural disturbance to sustain biodiversity in the longleaf pine (Pinus palustris) ecosystem: balancing complexity and implementation. For Ecol Manage 155: 347-356.

Palik BJ, Mitchell RJ, Houseal G, Pederson N, 1997. Effects of canopy structure on resource availability and seedling responses in a longleaf ecosystem. Can J For Res 27: 1458-1464. https://doi.org/10.1139/x97-081

Palik B, Mitchell RJ, Pecot S, Battaglia M, Pu M, 2003. Spatial distribution of overstory retention influences resources and growth of longleaf pine seedlings. Ecological Applications 13, 674–686 https://doi.org/10.1890/1051-0761(2003)013[0674:SDOORI]2.0.CO;2

Pecot SD, Mitchell RJ, Palik BJ, Moser EB, Hiers JK, 2007. Competitive responses of seedlings and understory plants in longleaf pine woodlands: separating canopy influences above and below ground. Can J For Res 37: 634-648. https://doi.org/10.1139/X06-247

Platt WJ, Evans GW, Davis MM, 1988. Effects of fire season on flowering of forbs and shrubs in longleaf pine forests. Oecologia 76: 353-363. https://doi.org/10.1007/BF00377029

R Project, Development Core Team, 2008. R: A language and environment for statistical computing. R foundation for statistical computing, Vienna, Austria. ISBN 3-900051-07-0.

Roach BA, 1974. What is selection cutting and how do you make it work—what is group selection and where can it be used? State University of New York Applied Forestry Research Institute AFRI Miscellaneous Report No. 5, 9 p.

Rodríguez G, 2007. Lecture notes on generalized linear models [online]. http://data.princeton.edu/wws509/notes/. [20 November 2014].

Rodriguez-Trejo DA, Duryea ML, White TL, English JR, McGuire J, 2003. Artificially regenerating longleaf pine in canopy gaps: initial survival and growth during a year of drought. For Ecol Manage 180: 25-36.

Samuelson LJ, Stokes TA, 2012. Leaf physiological and morphological responses to shade in grass-stage seedlings and young trees of longleaf pine. Forests 3: 684-699. https://doi.org/10.3390/f3030684

Schwarz GF, 1907. The longleaf pine in virgin forest- A silvical study. New York, John Wiley & Sons, London, NY.

South DB, Harris SW, Barnett JP, Hainds MJ, Gjerstad DH, 2005. Effect of container type and seedling size on early height growth of Pinus palustris seedlings in Alabama, USA. For Ecol Manage 204: 385-398.

USDA, 2014. Critical Conservation Area: Longleaf Pine Range. Soil Science and Resource Assessment, Resources Assessment Division. Beltsville, MD, USA.

Wahlenberg WG, 1946. Longleaf Pine: Its use, ecology, regeneration, protection, growth, and management. Charles Lathrop Pack Forestry Foundation, Washington, DC, USA.

Walker J, Peet RK, 1983. Composition and species diversity of pine-wiregrass savannas of the Green Swamp, North Carolina. Vegetatio 55: 163-179. https://doi.org/10.1007/BF00045019

Whitford KR, Mellican AE, 2011. Intensity, extent and persistence of soil disturbance caused by timber harvesting in jarrah (Eucalyptus marginata) forest on FORESTCHECK monitoring sites. Australian Forestry 74: 266-275. https://doi.org/10.1080/00049158.2011.10676371

9

Use of LiDAR data during multi-annual periods for estimating forestry variables

Manuel A. Valbuena[1], Esperanza Mateos[2] and Francisco Rodríguez[3]

[1]*Education Department of the Basque Government, IES Murguia BHI. C/Domingo de Sautu s/n. 01130 Murguía (Araba-Álava), Spain* [2]*University of the Basque Country UPV/EHU, Department of Chemical and Environmental Engineering, University of the Basque Country UPV/EHU. Paseo Rafael Moreno Pitxitxi, 3, 48013 Bilbao, Spain* [3]*Fora Forest Technologies SLL. C/ Oreste Camarca 4. 42004 Soria, Spain*

Abstract

Aim of study: To test the use of LiDAR data from a single acquisition in order to estimate volume over_bark variations in a 5-yr period of *Pinus radiata* D. Don.

Area of study: Province of Bizkaia in the Autonomous Community of the Basque Country (Spain).

Material and methods: Two field plot measurements were made in 2011 and 2015 and two wood volume models (one for each year) were fitted using the metric variables of the 2012 LiDAR points cloud. The models were applied to a 26.59 m raster covering the study area and the increase in volume at each pixel was calculated by subtraction.

Main results: The increase in estimated wood volume, when added to the volume of timber extracted in the area during the 5-yr period under consideration, yielded an average increase of 13.74 m^3 ha^{-1} yr^{-1}, which corresponds to the average growth of the *P. radiata* in that area. The harvest area estimated using this procedure largely coincides with the actual harvest area in the same period. The value of R^2 (85%) of the wood volume model for 2011 is similar to that obtained in other studies. However, as expected, the one obtained for the wood volume model for 2015 (80%) is significantly lower.

Research highlights: The increase in wood volume can be estimated using a single LiDAR flight and field data from the 5-yr period provided that data from plots subjected to this kind of harvest is included in the models.

Additional keywords: LiDAR forest inventory; wood volume increase; wood volume LiDAR model.

Abbreviations used: CCF (Canopy Cover Fraction); CV (Coefficient of Variation); Dm (Mean Diameter); G (basal area, m^2 ha^{-1}); Hm (mean height, m); Ho (top height, m); IFN4 (4th National Forest Inventory); LiDAR (Light Detection and Ranging); MAE (Mean Absolute Error); ME (Mean Error); RMSE (Root Mean Square Error); Vob (stem volume over bark, m^3 ha^{-1}).

Authors´ contributions: MAV conceived and designed the experiments and contributed to writing, data production, and analysis. EM wrote the introduction and reviewed the work. FR revised the statistical calculations and fitted models.

Funding: Gobierno Vasco (Projects SAI10/147- SPE10UN90); Vicerrectorado de Investigación de la Universidad del País Vasco-Euskal Herriko Unibertsitatea (UPV/EHU) (Project NUPV14/11).

Competing interests: The authors have declared that no competing interests exist.

Correspondence should be addressed to Manuel A. Valbuena: mavalbuena66@gmail.com; valbuena@irakasle.eus

Introduction

Over recent years, there has been increasing interest in the evaluation of the forestry variables, such as biomass and timber volume, especially within the context of the Kioto protocol (Lindner & Karjalainen, 2007; Zhang *et al.*, 2007). The quantification of these forestry parameters reveals the structure, operation and dynamics of forestry ecosystems and their assessment as an energy resource. For this reason, it is essential to determine these parameters in forestry management (Torres *et al.*, 2010; Pan *et al.*, 2011). In most European countries, the increase in these variables is greater than the amount harvested on an annual basis, converting forests into major carbon and biodiversity reserves (Mateos *et al.*, 2016). One of the key indicators for assessing forestry conditions is the estimation of forestry parameters, such as timber volume, but the usual methods for estimating this parameter require prolonged fieldwork (Song *et al.*, 2016). A large part of the forestry bibliography relating to this subject focuses on adjustment of timber volume equations by means of regression for geographical regions and specific tree species (Nilsson, 1996; Hyyppä *et al.*, 2001; Popescu *et al.*, 2003; Holmgren, 2004; Zianis *et al.*, 2005; Dalponte *et al.*, 2009; Latifi *et al.*, 2010; Magnussen *et al.*, 2010). The high cost of sampling, especially when this is done in inaccessible areas, limit in many cases the establishment of sufficient plots to determine real variability (Popescu, 2007).

The link between the parameters obtained in forestry inventories and LiDAR (*Light Detection And Ranging*) data through regression models offers an attractive approach to estimating forestry variables on a regional, continental or global scale (Li & Weiskittel, 2010; Næsset, 2011; Babcock *et al.*, 2015). Over recent decades, LiDAR technology has become an extremely powerful tool for estimating the biophysical properties of the vegetation. This is an active, non-destructive remote sensing method that offers a better alternative for the study of the structural variables of the canopy due to the penetrating capacity of the laser pulse through gaps in the vegetation (Leiterer *et al.*, 2015; Giannico *et al.*, 2016; Koenig & Höfle, 2016). The recording of multiple returns allows us to obtain sufficient three-dimensional, morphological information at a broad range of space and time scales of forestry variables such as the height of trees (Dean *et al.*, 2006; Popescu & Zhao, 2008), the basal area and the diameter of the crown (Giannico *et al.*, 2016), volume and aboveground biomass (Popescu, 2007; Jochem *et al.*, 2010; Zolkos *et al.*, 2013), and the density and classification of groups of forest species (Brandtberg, 2007; Zhang *et al.*, 2016). Moreover, the use of LiDAR data offers the possibility of reducing the number of required sampling plots (Maltamo *et al.*, 2007; Montaghi *et al.*, 2013) and the accuracy obtained with LiDAR is greater than that reached with traditional inventory techniques (Condes *et al.*, 2013; Watt *et al.*, 2013; Maltamo *et al.*, 2014).

Now that LiDAR technology has demonstrated its efficiency and accuracy in estimating most forestry variables, today the challenge is to try to take advantage of this tool in order to predict the annual growth of the plant structure at a large scale. The ease with which data can be updated allows us to readjust the estimations and quickly assess changes or disturbances in forest stands by estimating their development over time, or establish comparisons between different areas with a simple update of the forestry cartography (Riaño *et al.*, 2004). The bibliography contains numerous studies that used bi-temporal data from two LiDAR flights to estimate plant growth (Yu *et al.*, 2004, 2006; Næsset & Gobakken, 2005; Hopkinson *et al.*, 2008; Vepakomma *et al.*, 2011; Fekety *et al.*, 2014; Cao *et al.*, 2016; Song *et al.*, 2016). Some researchers have estimated the growth of trees in forests on the basis of the differences in height derived from LiDAR data in flights separated by a 2-yr interval (Næsset & Gobakken, 2005; Yu *et al.*, 2006; Vepakomma *et al.*, 2011; Cao *et al.*, 2016). These studies allowed the growth in different forestry parameters to be calculated, such as the basal area or volume (Næsset & Gobakken, 2005) as well as estimations of the vertical and lateral growth of trees (Vepakomma *et al.*, 2011). Other researchers have used multi-temporal LiDAR datasets to estimate the growth of forests (Yu *et al.*, 2006; Hopkinson *et al.*, 2008; Fekety *et al.*, 2014; Cao *et al.*, 2016; Song *et al.*, 2016). However, we have not found any previous studies that use the data from a single LiDAR flight to estimate the increase in timber volume over a period of several years by fitting models that relate the timber volume at different moments in time with the metrics of the heights of a single LiDAR points cloud obtained at the beginning of the period. In view of the high cost of the flight, the optimisation of the long-term use of LiDAR data gathered during a campaign may represent a considerable economic advantage.

Material and methods

This study assesses the application of data from the last LiDAR flight made over the Autonomous Community of the Basque Country in 2012 to evaluate the growth of volume over bark (Vob) during 5 years on stands of *Pinus radiata* D. Don (PR) through the comparison of estimated volumes by applying fitted models to relate separately the field data obtained with 5 years of difference with the metrics of the LiDAR points cloud at the beginning of the period.

The area based approach (ABA) for forest inventory applications using LiDAR has been utilised in this study, following the methodology applied by other researchers (Hopkinson *et al.*, 2008; Jochem *et al.*, 2010; Yu *et al.*, 2010). In order to get a better understanding of how LiDAR and the set of field data can be used to map Vob, we used two sets of field data gathered in 2011 and 2015 respectively. This involves analysing and validating the proposed Vob models, fitting the corresponding models for both sets of data in accordance with the metric variables of the LiDAR points cloud obtained in the 2012 LiDAR flight. The results will allow us to assess whether it is possible to estimate –with an acceptable margin of error– the growth in the timber volume based on field data obtained over a 5-yr period and the metrics of a single LiDAR flight made during this time could optimise the available resources and would represent a large decrease in the cost compared to traditional estimation methods.

Study area

The study area lies in the Province of Bizkaia in the Autonomous Community of the Basque Country

(ACBC) in the north of Spain. Its center is the municipality of Muxika and data from the surrounding municipalities between the coordinates 2° 30' – 2° 54' west and 43° 06' – 43° 24' north are also used. The average height is 230 m. The mean annual rainfall is 1277 mm and the mean annual temperature is 13.8 °C. Frosts are infrequent and no physiological drought is apparent. With regard to the silviculture applied in the area, the rotation cycle is, on average, 35 yrs, varying between 30 and 40 according to the site quality and the price of wood at the time when harvest takes place. Clear cutting are applied followed by repopulation with initial densities of 1300-1600 trees ha^{-1}. Between the 7th and 9th year, non-commercial thinning takes place, eliminating 500 trees ha^{-1} and pruning up to 2.5 m on the trees that have not been eliminated. Between the 12th and 15th year, the first commercial thinning is carried out, leaving between 500 and 700 trees ha^{-1}. Between the 17th and 20th year, the second commercial thinning is applied, leaving between 400 and 500 trees ha^{-1}. The third thinning is carried out between the 22nd and 25th year, leaving between 250 and 300 trees for the clear cutting. The total production is on average 550 m^3 ha^{-1} in one rotation, and therefore the average growth is 14.85 m^3 ha^{-1} yr^{-1}. Volumes and growths of 3153 ha occupied by *P. radiata* in the municipality of Muxika were estimated. The area in which data was gathered and results were obtained can be seen in Fig. 1.

2011 Field data

Two sets of data were used according to the year in which these were gathered. In order to fit the wood volume model from 2011, data was used from the Fourth National Forestry Inventory (IFN4) of the Province of Bizkaia (http://www.euskadi.eus/inventario-forestal-2011/web01-a3estbin/es/). Initially, all the plots included in the IFN4 and located within the boundaries of the municipality of Muxika and neighboring municipalities were selected. From these plots, 111 located within the area occupied by *P. radiata* were selected, according to the 2010 Forestry Map of the Basque Country (http://www.euskadi.eus/mapa-forestal-del-pais-vasco-ano-2010/web01-a2hestat/es/). From these 111 plots, 13 were eliminated as they had no data or did not have *P. radiata* as their main species. From the remaining 98, those that showed a clear incompatibility between the field data and the vegetation visible in the aerial photography were eliminated. This occurred because the coordinates of the centers of the plots were taken with GPS navigators that did not allow differential correction and had a potential error of more than 12 m. For the same reason, those plots in which the difference between the dominant height of the field trees and the 95th percentile of the height of the points of its LiDAR points cut was greater than 12 m were eliminated. Furthermore, those plots with a basal area (G) greater than 70 m^2 ha^{-1} were also eliminated as these were not considered possible for this species in this area. Following the application of these filters, only 8 additional plots were discarded. Therefore, 90 plots were used finally in this study. Their location can be seen in Fig. 1.

IFN4 plots are of variable radius, so that a tree will or will not be measured in accordance with its diameter and the distance to the center of the plot. For each selected

Figure 1. The area in which the study was carried out. The area in which field data were gathered is shown in green. The area in which the study was conducted and the results were obtained is shown in red. The points correspond to the location of the IFN4 plots selected for the study.

plot, the number of trees per hectare (N, trees ha^{-1}), the dominant height (Ho, m) defined as the mean height of the 100 thickest trees per hectare, the mean height (Hm, m), the mean diameter (Dm, cm), the basal area (G, m^2 ha^{-1}) and the wood volume over bark (Vob, m^3 ha^{-1}) were calculated. For the last of these calculations, the Excel add-in CUBIFOR (Rodríguez *et al.*, 2008) was used. With the aim of reducing data-gathering costs for subsequent years, a Vob tariff was adjusted in accordance with G and Ho. In this way, in order to calculate volumes in measured plots it is sufficient to measure both variables quickly and economically. The Vob estimated with this tariff will be considered as the field Vob for the entire study.

2015 Field data

In order to be able to assess the possibility of using LiDAR data from a single flight when estimating the forestry resources in subsequent years, 55 field plots were measured during the month of December, 2015. So as to prevent any lack of accuracy in locating the center of the plot from affecting the estimation of the final results, those areas with a sufficient level of uniformity were selected so that a displacement of up to 10 m in the center of the plot would have only a slight effect on the final result.

With the aim of locating uniform areas, two influential variables were selected, the first being Canopy Cover Fraction (CCF), as a percentage of the surface covered by the canopy of trees, and the second being the height of the vegetation. The data gathering area was divided with a 10 m-sided grid and in each 100 m^2 square and the CCF was estimated as a percentage of first returns, higher than 3 m over the total of first returns of the LiDAR points contained in it. For each grid, the typical deviation and the average of its CCF value and those of the 8 surrounding squares were calculated and those squares in which the coefficient of variation (CV) was not greater than 30% were selected if the CCF was lower than 70% and those in which the CV was greater than 10% if the CCF was greater than 70%. To estimate the height of the vegetation, the data gathering area was divided with a 10 m-sided grid and in each grid cell, the height of the 95th percentile of the heights of the LiDAR points contained in it was calculated. For each grid cell, the mean and typical deviation of its value and that of the 8 surrounding cells was calculated. Those grid cells in which the CV was not greater than 20% were selected if the 95th percentile was less than 8 m; those grid cells in which the CV was not greater than 10% were selected if the 95th percentile was between 8 and 12 m, and those in which the CV was not greater than 5% were selected if the 95th percentile was greater than 12 m. Only those grid cells that met both conditions were candidates for containing a plot center for data gathering. In accordance with these conditions, 55 points were found and located in the field with a GPS navigator. From each point, the dominant height was measured as the average of the thickest three trees within a radius of 9.77 m from the point that corresponds to a plot with a surface area of 300 m^2. The basal area was also measured by relascope inventory from the center of the plot. The measurement of these two variables allows the field work to be done cheaply and quickly. Later, the Vob was calculated for each plot applying the Vob tariff in accordance with Ho and G calculated with the plots from the IFN4. The values and statistics of the variables used in the measurements of the plots from the IFN4 (2011) and the new plots (2015) are shown in Table 1.

LiDAR Data

The LiDAR data used came from the flight over the Autonomous Community of the Basque Country in July, 2012, with a mean density of 1 pulse m^{-2} and up to 4 returns per pulse.

For the entire data-gathering area, those *las* format files that contained a selected plot from the IFN4 or some of the plots measured in the field in 2015 and all the sheets contained partially or completely in the

Table 1. Summary statistics of the stand variables of *Pinus radiata* used to develop a model to predict stem volume in plantations in Muxika (northern Spain)

	Year 2011 Plots (n = 90)			Year 2015 Plots (n=55)		
	Ho	G	Vob	Ho	G	Vob
Mean	23.53	28.53	273.88	26.06	27.65	278.78
Standard deviation	6.99	11.47	165.36	8.10	6.41	127.31
Minimum	7.84	3.66	11.68	12.70	15.00	92.57
Maximum	37.92	56.35	693.65	46.20	45.00	610.24
Range	30.08	52.69	681.97	33.50	30.00	517.67

H_0: dominant height (m). G: stand basal area (m^2 ha^{-1}). Vob: stand volume over bark (m^3 ha^{-1})

municipality of Muxika were selected. All the selected *las* format files were normalized by subtracting from the height Z of each point the corresponding height of the 1 m × 1 m Digital Terrain Model (DTM) of the grid cell published by the Basque Government (ftp://ftp.geo.euskadi.net/lidar/). Once the points clouds had been normalized, the corresponding cylinders were clipped to the projection of each one of the plots, the center of which was that of the IFN4 plots and radius 15 m. In this way, 90 sets of LiDAR points coinciding with the 90 IFN4 plots were obtained. For example, the cut of plot 559 can be seen in Fig. 2.

In order to make this clip, the PolyClipData tool of the FUSION program (http://forsys.sefs.uw.edu/fusion/fusionlatest.html) was used. Subsequently, the statistical parameters that describe each of the points clouds corresponding to plots were calculated. The results of the LiDAR points cloud statistics were compared to the results of the field measurements and volumes for each plot. In this way a table was obtained in which each row contains the field data and Vob of each IFN4 plot and the statistics corresponding to its LiDAR points cut. The selected plots and their data were separated in two groups of 60 and 30 plots respectively. The first of these was used to fit the model that relates the Vob with the statistics of the LiDAR points cloud and the second was used to validate this model. In order to separate the two groups, a random number between 0 and 100 was generated for each plot. Those plots with assigned numbers of under 65 were used to fit the model; those plots with a number over 65 were kept in order to validate the model. The same classification was used for adjusting the Vob tariff in accordance with G and Ho. The location of the plots of the model and of the validation group can be seen in Fig. 3.

The same procedure was followed using the LiDAR data for 2011 and the plots measured in the field in 2015. Fifty-five cuts of LiDAR points were obtained with their statistics and 55 Vob data calculated for each plot measured in 2015. Forty plots were selected to fit the model and the validation group was formed by the 15 plots. It should be pointed out that none of the plots measured in 2015 was located in areas in which clear cutting had been performed during the 5-yr period 2011-2015. This meant that the fitted model did not consider the possibility of volume 0. Some plots were indeed located in areas that had been subjected to thinning during the 5-yr period.

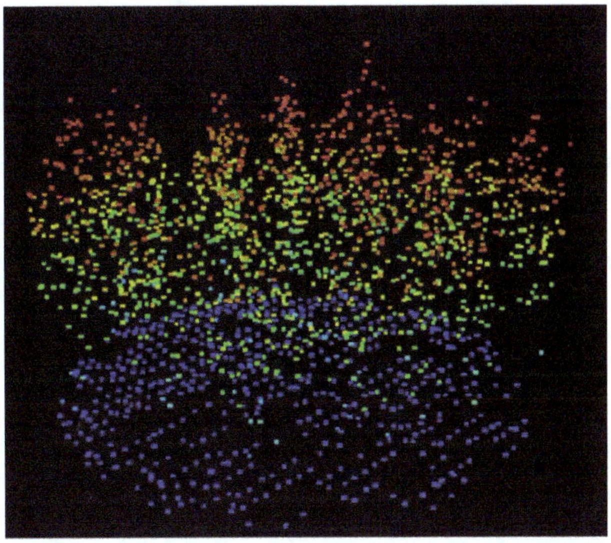

Figure 2. Normalized points cloud cut from plot 559 of the IFN4. The color of the points depends on their height above the ground.

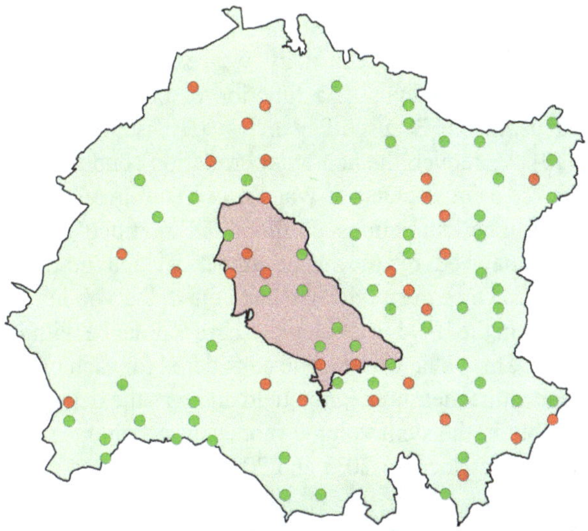

Figure 3. Location of the IFN4 plots used to fit the model (green points) and those used in the validation (red points)

Clear cutting and thinning data in Muxika during the period 2011-2015

Clear cutting and thinning data in the municipality of Muxika between 2011 and 2015 was gathered in order to be able to use this to estimate the growth in wood volume during the period 2011-2015. Data on harvest provided by the Provincial Council of Bizkaia (the administrative office that manages the forest stands of the Province of Bizkaia), can be seen in Table 2.

In order to adjust the Vob tariff per hectare in accordance with the stand variables G and Ho, 90 plots of the IFN4 were used. Of these, 60 were used to fit the model and 30 to validate this. Plots were assigned to the model group or the validation group according to the random number assigned to them. In order to build the tariff, the STATGRAPHICS (http://www.statgraphics.com) was used. Using the regression model selection option, Vob was used as a dependent variable and

Table 2. Clear cutting and thinnig data in Muxika between 2011 and 2015.

Year	Vob (m^3)	Clear cutting (m^3)	Thinning (m^3)	Total area (ha)	Clear cutting area (ha)	Thinning area (ha)
2011	11,813	5,746	6,067	199.34	18.35	180.99
2012	31,814	23,860	7,954	259.12	57.73	201.39
2013	21,663	18,755	2,908	132.31	53.95	78.36
2014	29,264	23,003	6,261	219.89	63.18	156.71
2015	41,070	33,391	7,679	257.48	87.5	169.98
Total	135,624	104,755	30,869	1,068.14	280.71	787.43

Ho, Ho^2, G, G^2 and Ho*G were used as independent variables (Table 3). The models were examined to determine whether all the terms should be retained in the final regression equations. Once the stepwise regression had been applied, the Mallows' Cp selection method was used as the best selection technique (Mallows, 1973). Only models whose parameters were significant at a given level (in our case, 5%) were taken into account.

The best model was chosen with the following indicators: adjusted determination coefficient (R^2), root mean square error (RMSE) and mean absolute error (MAE). The heterocedasticity was examined visually, by plotting residuals as a function of predicted values and using the Breusch-Pagan test (Breusch & Pagan, 1979). Moreover, the non-autocorrelation condition was checked using the Durbin-Watson test (Durbin & Watson, 1971). The estimations of the different fitted models were analyzed by means of numerical and graphical analysis. Once the model had been fitted, the Vob in each plot from the IFN4 was calculated in accordance with Ho and G. This value was the one considered for each plot as a Vob value measured in the field and was the dependent variable in the stem volume models in accordance with LiDAR statistics for 2011 and 2015.

Fitting the stem volume model based on 2011 LiDAR data (Vob_2011)

Linear models were used to establish empirical relationships between field measurements and LiDAR variables. In order to fit the Vob for 2011, the Vob calculated in each IFN4 plot based on the model fitted with the independent variables G and Ho was used as the dependent variable. In the STATGRAPHICS regression model selection procedure, all statistics relating to the height of the points obtained with the FUSION CloudMetrics order were included as independent variables. In order to fit the model, 60 plots from the IFN4 were used. The other 30 plots were used to validate the model. The highest model efficiency in accordance with the LiDAR metrics was obtained by ordering the proposed combinations according to the best Mallows Cp. The same goodness-of-fit test utilized for the models in the previous section was used. Likewise, for the selected models, a residuals analysis was made for the selected models, with the aim of identifying evidence of normality deviations, lack of adjustment and/or heterocedasticity.

Fitting the 2015 stem volume model (Vob_2015)

In order to fit the Vob for 2015, the Vob was used as dependent variable, calculated in each one of the 55 plots measured in the field in 2015 with the previously adjusted tariff of G and Ho. In order to maintain the uniformity of the results obtained, the model was fitted with the same independent variable as the Vob_2011 model, in other words, the 70 percentile of the height of the points cloud was used. In order to fit the model, 40 plots were used. The other 15 plots were used to validate the model.

Application of LiDAR data to Vob models

In order to apply the wood volume models obtained, the area occupied by *P. radiata* in Muxika (3,153 ha) was separated and a grid made up of 26.59-m grid cells was superimposed. Each grid cell had a surface area equal to a circular plot with a 15-m radius. Therefore, each

Table 3. Stem volume models evaluated

Equation	Model tested[1]	R^2	Mallows Cp	Breusch-Pagan Homoscedasticity test	Durbin-Watson Auto-correlation test
[1]	$Vob = \beta_0 + \beta_1 * H_0$	78	1721	7.28 (p<0.05)	2.07 (p >0.05)
[2]	$Vob = \beta_0 + \beta_1 * G$	89	872	2.04 (p>0.05)	1.76 (p >0.05)
[3]	$Vob = \beta_0 + \beta_1 * H_0 * G$	99	0.51	23.94 (p<0.05)	2.16 (p >0.05)

[1]Vob: stem volume over bark ($m^3\ ha^{-1}$). H_0: dominant height (m). G: basal area ($m^2\ ha^{-1}$). β_0, β_1: model parameters

grid cell was of 706.86 m^2. The cells that intersected the polygons corresponding to the surface areas of *P. radiata* according to the forestry map of the Basque Country, were singled out from the vector layer. In this way, a 26.59-m grid mask was obtained with the location of *P. radiata* forests in the municipality of Muxika. Parallel to this, the GridMetrics procedure of FUSION was applied to all the LiDAR files that intersected the municipality of Muxika. In order to apply GridMetrics, a grid cell size equal to that of the 26.59-m grid was used. The work process diagram can be seen in Fig. 4.

Results

Vob tariff per hectare

The lowest Mallows Cp value was 0.51. This corresponded to the model (3) that had a single independent variable: Ho*G (Table 3). This is a model with a high R^2 (0.99), although it presents heteroscedasticity ($p<0.05$) for the Breusch-Pagan test. This occurs because the higher the volume value, the greater the absolute error.

Vob_2011

The lowest Cp value was 3.28 and this corresponded to the model that had the value of the height of the percentile 70 (h_{70}) of the points cloud (Eq. [4], Table 4) as single independent variable. This model had a good R^2 (0.85) but was heterocedastic with a $p<$ 0.05 in the Breusch Pagan test. Once again, this occurs because the higher the volume value, the greater the absolute error.

Vob_2015

The stem volume model for 2015 based on LiDAR data is shown in Table 5.

The models obtained allowed us to estimate the total volume over bark with a different degree of accuracy, although in all cases it is quite high. The best regression

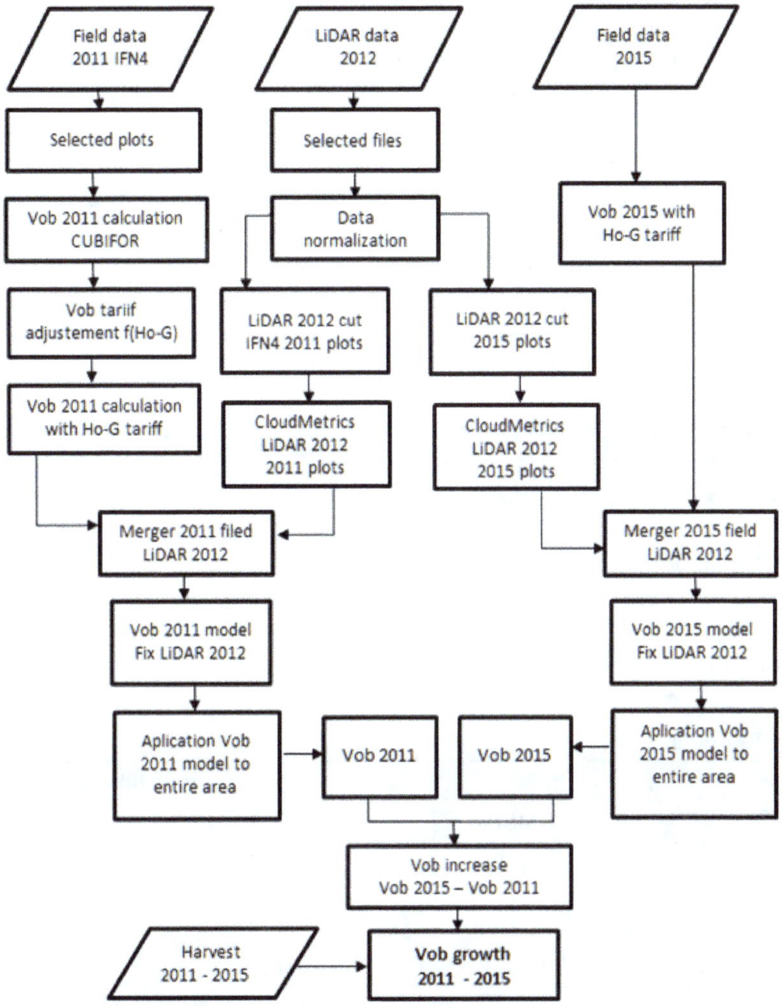

Figure 4. Work process diagram

Table 4. Stem volume models for 2011 based on LiDAR data

Equation	Model tested	R^2	Mallows Cp	Breusch-Pagan Homoscedasticity test	Durbin-Watson Auto-correlation test
[4]	$Vob\_2011 = \beta_0 + \beta_1 * h_{70}$	85	3.28	6 (p< 0.05)	1.71 (p> 0.05)
[5]	$Vob\_2011 = \beta_0 + \beta_1 * h_{40} + \beta_2 * h_{70}$	85	3.68	4.66 (p< 0.05)	1.77 (p> 0.05)
[6]	$Vob\_2011 = \beta_0 + \beta_1 * h_{75}$	85	4.52	6.31 (p< 0.05)	1.75 (p>0.05)
[7]	$Vob\_2011 = \beta_0 + \beta_1 * h_{80}$	85	5.89	6.69 (p< 0.05)	1.79 (p>0.05)

Vob: Stem volume over bark (m^3 ha^{-1}). h_{70}, h_{75}, h_{80} and h_{40} : percentiles 70, 75, 80 and 40 of the height of the points cloud. R^2: coefficient of determination. β_0, β_1 β_2: parameters of the models.

Table 5. Stem volume models for 2015 based on LiDAR data

Equation	Model tested	R^2	Mallows Cp	Breusch-Pagan Homoscedasticity test	Durbin-Watson Auto-correlation test
[8]	$Vob\_2015 = \beta_0 + \beta_1 * h_{70}$	80	--	3.50 (p>0.05)	2.12 (p>0.05)

Vob: Stem volume over bark (m^3 ha^{-1}). h_{70}: percentile 70 of the height of the points cloud. R^2: coefficient of determination. β_0, β_1: parameters of the models.

model corresponds to the Vob model based on stand variables (H_0, G). This model presents the highest value of the adjusted coefficient of determination (R^2_{adj} = 99), followed by the volume estimation model in accordance with LiDAR metrics (Vob_2011, R^2_{adj} = 85) and lastly, the model based on the field data obtained in 2015 (Vob_2015, R^2_{adj} = 80) (Table 6). Likewise, the lowest values of RMSE, MAE and ME corresponded to the best model (Vob) obtaining in this the RMSE values (13.61–12.70), MAE (10.04–9.86) and ME = (0–3.49) obtained in the adjustment plots and the validation, respectively (Table 6). Models Vob_2011 and Vob_2015 exhibited very similar values in RMSE (~ 60 m^3 ha^{-1}) and MAE (46 m^3 ha^{-1}). Residual plots for Vob prediction indicated that the proposed model was appropriate, as can be seem from the predicted versus the observed values for Vob in Fig. 5.

Each of the adjusted equations showed good fit statistics. All the parameters were significant at a confidence level of 95% (Table 7). In the graphical evaluation of the assumptions of normality and uniformity of the waste of the three fitted models, no evidence was found of any non-compliance with the assumptions.

The mean values of volume with bark of the three models are shown in Table 8.

Results of applying the Vob models to the entire surface area of Muxika

By applying the volume per hectare of wood with bark models to the entire surface area of Muxika for 2011 and 2015, the results shown in Table 9 were obtained.

The results of the distribution of the Vob for 2011 and 2015 and the difference between both can be seen in Fig. 6.

Annual increase in volume over bark

The difference between the Vob estimated for 2015 and that estimated for 2011 is 81,603 m^3 for the entire surface area of P. radiata in the municipality of

Table 6. Fits of the models

Model Equation	Fit					Validation			
	n	R^2 adj	RMSE (m^3 ha^{-1})	MAE	ME	n	RMSE (m^3 ha^{-1})	MAE	ME
[3]	60	99	13.61	10.04	0	30	12.70	9.86	3.49
[4]	60	85	62.54	46.24	0	30	68.00	51.75	-3.12
[8]	40	80	58.62	45.63	0	15	67.38	55.33	21.11

n: number of plots. R^2 adj : adjusted coefficient of determination. RMSE: root mean squared error. MAE: mean absolute error; ME: mean error

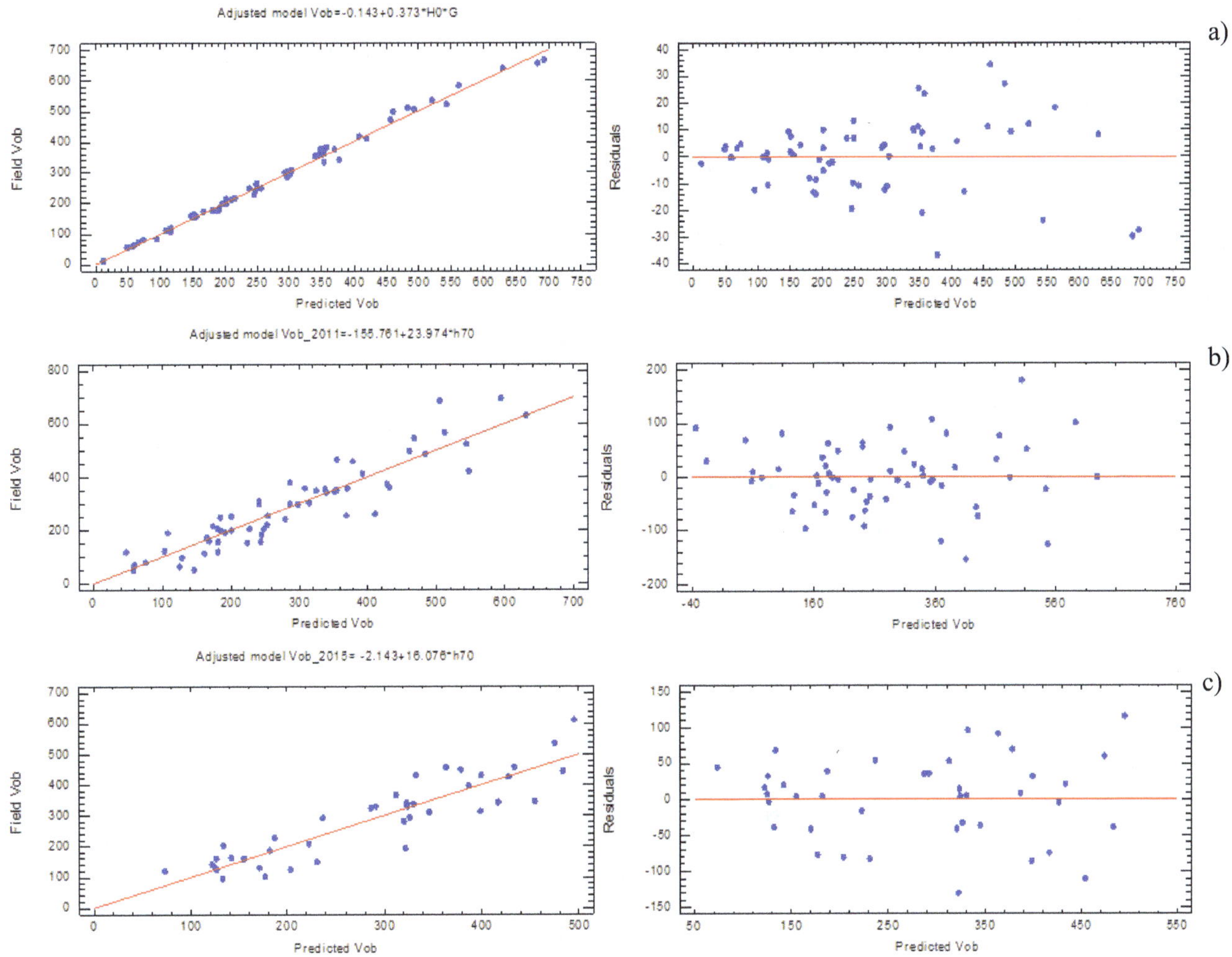

Figure 5. Chart showing models for stem volume and waste Vob model (a), Vob_2011 model (b) and Vob_2015 model (c)

Table 7. Parameter estimates, standard errors of the fitted models

Model	Dependent variable	Independent variable	Parameter estimate	Standard error	*t*-value	$p > \|t\|$
Vob (Eq. [3])	Vob	Intercept	-0.143	3.478	-0.041	0.967
		H_0*G	0.373	0.004	92.093	<0.001
Vob_2011 (Eq [4])	Vob_2011	Intercept	-155.761	24.675	-6.312	<0.001
		h_{70}	23.975	1.294	18.531	< 0.001
Vob_2015 (Eq. [8])	Vob_2015	Intercept	-2.143	24.176	-0.088	0.930
		H_{70}	16.076	1.251	12.853	<0.001

Vob: Stem volume (m^3 ha^{-1}); G: Basal area (m^2 ha^{-1}); Ho: Dominant height (m); h_{70}: 70th percentile of points height for cell.

Table 8. Mean stem volume in the models analyzed

Model	Fit		Validation	
	n	m^3 ha^{-1}	n	m^3 ha^{-1}
Vob	60	276.31	30	269.03
Vob_2011	60	276.31	30	269.03
Vob_2015	40	284.84	15	262.60

Table 9. Application of the Vob models to the entire surface area of Muxika

Year	Vob (m^3 ha^{-1})	Mean Vob (m^3 ha^{-1})	Vob increase 2011-2015 (m^3 ha^{-1})	Annual mean increase Vob (m^3 ha^{-1} yr^{-1})
2011	712021	225.80	-	-
2015	793084	251.51	81603	5.14

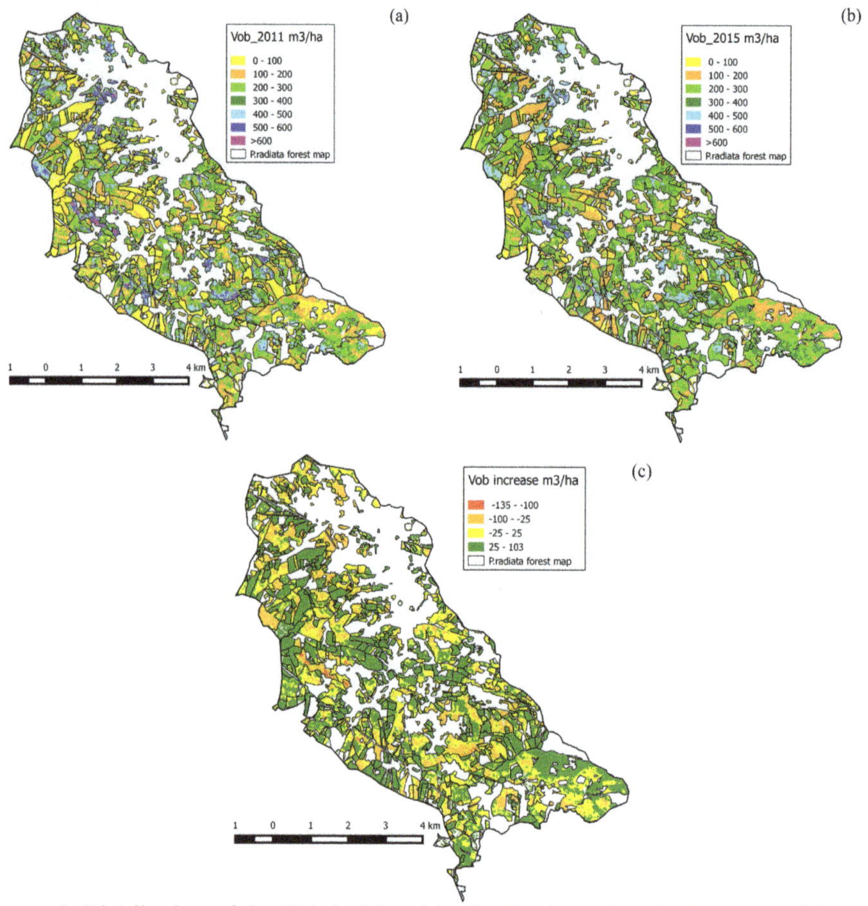

Figure 6. Distribution of the Vob in 2011 (a), distribution of the Vob in 2015 (b), and difference 2015 Vob minus 2011 Vob (c).

Muxika. Therefore, the annual mean growth of the volume over bark (ΔVob) obtained during the 5-yr period was 5.14 m^3 ha^{-1} yr^{-1} (Table 9).

The forestry harvests data shows that the wood volume extracted during the period 2011-2015 was 135,624 m^3 (Table 2) and therefore, on average, 8.60 m^3 ha^{-1} were extracted. The estimated mean growth for the period 2011 - 2015 is 13.74 m^3 ha^{-1} yr^{-1} , this being the value obtained by adding the estimated mean growth in the aforementioned period (ΔVob = 5.14 m^3 ha^{-1} yr^{-1}) and the mean volume extracted 8.6 m^3 ha^{-1}.

As can be seen in Fig. 7, the fitted models predict the greatest increases for Vob of between 0 and 150 m^3 ha^{-1}. The increases diminish gradually for Vob values close to 300 m^3 ha^{-1} and become negative above that value.

Discussion

Both the R^2 and the RMSE of the model fitted for the Vob of 2011 are similar to those published by other authors under similar conditions (Packalen *et al.*, 2011; Stone *et al.*, 2011; Tonolli *et al.*, 2011; Watt *et al.*, 2013). The values of R^2 and RMSE of the Vob model for 2015 are more unfavorable than those obtained under similar working conditions. This is due to the fact that in the preparation of the Vob_2011 model, the field data and the LiDAR data were almost contemporary, *i.e.*, they were taken just a few months apart. However, for the fit of the Vob_2015 model, the field data is more than four years older than the LiDAR data. This time difference causes a reduction in the ability to predict the fitted model, even more so if we take into consideration that *P. radiata* is a fast-growing species and is normally

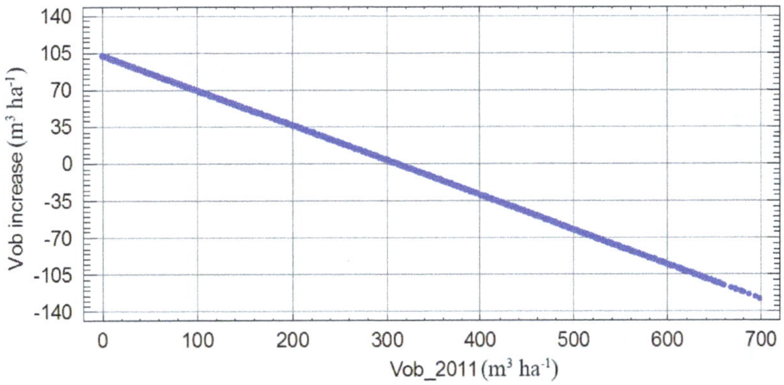

Figure 7. Vob increase versus Vob_2011

subjected to intensive forestry regimes that bring about considerable changes within a short time span.

With regard to adapting the results obtained, the estimated mean growth being 13.74 m^3 ha^{-1} yr^{-1}, the volume will have increased 68.7 m^3 ha^{-1} on average during the period 2011-2015. Therefore, in those places where harvest was more than 68.7 m^3 ha^{-1}, the growth has failed to offset the volume extracted and as a result will have less volume in 2015 than in 2011. These are the grid cells with a negative growth value. On the other hand, the mean volume extracted in a thinning operation according to Table 2 is 39.2 m^3 ha^{-1}. It is for this reason that the areas that were subjected to thinning between 2011 and 2015 must correspond to cells with a volume difference of less than 29.5 m^3 ha^{-1} as the 68.7 m^3 of growth in the 5 yrs, minus the 39.2 m^3 extracted in thinning operations will have allowed this amount to accumulate on average in these areas. Those grid cells with a volume difference of under that amount add up to 1,552 ha. This is the surface area in which, according to the fitted models, some kind of harvest took place during 2011 and 2015 and coincides with those contributed as real registered harvest: 1068.1 ha. As shown in Table 1, the minimum Vob of the plots measured in the field data taken in 2015 was 90.84 m^3 ha^{-1} and, therefore, no plot in which clear cutting had taken place was measured. Consequently, the 2015 model cannot detect the clear cutting that took place during this period and assimilates these to intensive thinning, and therefore the surface area of harvest estimated by the models includes both clear cutting and thinning. This circumstance must be borne in mind for subsequent applications of the method and those plots subjected to clear cutting during the period of use of LiDAR data. It can be seen, moreover, that the areas in which the models predict greater harvest intensity correspond to those that had a higher volume in 2011. These areas are the ones in which third thinnings and clear cuttings (being the harvest of greater weight) are normally carried out. Both the assimilation of clear cutting to intensive thinning and the fact that the areas that were actually thinned or clearcut after July 2012 are captured in the raster models, which are based on a Lidar dataset from July 2012, can be explained when analyzing the diagram resulting from a comparison of Vob_2011 and the growth between 2011 and 2015 (Fig. 7).

One of the aspects that must be clarified in subsequent studies is the development in growth in annual periods between both ends of the 5-yr interval. In this study it has not been possible to deal with this question due to a lack of field data in this intermediate period. Moreover, due to a lack of funding, locating the IFN plots was not possible for the second gathering of field data and for this reason the samples for 2011 and 2015 were different. Even though this circumstance might reduce the validity of the results obtained, it was decided to begin the study to verify whether this methodology could be used in subsequent studies with higher funding levels. Once sufficient funds are available, field data must be taken on an annual basis and in the same plot sample.

With regard to the multi-annual use and estimation of growth using a single LiDAR flight, no published references have been found concerning similar methods. This is the main contribution of this paper: the use of LiDAR data in successive years through the updating of the mass conditions with fast, reliable and cost-effective annual field data gathering. In this way, the investment involved in LiDAR data gathering is justifiable as this can be amortized in several years.

Fitted models that relate the descriptive variables of the points cloud of a LiDAR flight with the wood volume field data obtained in field plots with a 5-yr interval, provides a good approximation to the growth values produced in masses of *P. radiata* in this region (13.74 m^3 ha^{-1} yr^{-1}), as well as an estimation of the surface area subjected to harvest in this period. This allows the multi-annual use of LiDAR data gathered in a campaign with the consequent decrease in costs. To ensure the highest levels of accuracy in the estimation of the growth and volume of wood during the years following the gathering of LiDAR data, it is necessary to include in the sample of plots to be measured those in which both thinning and clear cutting operations have been carried out.

Acknowledgements

Our sincere thanks to Hazi Fundazioa for providing essential data for this study.

References

Babcock C, Finley AO, Bradford JB, Kolka R, Birdsey R, Ryan MG, 2015. LiDAR based prediction of forest biomass using hierarchical models with spatially varying coefficients Remote Sens Environ 169: 113-127. https://doi.org/10.1016/j.rse.2015.07.028

Brandtberg T, 2007. Classifying individual tree species under leaf-off and leaf-onconditions using airborne LiDAR. ISPRS J Photogramm. Remote Sens 61: 325-340. https://doi.org/10.1016/j.isprsjprs.2006.10.006

Breusch TS, Pagan AR, 1979. A simple test for heteroscedaticity and random coefficient variation. Econometrica 47: 1287-1294. https://doi.org/10.2307/1911963

Cao L, Coops NC, Innes JL, Sheppard SR, Fu L, Ruan H, She G, 2016. Estimation of forest biomass dynamics in subtropical forests using multi-temporal airborne LiDAR data. Remote Sens Environ 178: 158-171. https://doi.org/10.1016/j.rse.2016.03.012

Condes Ruiz S, Fernández-Landa A, Rodríguez F, 2013. Influencia del inventario de campo en el error de muestreo obtenido en un inventario con tecnología Lidar. Actas del 6º Congreso Forestal Español, Vitoria-Gasteiz, Spain.

Dalponte M, Coops N, Bruzzone L, Gianelle D, 2009. Analysis on the use of multiple returns LiDAR data for the estimation of tree stems volume. IEEE J-STARS 2: 310-318. https://doi.org/10.1109/JSTARS.2009.2037523

Dean TJ, Cao QV, Roberts SD, Evans DL, 2009. Measuring heights to crownbase and crown median with LiDAR in a mature even-aged loblolly pine stand. Forest Ecol Manag 257: 126-133. https://doi.org/10.1016/j.foreco.2008.08.024

Durbin J, Watson GS, 1971. Testing for serial correlation in least squares regression. III Biometrika 58: 1-19. https://doi.org/10.2307/2334313

Fekety PA, Falkowski MJ, Hudak AT, 2014. Temporal transferability of LiDAR-based imputation of forest inventory attributes. Can J Forest Res 45 (4): 422-435. https://doi.org/10.1139/cjfr-2014-0405

Giannico V, Lafortezza R, John R, Sanesi G, Pesola L, Chen J, 2016. Estimating stand volume and above-ground biomass of urban forests using LiDAR. Remote Sens 8 (4): 339. https://doi.org/10.3390/rs8040339

Holmgren J, 2004. Prediction of tree height basal area and stem volume in forest stands using airborne laser scanning. Scand J Forest Res 19: 543-553. https://doi.org/10.1080/02827580410019472

Hopkinson C, Chasmer L, Hall RJ, 2008. The uncertainty in conifer plantation growth prediction from multi-temporal lidar datasets. Remote Sens Environ 112 (3): 1168-1180. https://doi.org/10.1016/j.rse.2007.07.020

Hyyppä J, Kelle O, Lehikoinen M, Inkinen M, 2001. A segmentation-based method to retrieve stem volume estimates from 3-D tree height models produced by laser scanners. IEEE T Geosci Remote 39 (5): 969-975. https://doi.org/10.1109/36.921414

Jochem A, Hollaus M, Rutzinger M, Höfle B, 2010. Estimation of aboveground biomass in alpine forests: A semi-empirical approach considering canopy transparency derived from airborne LiDAR data. Sensors 11 (1): 278-295. https://doi.org/10.3390/s110100278

Koenig K, Höfle B, 2016. Full-waveform airborne laser scanning in vegetation studies-A review of point cloud and waveform features for tree species classification. Forests 7 (9): 198. https://doi.org/10.3390/f7090198

Latifi H, Nothdurft A, Koch B, 2010. Non-parametric prediction and mapping of standing wood volume and biomass in a temperate forest: application of multiple optical/LiDAR-derived predictors. Forestry 83 (4): 395-407. https://doi.org/10.1093/forestry/cpq022

Leiterer R, Furrer R, Schaepman ME, Morsdorf F, 2015. Forest canopy-structure characterization: A data-driven approach. Forest Ecol Manag 358: 48-61. https://doi.org/10.1016/j.foreco.2015.09.003

Li R, Weiskittel AR, 2010. Comparison of model forms for estimating stem taper and volume in the primary conifer species of the North American Acadian Region. Ann For Sci 67 (3): 302. https://doi.org/10.1051/forest/2009109

Lindner M, Karjalainen T, 2007. Carbon inventory methods and carbon mitigation potentials of forests in Europe: a short review of recent progress. Eur J Forest Res 126: 149-156. https://doi.org/10.1007/s10342-006-0161-3

Magnussen S, Næsset E, Gobakken T, 2010. Reliability of LiDAR derived predictors of forest inventory attributes: A case study with Norway spruce. Remote Sens Environ 114 (4): 700-712. https://doi.org/10.1016/j.rse.2009.11.007

Mallows CL, 1973. Some comments on CP. Technometrics 15 (4): 661-675.

Maltamo M, Packalén P, Peuhkurinen J, Suvanto A, Pesonen A, Hyyppä J, 2007. Experiences and possibilities of ALS based forest inventory in Finland. Proc ISPRS Workshop on Laser Scanning, Spoo, Finland. pp: 270-279.

Maltamo M, Næsset E, Vauhkonen J, 2014. Forestry applications of airborne laser scanning. Concepts and case studies. Manag For Ecosys 27: 460.

Mateos E, Garrido F, Ormaetxea L, 2016. Assessment of biomass energy potential and forest carbon stocks in Biscay (Spain). Forests 7 (4): 75-89. https://doi.org/10.3390/f7040075

Montaghi A, Corona P, Dalponte M, Gianelle D, Chirici G, Olsson H, 2013. Airborne laser scanning of forest resources: an overview of research in Italy as a commentary case study. Int J Appl Earth Obs 23: 288-300. https://doi.org/10.1016/j.jag.2012.10.002

Næsset E, Gobakken T, 2005. Estimating forest growth using canopy metricsderived from airborne laser scanner data. Remote Sens Environ 96: 453-465. https://doi.org/10.1016/j.rse.2005.04.001

Næsset E, 2011. Estimating above-ground biomass in young forests with airborne laser scanning. Int J Remote Sens 32: 473-501. https://doi.org/10.1080/01431160903474970

Nilsson M, 1996. Estimation of tree heights and stand volume using an airborne lidar system. Remote Sens Environ 56: 1-7. https://doi.org/10.1016/0034-4257(95)00224-3

Packalén P, Mehtätalo L, Maltamo M, 2011. ALS-based estimation of plot volume and site index in a eucalyptus plantation with a nonlinear mixed-effect model that accounts for the clone effect. Ann Forest Sci 68 (6): 1085-1092. https://doi.org/10.1007/s13595-011-0124-9

Pan Y, Birdsey R, Fang J, Houghton R, Kauppi P, Kurz W, Phillips O, Shvidenko A, Lewis S, *et al.* A large and persistent carbon sink in the world's forests. Science 333: 988-993. https://doi.org/10.1126/science.1201609

Popescu SC, 2007. Estimating biomass of individual pine trees using airborne Lidar. Biomass Bioenerg 31: 646-655. https://doi.org/10.1016/j.biombioe.2007.06.022

Popescu SC, Zhao K A, 2008. Voxel-based LiDAR method for estimating crownbase height for deciduous and pine trees. Remote Sens Environ 112: 767-781. https://doi.org/10.1016/j.rse.2007.06.011

Popescu SC, Wynne RH, Nelson RF, 2003. Measuring individual tree crown diameter with lidar and assessing its influence on estimating forest volume and biomass. Can J Remote Sens 29 (5): 564-577. https://doi.org/10.5589/m03-027

Riaño D, Chuvieco E, Condés S, González-Matesanz J, Ustin SL, 2004. Generation of crown bulk density for Pinus sylvestris L from lidar. Remote Sens Environ 92 (3): 345-352. https://doi.org/10.1016/j.rse.2003.12.014

Rodríguez F, Broto M, Lizarralde I, 2008. CubiFOR: Herramienta para cubicar clasificar productos y calcular biomasa y CO2 en masas forestales de Castilla y León. Montes 95: 33-39.

Song Y, Imanishi J, Sasaki T, Ioki K, Morimoto Y, 2016. Estimation of broad-leaved canopy growth in the urban forested area using multi-temporal airborne LiDAR datasets. Urban For Urban Gree 16: 142-149. https://doi.org/10.1016/j.ufug.2016.02.007

Stone C, Penman TD, Turner R, 2011. Determining an optimal model for processing lidar data at the plot level: results for a Pinus radiata plantation in New South Wales Australia. NZ J For Sci 41: 191-205.

Tonolli S, Dalponte M, Vescovo L, Rodeghiero M, Bruzzone L, Gianelle D, 2011. Mapping and modeling forest tree volume using forest inventory and airborne laser scanning. Eur J For Res 130 (4): 569-577. https://doi.org/10.1007/s10342-010-0445-5

Torres AB, Marchant R, Lovett JC, Smart JC, Tipper R, 2010. Analysis of the carbon sequestration costs of afforestation and reforestation agroforestry practices and the use of cost curves to evaluate their potential for implementation of climate change mitigation. Ecol Econ 69 (3): 469-477. https://doi.org/10.1016/j.ecolecon.2009.09.007

Vepakomma U, St-Onge B, Kneeshaw D, 2011. Response of a boreal forest to canopy opening: assessing vertical and lateral tree growth with multi-temporal LiDAR data. Ecol Appl 21: 99-121. https://doi.org/10.1890/09-0896.1

Watt M, Meredith A, Watt P, Gunn A, 2013. Use of LiDAR to estimate stand characteristics for thinning operations in young Douglas-fir plantations. NZ J For Sci 43 (18): 1-10. https://doi.org/10.1186/1179-5395-43-1

Yu X, Hyyppä J, Kaartinen H, Maltamo M, 2004. Automatic detection of harvested trees and determination of forest growth using airborne laser scanning. Remote Sens Environ 90: 451-462. https://doi.org/10.1016/j.rse.2004.02.001

Yu X, Hyyppä J, Kukko A, Maltamo M, Kaartinen H, 2006. Change detection techniques for canopy height growth measurements using airborne laser scanner data Photogramm. Eng Remote Sens 72: 1339-1348. https://doi.org/10.14358/PERS.72.12.1339

Yu X, Hyyppä J, Holopainen M, Vastaranta M, 2010. Comparison of area-based and individual tree-based methods for predicting plot-level forest attributes. Remote Sens 6: 1481-1495. https://doi.org/10.3390/rs2061481

Zhang J, Ge Y, Chang J, Jiang B, Jiang H, Peng C, Zhu J, Yuan W, Qi L, Yu S, 2007. Carbon storage by ecological service forests in Zhejiang Province subtropical China. For Ecol Manag 245: 64-75.

Zhang Z, Kazakova A, Moskal LM, Styers DM, 2016. Object-based tree species classification in urban ecosystems using LiDAR and hyperspectral data. Forests 7 (6): 122. https://doi.org/10.3390/f7060122

Zianis D, Muukkonen P, Mäkipää R, Mencuccini M, 2005. Biomass and stem volume equations for tree species in Europe. Silva Fennica Monographs 4, 63 pp.

Zolkos SG, Goetz SJ, Dubayah RA, 2013. Meta-analysis of terrestrial aboveground biomass estimation using LiDAR remote sensing. Remote Sens Environ 128: 289-298. https://doi.org/10.1016/j.rse.2012.10.017

10

Temporal optimisation of fuel treatment design in blue gum (*Eucalyptus globulus*) plantations

Ana Martín[1], Brigite Botequim[2]*, Tiago M. Oliveira[3], Alan Ager[4], Francesco Pirotti[5]

[1] *Sustainable Forest Management Research Institute University of Valladolid-INIA, Avda de Madrid 44, 3004 Palencia, Spain.* [2] *Centro de Estudos Florestais (CEF), Instituto Superior de Agronomia, Universidade de Lisboa (UL). Edificio Mário de Azevedo Gomes Tapada da Ajuda, n/s, 1349 - 017, Lisboa, Portugal.* [3] *The Navigator Company, Forest Protection Department, Apartado 55, 2901-861 Setúbal, Portugal.* [4] *USDA Forest Service, Rocky Mountain Research. 5775 Highway 10 West, Missoula, MT 59808 USA.* [5] *Università degli Studi di Padova, Dipartimento Territorio e Sistemi Agro-forestali. Viale dell'Universita, 16, 35020, Padova, Italy*

Abstract

Aim of study: This study was conducted to support fire and forest management planning in eucalypt plantations based on economic, ecological and fire prevention criteria, with a focus on strategic prioritisation of fuel treatments over time. The central objective was to strategically locate fuel treatments to minimise losses from wildfire while meeting budget constraints and demands for wood supply for the pulp industry and conserving carbon.

Area of study: The study area was located in Serra do Socorro (Torres Vedras, Portugal, covering ~1449 ha) of predominantly *Eucalyptus globulus* Labill forests managed for pulpwood by The Navigator Company.

Material and methods: At each of four temporal stages (2015-2018-2021-2024) we simulated: (1) surface and canopy fuels, timber volume (m^3 ha^{-1}) and carbon storage (Mg ha^{-1}); (2) fire behaviour characteristics, i.e. rate of spread (m min^{-1}), and flame length (m), with FlamMap fire modelling software; (3) optimal treatment locations as determined by the Landscape Treatment Designer (LTD).

Main results: The higher pressure of fire behaviour in the earlier stages of the study period triggered most of the spatial fuel treatments within eucalypt plantations in a juvenile stage. At later stages fuel treatments also included shrublands areas. The results were consistent with observations and simulation results that show high fire hazard in juvenile eucalypt stands.

Research highlights: Forest management planning in commercial eucalypt plantations can potentially accomplish multiple objectives such as augmenting profits and sustaining ecological assets while reducing wildfire risk at landscape scale. However, limitations of simulation models including FlamMap and LTD are important to recognise in studies of long term wildfire management strategies.

Keywords: Eucalypt plantations; fire hazard; FlamMap; fuel treatment optimisation; Landscape Treatment Designer; wildfire risk management.

Funding: The authors received no specific funding for this research.

Competing interests: The authors have declared that no competing interests exist.

Correspondence should be addressed to Brigite Botequim: bbotequim@isa.ulisboa.pt

Introduction

Portugal is covered by 3.2 million hectares of forest, representing 34.5% of the total mainland area, and ranking eighth in Europe as highest country with forestlands (ICNF, 2013). Portuguese plantations are dominated by eucalypt (*Eucalyptus globulus* Labill) encompassing 812 thousand hectares (26% of the whole forested area), and constitute the main forest cover type in continental Portugal (ICNF, 2013). Eucalypt plantations are an important economic resource and support a large and significant pulp and paper industry, one of the key businesses in Portugal (Soares *et al.*, 2007). The culture of eucalypt has many other potential benefits including carbon sequestration (Botequim *et al.*, 2013). However, commercial eucalypt plantations are inherently flammable due to the nature and accumulation of their litter and bark fuels with high levels of biomass, and thus prone to intense wildfires (Fernandes, 2009).

Portugal has the highest incidence of wildfires events in the Mediterranean basin (Pereira *et al.*, 2014). Eucalypt plantations accounted for 35.9% of all Portuguese forest burned during 1996 - 2012 (Mateus &

Fernandes, 2014). These wildfires pose a significant impact on revenues and costs associated with eucalypt management scheduling (Ferreira *et al.*, 2012). However, appropriate management activities can reduce fuel loadings and alter forest structure to change wildfire occurency and reduce wildfire losses (Botequim *et al.*, 2013; Marques *et al.*, 2011). In particular, fuel reduction methods for modifying fire behaviour have become a central management issue and are actively practiced by fire and forest managers in eucalypt stands to reduce wildfire risk (Botequim *et al.*, 2013; Fernandes, 2015).

Wildfire simulation methods provide a framework to quantitatively measure performance of the fuel treatments (Ager *et al.*, 2010), and evaluate the effects of the fuel treatment planning on fire behaviour before and after the fuel treatment design. Large recent wildfires have led to an increased interest in decision support tools to mitigate wildfire risk. These include growth and yield modeling, wildfire behaviour models and optimisation approaches to assist managers in planning and evaluating management decisions to support hazard-reduction fuel practices (Ager *et al.*, 2012; Bradstock *et al.*, 2012; Vogler *et al.*, 2015).

Landscape-scale fuel treatments decrease fire intensity and fire growth rate, hence resulting in smaller and less severe fires (Fernandes, 2015). Fuel reduction treatments (e.g. thinning and prescribed burning) have long been identified as key to decrease fire size and fire severity (Agee & Skinner, 2005), and are a vital part of forest management. However, with limited resources, it is necessary to prioritise where, when, and how fuel treatments should be applied across the landscape. Finney *et al.* (2006) addressed how fuel treatments placed in random and optimal spatial patterns affect fires behaviour. They found that strategic allocation of fuel treatments reduced the predicted growth rates of simulated fires under unfavorable weather conditions more effectively than random placement. Indeed, fire risk and wildfire damage can be reduced by removing or reducing fuels in strategic locations.

Where should the forest manager spatially invest in prevention? Which stands should be assumed for fire prevention treatment management? This study was conducted to monitor trends in wildfire hazard through strategic fuel treatment locations into a spatial-temporal analysis (2015-2018-2021-2024). The main interest was set at reducing the impact of wildfire as determined by fire hazard thresholds (rate of spread, m min^{-1}), and flame length (m) while considering wood supply (m^3 ha^{-1}) for the pulp industry and demands of carbon values (Mg ha^{-1}) with an annual budget constraint (or area treated as surrogate) for the prevention activities in Serra do Socorro. The current framework was designed taking into account several decision support tools used in the United States for wildfire risk management (e.g. FlamMap – Finney, 2006 and Landscape Treatment Designer, LTD - Ager *et al.*, 2012). We used these tools in a three-tiered approach in Serra do Socorro (Centre Portugal). This research resulted in: (i) improved understanding of the relative role of biometric variables (e.g. forest fuel load and stand structure) in fire behaviour characteristics, with the challenge of incorporating such knowledge into fire-management planning; and (ii) information to help policy makers define a scientific approach to spatially and temporally prioritise fuel management interventions for minimising fire impacts, thus protecting eucalypt areas from burning without ecological and commercial timber values losses.

Material and Methods

The study area

The study area is located in Serra do Socorro (Torres Vedras, central Portugal) and covers ~1449 hectares (Fig. 1). Serra do Socorro is the steepest landscape in the municipality of Torres Vedras (Gabinete Técnico Florestal, Torres Vedras, 2015-2019). Topography is dominated by uplands and lowlands ranging from 51 meters to a maximum of 389 meters elevation, with an average value of 168 meters.

The area is characterised by a strong Atlantic wind influence and consequently, throughout the whole year, atmospheric humidity remains high which is especially noticeable in summer when compared with other regions in Portugal (more than 70% relative humidity). In terms of temperature, the maximum (26 °C) is reached during August. Precipitation values, in the fire season (especifically during July and August), do not reach 7 mm as average value per month (values are given for the historical period 1961-1990) (Gabinete Técnico Florestal, Torres Vedras, 2015-2019).

The main forest land cover type is eucalypt plantations (175 ha) surrounded by small agriculture, shrublands and uncultivated land. These plantations are managed by The Navigator Company, the leading paper company in Europe. The site presents the following land use classes: non burnable areas (5.9%), blue gum plantations (12%), extensive agricultural lands and pasture lands (56.7 %), shrublands (24%), pine stands (0.24%).

The municipality of Torres Vedras is considered under a low intensity fire regime with a long fire season. Fire history in Torres Vedras is defined by a significant number of events, mainly in summer season (maximun in October) and small size (less than 1

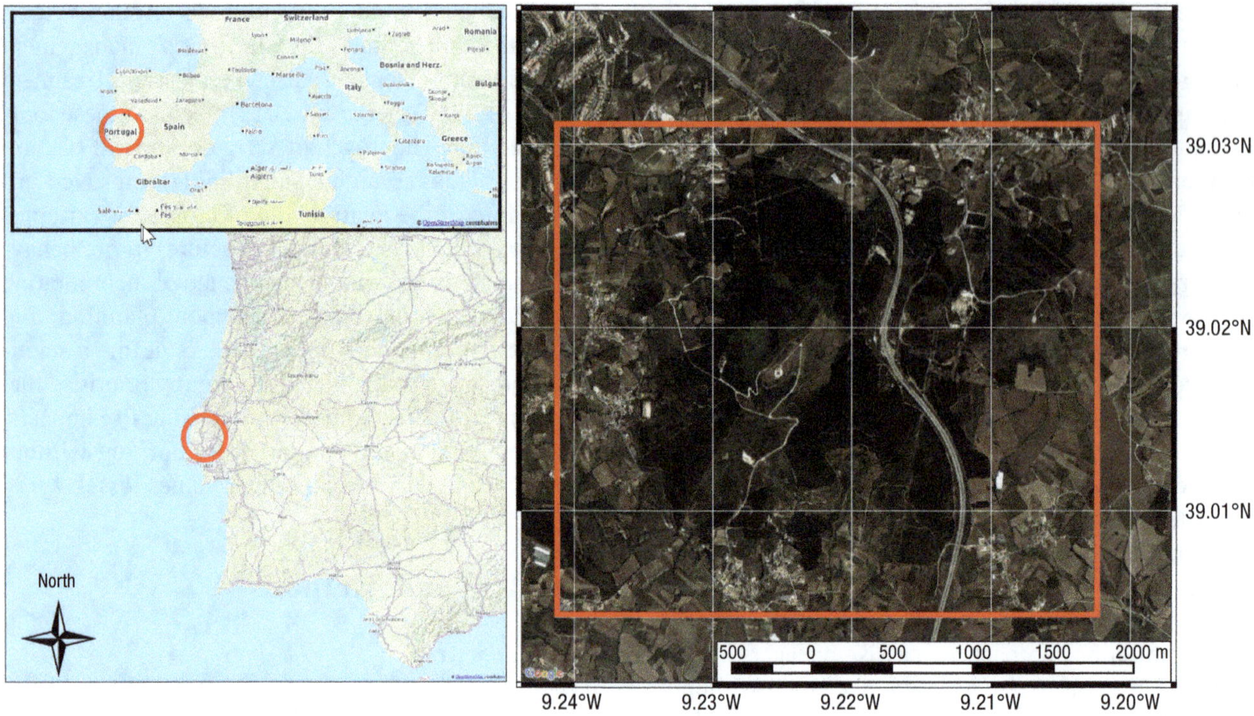

Figure 1. Location of the study area.

hectare) (Gabinete Técnico Florestal, Torres Vedras, 2015-2019). Those autumn and winter fires mainly affect shrublands and farming areas (Marques *et al.*, 2014). Social and demographical factors also play a determining role in Serra do Socorro fire regime, given the strong links between population and agricultural activities and the common use of fire irrespective of season (Raínha & Fernandes, 2002).

Methods

Changes in fuels, carbon, timber volume and fire behaviour, were explored in response to fuel treatments over time for the present (2015, t=0) and three future periods (2018-2021-2024, t=1, t=2, t=3, respectively). A three-tiered strategy was applied as follows.

1. *Dynamic fuel trends*

Field inventory records (year 2013) and data from the growth simulation model (Globulus Model 3, Tomé *et al.*, 2006) were considered for analyses at 3-year intervals in order to conduct the fire behaviour simulation and obtain timber volume and total biomass over time - converted to carbon storage value multiplying by a factor of 0.5 - Kyoto protocol.

A set of 18 customised fuel models developed for Portugal by Fernandes *et al.* (2009), and Cruz (2005) for the case of eucalypt slash were assigned to obtain the surface fuel map of the study area (2015, t=0). Urban areas and structures, water streams, irrigated agricultural lands, roads and forest roads were classified as non-burnable (fuel model 99); short grass (fuel model 232) was assumed for greenhouses, golf facilities, isolated homes, vineyards, agro-forestry systems, fruit orchards, olive groves and natural pastures; in high grass fuel type (fuel model 231) were included non-irrigated agricultural lands and others natural pastures. Besides that, two more classes were introduced to define a region with riparian vegetation (fuel model 221) and small patches with pine forest of *Pinus pinaster* Ait. and *Pinus pinea* L. (227 and 213). The percentage of canopy cover data (84%,60%,44%,22%) assigned to each stand in the current landscape came from previous analysis of the National Forest Inventory blue gum plots (Oliveira *et al.*, 2016) associated with field visits, expert local knowledge and further supported by visual interpretation of georeferenced digital photography.

Dynamic fuel projections (surface fuel types change over time assess changes over time due to planting operations, harvesting, and vegetation regrowth) were defined by fuel model descriptions in field visits and fuel load characterisation based on the dominant vegetation, structures and fuel loads (Identification fuel model key and forest fuel model guideline for Portugal, Fernandes *et al.*, 2009), and expert knowledge provided by The Navigator Company forest protection department.

Canopy characteristics followed equations in Cruz & Viegas (1998) and Soares & Tomé (2001) for canopy bulk density (CBD, kg m^{-3}) and canopy base height (CBH, m) based on the previously calculated biometric data. Canopy cover (CC) percentage was estimated over time based on local experience and inputs from fire managers. A mean CC value was assigned to the plot and classified into intervals (i.e. increased to the next canopy cover percentage category every period until the final cut at 12 years old).

2. Spatial fire behaviour simulations and fire metrics

FlamMap5 (Finney, 2006) was used to explore fire behaviour characteristics and fire spread and intensity under constant wind speed and fuel moisture contents. We analysed FlamMap outputs for flame length (FL, m), fireline intensity (FLI, kW m^{-1}), rate of spread (ROS, m min^{-1}). The equations for these parameters are described in Ager *et al.* (2011).

FlamMap uses a gridded input file containing information on fuel and topography. The latter information consisted of elevation (meters), aspect (degrees) and slope (degrees) extracted from a digital elevation model (DEM) at 30 m resolution. Fuel-related data themes consisted of calibrated fuel models (Fernandes *et al.*, 2009), canopy closure, stand height, canopy bulk density and canopy base height derived from inventory records, as well as estimated by growth simulation models and canopy equations. FlamMap simulations were performed for each time period based on temporal projections in surface and canopy fuels as described above.

Wind speed, direction and fuel moisture for the FlamMap simulations were derived from local weather data. Data records from the Dois Portos weather station (ca. 7 km from the study area) were used to calculate fuel moisture for high fire risk season (May-October) for twelve years observation (2001-2012). For fuel moisture, a value of 7% was used for the 1-hour fuels corresponding to air temperature of 20-35 °C and a relative humidity of 23-35%. These values represented 97th percentile weather during fire season. Values for the 10 and 100-hour dead fuel size classes were computed adding 1% and 3% to the 1-hour dead fuel size, respectively. For live fuel moisture content, we assumed 85-95% for shrubs and canopy fuelbeds, respectively. The simulaton was carried out according 10-m open values for wind speed of 32 km h^{-1} aligned with predominant winds from Northwest, azimuth direction of 320 degrees, generating the maximum fire potential at any given cell.

3. Optimal treatment locations

For modeling optimal fuel treatment locations we used the Landscape Treatment Designer (LTD) (Ager *et al.*, 2012, Ager *et al.*, 2013, Vogler *et al.*, 2015). LTD prioritised fuel treatments bases on fire behaviour characteristics, highly valuable resources and economic constraints. LTD creates a sequence of spatially defined project areas that maximises the objectives with the provided constraints and treatment thresholds. The specific variables modeled were fire hazard (FL, ROS), timber volume (m^3 ha^{-1}), total biomass as carbon storage (Mg ha^{-1}). The treatment thresholds were based on fire potentials conditions and fuel treatments were triggered when a stand exceeded a ROS of 10 meters per minute and a FL of 1.5 meters. Fire behaviour above these thresholds generally exceed firefighting capabilities. Priority stands for treatment were defined by timber volume and total biomass as carbon storage (Mg ha^{-1}). The treatments were constrained to a percentage of the total area of the landscape, 70 ha with a buffer of 10 ha extra "slack variable", as a surrogate for the seasonal budget available for this area (i.e fuel treatment allowance level intensity about 4% for the whole landscape and 40% of the area within eucalypt plantations).

LTD can create both aggregated and non-aggregated project areas. LTD creates a map of treatment priority resulting in a sequence of project areas and respective priorities (ranking in priority planning, e.g. the project labelled "1" represents the highest priority planning area).

Results and discussion

Assessing fire patterns over time

We focused on changes in two fire metrics, flame length (FL) and rate of spread (ROS) patterns over time in response to treatments. The maximum FL was observed at the earlier time periods and decreased thereafter. The maximum FL was observed in year 2018 (t2) presumably due to eucalypt stand structure in young plantations where most trees are between three and five years old. The highest ROS was observed for young eucalypt plantations and in the adjoining shrublands. We observed decreasing ROS with increasing stand age.

Optimising fuel treatment locations at landscape level

LTD creates both aggregated (Fig. 2) and dispersed treatment schedules (Fig. 3), each representing a different spatial treatment scenario. It might be assumed that large land owners such as The Navigator Company would focus on aggregate fuel treatments while forest owners might prefer non-aggregated treatments.

The higher fire risk levels of the periods t_0 and t_1 (Fig. 2a, 2b; 3a ,3b) resulted in the allocation of most of the fuel treatments within plantations in a juvenile stage. This is consistent with the previous fire behaviour outputs by FlamMap in terms of FL and ROS, which were largest in the young eucalypt stands. In t_2 and t_3 (Fig. 2c, 2d; 3c, 3d) fuel treatments were allocated by LTD to both, within eucalypt plantations and outside plantations, in shrublands areas. Young eucalypt plantations often have high density, crowns are low and closer to the ground and have shrubs understory and thus generate a continuous fuel layer and lead to higher fire hazard (Botequim *et al.*, 2013).

The decreasing trend in the number of treatment projects over time was consequence of the forest fire hazard status through time (e.g. the strategic placement of project areas over time in Fig 3a,3b presented projects 1, 2, 3 and 4 while in Fig. 3c, 3d presented only projects 1 and 2). The number of resulting projects decrease as fire metric values decrease over time (t2, t3) and do not surpass the defined thresholds thus do not trigger fuel treatments operations.

The output from fire simulation and strategic allocation of fuel treatment projects from LTD were consistent with experiences from local managment and current silviculture practices performed by the The Navigator Company to manage wildfire risk.

The LTD program selected areas for treatments based on fire hazard and fire risk. In Portugal, a typical eucalypt full rotation may include up to 2 or 3 coppice cuts, each cut being followed by a stool thinning that may leave an average number of shoots per stool ranging from 1 to 2. Harvest ages typically range from 10 to 12 (Soares & Tomé, 2001). Prescriptions further include several shrub cleanings over a rotation (i.e., 1 to 3 fuel treatments per cycle - Duncker *et al.*, 2012).

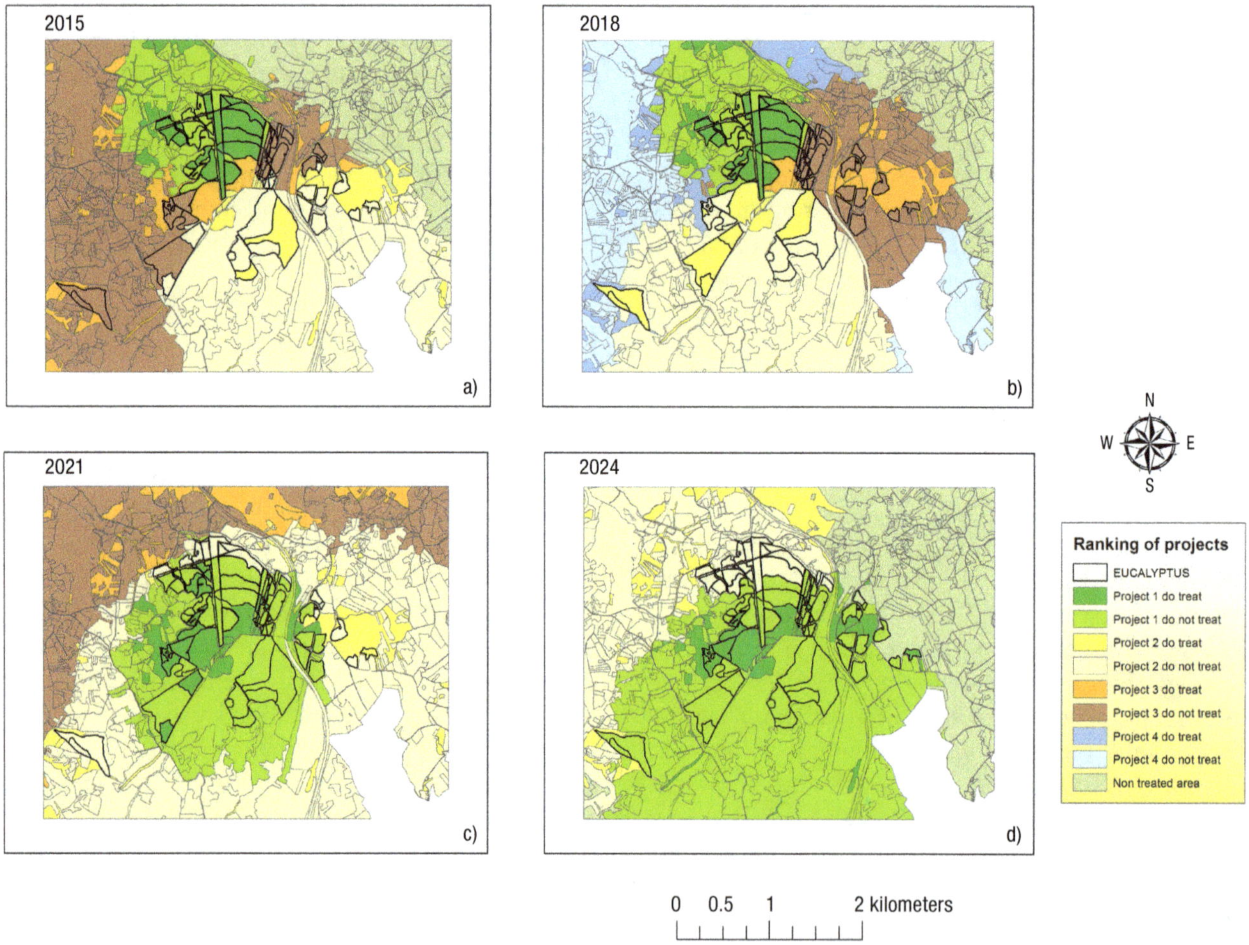

Figure 2. Ranking of projects (aggregate option) to prioritise the entire study area into a sequence of project areas. The project labelled with numbers from 1 to 4 represent respectively the highest to lowest priority planning area for the given objective. Areas labelled as "do treat" were selected for treatment based on the expected flame length and rate of spread exceeding the threshold; the other areas are within the project areas but were not selected for treatment.

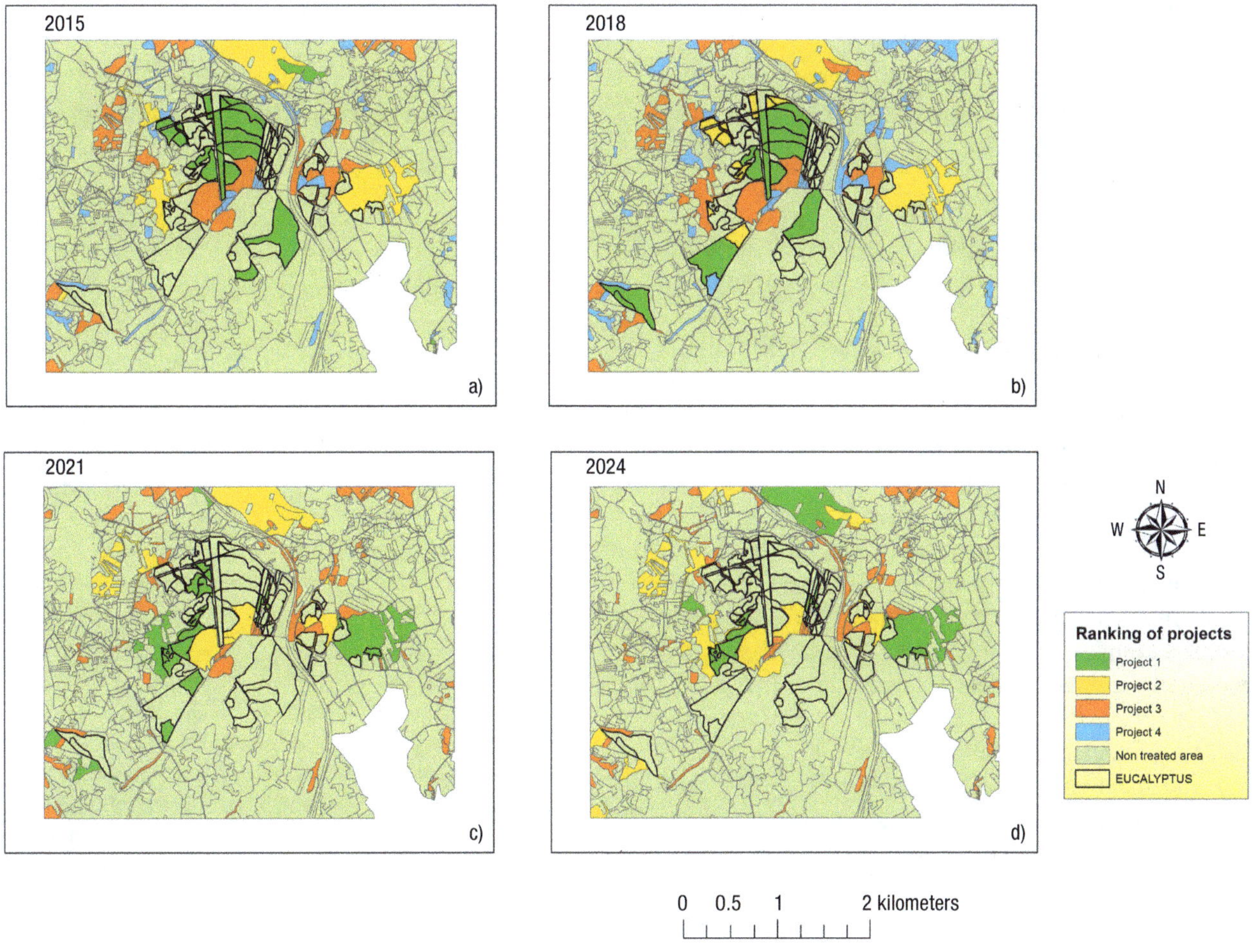

Figure 3. Ranking of projects (non-aggregate option) to prioritise the entire study area into a sequence of project areas. The project labelled with numbers from 1 to 4 represent respectively the highest to lowest priority planning area for the given objective.

The specific fuel treatments, including slash removal and prescribed burning would be carried out in these areas according to local conditions to modify fuel structure and fuel load changes. In our study area, prescribed burning as fuel management is the treatment preferred for the removal of accumulated fuel loads in blue gum plantations (i.e., after stool selection) (Pinto *et al.*, 2013). Other punctual operations depending on the stand status might include mechanical treatments used separately or in concert with prescribed fire to reduce fuel loads and alter fuel structure and thus minimise fire impacts (Fernandes *et al.*, 2011). Nevertheless, several studies suggest that the final size of individual fires can be impacted by landscape treatment rates as low as 5% (e.g. Cochrane *et al.*, 2012).

Contribution to fire-management strategies

The methods we used can potentially have wider application to explore fuel and fire management options in eucalypt plantations. By coupling FlamMap and LTD to examine temporal changes in plantations we were able to add a temporal analyses to the strategic implementation of fuel management. This temporal aspect was made possible by pre-existing research that led to the development of growth and yield model, canopy equations and fuel models descriptions.

Given the fire risk and cost of performing forest and shrublands fuel treatments across large areas, effective allocation of resources is critical. It is very challenging to strategically select fuel treatment locations on space and time with heterogeneous landscape i.e., topography, fuel models, and highly valuables resources susceptible to potential fire damages (Wei & Long, 2014). In this sense, we demonstrated a new approach for prioritising forest management through time in eucalypt stands in Serra do Socorro.

The framework of our method joined models of forest stand dynamics with optimisation models to provide a new decision support system for locating areas for fuel treatments. The study demonstrates new ways to develop efficient risk mitigation practices focusing on spatial-time-investment fuel treatment strategies in fire-prone landscape. The simulation methods also provide a way to communicate wildfire risk to the diverse array of landowners and other stakeholders in Serra do Socorro and adjacent areas in Central Portugal. Thus the methodology can be implemented as part of operational-planning decision-making and helps efficient forest management when answering spatial-temporal question such as where and when fuel treatments are required under the desired objectives.

The literature provides little guidance for determining the required fire behaviour knowledge (e.g. the use of crown fire models developed for conifers and the spotting potential of eucalypts) for an effective protection of eucalypt plantations (Fernandes *et al.*, 2011). Indeed, the limitations of simulation models has been suggested by several authors (Alexander & Cruz, 2013; Cruz & Alexander, 2010). A large number of observations obtained from carrying out experimental simulations studies (e.g. FlamMap5) showed that, in many cases, they are instrumental to predict fire behaviour characteristics and can quantitatively replicate large wildfire events, in terms of predicted burned area (Finney *et al.*, 2011), and size and shape of perimeters (Ager *et al.*, 2014a; Salis *et al.*, 2014). Wildfire simulations should be regarded as general indicators of wildfire exposure and fuel management planning for prioritising fire protection efforts (Salis *et al.*, 2013). We attempted to minmise errors with the application of FlamMap by using calibrated fuel models for the study area (Fernandes *et al.*, 2009). We acknowledge that even if the strategic location of fuel treatments projects are in line with local manager experiences and silviculture practices, uncertainties in the the fuel model and weather data input (i.e., wind conditions) contribute to variability in our modeled outputs (Oliveira *et al.*, 2016).

In future work we will conduct sensitivity analyses to test a wide range of scenarios that vary in terms of intensity of area treated (hectares). We also suggest that future studies include social and demographic factors since these play a role in Serra do Socorro fire regime as evidenced by the strong links between population and agricultural activities and the common use of fire irrespective of season (Raínha & Fernandes, 2002). Coupling mitigation efforts for both social and biophysical risk factors in terms of human ignitions and fuel management is likely the most efficient mitigation strategy for wildfire risk management in fire prone eucalypt plantations in Portugal. Moreover, a better understanding of wildfire risk transmission between eucalypt plantations and surrounding communities (Ager *et al.*, 2014b; Haas *et al.*, 2015; Oliveira *et al.*, 2016) will facilitate the development of strategic fuel treatment plans aimed at reducing future wildfire impacts.

Acknowledgements

The study was framed under the thesis work of the Erasmus Mundus master MEDfOR - Mediterranean Forestry and Natural Resources Management (2013/2015). We gratefully acknowledge Dr. Paulo Fernandes (Universidade de Trás-os-Montes e Alto Douro) and Dr. Margarida Tomé (Universidade de Lisboa) for comments on an earlier version of the manusctript, Dr. Paula Soares for the use of the software Globulus to model the time-line growth of the inventory plots, and The Navigator Company for supplying the inventory databases and support in field visits and Filipa Araújo from the Camara Municipal de Torres Vedras for the spatial dataset.

References

Agee JK, Skinner CN, 2005. Basic principles of forest fuel reduction treatments. Forest Ecol Manag 211(1-2): 83–96. http://dx.doi.org/10.1016/j.foreco.2005.01.034.

Ager A, Vaillant N, Finney M, 2010. A comparison of landscape fuel treatment strategies to mitigate wildland fire risk in the urban interface and preserve old forest structure. Forest Ecol Manag 259: 1556–1570. http://dx.doi.org/10.1016/j.foreco.2010.01.032.

Ager A, Vaillant N, Finney M, 2011. Integrating Fire Behavior Models and Geospatial Analysis for Wildland Fire Risk Assessment and Fuel Management Planning. J Combustion 2011: 1–19. http://dx.doi.org/10.1155/2011/572452.

Ager AA, Vaillant NM, Owens DE, Brittain S, Hamann J, 2012. Overview and example application of the Landscape Treatment Designer. Gen. Tech. Rep. PNW-GTR-859, USDA Forest Service, Pacific Northwest Research Station, Portland, OR.

Ager AA, Vaillant NM, McMahan A, 2013. Restoration of fire in managed forests: a model to prioritize landscapes and analyze tradeoffs. Ecosphere 4: 29. http://dx.doi.org/10.1890/ES13-00007.1.

Ager AA, Day MA, McHugh CW, Short K, Gilbertson-Day J, Finney MA, Calkin DE, 2014a. Wildfire exposure and fuel management on western US national forests. J. Environ Manag 145: 54-70. http://dx.doi.org/10.1016/j.jenvman.2014.05.035.

Ager AA, Day MA, Finney MA, Vance-Borland K, Vaillant NM, 2014b. Analyzing the transmission of wildfire exposure on a fire-prone landscape in Oregon, USA. Forest Ecol Manag 334: 377-390. http://dx.doi.org/10.1016/j.foreco.2014.09.017.

Alexander ME, Cruz MG, 2013. Are the applications of wildland fire behaviour models getting ahead of their evaluation again? Environ Modell Softw 41: 65- 71. http://dx.doi.org/10.1016/j.envsoft.2012.11.001.

Botequim B, Garcia-Gonzalo J, Marques S, Ricardo A, Borges JG, Tomé M, Oliveira MM, 2013. Developing wildfire risk probability models for *Eucalyptus globulus* stands in Portugal. iForest 6: 217-227.

Bradstock RA, Cary GJ, Davies I, Lindenmayer DB, Price OF, Williams RJ, 2012. Wildfires, fuel treatment and risk mitigation in Australian eucalypt forests: insights from landscape-scale simulation. J Environ Manag 105: 66–75. http://dx.doi.org/10.1016/j.jenvman.2012.03.050.

Cochrane M, Moran C, Wimberly M, Baer A, Finney M, Beckendorf K, Eidenshink J, Zhu Z, 2012. Estimation of wildfire size and risk changes due to fuels treatments. Int J Wildland Fire 21: 357-367. http://dx.doi.org/10.1071/WF11079.

Cruz MG, 2005. Guia fotográfico para identificação de combustíveis florestais Região Centro Portugal. Centro de Estudos sobre Incêndios Florestais. (p. 38). ADAI, Coimbra, Portugal.

Cruz MG, Viegas DX, 1998. Crown fuel dynamics in bluegum eucalyptus (Eucalyptus globulus Labill.) plantations fuel complex: Implications on extreme fire behavior phenomena. In Proceeding of 3rd International Conference on Forest Fire Research-14th Conference on Fire and Forest Meteorology, Luso-Coimbra, Portugal.16/20 November, pp. 2089–2109.

Cruz MG, Alexander ME, 2010. Assessing crown fire potential in coniferous forests of western North America: a critique of current approaches and recent simulation studies. Int J Wildland Fire 19: 377-398. http://dx.doi.org/10.1071/WF08132.

Duncker P, Barreiro S, Hengeveld GM, Lind T, Mason WL, Ambrozy S, Spiecker H 2012. Classification of forest management approaches: a new methodological framework and its applicability to Europan forestry. Ecology Society 17(4): 51. http://dx.doi.org/10.5751/ES-05262-170451.

Ellis PF, 2000. The Aerodynamic and Combustion Characteristics of eucalypt bark. A Firebrand study. https://digitalcollections.anu.edu.au/handle/1885/49422.

Fernandes P, 2009. Combining forest structure data and fuel modelling to classify fire hazard in Portugal. Ann For Sci 66 (415). http://dx.doi.org/10.1051/forest/2009013.

Fernandes P, Gonçalves H, Loureiro C, Fernandes M, Costa T, Cruz MG, Botelho H, 2009. Modelos de Combustível Florestal para Portugal. In Actas do 6o Congresso Florestal Nacional. SPCF, Lisboa, Portugal. pp. 348–354.

Fernandes PM, Loureiro C, Palheiro P, Vale-Gonçalves H, Fernandes M.M, Cruz MG, 2011. Fuels and fire hazard in blue gum (*Eucalyptus globulus*) stands in Portugal. Boletín del CIDEU 10: 53-61.

Fernandes PM, 2015. Empirical Support for the Use of Prescribed Burning as a Fuel Treatment. Current Forestry Reports 1(2): 118-127. http://dx.doi.org/10.1007/s40725-015-0010-z.

Ferreira L, Constantino MF, Borges G, Garcia-Gonzalo J, 2012. A Stochastic Dynamic Programming Approach to Optimize Short- Rotation Coppice Systems Management Scheduling : An Application to Eucalypt Plantations under Wildfire Risk in Portugal. Forest Science 58(4): 353–365. http://dx.doi.org/10.5849/forsci.10-084.

Finney MA, 2006. An Overview of FlamMap Fire Modeling Capabilities. In Fuel Management-How to Measure Success: Conference Proceedings RMRS-P-41. 28-30 March, Portland, Oregon. pp. 213–220.

Finney MA, Seli RC, Mchugh CW, Ager AA, Bahro B, Agee JK, 2006. Simulation of Long-Term Landscape-Level Fuel Treatment Effects on Large Wildfires. In USDA Forest Service Proceeding RMRS-P-41, pp. 125–147.

Finney MA, Grenfell IC, McHugh CW, Seli RC, Trethewey D, Stratton RD, Brittain S, 2011. A method for ensemble wildland fire simulation. Environ Model Assess 16: 153-167. http://dx.doi.org/10.1007/s10666-010-9241-3.

Gabinete Técnico Florestal, 2015. Plano Municipal de Defesa da Floresta Contra Incêndios (PMDFCI) 2015-2019. Forest Fires Municipal Plan in Torres Vedras, 2015-2019. http://www.cm-tvedras.pt/assets/upload/documentos/2015/10/05/pmdfci_torres_vedras_2015_2019.pdf.

Haas JR, Calkin DE, Thompson MP, 2015. Wildfire risk transmission in the Colorado Front Range, USA Risk Analysis 35(2): 226–240. http://dx.doi.org/10.1111/risa.12270.

ICNF, 2013. IFN6 - Áreas dos usos do solo e das espécies florestais de Portugal continental. Instituto da Conservação da Natureza e das Florestas, Lisboa, Portugal, pp. 34.

Kalabokidis K, Palaiologou P, Finney M, 2013. Fire Behavior Simulation in Mediterranean Forests Using the Minimum Travel Time Algorithm. In Proceeding of 4th Fire Behavior and Fuel Conference (p. 25). Missoula, Montana, USA: Internat Ass Wildland Fire.

Marques S, Garcia-Gonzalo J, Borges JG, Botequim B, Oliveira MM, Tomé J, Tomé M, 2011. Developing post-fire *Eucalyptus globulus* Labill stand damage and tree mortality models for enhanced forest planning in Portugal. Silva Fenn 45(1): 69-83. http://dx.doi.org/10.14214/sf.32.

Marques AL, Pacheco AP, Claro J, Oliveira TM, 2014. FIRE-ENGINE, from empirical studies to system design and policy. A cohesive tool set: aplicação ao município de Torres Vedras. p. 142.

Mateus P, Fernandes PM, 2014, Forest Fires in Portugal: Dynamics, Causes and Policies. In: Reboredo, F. Forest Context and Policies in Portugal, Present and Future Challenges. World Forests, Vol.19. Springer. Pp: 219-236.

Oliveira TM, Barros AM, Ager A, Fernandes P, 2016. Assessing the effect of a fuel break network to reduce burnt area and wildfire risk transmission, Int J Wildfire Fire 25(6): 619-632. http://dx.doi.org/10.1071/WF15146.

Pereira MG, Aranha, J, Amraoui M, 2014. Land cover fire proneness in Europe. Forest Systems, 23(3): 598–610. http://dx.doi.org/10.5424/fs/2014233-06115.

Pinto A, Fernandes PM, Espinosa-Prieto J, Loureiro C, 2013. FIREglobulus: Experimental Study of Fire Behaviour and Effects in Blue Gum Plantations. Silva Lusitana 21, Junho.

Raínha M, Fernandes, PM, 2002. Using the Canadian Fire Weather Index (FWI) in the Natural Park of Montesinho, NE Portugal: calibration and application to fire management.

Salis M, Ager AA, Arca B, Bacciu V, Duce P, Spano D, 2013. Assessing exposure of human and ecological values to

wildfire in Sardinia, Italy. Int J Wildland Fire 22: 549–565. http://dx.doi.org/10.1071/WF11060.
Salis M, Ager A, Finney M, Alcasena F, Arca B, Munoz O, Santoni P, Spano D, 2014. Application of simulation modeling for wildfire risk assessment and management. In VII International Conference on Forest Fire Research. (Ed. D Viegas). http://dx.doi.org/10.14195/978-989-26-0884-6_181.
Scott JH, Reinhardt ED, 2001. Assessing crown fire potential by linking models of surface and crown fire behavior. Research Paper RMRS–29. USDA Forest Service, Rocky Mountain Research Station, Fort Collins, Colorado, USA, p. 59.
Soares P, Tomé M, 2001. A tree crown ratio predicion equation for eucalypt plantations. Ann For Sci 58(2): 193-202. http://dx.doi.org/10.1051/forest:2001118.
Soares P, Tomé M, Pereira JS, 2007. A produtividade do eucaliptal. In: Alves, A.M., Pereira, J.S., Silva, J.M.N. (Eds.), O Eucaliptal em Portugal – Impactes Ambientais e Investigação Científica. ISAPress, Lisbon, pp. 27–60.
Stratton RD, 2006. Guidance on Spatial Wildland Fire Analysis : Models, Tools, and Techniques. Gen. Tech. Rep. RMRS-GTR-183. Fort Collins, CO: U.S. Department of Agriculture, Forest Service, Rocky Mountain Research Station.
Tomé M, Oliveira T, Soares P, 2006. O modelo GLOBULUS 3. 0 dados e equações (p. 23). Universidade Técnica de Lisboa - Instituto Superior de Agronomia Departamento de Engenharia Florestal (Portugal).
Van Wagner CE, 1977. Conditions for the start and spread of crown fire. Can J For Res 7: 23-34. http://dx.doi.org/10.1139/x77-004.
Vogler, KC, Ager AA, Day MA, Jennings M, Bailey JD, 2015. Prioritization of forest restoration projects: tradeoffs between wildfire protection, ecological restoration and economic objectives. Forests, 6(12): 4403-4420. http://dx.doi.org/10.3390/f6124375.
Wei Y, Long Y, 2014. Schedule fuel treatments to fragment high fire hazard fuel patches. MCFNS, 6(1): 1–10.

11

A forest simulation approach using weighted Voronoi diagrams: An application to Mediterranean fir *Abies pinsapo* Boiss stands

Begoña Abellanas[1,*], Manuel Abellanas[2], Arne Pommerening[3], Dolores Lodares[2], Simón Cuadros[1]

[1]*Departamento de Ingeniería Forestal. Universidad de Córdoba. Campus de Rabanales. Córdoba. Spain.* [2] *Dpto. Matemática Aplicada. Facultad de Informática. Universidad Politécnica de Madrid. Campus de Montegancedo s/n. Boadilla del Monte. Spain.* [3]*Swedish University of Agricultural Sciences SLU. Faculty of Forest Sciences. Department of Forest Resource Management. Skogsmarksgränd. Umeå. Sweden*

Abstract

Aim of study: a) To present a new version of the forest simulator Vorest, an individual-based spatially explicit model that uses weighted Voronoi diagrams to simulate the natural dynamics of forest stands with closed canopies. b) To apply the model to the current dynamics of a Grazalema *pinsapo* stand to identify the nature of its competition regime and the stagnation risks it is currently facing.

Area of study: Sierra del Pinar de Grazalema (S Spain)

Material and methods: Two large plots representative of Grazalema *pinsapo* stands were used to fit and validate the model (plus 6 accesory plots to increase the availability of mortality data). Two inventories were carried out in 1998 and 2007 producing tree size and location data. We developed a forest simulator based on three submodels: growth, competition and mortality. The model was fitted, evaluated and validated for Grazalema plots. The simulation outputs were used to infer the expected evolution of structural diversity of forest stands.

Main results: Vorest has proved to be a good tool for simulating dynamics of natural closed stands. The application to Grazalema *pinsapo* stands has allowed assessing the nature of the main processes that are driving its development pathway. We have found that the prevailing size-asymmetric competition dominates the self-thinning process in small-sized trees. At the same time, there is an active tree-size differentiation process.

Research highlights:

- Vorest has proved to be a good tool for simulating natural stands with closed canopies.
- The Grazalema *pinsapo* stand under consideration is currently undergoing a natural process of differentiation, avoiding long-term stagnation.

Keywords: Vorest; stand dynamics; individual-based forest model; spatially explicit forest model; *pinsapo*.

Funding: Spanish National Research Project from the Ministry of Science and Education: MTM2008-05043/MTM. Research contract from Andalusian Government (Junta de Andalucía. Consejería de Medio Ambiente).

Competing interests: The authors have declared that no competing interests exist.

Correspondence should be addressed to Begoña Abellanas: ir1aboab@uco.es

Introduction

Abies pinsapo Boiss belongs to the Mediterranean group of firs that are the most ancient and less evolutionarily developed European species of this genus. Together with the Algerian species *Abies numidica* and the Turkish *Abies cilicica, Abies pinsapo* seems to be one of the first species to differentiate from the hypothesised common tertiary ancestor of all current Mediterranean species of this genus, including the widely spread *Abies alba* (Linares, 2011; Xiang *et al.*, 2009). Some authors such as Alba-Sánchez *et al.*, (2010) and Linares (2011) suggest an ancient segregation of the two Iberian *Abies* species: *A. alba* and *A. pinsapo* that could have remained isolated throughout the whole Quaternary. *Abies pinsapo* has occurred since then in two reduced areas on both sides of the Strait of Gibraltar, i.e. on European and African shores, between 1000 and 1800 m a.s.l. on the northern side, and between 1500 and 2150 m on the southern one. The genetic isolation and the limited area to which these populations have spread, aggravated by the intense habitat fragmentation as a result of ancient human intervention in the Mediterranean basin (Linares, 2011), have caused its current extremely low genetic diversity (Scaltsoyiannes *et al.*, 1999) and its current status as endangered species. The small size of *Abies pinsapo* populations presents a considerable extinction risk. There are only three small areas of *pinsapo* forest in Spain, with a total extension of 3750 ha, from which less than 50% (1758 ha) are actually forests with closed canopies (MAGRAMA, 2015). Grazalema is one of these areas with nearly 500 ha of pure dense *pinsapo* stands that are the object of this study. The consequences of climate change involving an increase of temperature and a decrease of rainfall are likely to further reduce the chance of survival of these unique stands (Linares *et al.*, 2009). Conservation management of these ecosystems requires a thorough understanding of its current dynamics that were heavily influenced by past human disturbances and subsequent conservation attempts (Linares & Carreira, 2009). It is assumed that the regeneration dynamics of pure *Abies* stands mostly rely on disturbance events opening up potential growing space (Morin, 1994). *Abies* dynamics involve long periods of biomass accumulation and increasing stocking (Matic, 2001) followed by eventual disturbance events (mainly pests and diseases) that provide sporadic opportunities for new cohorts of trees (Antos & Parish, 2002). Gap processes and patch dynamics have also been shown to play an important role in shade-tolerant stands that have been undisturbed for a long time, leading to a slower and more continuous turnover of the canopy (McCarthy, 2001; Antos & Parish, 2002). The dynamics of stands in intermediate overstocked periods are hardly understood. Competitive interaction and resulting differentiation and mortality processes are crucial issues for the dynamics and the stand structure at these developmental stages (Gray & He, 2009). Competition and also growth in dense forest stands are highly dependent on resource partitioning among trees. Trees acquire growth resources gathering them from their surrounding space and therefore resource partitioning has much to do with space partitioning.

Voronoi diagrams divide space into cells according to the closest distance to a tree location. For every tree location, its Voronoi region contains the points of the stand area that are closer to it than to any other tree location. The original Voronoi diagrams, developed by Georgy Voronoi in the early 1900s, consider Euclidean space. Since then, many generalizations of the Voronoi diagrams have been developed considering different spaces, different points or different metrics. The book by Okabe *et al.* (2000) provides a good overview of the subject. Our model considers tree locations and the corresponding Voronoi cells are weighted by the size of each tree. We use different distance functions for computing distances between different trees. This way, we can simulate the different ability of neighboring trees to explore the surrounding space and can ensure an adequate partitioning of space among them.

For forest stands with closed canopies, the more size-symmetric the competition the higher the risk of stagnation (Oliver & Larson, 1996), especially if drought episodes become more frequent as it has been hypothesized for *pinsapo* stands in Spain (Linares, 2011). Over the last decades Spanish *pinsapo* stands have undergone a dramatic change in their dynamics due to a significant shift in management, i.e. from intensive timber management to strict conservation. Currently, after more than fifty years of almost total suppression of traditional ways of forest use including firewood, charcoal and livestock grazing, they have reached quite a high level of biomass accumulation (Coca *et al.*, 2001; Linares *et al.*, 2009) and there is no recruitment of small trees. On the other hand there is abundant advance regeneration in the understory waiting for opportunities (Arista, 1995; Cuadros *et al.*, 2003).

Size-symmetric competition in forest stands has been defined as a situation where each individual competes and obtains growth resources in proportion to its size whilst size-asymmetric competition is defined as a situation where larger individuals have a disproportionate competitive effect on smaller individuals leading to the mortality of smaller ones (Weiner, 1990; Hara, 1992; Lundqvist, 1994; Bauer *et al.*, 2004).

Size-asymmetric competition is often associated with even-aged stands and size-symmetric competition with uneven-aged ones (Lundqvist, 1994). In even-aged stands, canopy closure results in the extinction of light below canopy level and therefore in the suppression and death of slightly smaller individuals that remain below this light threshold (Weiner & Damgaard, 2006; Lundqvist, 1994; Oliver & Larson, 1996). However, in uneven-aged forests, the canopy is not totally closed at any height because otherwise there would not be enough light for smaller tree cohorts to develop. As a consequence self-thinning is not limited to small trees, rather it is distributed over the entire size range (Lundqvist, 1994). The nature of competition has much to do with stand structure and mostly with light extinction due to canopy closure. Nevertheless, size-asymmetric competition requires that some previous size differentiation among trees has established before canopy closure (Oliver & Larson, 1996). Otherwise, the competition among equal sized competitor trees would lead to an equal partitioning of resources (Uchmansky, 2000). In a forest stand with a closed canopy, this can lead to a high risk of stagnation, because many trees will approach light starvation as a result of insufficient growing space due to a lack of self-thinning (Oliver & Larson, 1996). This is the risk that threatens most artificially established dense even-aged pure stands. Silvicultural thinning is a way to avoid stagnation in such cases, just doing what nature is not able to do because of extreme homogeneity.

The rapid restocking that Grazalema *pinsapo* stands have undergone in the last decades and the resulting canopy closure achieved poses the question about the nature of the dynamics that they are currently experiencing and their needs for silvicultural treatments. Size differentiation and self-thinning processes may be most instrumental in maintaining the stand in a healthy good condition. An absence of these processes would require silvicultural interventions in order to avoid stagnation and decline.

The objective of this paper is both to present an improved version of the growth model Vorest (Abellanas *et al.*, 2007; Abellanas *et al.*, 2012) that allows to deterministically simulate the natural dynamics of forests stands with closed canopies and to apply it for diagnosing the nature of the dynamics that the studied Grazalema *pinsapo* stand is currently undergoing. Our aim is to identify the nature of the prevailing competition regime and the level of stagnation risk this Grazalema *pinsapo* stand it is potentially confronted with. Our findings inform what silvicultural management is required for conserving these unique stands.

Materials and methods

Study site and experimental layout

The study site is a 460 ha natural stand of *Abies pinsapo*, which occupies the steep slopes (average gradient of 50%) of the northern side of the range *Sierra del Pinar de Grazalema*, located in the south of Spain (36° 46' N; 5° 23' W).

Pinsapo stands typically grow on weathered limestone soils and, due to the exposition of these mountains facing the coast, there are rainfall averages of 2,133 mm/year, being the wettest location in the Iberian Peninsula. Nevertheless Mediterranean climate prevails with a distinct summer drought period (accumulated average rainfall in July and August is less than 10 mm) and high rainfall variability between years of approximately 500 mm per year.

Grazalema is, as a whole, an irregular forest with wide size ranges (in terms of diameter at breast height, or *dbh,* of trees), although the current high overstory density is effectively preventing the recruitment of new cohorts of small trees.

The main monitoring layout includes two 2100m^2-sized rectangular plots, representative of the Grazalema *pinsapo* stand, one of them, plot F, was used to fit and evaluate the model and the other one (plot V) to validate it. Additionally we analyzed mortality data from another six similar plots (M1 to M6) in the same stand. All plots were surveyed in 1998 and again in 2007 so that the data span a period of 9 years (Coca *et al.*, 2001; Cuadros *et al.*, 2003; Abellanas *et al.*, 2005; Cuadros *et al.*, 2005). Table 1 shows the main stand characteristics of experimental plots. Among the variables measured in both surveys, we studied *dbh*, tree locations and mortality. The usefulness of this stand for modelling lies in the absence of silvicultural interventions during the last decades, i.e. the only type of mortality that the stand has undergone is natural mortality.

The increase in mean and maximum stand *dbh* values over nine years reflects the rarity of in-growth and the prevalence of mortality by suppression over ageing or external factors.

Table 1. Stand characteristics of the monitoring plots in the Sierra del Pinar de Grazalema. N98, N07: Number of trees per ha in 1998 and 2007, respectively; G98, G07: Basal area (m^2/ha) in 1998 and 2007, respectively; Dg98, Dg07: Mean quadratic stand diameter at breast height (cm) in 1998 and 2007, respectively; H98, H07: Mean stand height (m) in 1998 and 2007. St. dev. is standard deviation.

Plot	N98	N07	G98	G07	Dg98	Dg07	H98	H07
F	377.8	292.6	32.5	28.2	33.09	35.03	11.31	12.02
V	325.9	255.6	42.5	37.7	40.76	43.32	14.87	16.20
M1	218.5	166.7	37.9	34.3	47.01	51.21	15.01	16.87
M2	555.6	400.0	46.8	36.3	32.75	34.01	13.83	14.92
M3	500.0	392.6	43.5	36.6	33.30	34.46	12.73	13.28
M4	722.2	466.7	40.7	30.8	26.78	28.97	9.56	10.46
M5	577.8	374.1	41.6	35.1	30.28	34.58	12.60	15.10
M6	885.2	551.9	46.8	39.7	25.94	30.26	11.75	14.08
Mean	520.4	362.5	41.5	34.8	33.7	36.5	12.7	14.1
St. dev.	216.5	121.8	4.7	3.7	7.1	7.3	1.9	2.1

VOREST model description

The model description follows a reduced version of the ODD (overview, design concepts and details) standard protocol for describing individual- and agent-based models proposed by Grimm *et al.* (2006, 2010).

Purpose

The purpose of the model is to simulate the current space-time development of natural forest stands with closed canopies and no significant, active recruitment. The main driving forces are growth, competition, and mortality.

Entities, state variables and scales

The model handles two hierarchical levels: The lower level consists of individual trees and the upper level of forest stands (plots). The basic entities are individual trees that are characterized by the attributes and state variables specified in Table 2. The time resolution is one year (one time step of the simulation represents one year), according to the growth rate of trees in temperate zones. The model time horizon is between one and a few decades.

The spatial resolution is one decimeter. The model landscape corresponds to forest plots or stands between 1,000 m^2 and a few hectares. Tree locations are measured in meters (to the nearest decimeter) and *dbh* of trees in centimeters (to the nearest half millimeter).

Process overview and scheduling

Vorest overview

VOREST is a simulation model implemented as a computer program in C++ that consists of three submodels, which correspond to the three basic processes governing the main dynamics of the stands under consideration: growth, competition and mortality.

Table 2. Entities, attributes and state variables of the Vorest model.

Entities	Attributes	State variables
Individual trees	– Coordinates (*x*, *y*) – Identification number (number) – Tree species (species) – Species-specific parameters (growth, PGS and mortality functions)	– Size (*DBH*) – Occupied growing space (OGS) – Available growing space (AGS) – Potential growing space (PGS) – Status (live or dead)

Tree growth is considered in the model in terms of diameter increment (*id*). The model relates tree growth to its capacity to acquire new growing space and estimates it by the area available to access solar radiation, assuming that light is the most limited resource in this stand and therefore the one that dominates the competition among tree neighbors.

When the main forest canopy is completely closed, the growing space is shared between the neighboring trees and typically most trees occupy a smaller growing space than they would do if they grew in the open. This behaviour leads to a corresponding reduction in tree growth compared to an open-grown tree of the same size (*dbh*).

The area that an open grown tree of the same *dbh* can explore for light we refer to as *potential growing space* (PGS) (or *zone of influence)* of a tree and we estimate this area from the crown projection area of an open grown tree with the same *dbh*. Some studies on crown size and diameter growth support the notion that growth follows the acquisition of canopy space (Thorpe *et al.*, 2010).

We define the *available growing space* (AGS) of a tree as the area in its surroundings not occupied by neighboring trees and therefore available to that tree to search for light. Finally, we define *occupied growing space* of a tree (OGS) as the portion of its *available growing space* (as a subset of its *potential growing space*) that the tree actually occupies in terms of crown biomass.

VOREST estimates the *available growing space* (AGS) of each tree by the area of the multiplicatively weighted Voronoi region assigned to it in the Section Competition model (Römisch, 1996; Sterba & Zingg, 2001) and computes the *occupied growing space* (OGS) by limiting the extent of the Voronoi region (AGS) by the range of its *potential growing space* (PGS) or *zone of influence* (radius in short) according to its size (*dbh*).

The Voronoi diagram subdivides space acording to the proximity to a given set of points (Okabe *et al.*, 2000). Our points of interest are tree locations and the Voronoi regions correspond to the space closer to each of them rather than to another tree location. For taking into account size differences between trees, the model uses weighted Voronoi diagrams (Aurenhammer & Edelsbrunner, 1984). In multiplicatively weighted Voronoi diagrams, Euclidean distances are modified by a factor that depends on the tree to which a distance is measured. In our case, the larger the tree, the smaller the factor. In this way, the Voronoi partitioning simulates the fact that larger trees have access to more resources. As potential growing space (PGS) or zone of influence, (maximum spatial range that a tree can explore depending on its size), we consider the area that a tree of the same size would occupy when growing in the absence of competition. The weights used to construct the weighted Voronoi diagrams rely on the relative size of the trees. The zone of influence (PGS) corresponds with the crown projection area of an open-grown tree, which is estimated on the basis of its *dbh* using allometric models from the literature (Krajiceck *et al.*, 1961, Hasenauer, 1997; Ek, 1974; Bella, 1967, Farr *et al.*, 1989, Leech, 1984, Smith *et al.* 1992; Paine & Hann, 1982).

We call *relative area* (*RelArea*) the ratio of the occupied growing space (OGS) of a tree and its potential growing space (PGS) and it is used as a surrogate for the growing capacity of a tree.

Vorest schedule

In each simulation step, the following processes are run in the following order:

1. Synchronous size (*dbh*) updating of all trees according to the defined growing function.
2. Identification of natural neighbors of each tree and updating the weights of each tree.
3. Computing the range of the potential growing space (PGS) of each tree according to its current size.
4. Synchronous simulation of natural mortality.
5. Calculation of the occupied growing space (OGS) for each tree.

Design concepts

a) Basic principles

The basic principles underpinning the model can be summarized as follows: The growth of each tree within the stand is calculated by applying a reduction coefficient to its potential growth to incorporate the effect of competition (Newnham, 1964; Botkin *et al.*, 1972; Pretzsch, 2009, Pommerening *et al.*, 2011). The potential growth of a tree (in analogy to that of an open-grown tree) is estimated by the growth of dominant trees of the stand (Pretzsch *et al.*, 2002; Pommerening *et al.*, 2011). Competition is simulated by the reduction in the occupied growing space (OGS) compared to the potential one (PGS), an effect caused by its neighbors.

Natural mortality occurs after a period of reduced or zero growth, which is related to the size of the tree.

b) Emergence

The model simulates the growth of trees and their resulting size, survival or mortality and the space they occupy. A number of emergent properties, some highly relevant to forest management, are derived from the simulation, such as the spatial structure of the stand, both in terms of tree locations and dimensional aspects, as well as of population characteristics such as size distribution or stocking and the stand self-thinning dynamics.

c) Interaction

The main interaction between trees is competition. The partitioning of growing space between trees is simulated by linking resource availability with physical space. The area actually assigned to a tree can increase or decrease during the simulation, consistently varying its growing rate.

d) Stochasticity

Currently, there is no stochasticity in the model. All simulated processes are deterministic.

Initialization

The input data for the initialization are provided in a text file (ASCII). The file includes the coordinates, species and initial diameter at breast height (*dbh*) of every tree. The header rows of this file also include the initial number of trees and function parameters (growth, mortality and allometric functions). These data are used in each run of the simulation. The length of the simulation period is optional and can be specified by the user during the execution of the program.

Sub models: Description

Growth model

The growth model is based on the potential modifier method (Newnham, 1964; Botkin *et al.*, 1972; Pretzsch, 2009; Pommerening *et al.*, 2011). Potential diameter increment (*dbh*) curves have been derived from dominant trees of the stands. The growth model used to fit potential diameter increment functions is the first derivative of the Chapman-Richards function (Pretzsch *et al.*, 2002; Pommerening *et al.*, 2011; Pommerening & Särkkä, 2013):

$$id_{i,t}^{pot} = A \cdot \left(1 - \exp\left(-k \cdot DBH_{i,t}\right)\right)^{(p-1)} \cdot k \cdot p \cdot \exp\left(-k \cdot DBH_{i,t}\right) \quad [1]$$

Where $id_{i,t}^{pot}$ is the potential diameter increment of tree i at time t; $DBH_{i,t}$ is the diameter at breast height of tree i at time t and A, k, p are model parameters.

Actual increment is simulated as a fraction of this potential increment depending on the competitive status of the tree.

Competition model

Competition is related to the magnitude of the reduction in growing space that a tree undergoes due to the presence of adjacent trees that act as competitors. Vorest considers the "natural neighbors" as competitors of a subject tree as defined by the weighted Voronoi diagrams (see Figure 1), unless the radii of their potential growing space (PGS) do not reach that of the target tree.

As a surrogate for potential growing space (PGS), Vorest incorporates functions published in the literature (Hasenauer, 1997) to estimate the radius of the crown projections of open grown trees as a function of *dbh* for several species. The occupied growing space (OGS) of each tree in each time period is simulated by the weighted Voronoi region that corresponds to it (AGS), limited by the range of its potential growing space, PGS. The weights used to construct the Voronoi diagrams are dependent on the relative sizes of the subject tree and its neighbors as shown in equation [2].

$$w_{i,t} = DBH_{i,t} \cdot \left(\frac{c}{\bar{d}_{1,t}} + \frac{g}{\bar{d}_{2,t}} \right) \qquad [2]$$

where $\bar{d}_{1,t}$ and $\bar{d}_{2,t}$ are the arithmetic mean of the stem diameters of the nearest neighbors and the arithmetic mean stand diameter, respectively. The weight $w_{i,t}$ of a tree i at time t, is dependent on the relative size of the tree i ($DBH_{i,t}$) in relation to the mean *dbh* of its neighbors at time t (local component) and on the *dbh* of tree i relative to the mean stand *dbh* at time t (global component). c and g are scaling parameters satisfying $c + g = 1$, $c >> g$ (we used $c = 0.8$ and $g = 0.2$).

The individual distance function assigned to each tree is continuously adapted at each time step to generate the weighted Voronoi diagrams. This individual distance function is derived from the Euclidean distance divided by the weight assigned to the tree ($w_{i,t}$) according to equation [2]. Thus, there is a positive relationship between the extent of the area of each tree estimated as AGS and its relative size in relation to its neighbors.

At each time step the weighted Voronoi diagrams define two issues: Which are the nearest neighbors of each tree (Figure 1) and the relative load of competition among them. The use of individual distance functions allows simulating the competition signals sent out by

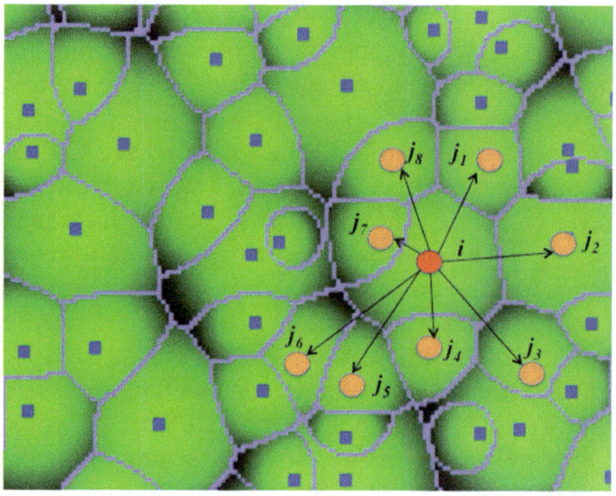

Figure 1. Natural neighbors (j_x) of a tree i at time t according to the weighted Voronoi diagram approach.

trees as an isotropic process. Evidence of directional growth factors could not be confirmed in this study. The competition load of a tree i at time step t is incorporated in the model as the ratio of OGS and PGS:

$$RelArea_{i,t} = \frac{OGS_{i,t}}{PGS_{i,t}} \qquad [3]$$

$OGS_{i,t}$ is the occupied growing space of tree i in year t, computed as the area of the weighted Voronoi region of the tree i restricted by the range of its zone of influence (radius) at time t and $PGS_{i,t}$ is the potential growing space of tree i in year t estimated as the crown projection area of an open grown tree of the same *dbh*.

If all neighbors of a tree are sufficiently far from a target tree (i.e. when the zones of influence (PGS) of every neighbor do not overlap with that of the target tree), the occupied growing space (OGS) of the tree equals the potential one (PGS) and thus relative area (*RelArea*) equals one, indicating an open-grown tree. Vorest uses relative area (*RelArea*) [3] to define the final growth function in equation [4]:

$$id_{i,t} = id_{i,t}^{pot} \cdot \left[1 - \nu \cdot (1 - RelArea_{i,t}) \right] \qquad [4]$$

$id_{i,t}$ is the diameter increment of tree i at time step t and ν is a specific local model parameter.

Mortality model

Natural mortality can be either the consequence of sustained suppression or of ageing. In any case it is the ultimate result of poor growing conditions (Bauer *et al.*, 2004). Vorest implements a memory function for simulating the death of a tree depending on its relative increment of the previous five years (Berger *et al.*, 2004; Murphy & Pommerening, 2010; Pommerening *et al.*, 2011, Pommerening & Särkkä, 2013) in relation to its size (*dbh*) according to equations [5] and [6].

$$pd^5_{i,t} = \frac{DBH_{i,t} - DBH_{i,t-5}}{DBH_{i,t}} \quad [5]$$

$pd^5_{i,t}$ is the relative *dbh* increment of tree *i* during the five year period previous to year *t*; $DBH_{i,t}$ is the *dbh* of tree *i* in year *t* and $DBH_{i,t-5}$ denotes the *dbh* of tree *i* in year $t-5$. Tree *i* dies in year *t* if

$$pd^5_{i,t} \leq \frac{\alpha}{DBH^{\beta}_{i,t}} \quad [6]$$

α and β are model parameters.

Figure 2 illustrates the simulation process of this version of Vorest.

Graphical output

Apart from various estimations, Vorest also provides a graphical simulation output, which shows the spatial distribution of living and dead trees and the growing space occupied and available in the stand. We addressed the spatial edge effect by periodic edge correction methods (Illian *et al.*, 2008) constructing a simulated edge (referred to as plus-sampling, see Figure 3).

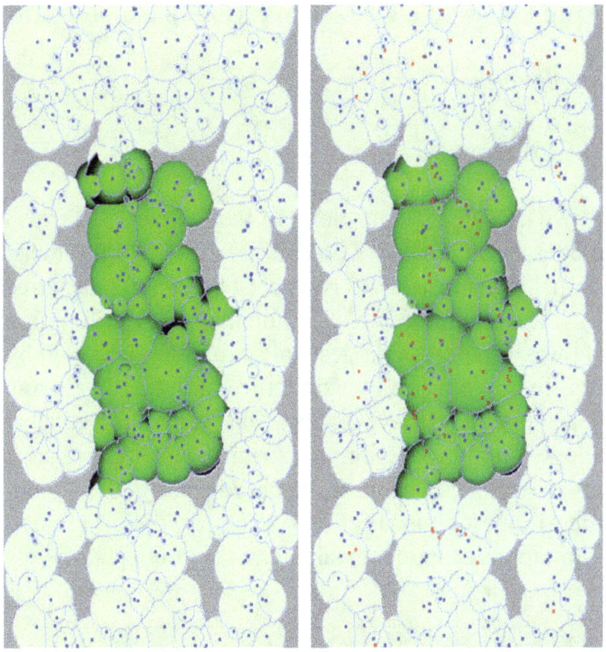

Figure 3. Graphical output of Vorest. Left: Plot F before the start of the simulation, right: Plot F after nine years of simulation. Blue dots are live trees and red dots are dead trees. Green bound regions are occupied growing spaces (OGS) assigned to live trees: Darker green areas for plot trees and pale green areas for constructed edge trees. Interspersed grey regions represent unoccupied growing space.

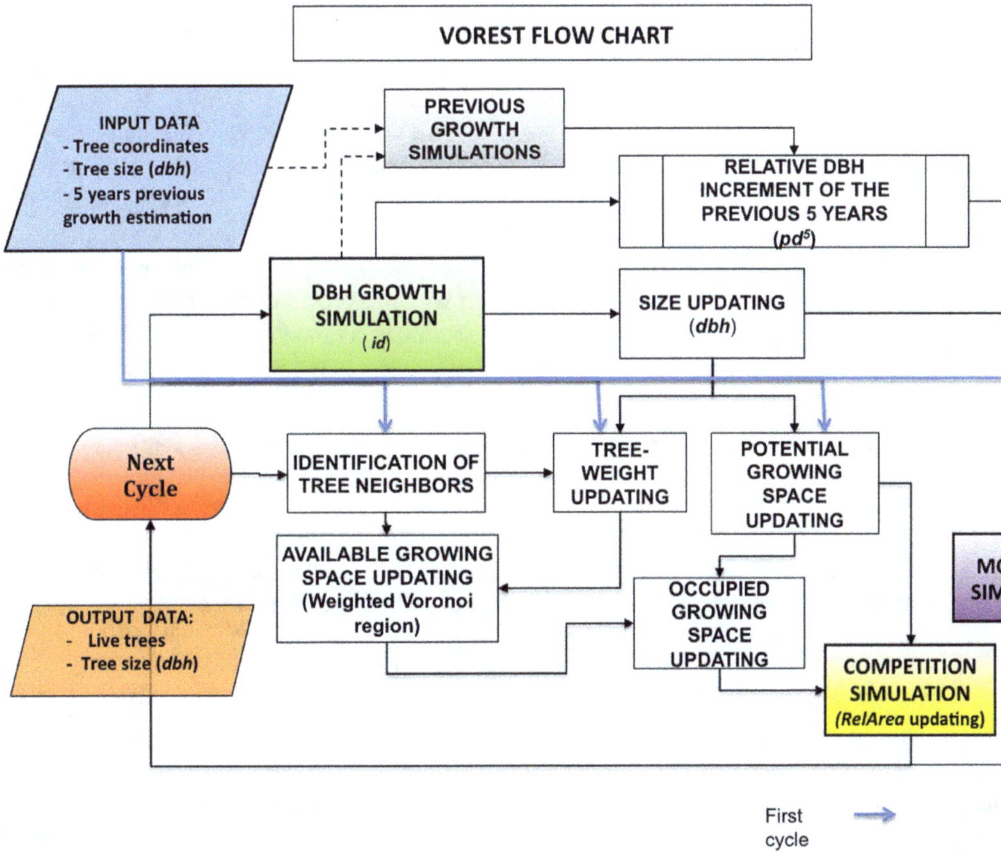

Figure 2. Vorest flowchart for natural stands with closed canopies.

Application to a Grazalema *Abies pinsapo* stand

Model fitting

a) Growth model

We used Grazalema plot F to fit the potential growth model. From measured *dbh* values (1998, 2007), mean annual increment values have been derived for each tree by linear interpolation. The potential growth function [1] was fitted to the largest values of *dbh* increment by nonlinear regression using the statistical package quantreg of R (R Development Core Team, 2011), which allows fitting a function to a specified quantile of a set of data points. In our case, it was adjusted to 0.99 quantile of data points (Figure 4). The regression yielded estimations of model parameters A, k and p [1].

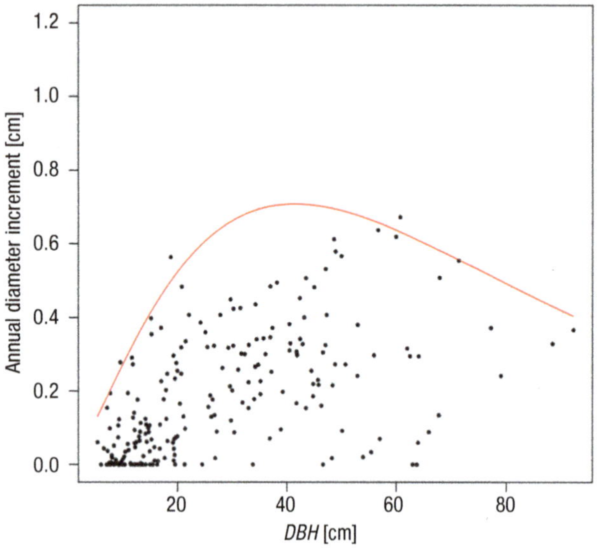

Figure 4. Annual diameter increments of plot F computed by linear interpolation from *dbh* values in 1998 and 2007. The envelope curve represents the potential *dbh* increment, which has been obtained by fitting the first derivative of Chapman-Richards function to percentile 99 of the growth data fitted by the quantreg R package.

b) Competition model

To fit the competition model we needed to estimate initial values of *RelArea* and according to its definition in [3] also of PGS and OGS. To estimate crown radius we used the parameter values proposed by Hasenauer (1997) for fir (Table 3).

To fit the final growth model [4] an auxiliary iteration of Vorest was run in plot F with the sole purpose of computing the initial tree AGS (i.e. the weighted Voronoi regions) and PGS areas that allowed obtaining initial values of OGS, PGS and *RelArea* for all trees but without simulating any growth. With these initial values, the growth model was fitted using nonlinear regression and least-square methods as implemented in the R software.

c) Mortality model

As natural mortality is a sporadic event in forest stands and our time series only spanned nine years, we added mortality data of plots M1 to M6 to those of plot F in order to increase the availability of mortality data.

A threshold value of relative *dbh* increment of the previous five years (pd^5) typical of live trees of the species *Abies pinsapo* has been investigated (Berger *et al.*, 2004; Pommerening *et al.*, 2011; Pommerening & Särkkä, 2013). Figure 5 shows the density functions of pd^5 for dead and live trees.

We can see that pd^5 does not allow to sharply split dead and live trees, even if the average differences are statistically significant (mean $pd^5 = 0.017$ for dead trees and 0.033 for live trees; $p = 0.0069$). This is likely so because firs are very shade tolerant species that can survive a very long time in the shade of an overstory. The fate of these suppressed trees depends on the creation of canopy openings around them, which would allow them to resume active growth. In a previous study of the same plots (Cuadros *et al.*, 2005), a random sample of 32 live trees (trees with more than 1.5 m height and less than 5 cm *dbh*) were

Table 3. Vorest model parameters for *Abies pinsapo* in Grazalema (numbers in square brackets refer to the equations to which the parameters apply. τ is the percentile used to fit the potential growth function; see Figure 4). *SE* – standard error.

Submodel	Function	Parameter	Fitted value
Growth	Potential *dbh* increment (id^{pot}) ($\tau = 0.99$) [1]	A	69.2108 (*SE* = 15.38)
		k	0.022 (*SE* = 0.01)
		p	2.492(*SE* = 0.83)
	dbh increment (*id*) [4]	ν	0.6522 (*SE* = 0.03)
Competition	Range of zone of influence (radius)	$a^{(*)}$	0.092
		$b^{(*)}$	0.538
Mortality	Survival threshold [6]	α	0.18(*SE* = 0.05)
		β	0.80 (*SE* = 0.10)

(*) Hasenauer, 1997 (values for *Abies alba*).

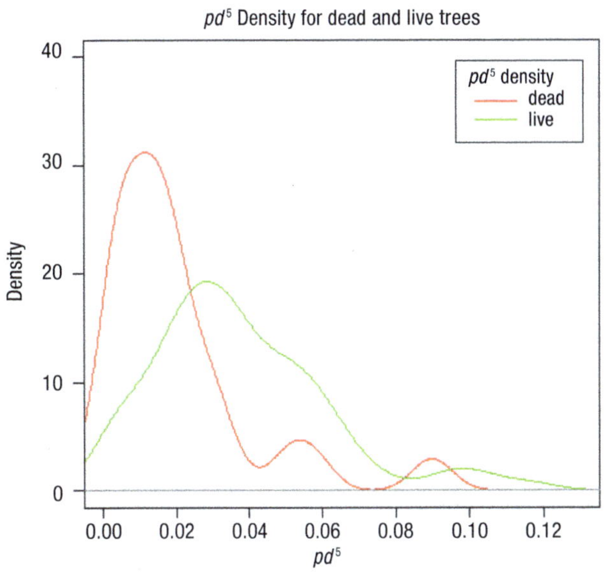

Figure 5. Density functions for relative *dbh* increments of the previous 5 years (pd^5) for live (green line) and dead (red line) trees.

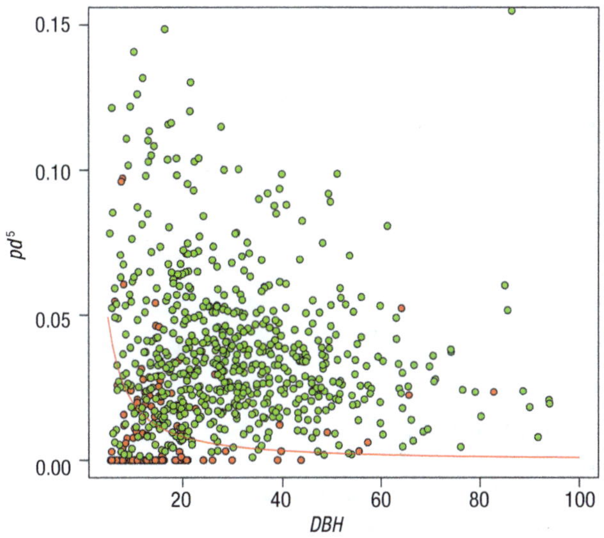

Figure 6. pd^5 (relative *dbh* increment of the previous five years) versus the *dbh* values for live (green dots) and dead (red dots) trees. The red curve is a threshold mortality function fitted to the 0.05 quantile of live trees.

cut and dated by tree-ring counting at the stem base. The average age in this stand is 45 years indicating a mean *dbh* increment of 1.1 mm/year. We estimated a maximum value of pd^5 of these suppressed live trees of 0.022.

We analysed the relationship between pd^5 and *dbh* for live and dead trees. Figure 6 shows that most dead trees are below 20 cm *dbh* and that the lower the *dbh* the higher is the pd^5 value of dead trees. We realized that a fixed value of pd^5 as mortality threshold would not fit well to this condition and consequently defined a decreasing threshold survival function for pd^5 as in [6] producing lower pd^5 mortality values for trees with larger *dbh* values.

We have fitted α and β parameters of this function using again the statistical package quantreg of R based on a quantile of 0.05 of the data of live trees (Figure 6).

Model evaluation and validation

The model performance was analyzed comparing simulated versus actual data both for growth and mortality.

To evaluate and validate growth simulation, goodness of fit has been assessed with the statistics root mean squared error [7] and fit index [8]:

$$RMSE = \sqrt{\frac{\sum_{i=1}^{N}\left(DBH_i - \widehat{DBH_i}\right)^2}{N-p}} \qquad [7]$$

$$FI = 1 - \frac{(N-1)\sum_{i=1}^{N}\left(DBH_i - \widehat{DBH_i}\right)^2}{(N-p)\sum_{i=1}^{N}\left(DBH_i - DBH\right)^2} \qquad [8]$$

The predictive ability of the growth model has been evaluated by statistical estimates of bias (mean error, ME) [9] and precision (mean absolute error, MAE) [10]:

$$ME = \frac{\sum_{i=1}^{N}\left(DBH_i - \widehat{DBH_i}\right)}{N} \qquad [9]$$

$$MAE = \frac{\sum_{i=1}^{N}\left|DBH_i - \widehat{DBH_i}\right|}{N} \qquad [10]$$

DBH_i and $\widehat{DBH_i}$ are the observed and the estimated *dbh* values of tree *i*, respectively, at the end of each simulation period. *DBH* is the mean *dbh* of the plot at the same time. *N* is the number of live trees and *p* is the number of parameters of the model.

The growth model is also assessed graphically by means of plotting expected *versus* observed *dbh*. The same assessment procedure is used for the evaluation and validation. The data of the plot used to fit the model were considered for evaluation (plot F) and data from an external plot for validation (plot V).

To evaluate and validate the mortality model two similar statistics were used: Mean deviation (MD) [11] and mean absolute deviation (MAD) [12] for each simulated period (Nunes *et al.*, 2011):

$$MD = \frac{\sum_{i=1}^{N}(y_i - p_i)}{N} \qquad [11]$$

$$MAD = \frac{\sum_{i=1}^{N}|y_i - p_i|}{N} \qquad [12]$$

y_i is a dummy variable that takes the value 0 if tree i actually dies during the simulated period and takes the value 1 otherwise. p_i is another dummy variable that accounts for the simulated death or survival of trees. Comparing the number of dead trees in the simulation with the actual mortality in the same period is another global assessment of the mortality model.

Simulation of *pinsapo* stand dynamics and structure

In order to answer the initial question about the current stagnation risk of Grazalema *pinsapo* stands, we have projected the development of both plots, F and V, with Vorest 60 years forward. The aim of this simulation was to analyse the current trend of forest structure rather than to project the stand very far into the future. We are particularly interested in size differentiation, since differentiation is a natural strategy that dense forest stands adopt to avoid stagnation (Oliver & Larson, 1996).

From the simulated stands we have derived *dbh* and basal area distributions for consecutive simulation periods and then graphically analyzed the expected stand structure evolution. The expected evolution of stand structural diversity has also been assessed computing the Shannon diversity index (H') [13] for 10cm-diameter classes and for the same simulation periods.

$$H' = -\sum_{i=1}^{R} p_i \cdot \ln p_i \qquad [13]$$

p_i is the proportion of trees belonging to the ith 10 cm-diameter class and R is the number of classes.

Results

Model parameters

Table 3 shows the fitted values obtained for the model parameters.

Graphical output

Figure 3 provides a graphical output of the initial and the final nine-year simulation of plot F.

Model evaluation

The values of the evaluation statistics for the *dbh* estimations are shown in Table 4. Figure 7 shows graphically the goodness of fit for the period of time simulated (9 years) and the actual and simulated *dbh* distributions for plot F. For the mortality evaluation we obtained values of MD = -0.01 and MAD = 0.21. The number of surviving trees at the end of the simulation (nine years) was 80 while the actual number of survivors in 2007 in plot F was 79.

Table 4. Growth model evaluation (plot F). The statistics refer to *dbh* estimates. Definition of statistics explained in the text.

RMSE (cm)	*ME* (cm)	*MAE* (cm)	Fit index
1.69008	-0.34973	1.20346	0.99179

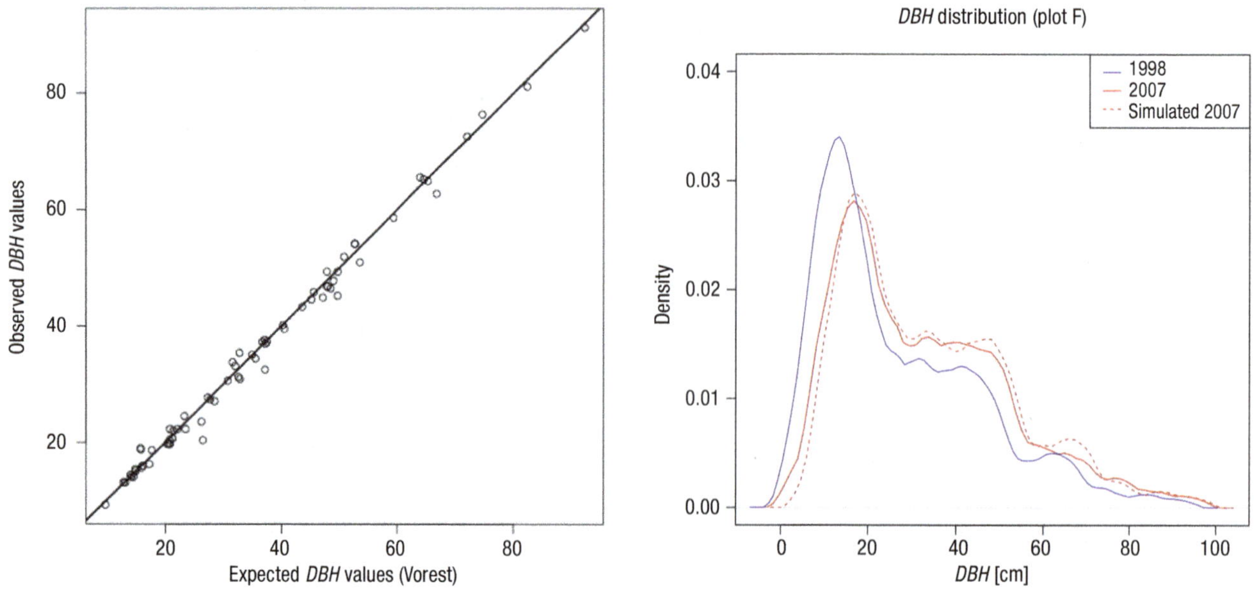

Figure 7. Growth model evaluation (plot F) involving a nine-year simulation period. Observed vs expected *dbh* values of 2007 (left) and actual and simulated *dbh* distributions (right).

Model validation

The data of another plot (plot V), not used to fit any of the submodels, was included in the validation of the model. The plot consisted of 127 trees at the beginning of the simulation in 1998. Table 5 presents the values of growth validation statistics. Figure 8 shows the goodness of fit for plot V and the period of time simulated (9 years) along with the corresponding observed and simulated *dbh* distributions.

For the validation of the mortality model we obtained values of $MD = -0.04$ and $MAD = 0.15$. The number of surviving trees at the end of simulation period (9 years) was 108 while the actual number of survivors in 2007 in plot V was 103.

Forecasting stand structure in Grazalema *pinsapo* stands

Figure 9 shows the expected evolution of diameter distributions and relative basal area distributions by diameter classes (for plots F and V).

The observed trend of the diameter distributions is similar in two ways: There is a marked decrease in the initial peak located around 15 cm, and there is a progressive expansion of the frequencies along an increasing interval towards the upper classes. The initial distributions reflect the presence of two distinct cohorts with a mean *dbh* of about 40 cm, the oldest and less numerous one, and around 15 cm, the youngest and most frequent one. The latter most likely derived from the rapid restocking which was the result of putting the stands under strict protection during the last decades. There are some big trees with *dbh* values greater than 50 cm but their frequencies are quite low. In the basal area distributions (Figure 9, bottom), we can see a trend towards an increasing relative contribution of progressively larger diameter classes to the whole basal area of the stand. The expected evolution of size differentiation has also been assessed with the Shannon index applied to 10cm-diameter classes (Figure 10). The simulation indicates a progressive increase in the Shannon index value involving a subsequent increase in size diversity with time. This tendency points towards an evident differentiation process in the size distribution.

Table 5. Growth model validation (plot V). Statistics refer to *dbh* estimations. 9-year simulation period.

RMSE (cm)	*ME* (cm)	*MAE* (cm)	Fit index
1.39042	0.13950	1.08380	0.99338

Discussion

Vorest has proved a good tool to simulate current Grazalema *pinsapo* dynamics through an appropriate simulation of self-thinning of dominated trees and by continuously re-adjusting the growing capacity of both, previously dominant and newly promoted trees. The growing space allocation among trees using weighted Voronoi diagrams has proved to be a good approach to simulate the mainly size-

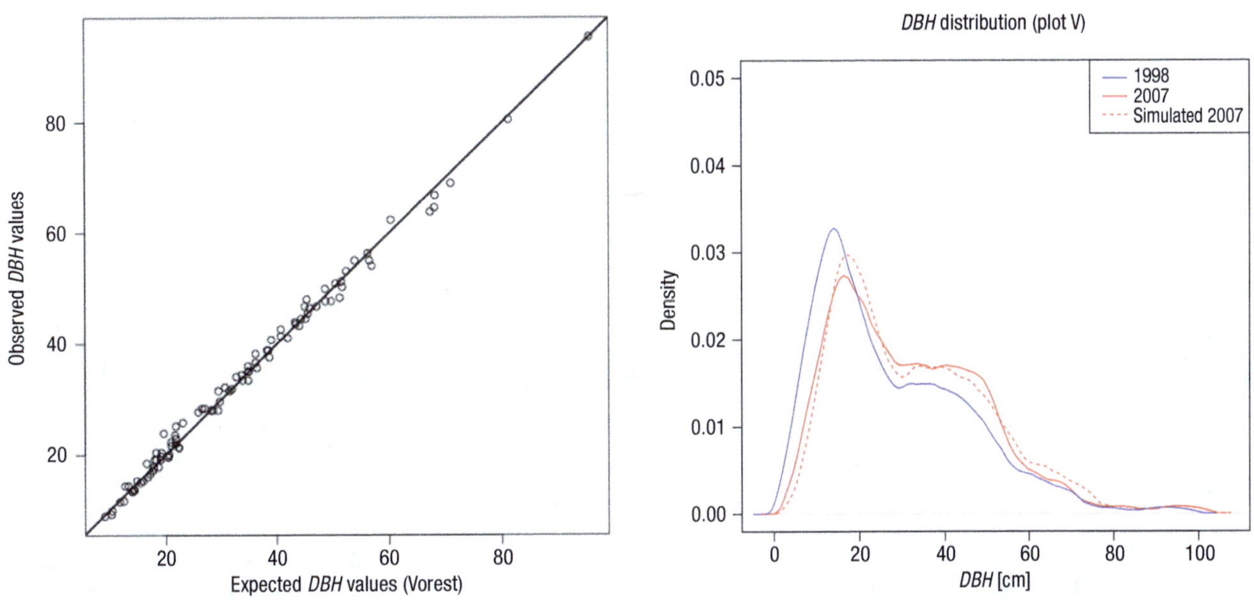

Figure 8. Growth model validation (plot V, nine-year simulation period). Left: Observed vs expected *dbh* values of 2007. Right: Actual and simulated *dbh* distributions.

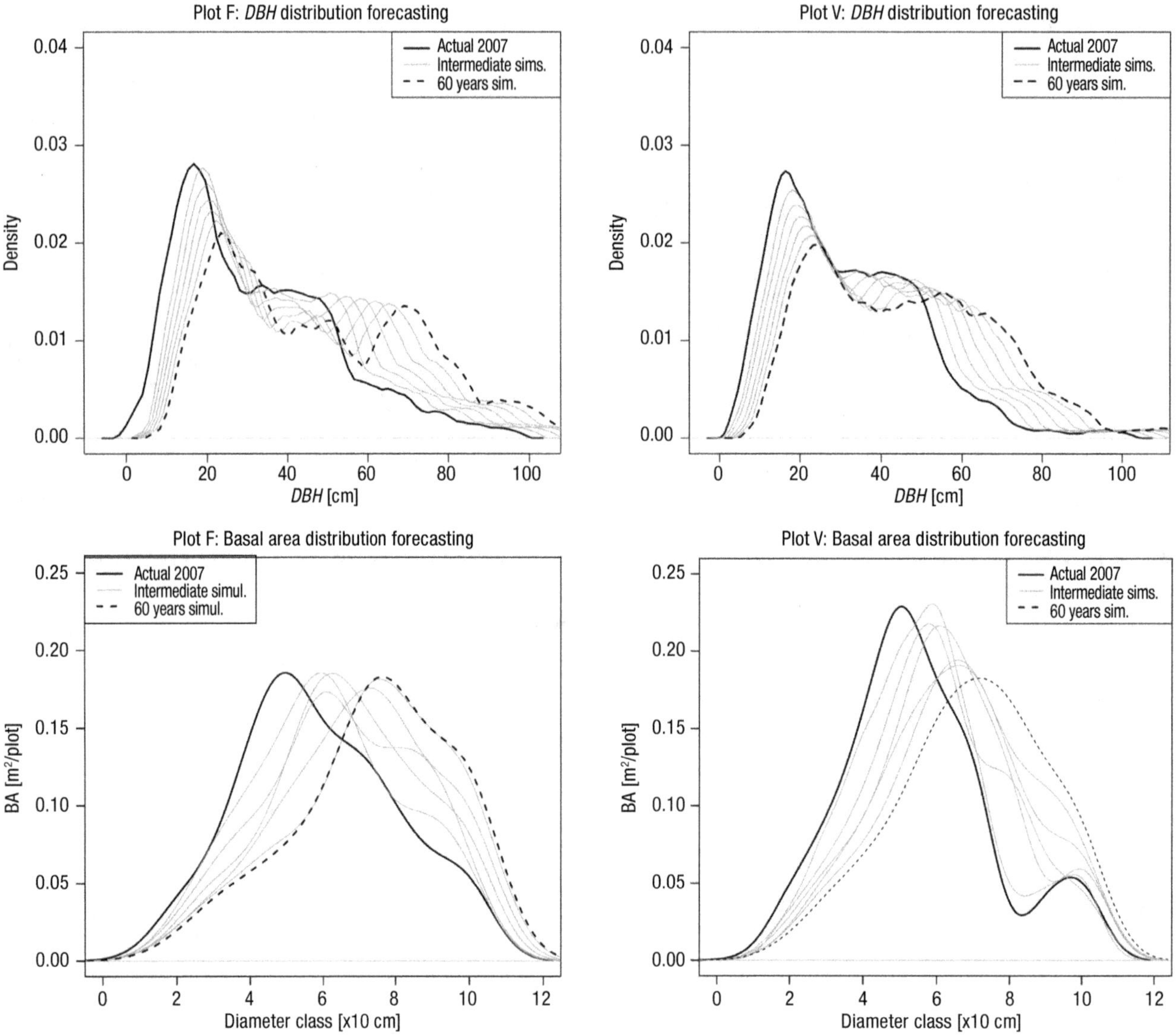

Figure 9. Simulation of *dbh* distributions (top) and *basal area* distributions by *dbh* classes (bottom), 60 year into the future. Simulation using the observed data from 2007 of plots F (left) and V (right).

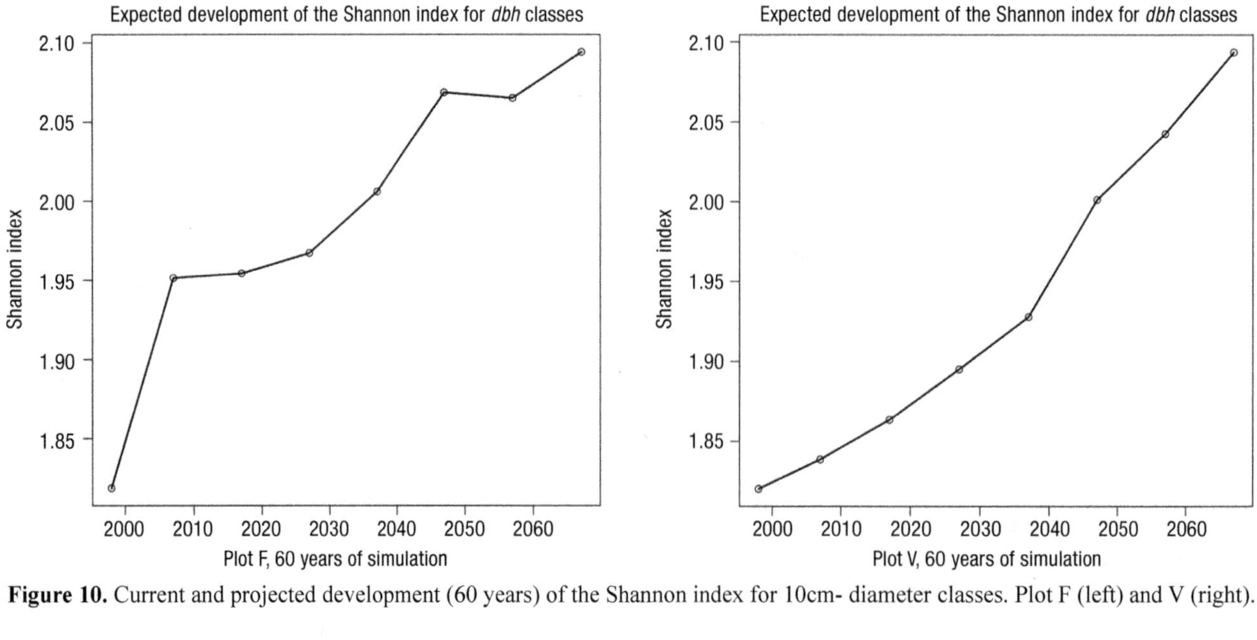

Figure 10. Current and projected development (60 years) of the Shannon index for 10cm- diameter classes. Plot F (left) and V (right).

asymmetric competition that dominates differentiation and self-thinning in the Grazalema *pinsapo* stands.

Nevertheless, the version of Vorest presented in this paper deals with the specific dynamic stage that the Grazalema *pinsapo* stand is currently undergoing based on its history and is not supposed to cover extensive simulation periods. As stated by Antos & Parish (2002), in this kind of forests, variation in dynamics and in the processes controlling changes can be large in the long term.

The model presented here has considered a stand development stage involving a closed canopy with a high level of competition that is currently blocking new recruitment. The aim of the simulated 60 years of stand development was to elucidate the level of stability we can expect in the stand in the near future without the intervention of silvicultural treatments or other kind of disturbances.

The long term dynamics of fir forests are currently much debated in many countries. It is generally assumed that the shade tolerance and the branching pattern provide the genus with a high competitive advantage. *Abies* dynamics involve long periods of biomass accumulation and increasing stocking (Matic, 2001). Both dominated stems and seedlings, grow comparatively slowly and can persist for very long periods until their conditions improve (Antos *et al.*, 2000; Antos & Parish, 2002). Consistent with this statement, we have found very low mortality related growth-rate threshold values for *pinsapo* trees.

Seedling banks play a major role in the regeneration process (Morin & Laprise 1997; Stewart *et al.*, 1991; Woods, 1984; Antos & Parish, 2002). The recruitment of advance regeneration can occur as a result of major disturbances affecting canopy cover as well as of less catastrophic events leading to patch dynamics and gap processes or to even almost subtle endogenous events affecting only one or a few trees (Antos & Parish, 2002).

We have shown that, in our case, high levels of stocking together with high frequencies in lower diameter classes do not cause stagnation and the dynamics lead to an increasing size diversity. In an old growth spruce-fir stand, Antos & Parish (2002) have also found that there is no need for major disturbances to maintain a long-term stability in this kind of forest. In their case study, it probably was due to the low density of the subalpine forest that there was no sign of stagnation features after a long time period without exogenous disturbing events. In our case and for a limited time scale, the observed trend is probably the result of the presence of a few larger trees acting as seed trees.

The presence of one older cohort in the Grazalema *pinsapo* stand, has probably made the increasing size evenness possible, blurring the known effect towards bimodality that is the typical consequence of prevailing size-asymmetric competition (Weiner, 1986; Bauer *et al.*, 2004; Eichhorn, 2010). On the other hand the simulation has suggested a likely decline in mortality rates for the next decades (not shown) that together with the reduction in basal area and number of trees per hectare could indicate a global decrease in competition load.

Pinsapo stands of Grazalema are currently undergoing a process of mortality by suppression caused by a relatively sudden restocking based on the radical change of management imposed a few decades ago by the establishment of strict protection rules. However, our simulation has also indicated a sufficient promotion of some of the youngest cohort trees that progressively colonise the upper diameter classes, thus changing stand biomass distribution towards a more uniform partitioning among the size classes and a simultaneous increase of the weight of intermediate and larger trees. This trend allows expecting an increasing stability of the stand in the near future even in the absence of exogenous disturbances.

Conclusions

Using the present version of Vorest we have found that weighted Voronoi diagrams are a good method of modelling resource partitioning among trees and provide correct feedback loops steering the growth-competition processes in conditions of high stand density.

The simulation with Vorest has shown that there is enough asymmetry in competition relationships among *pinsapo* trees to promote an active self-thinning. But due to the high shade tolerance of firs and likely due to the presence of only few interspersed big trees of previous cohorts, these dynamics have also entered an increasing size differentiation process.

The slow suppression process is typical of shade tolerant species and in our case a comparatively low mortality-growth threshold has been confirmed.

Our simulation suggests a likely decline in competition pressure that probably can allow for new episodes of recruitment to occur, even if no catastrophic regeneration episodes take place.

Our study also revealed that there is currently a low risk of stagnation in the *pinsapo* stands of Grazalema, due to an increase in structural diversity that tends to prosper because of self-thinning and the promotion of younger cohort trees by old seed trees.

As other authors (e.g. Antos & Parish, 2002) have shown, the absence of disturbances not necessarily leads to a stagnation of stands with closed canopies, if size differentiation is encouraged.

It has frequently been demonstrated that tree growth in stands with closed canopies is largely a function of local neighborhood competition and that the nature of this competition is mainly size-asymmetric (Oheimb *et al.*, 2011; Grabarnik & Särkkä, 2011). The nature of competition and not only its intensity has proved an important factor driving forest dynamics and the resulting structure (Toda *et al.*, 2010). As Bauer *et al.* (2004) have shown, it seems that size-asymmetric competition is indeed a general outcome of tree neighborhood interactions.

References

Abellanas B, Coca M, Cuadros S, Oliet J, 2005. Análisis de la diversidad estructural del pinsapar puro en la sierra de grazalema. Influencia sobre la dinámica de la regeneración. Actas del IV Congreso Forestal Español. SECF, 4CFE05-004-T1: 115.

Abellanas B, Abellanas M, Vilas C, 2007. Vorest: Modelización de bosques mediante diagramas de Voronoi. Actas XII Encuentro de Geometría Computacional: 2249-256. Valladolid, España.

Abellanas B, Abellanas M, Pommerening A, Lodares D, 2012. Vorest: un modelo de simulación basado en diagramas de Voronoi. Cuadernos de la Sociedad Española de Ciencias Forestales, 34: 27-37.

Alba-Sánchez F, López-Sáez JA, Benito-De Pando B, Linares JC, Nieto-Lugilde D, López-Merino L, 2010. Past and present potential distribution of the Iberian Abies species: a phytogeografic approach using fossil pollen data and species distribution model. Diversity Distrib 16: 214-218. http://dx.doi.org/10.1111/j.1472-4642.2010.00636.x.

Antos JA, Parish R, Conley K, 2000. Age structure and growth of the tree-seedling bank in subalpine spruce-fir forests of south- central British Columbia. Am Midl Nat 143: 342–354. http://dx.doi.org/10.1674/0003-0031(2000)143[0342:ASAGOT]2.0.CO;2.

Antos JA, Parish R, 2002. Structure and dynamics of a nearly steady-state subalpine forest in south-central British Columbia, Canada. Oecologia 130: 1, 126-135. http://dx.doi.org/10.1007/s004420100787.

Arista M, 1995. The structure and dynamics of an Abies pinsapo forest in Southern Spain. Forest Ecol Manag 74: 81-89. http://dx.doi.org/10.1016/0378-1127(94)03507-S.

Aurenhammer F, Edelsbrunner H, 1984. An optimal algorithm for constructing the weighted Voronoi diagram in the plane. Pattern Recognition 17(2): 251-257. http://dx.doi.org/10.1016/0031-3203(84)90064-5.

Bauer S, Wyszomirski T, Berger U, Hildebrandt H, Grimm V, 2004. Asymmetric competition as a natural outcome of neighbor interactions among plants: results from the field-of_neighborhood modelling approach. Plant Ecol 170: 135–145. http://dx.doi.org/10.1023/B:VEGE.0000019041.42440.ea.

Bella IE, 1967. Crown width/diameter relationships of open-growing jack pine on four site types in Manitoba, Canada. Dep. Forest & Rural Development Research Notes 23(1): 5-6.

Berger U, Hildenbrandt H, Grimm V, 2004. Age-related decline in forest production: modelling the effects of growth limitation, neighbourhood competition and self-thinning. J Ecol 92: 846-853. http://dx.doi.org/10.1111/j.0022-0477.2004.00911.x.

Botkin DB, Janak JF, Wallis JR, 1972. Some ecological consequences of a computer model of forest growth. J Ecol 60: 849-872. http://dx.doi.org/10.2307/2258570.

Coca M, Abellanas B, Cuadros S, Oliet JA, Miguel A, 2001. Caracterización estructural del pinsapar de la Sierra de Grazalema. Actas del III Congreso Forestal Español. Ed. Junta de Andalucía, TRAGSA, SECF: 257-264.

Cuadros S, Oliet JA, Abellanas B, Coca M, Padrón E, 2003. La regeneración en el pinsapar de la Sierra de Grazalema. II: Estructura y dinámica del regenerado consolidado en el pinsapar puro. Cuadernos de la Sociedad Española de Ciencias Forestales 15(2): 27-32.

Cuadros S, Ramírez A, Abellanas B, 2005. Epidometría basada en análisis de imagen y apoyada en un sig. Cuadernos de la Sociedad Española de Ciencias Forestales 19: 85-90.

Eichhorn MP, 2010. Spatial organisation of a bimodal forest stand. J Forest Res 15: 391–397. http://dx.doi.org/10.1007/s10310-010-0200-2.

Ek AR, 1974. Dimensional relationships of forest and open-grown trees in Wisconsin.Univ. Wisconsin Forest Research Notes 181. 7pp.

Farr WA, Demars DJ, Dealy JE, 1989. Height and crown width related to diameter for open-grown western hemlock and Sitka spruce. Can J Forest Res 19: 1203-1207. http://dx.doi.org/10.1139/x89-181.

Grabarnik P, Särkkä A, 2011. Modelling the spatial and space-time structure of foreststands: How to model asymmetric interaction between neighbouring trees. Procedia Environ Sci 7: 62–67. http://dx.doi.org/10.1016/j.proenv.2011.07.012.

Gray L, He F, 2009. Spatial point-pattern analysis for detecting density-dependent competition in a boreal chronosequence of Alberta. Forest Ecol Manag 259: 98-106. http://dx.doi.org/10.1016/j.foreco.2009.09.048.

Grimm V, Berger U, Bastiansen F, Eliassen S, Ginot V, Giske J, Goss-Custard J, Grand T, Heinz SK, Huse G, 2006. A standard protocol for describing individual-based and agent-based models. Ecol Modelling 198: 115-126. http://dx.doi.org/10.1016/j.ecolmodel.2006.04.023.

Grimm V, Berger U, Deangelis DL, Polhill JG, Giske J, Railsback SF, 2010. The ODD protocol: A review and first update. Ecol Modelling 221: 2760-2768. http://dx.doi.org/10.1016/j.ecolmodel.2010.08.019.

Hara T, 1992. Effects of the mode of competition on stationary size distribution in plant populations. Ann Bot 69: 509-513.

Hasenauer H, 1997. Dimensional relationships of open-grown trees in Austria. Forest Ecol Manag 96: 197-206. http://dx.doi.org/10.1016/S0378-1127(97)00057-1.

Illian J, Penttinen A, Stoyan H, Stoyan D, 2008. Statistical Analysis and Modelling of Spatial Point Patterns. John Wiley &Sons, Ltd. West Sussex. 534 pp.

Krajiceck JE, Brinkman KA, Gingrich SF, 1961. Crown competition- a measure of density. Forest Sci 7: 35-42.

Leech JW, 1984. Estimating crown width from diameter at breast height for open-grown radiate pine trees in South Australia. Aust Forest Res 14: 333-337.

Linares JC, Camarero JJ, Carreira JA, 2009. Interacting effects of changes in climate and forest cover on mortality and growth on the southernmost European fir forests. Global Ecol Biogeogr 18: 485–497. http://dx.doi.org/10.1111/j.1466-8238.2009.00465.x.

Linares JC, Carreira JA, 2009. Temperate-like stand dynamics in relict Mediterranean-fir (Abies pinsapo, Boiss.) forests from southern Spain. Ann Forest Sci 66: 610:1-10. http://dx.doi.org/10.1051/forest/2009040.

Linares JC, 2011. Biogeography and evolution of Abies (Pinaceae) in the Mediterranean Basin: the roles of long-term climatic change and glacial refugia. J Biogeogr 38: 619-630. http://dx.doi.org/10.1111/j.1365-2699.2010.02458.x.

Lundqvist L, 1994. Growth and competition in partially cut sub-alpine Norway spruce forests in northern Sweden. Forest Ecol Manag, 65: 115-122. http://dx.doi.org/10.1016/0378-1127(94)90163-5.

MAGRAMA, 2015. 3° Inventario Forestal Nacional de España (IFN3). Ministerio de Agricultura, Alimentación y Medio Ambiente de España. http://www.magrama.gob.es/es/biodiversidad/servicios/banco-datos-naturaleza/informacion-disponible/ifn3.aspx.

Matic S, 2001. Silver fir (Abies alba Mill) in Croatia. Ed: Akademija Sumarskih Znanosti – Hrvatske Sume, p.o. Zagreb. 895 pp.

McCarthy J, 2001. Gap dynamics of forest trees: a review with particular attention to boreal forests. Environ Rev 9, 1: 1-59. http://dx.doi.org/10.1139/a00-012.

Morin H, 1994. Dynamics of balsam fir forests in relation to spruce budworm outbreaks in the Boreal Zone of Quebec. Can J Forest Res 24, 4: 730-741. http://dx.doi.org/10.1139/x94-097.

Murphy ST, Pommerening A, 2010. Modelling the growth of Sitka spruce (Picea sitchensis (Bong)carr.) in Wales using Wenk's model approach. Allg Forst J Ztg, 181(1/2): 35-43.

Newnham RM, 1964. The development of a stand model for Douglas-fir. PhD Thesis. Univ. British Columbia. Vancouver. Canada. 201 pp.

Nunes L, Tomé J, Tomé M, 2011. Prediction of annual tree growth and survival for thinned and unthinned even-aged maritime pine stands in Portugal from data with different time measurement intervals. Forest Ecol Manag 262: 1491-1499. http://dx.doi.org/10.1016/j.foreco.2011.06.050.

Oheimb G, Lang AC, Bruelheide H, David H, Forrester I, Wäsche I, Yu M, Härdtle W, 2011. Individual-tree radial growth in a subtropical broad-leaved forest: The role of local neighbourhood competition. Forest Ecol Manag 261: 499-507. http://dx.doi.org/10.1016/j.foreco.2010.10.035.

Okabe A, Boots B, Sugihara K, Chiu SN, 2000. Spatial Tessellations – Concepts and Applications of Voronoi Diagrams. 2nd edition. John Wiley, 671 pp. http://dx.doi.org/10.1002/9780470317013.

Oliver CD, Larson BC, 1996. Forest Stand Dynamics. Ed: John Wiley and Sons, New York. 544 pp.

Paine DP, Hann DW, 1982. Maximum crown-width equations for southwestern Oregon tree species. School of Forestry Oregon State Univ. Forest Res Paper 46: 1-20.

Pommerening A, LeMay V, Stoyan D, 2011. Model-based analysis of the influence of ecological processes on forest point pattern formation- A case study. Ecol Modelling, 222: 666-678. http://dx.doi.org/10.1016/j.ecolmodel.2010.10.019.

Pommerening A, Särkkä A, 2013. What mark variograms tell about spatial plant interactions. Ecol Modelling, 251: 64-72. http://dx.doi.org/10.1016/j.ecolmodel.2012.12.009.

Pretzsch H, Biber P, Ďurský J, 2002. The single tree-based stand simulator SILVA: construction, application and evaluation. Forest Ecol Manag, 162(1): 3-21. http://dx.doi.org/10.1016/S0378-1127(02)00047-6.

Pretzsch H, 2009. Forest dynamics, growth and yield. From measurement to model. Spriger-Verlag. Berlin. 664 pp.

R Development Core Team, 2011. R: A language and environment for statistical computing. R Foundation for Statistical Computing, Vienna, Austria. http://www.R-project.org/.

Römisch K, 1996. Durchmesserwachstum und ebene Bestandesstruktur am Beispiel der Kiefernversuchsfläche Markersbach. In: Hempel, G. (Ed.), 8. Tagung der Sektion Forstliche Biometrie und Informatik, pp. 84–103. Biotechnische Fakultät, Abteilung Forstwirtschaft, Ljubljana, Slovenia.

Scaltsoyiannes A, Tsaktsira M, Drouzas AD, 1999. Allozyme differentiation in the Mediterranean firs (Abies, Pinaceae).A first comparative study with phylogenetic implications. Plant Syst Evol, 216(3/4): 289-307. http://dx.doi.org/10.1007/BF01084404.

Smith WR, Farrar RM, Murphy RA, Yeiser JL, Meldahl RS, Kush JS, 1992. Crown and basal area relationships of open-grown southern pines for modeling competition and growth. Can J Forest Res 22: 341-347. http://dx.doi.org/10.1139/x92-044.

Sterba H, Zingg A, 2001. Target diameter harvesting- a strategy to convert even-aged forests. Forest Ecol Manag 152(1-3): 95-105. http://dx.doi.org/10.1016/S0378-1127(00)00700-3.

Stewart GH, Rose AB, Veblen TT, 1991. Forest development in canopy gaps in old-growth beech (Nothofagus) forests, New Zealand. J Veg Sci 2: 679–690. http://dx.doi.org/10.2307/3236178.

Thorpe SC, Astrup R, Trowbridge A, Coates KD, 2010. Competition and tree crowns: A neighborhood analysis of three boreal tree species. Forest Ecol Manag 259: 1586–1596. http://dx.doi.org/10.1016/j.foreco.2010.01.035.

Toda M, Yokozawa M, Emori S, Hara T, 2010. More asymmetric tree competition brings about more evapotranspiration and less run off from the forest ecosystems: A simulation study. Ecol Modelling 221: 2887-2898. http://dx.doi.org/10.1016/j.ecolmodel.2010.08.025.

Uchmansky J, 2000. Resource partitioning among competing individuals and population persistence: an individual-based model. Ecol Modelling 131:21–32. http://dx.doi.org/10.1016/S0304-3800(00)00242-8.

Weiner J, 1986. How competition for light and nutrients affects size variability in Ipomoea tricolor populations. Ecology, 67: 1425–1427. http://dx.doi.org/10.2307/1938699.

Weiner J, 1990. Asymmetric competition in plant populations. Trends Ecol Evol 5: 360-364. http://dx.doi.org/10.1016/0169-5347(90)90095-U.

Weiner J, Damgaard C, 2006. Size-asymmetric competition and size-asymmetric growth in a spatially explicit zone-of-influence model of plant competition. Ecol Res 1: 707–712. http://dx.doi.org/10.1007/s11284-006-0178-6.

Woods KD, 1984. Patterns of tree replacement: canopy effects on understory pattern in hemlock – northern hardwood forests. Vegetatio 56: 87–107.

Xiang QP, Xiang QY, Guo YY, Zhang XC, 2009. Phylogeny of Abies (Pinaceae) inferred from nrITS sequence data. Taxon 58(1): 141-152.

12

Rapid discrimination of wood species from native forest and plantations using near infrared spectroscopy

Fernanda M. G. Ramalho[1], Jéssica M. Andrade[2] and Paulo R. G. Hein[1]

[1]*Universidade Federal de Lavras, Dept. de Ciências Florestais, 37200-000, Lavras, Brazil.* [2]*Universidade Federal de Lavras, Dept. de Engenharia, 37200-000, Lavras, Brazil.*

Abstract

Aim of study: To verify how well near infrared (NIR) spectroscopy is able to discriminate wood specimens from natural and planted forests. This study was carried out using tropical trees from Brazil.

Area of study: Wood specimens coming from Lavras (21°10′S, 44°54′W), Paraopeba (19°16′S, 44°24′W) and Belo Oriente (19°17′S, 42°23′W) cities, Minas Gerais state, southeastern Brazil were insvetigated.

Material and methods: NIR spectra were recorded in the radial surface of wood specimens of four native species (*Cedrela* sp., *Apuleia* sp., *Aspidosperma* sp. and *Jacaranda* sp.) and two commercial clones (*Eucalyptus* for bioenergy and pulp & paper).

Main results: The principal component analysis (PCA) of spectral information revealed that it is possible to distinguish wood from planted and native forests. The dispersion of scores in the graphic formed by the first and second principal component formed two groups allowing differentiating very clearly the Eucalyptus clones from the native woods. The partial least squares discriminant analysis (PLS-DA) allowed the prediction of group of species with a high degree of correct classification. The PLS-DA models performed from untreated NIR spectra obtained 86 to 100% accuracy for the natural wood species.

Research highlights: From PLS-DA of treated NIR spectra, no *Eucalyptus* wood sample was classified as a natural forest species and vice versa. NIR technique associated with multivariate statistics are promising to discriminate wood specimens from native or planted forests and thus identify frauds.

Additional keywords: illegal logging; forest exploitation; wood identification; timber classification.

Abbreviations used: 1d (First Derivative); NIR (Near Infrared); PCA (Principal Component Analyses); PLS-DA (Partial Least Squares Discriminant Analyses); PLS-R (Partial Least Squares Regression); RER (Range Error Ratie); RMSE (Root Mean Square Error); RPD (Ratio of Performance to Deviation); SNV (Standard Normal Variate).

Authors′ contributions: FMGR: Elaboration of the manuscript, statistical analysis and acquisition of the spectra. JMA: Acquisition of spectra. PRGH: Coordination of the research project, preparation of the manuscript and statistical analysis. All authors read and approved the final manuscript.

Funding: CNPq (National Council for Scientific and Technological Development, Brazil, grant number 405085/2016-8); CAPES (Higher Education Personnel Improvement Coordination, Brazil); and FAPEMIG (Foundation for Research Support of the State of Minas Gerais, Brazil, grant number APQ-00509-14). FMG Ramalho was supported by CAPES grants while PRG Hein was supported by CNPq (process no. 303675/2017-9) grants.

Competing interests: The authors have declared that no competing interests exist.

Correspondence should be addressed to Paulo R. G. Hein: paulo.hein@dcf.ufla.br

Introduction

Deforestation and illegal logging in natural forests are a global concern because of the threats they pose to the rich biodiversity of some regions and because of their contributions towards climate change (Brito & Barreto, 2006). In Brazil, despite the enforcement against illegal logging has improved, the deforestation is still increasing. One of the great challenges is to identify illegal from legal wood. According to Yang *et al.* (2015) the traditional methods for wood identification are time consuming, laborious and require expertise because it involves physical, anatomical, and visual aspects of wood species.

Discriminate natural and planted forests is to perceive differences between them and to classify them in function of these differences. The non-destructive and fast separation of wood species could improve the control in terms of monitoring of forest exploitation and wood transportation. Near infrared (NIR) spectroscopy has been successfully used for the nondestructive measurement of organic materials such as agricultural products or foods (Tsuchikawa, 2007). NIR spectroscopy with aid of multivariate statistics and computational systems is useful not only for quantitative but also for qualitative applications, including classification of wood materials (Tsuchikawa & Kobori, 2015).

Many studies have shown that it is feasible to identify wood species by NIR spectroscopy coupled with multivariate analysis proposing a new method to rapidly identify or classify wood samples. Promising classifications based on NIR spectra recorded fromdifferent wood samples of the same genera were reported in literature for *Eucalyptus* (Michell & Schimleck, 1998), *Pinus* (Espinoza *et al.*, 2012), *Larix* (Gierlinger *et al.*, 2004) and *Quercus* (Adedipe *et al.*, 2008). In *Eucalyptus*, Schimleck *et al.* (1996) recorded NIR spectra from wood meals and discriminated woods samples by PCA (Principal Component Analyses), showing that it was possible to differentiate wood from different *Eucalyptus* species (*E. globulus, E. grandis* and *E. nitens*) and provenances (Victoria and Tasmania), and woods samples from the same species of *Eucalyptus* on different sites. Few studies presented models for classification of wood from different genera. For instance, Brunner *et al.* (1996) classified twelve different species of wood according to their NIR spectra suggesting that NIR would be able to distinguish wood samples of the same species but from different origins. Russ *et al.* (2009) have successfully classified seven hardwood species by PCA calculated from their NIR spectra recorded from wet and dried chips. Yang *et al.* (2012) applied the technique in eight rosewood samples and clearly classified them in eight categories using three-dimensional PCA scores plot. These studies have shown promising findings in regard to the ability of NIR spectroscopy associated with multivariate analysis for classifying wood samples.

This approach could be useful in situations where wood species are visually and anatomically similar and can be improperly commercialized. In regard to this issue, Pastore *et al.* (2011) have investigated four native rainfall forest tree species by NIR spectroscopy. They reported interesting classifications for four native species and showed that it is possible to perform safe discriminations of *Swietenia* (mahogany), *Carapa* (andiroba or crabwood), *Cedrela* (cedar) and *Micropholis* (Curupixá) woods with a very low probability of misclassifying a sample. Cooper *et al.* (2011) used the same approach for differentiating certain wood species groups. They successfully distinguished true firs (balsam fir (*Abies balsamea* (L.) Mill.) and sub alpine fir (*Abies lasiocarpa* (Hook.) Nutt.)) from pine and spruce in eastern and western spruce-pine-fir, respectively. Yang *et al.* (2015) also applied the same approach and successfully discriminated three wood genera (*Pometia, Instia* and *Couratari*) from Beijing wood market.

Here, wood species from native forest and commercial plantations were investigated in order to verify how well NIR spectroscopy is able to discriminate wood specimens from natural and planted forests. This technology would be useful as a tool to be used by control agents for supervising and monitoring illegal logging and deforestation.

Material and methods

Biological material and specimens preparation

Wood specimens from natural forest (cerrado biome) and commercial plantations typically found in Brazil were used in this study. The wood specimens from natural forests belongs to the following genera: *Cedrela* sp. (Cedar, C), *Apuleia* sp. (Garapa, T), *Aspidosperma* sp. (Peroba, P) and *Jacaranda* sp. (Rosewood, J). The natural forest is located at Lavras municipality, Minas Gerais state, southeastern Brazil (21°10′S, 44°54′W).

On the other hand, *Eucalyptus* wood represented the planted forests, the main genera cultivated for industrial application in Brazil. In order to represent the large variability in terms of genetic material used in reforestation in Brazil, *Eucalyptus* clones commercially used by two forestry companies were selected: i) Vallourec (V&M do Brasil) that utilizes hybrids of *Eucalyptus grandis* × *E. urophylla* at 6.5 years old located at Paraopeba city (19°16′S, 44°24′W) with focus on charcoal for bioenergy purposes and ii) Cenibra Nipo Brasileira S.A. that utilizes hybrids of *Eucalyptus grandis* × *E. urophylla* at 6 years old located at Belo Oriente (19°17′S, 42°23′W) state of Minas Gerais, Brazil, for cellulosic pulp production.

The total wood sampling was composed by six vegetal materials: 4 genera from native forest and 2 different types of *Eucalyptus*. The wood specimens used in this study came from the same experiment described in Ramalho *et al.* (2017) who investigated the charcoal produced from these wood specimens.

Ninety wood specimens with nominal dimensions of 35 mm × 35 mm × 100 mm (R × T × L) were removed from the first log of felled trees of each vegetable material, except for *Cedrela, Aspidosperma* and *Apuleia* genera that was represented by 45, 88 and 87 specimens, respectively. The total sampling was composed by 490 clear specimens [(90 · 3) + 45 + 88 + 87)] free of defects and well defined tangential-longitudinal and radial-longitudinal surfaces.

Specimens were conditioned in a room set for 20°C and 65% relative humidity to maintain the equilibrium moisture content (EMC) of ~14% prior to NIR spectra recordings.

NIR spectra acquisition

The near-infrared spectra were acquired using the Bruker spectrometer (model MPA, Bruker Optik GmbH, Ettlingen, Germany) based on a Fourrier transform and equipped with an integrating sphere. Spectral acquisition was performed within the 12,500 - 3,600 cm^{-1} with a spectral resolution of 8 cm^{-1} in diffuse reflection mode. Each spectrum represented the average of 32 scans. NIR spectra were recorded in the center of the two sides of the longitudinal-radial surfaces of each wood samples and averaged. Previous papers have indicated that the radial surface is more suitable for recording NIR spectra because is more representative (*e.g.* Hein *et al.*, 2009; Costa *et al.*, 2018). First derivatives (13-point filter and a second order polynomial) using the Savitzky & Golay (1964) algorithm and Standard Normal Variate (SNV) transformations were applied on the NIR spectra to enhance the quality of the information and analyses.

Multivariate statistical analysis

PCA were calculated for compressing the main information in a set of variables into a lower number of new variables by means of the Chemoface (Nunes *et al.*, 2012, v. 1.61, http://ufla.br/chemoface/) software. The PCA were calculated using full model size and maximum of eight principal components. For cross-validations 8 segments of 60 samples per segment were adopted while for test set validation were done using 2/3 of samples in calibration set and the remaining 1/3 in test set. Martens' uncertainty test (Westad & Martens, 2000) was used to select the wavenumbers with significant x-loadings. Outliers were identified from the analyses of residuals x-variances and leverage value plot. The number of latent variables (LVs) adopted for each model corresponded to the first minimal residual x-variance.

Partial least squares discriminant analyses (PLS-DA) were carried out for developing models for classifying wood species. The PLS-DA classification method, which is based on the PLS-R (partial least squares regression) approach, correlated two blocks of variables, X (independent variables) and Y (dependent variables). The independent variables (X) were the NIR spectra matrix while the dependent variables (Y) were the classes of wood species. For evaluating PLS-R regressions, the root mean square error (RMSE), ratio of performance to deviation (RPD) and range error ratio (RER) were used. As described in Costa *et al.* (2018), the RMSE measures the efficiency of the calibration model to predict the property of interest in many unknown sample, while the RPD is a way to identify the accuracy of the calibration and is the ratio between the standard deviation of the reference values and RMSE. The RER is the ratio between range of reference data and standard error of prediction. According the AACC (1999) any model that has a RER $\geq$ 4 is qualified for screening calibration. When RER $\geq$ 10 the model is acceptable for quality control, and if RER $\geq$ 15 the model is very good for research quantification.

The samples were grouped in six different classes: EC (*Eucalyptus* clones commercially used by Cenibra Nipo Brasileira S.A.), EV (*Eucalyptus* clones commercially used by Vallourec do Brasil), C (*Cedrela* sp. from natural forest), J (*Jacaranda* sp. from natural forest), P (*Aspidosperma* sp. from natural forest) and T (*Apuleia* sp. from natural forest). The classification of the wood specimens into classes of vegetal material was carried out through PLS-DA and the percentages of correct predictions were analyzed. The PLS-DA models were validated using cross- and test set validation methods. The cross-validation was carried out using eight segments while for the independent test set validation 2/3 of the samples were used in calibration set and 1/3 in validation set.

Results

NIR spectral signatures recorded from wood of natural forest and commercial plantations are shown in Fig. 1. Each NIR spectrum represents the average of NIR spectra taken from several wood specimens. In short, it is not possible to distinguish materials from their NIR signature because no interpretation can be properly done from such visual analysis (Fig. 1). Thus, multivariate statistical analysis was carried out for distinguishing vegetal material from their NIR spectra.

PCA were carried out from untreated and mathematically treated NIR spectra in order to reduce and concentrate the information in few principal components (PCs). Table 1 displays the cumulative explained variance taken into account for each PC. For instance, the first PC of the untreated NIR spectra is sufficient for explaining 98.08% of the variance between specimens. The scatter plot of the PC 1 and PC 2 from the PCA of the untreated NIR spectra is shown in Fig. 2A. Fig. 2B shows the first and second PC of the NIR-based scores after normalization (SNV) followed by first derivative of NIR spectra recorded from natural and planted wood specimens. From this analysis, two *Eucalyptus* types formed a clear group while wood

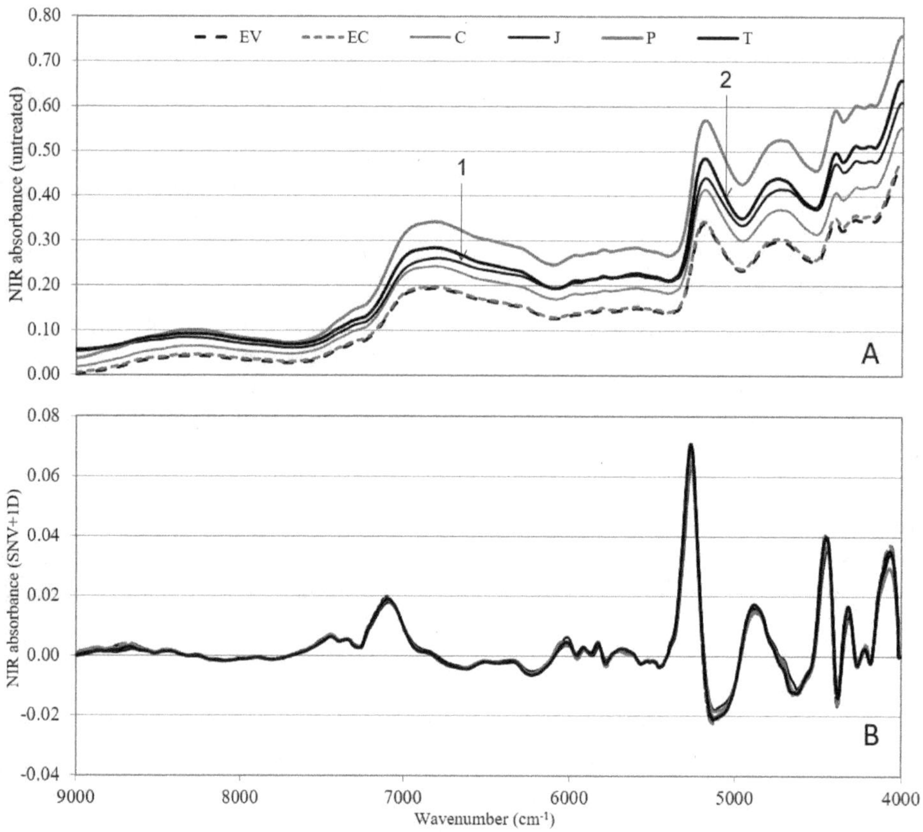

Figure 1. Original (A) and treated (B) NIR spectra recorded from wood of native forest trees (for abbreviations see text) and commercial plantations.

specimens from natural forests were graphically organized independently.

As in PCA approaches, PLS-DA were performed from treated NIR spectrum to verify is even better classifications could be achieved. The statistics associated to the preliminary PLS-R used in the following PLS-DA models are presented in Table 2, and the general graph of the regression coefficients is shown in Fig. 3. These preliminary PLS-R models developed for estimating continuous values in each of six categories (EC, EV, C, J, T, P) taking into account the reference data 1 (for specimen belonging to the correct category) and 0 (for specimen belonging to another category).

Table 1. Explained X-variance (in %) for each principal component by treatment applied in NIR spectra for discriminating wood from native and plantations tree species.

	Untreated	1d	SNV	SNV+1d
PC 1	98.08	69.28	70.38	63.113
PC 2	99.49	87.96	92.01	76.485
PC 3	99.75	91.28	96.41	85.116
PC 4	99.92	94.80	98.26	91.461
PC 5	99.95	95.94	98.73	94.104

1d: first derivative. SNV: standard normal variate.

The correct classifications of native species using the untreated spectra (for both cross-validation and test set) obtained values above 99%, except for the species named T (Table 3). Based on NIR spectra treated by SNV and first derivative (1d) method the PLS-DA yielded classification 100% corrects for all native species both for cross- and test set validation (Table 4).

Discussion

NIR spectra signature

Fig. 1 displays NIR spectra signatures taken from 9,000 to 4,000 cm^{-1} on the radial surface of wood specimens. According to Brereton (2003), this absorption bands correspond to the specific functional groups present in wood samples. Most band assignments for wood has been presented by Schwanninger *et al.* (2011), in which they described band locations in wavenumber, the component likely to absorb at this band location and also the bond vibration.

Here, after treatment of the spectra using standard normal variate and first derivative (SNV + 1d), the existence of the absorption bands became more evident at certain wavenumbers (Fig. 1B), although the peaks

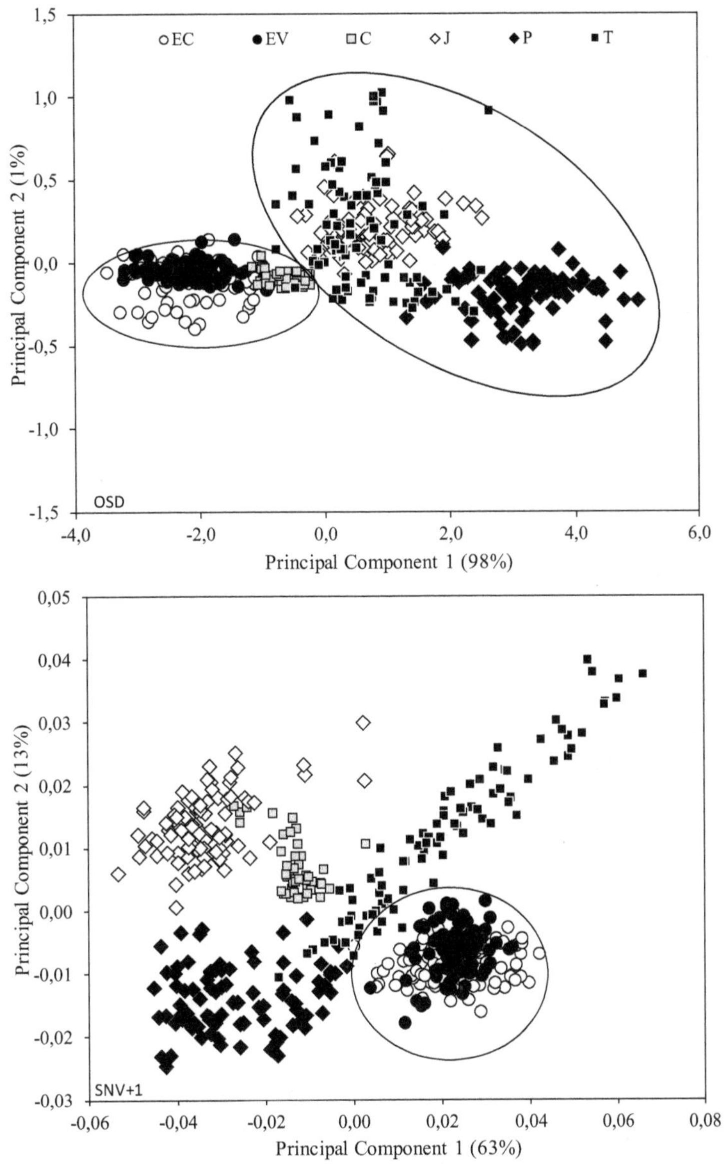

Figure 2. Two-dimensional scatter plot of the first and second PC of the PCA scores of untreated NIR spectra (Model OSD, original spectra data) (A) and after normalization (SNV) and first derivative (1d) of NIR spectra recorded from the surface of native and planted wood samples (Model SNV+1d) (B).

are inflection on raw spectra. From a visual analysis, small differences between NIR spectra of native woods can be observed in terms of intensity of the absorption bands at certain wavenumbers (indicated by narrows). For instance, line J (black fine line) decreases less intensely than lines T, P and C (native woods) in the zones indicates by the narrows. However, no difference can be detected between NIR spectra measured from *Eucalyptus* wood (dotted lines).

Principal component analyses (PCA)

An exploratory analysis using NIR spectroscopy is useful and important for wood science and industry as well as quantitative analysis (Tsuchikawa, 2007). PCA was performed with the treated and untreated spectra. The mathematical treatment did not improve the results in terms of increasing the explained variance (Table 1) but from a visual analysis of the score plots, treatment on NIR seems to clarify the separation of different types of wood.

It was possible to identify two wood groups with the PCA of the untreated NIR spectra (Fig. 2A). These clusters were formed because each vegetal material presents different characteristics and properties, such as the chemical composition and anatomical features (Kollmann & Coté, 1968). Thus, the two types of *Eucalyptus* wood (EC and EV) clearly could be

Table 2. Statistics associated to the preliminary PLS-R cross-validations and test set validations for the PLS-DA classifications without treatment.

Model	R^2cv	RMSECV	R^2p	RMSEP	RPD	RER
EC	0.457	0.285	0.468	0.279	1.04	3.58
EV	0.491	0.276	0.415	0.287	1.35	3.48
C	0.862	0.141	0.881	0.139	2.79	7.19
J	0.949	0.088	0.945	0.082	4.73	12.20
P	0.913	0.114	0.900	0.119	3.23	8.40
T	0.862	0.142	0.839	0.154	2.48	6.49

Model abbreviations: see text. R^2cv, R^2p: coefficient of determination for cross and test set validation, respectively. RMSECV, RMSEP: root mean square error for cross validation and prediction, respectively. RPD: ratio of performance to deviation. RER: range error ratio.

distinguished from most of wood specimens from natural forests (J, P and T). However, the samples of *Cedrela* (C, gray squares) are mixed with *Eucalyptus* specimens in this exploratory analysis done from untreated NIR spectra.

PCA of the treated spectra promoted the separation of the species into groups according to the genetic material (Fig. 2B). *Cedrela* samples were correctly grouped as native species. NIR spectroscopy is very sensitive to variations in chemical properties and it is difficult to discriminate materials with chemical or anatomical similarity. Wood specimens from natural forests clearly differ from commercial *Eucalyptus* samples for any reasons, such as density, color, grain. In this study, all specimens present same moisture and were processed using the same procedure, so the differences captured by NIR sensor is due to intrinsic properties of each material. In short, NIR signature is able to perceive these small differences and mathematical treatments can make this information clearer. As pointed out by Yang *et al.* (2015) the non-destructive and fast separation of wood species can improve the speed of production and enable the processes to be more efficient.

In both Figs. 2A and 2B, the two types of *Eucalyptus* (EC, managed for pulp and paper industry and EV, managed for bioenergy) were grouped together and no difference between them could was highlighted.

Partial least squares discriminant analysis (PLS-DA)

In order to generate PLS-DA models capable of classifying the wood specimens by vegetal material based on NIR spectra, preliminary PLS-R models were calibrated and tested by cross-validation. The models from natural forest presented better values of statistical parameters than the models of planted forests (Table 2). The RER of PLS-R models for native woods were above 6 while the models for *Eucalyptus* were approximately 3.5 (Table 2). The PLS-R model for rosewood (J) wood presented better statistics (higher RPD and RER and lower RMSEP) and consequently the wood specimens were classified better by PLS-DA.

The classification of wood specimens from PLS-DA models both for the cross-validation and the test set yielded findings similar to the results observed by the PCAs (Table 3). The PLS-DA models presented a large percentage of correct classifications for wood specimens, except for the *Eucalyptus* that were confused between them. However, the misclassifications between *Eucalyptus* type specimens do not devalue the method since the main goal of the research was to develop models able to distinguish between the two large groups: natural (native species) and planted forests (*Eucalyptus*). Moreover, only one specimen from natural forest (Ceder) was confounded

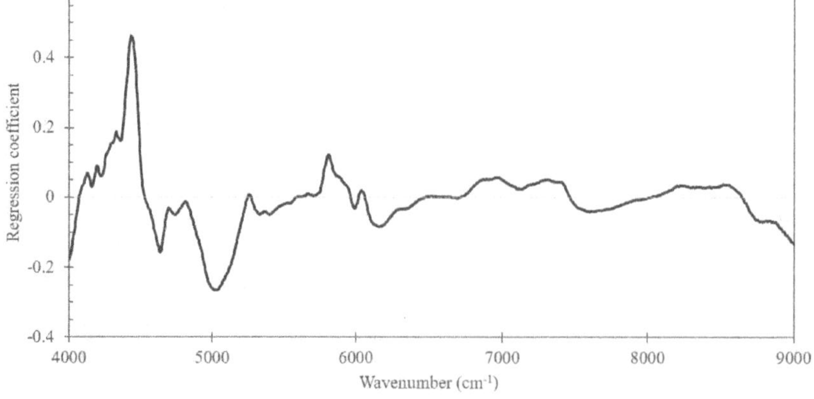

Figure 3. Absorption bands and regression coefficients of the models to predict the species classes.

Table 3. Classification of wood samples according to vegetal material by means of PLS-DA using cross-validation and test set from untreated NIR spectra.

Validation	Class of vegetal material predicted by NIR							N°	% Success
	Material	EC	EV	C	J	P	T		
Cross-validation	EC	63	27					90	70
	EV	17	73					90	81
	C		1	44				45	99
	J				89		1	90	99
	P					87	1	88	99
	T					7	80	87	92
Test set	EC	21	9					30	70
	EV	7	23					30	77
	C			15				15	100
	J				30			30	100
	P					29		29	100
	T					4	25	29	86

Table 4. Classification of wood samples according to vegetal material by means of PLS-DA using independent validation from treated NIR spectra by SNV+1d.

Validation	Class of vegetal material predicted by NIR							N°	% Success
	Material	EC	EV	C	J	P	T		
Cross-validation	EC	83	7					90	92
	EV	67	23					90	26
	C			45				45	100
	J				90			90	100
	P					88		88	100
	T						87	87	100
Test set	EC	22	8					30	73
	EV	20	10					30	33
	C			15				15	100
	J				30			30	100
	P					29		29	100
	T						29	29	100

with *Eucalyptus*, indicating that this native wood has characteristics similar, in some extent, to those of planted forest.

Among the specimens of natural forests there were incorrect classifications between T and P specimens (Table 3). Seven from 87 specimens belonging to *Apuleia* sp. group, T) was confounded and classified as *Aspidosperma* sp. wood specimens by the cross-validated models. In the model validated by test set, 14% of the T samples were misclassified as P wood samples. Besides obtaining high index of correct classification, the native species J (*Jacaranda* sp.), P (*Aspidosperma* sp.) and T (*Apuleia* sp.) were not confused with *Eucalyptus*. To sum up, the classification of native wood samples from PLS-DA carried out using untreated NIR spectra present high percentage of success (86% to 100%) indicating good efficiency of the classification models.

The PLS-DA models of Table 4 shown to be robust for classifying specimens from natural and planted forests; In this analysis, no specimen of planted forest was classified as a wood specimen from natural forest and vice versa. These results indicate that mathematical treatment can improve the signal quality of NIR spectra, allowing good classification by PLS-DA.

Again, these PLS-DA models (Table 4) present faults for classifying the varieties of the *Eucalyptus* genus. Compared to classifications of Table 3, the EV became even more mixed with EC resulting in the higher rate of misclassification (74%) while only 7 of 90 specimens of EC was confounded with EV. It is likely that these misclassifications occur because these *Eucalyptus* samples are very similar from a chemical point of view although no chemical characterization was performed in this study.

In short, the aim of distinguishing vegetal material, especially from planted or planted forests was successfully achieved. The similar findings obtained from PCAs and PLS-DAs show that it is possible to use NIR spectra for distinguishing different wood species, especially for separating unknown wood samples from natural to planted wood, but with some limitations.

Limitations

The specimens of *Cedrela* sp. wood had a greater number of samples mistakenly grouped as *Eucalyptus* wood in the PCA carried out from untreated NIR spectra. This deportment shows that the wood of some native species can be mistaken with *Eucalyptus* woods. The chemical similarity between wood from natural and planted forests can be an barrier in implementing this approach in real condition, since the chemical composition of some woods can be suchlike. This work is an approach to a preliminary study to develop a system for rapid identification of the type of wood that has been transported and sold illegally in many countries.

According to Yang *et al.* (2015) the identification of wood species is of great significance for industrial utilization as well as the quality of the final product. Moreover, this technology can be used for monitoring illegal logging. The inspection agent can record a spectrum of the sample of unknown species and compare with the database. Thus, depending on the group in which the spectrum of wood sample frame, it will be possible to classify it as originating from native or planted forest.

More studies applying this approach may include a wide range of planted and native forest species in models. Moreover, models can be developed with wood samples in different moisture content and with different surface qualities in order to generate a comprehensive database that can be efficient when used to identify the source of wood.

This study demonstrates that NIR spectroscopy has sensibility for clearly separating *Eucalyptus* from native wood and can be used for quickly and reliably identify the origin of unknown wood samples.

Concluding remarks

PCA based on untreated NIR spectra is able to clearly distinguish *Eucalyptus* wood samples from *Jacaranda* (J), *Aspidosperma* (P) and *Apuleia* (T) genus; however, most of specimens of *Cedrela* (C) wood were mixed with *Eucalyptus.* PCA of treated NIR spectra by normalization (SNV) and first derivative clearly showed two groups formed by wood samples from *Eucalyptus* and natural forests.

PLS-DA models carried out from untreated NIR spectra presented a large percentage of correct classifications (86 to 100%) for wood specimens of natural forest, except for the *Eucalyptus* samples that were confused between the two varieties. PLS-DA models based on treated NIR signatures are robust for classifying wood specimens from natural and planted forests and no specimen of planted forest was classified as specimen of natural forest and vice versa.

Acknowledgements

The author expresses your special thanks to the Wood Science and Technology Laboratory of the Federal University of Lavras (UFLA, Brazil) for supporting the experimental work.

References

AACC, 1999. Method 39-00, 15. American Association of Cereal Chemists.

Adedipe OE, Dawson-Andoh AB, Slahor J, Osborn AL, 2008. Classification of red oak (Quercus rubra) and white oak (Quercus alba) wood using a near infrared spectrometer and soft independent modelling of class analogies. J Near Infrared Spectrosc 16: 49-57. https://doi.org/10.1255/jnirs.760

Brereton RG, 2003. Chemometrics: data analysis for the laboratory and chemical plant. J. Wiley, Chichester, England. https://doi.org/10.1002/0470863242

Brito B, Barreto P, 2006. Enforcement against illegal logging in the Brazilian Amazon. 4th IUCN Academy of Environmental Law Colloquium. IUCN. 27 pp.

Brunner M, Eugster R, Trenka E, Berganmin-Strotz L, 1996. FT-NIR spectroscopy and wood identification. Holzforschung 50: 130-134. https://doi.org/10.1515/hfsg.1996.50.2.130

Cooper PA, Jeremic D, Radivojevic S, Ung YT, Leblon B, 2011. Potential of near-infrared spectroscopy to characterize wood products. Can J For Res 41: 2150-2157. https://doi.org/10.1139/x11-088

Costa EVS, Rocha MFV, Hein PRG, Amaral E, Santos LM, Brandao LEVS, Trugilho PF, 2018. Influence of spectral acquisition technique and wood anisotropy on the statistics of predictive NIR-based models for wood density. J Near Infrared Spectrosc 26: 1. https://doi.org/10.1177/0967033518757070

Espinoza JA, Hodge GR, Dvorak WS, 2012. The potential use of near infrared spectroscopy to discriminate between different pine species and their hybrids. J Near Infrared Spectrosc 20: 437-447. https://doi.org/10.1255/jnirs.1006

Gierlinger N, Schwanninger M, Wimmer R, 2004. Characteristics and classification of Fourier-transform near infrared spectra of the heartwood of different larch species (Larix sp.). J Near Infrared Spectrosc 12: 113-119. https://doi.org/10.1255/jnirs.415

Hein PRG, Lima JT, Chaix G, 2009. Robustness of models based on near infrared spectra to predict the basic density in *Eucalyptus urophylla* wood. J Near Infrared Spectrosc 17: 141-150. https://doi.org/10.1255/jnirs.833

Kollmann FR, Coté WA, 1968. Principles of wood science and technology. Springer-Verlag, Berlin. https://doi.org/10.1007/978-3-642-87928-9

Michell AJ, Schimleck LR, 1998. Further classification of eucalypt pulp woods using principal components analysis of near-infrared spectra. Appita J 51: 127-131.

Nunes CA, Freitas MP, Pinheiro ACM, Bastos SC, 2012. Chemoface: A novel free user-friendly interface for chemometrics. J Braz Chem Soc 23: 2003-2010. https://doi.org/10.1590/S0103-50532012005000073

Pastore TCM, Braga JWB, Coradin VTR, Magalhães WLE, Okino EYA, Camargos JAA, De Muniz GIB, Bressan OA, Davrieux F, 2011. Near infrared spectroscopy (NIRS) as a potential tool for monitoring trade of similar woods: Discrimination of true mahogany, cedar, andiroba, and curupixá. Holzforschung 65: 73-80. https://doi.org/10.1515/hf.2011.010

Ramalho FMG, Hein, PRG, Andrade JM, Napoli A, 2017. Potential of near infrared spectroscopy for distinguishing charcoal produced from planted and native wood for energy purpose. Energy Fuels 31: 1593-1599. https://doi.org/10.1021/acs.energyfuels.6b02446

Russ A, Fiserova M, Gigac J, 2009. Preliminary study of wood species identification by NIR spectroscopy. Wood Res 54: 23-32.

Savitzky A, Golay MJE, 1964. Smoothing and differentiation of data by simplified least-squares procedures. Anal Chem 36: 1627-1639. https://doi.org/10.1021/ac60214a047

Schimleck LR, Michell AJ, Vinden P, 1996. Eucalypt wood classification by NIR spectroscopy and principal components analysis. Appita J 49: 319-324.

Schwanninger M, Rodrigues JC, Fackler K, 2011. A review of band assignments in near infrared spectra of wood and wood components. J Near Infrared Spectrosc 19: 287-308. https://doi.org/10.1255/jnirs.955

Tsuchikawa S, 2007. A review of recent near infrared research for wood and paper. Appl Spectrosc Rev 42: 43-71. https://doi.org/10.1080/05704920601036707

Tsuchikawa S, Kobori H, 2015. A review of recent application of near infrared spectroscopy to wood science and technology. J Wood Sci 61: 213-220. https://doi.org/10.1007/s10086-015-1467-x

Westad F, Martens H, 2000. Variable selection in near infrared spectroscopy based on significance testing in partial least square regression. J Near Infrared Spectrosc 8: 117-124. https://doi.org/10.1255/jnirs.271

Yang Z, Jiang ZH, Lu B, 2012. Investigation of near infrared spectroscopy of rosewood. Spectrosc Spectr Anal 32: 2405-2408.

Yang Z, Liu Y, Pang X, Li K, 2015. Preliminary investigation into the identification of wood species from different locations by near infrared spectroscopy. BioResources 10: 8505-8517. https://doi.org/10.15376/biores.10.4.8505-8517

13

Comparison of height-diameter models based on geographically weighted regressions and linear mixed modelling applied to large scale forest inventory data

María Quirós-Segovia[1]*, Sonia Condés-Ruiz[2] and Karel Drápela[1]

[1] *Department of Forest Management and Applied Geoinformatics, Faculty of Forestry and Wood Technology, Mendel University in Brno, Zemědělská 3, 61300 Brno, Czech Republic.* [2] *Department of Natural Systems and Resources, School of Forestry, Technical University of Madrid, Ciudad Universitaria s/n, 28040 Madrid, Spain*

Abstract

Aim of the study: The main objective of this study was to test Geographically Weighted Regression (GWR) for developing height-diameter curves for forests on a large scale and to compare it with Linear Mixed Models (LMM).

Area of study: Monospecific stands of *Pinus halepensis* Mill. located in the region of Murcia (Southeast Spain).

Materials and Methods: The dataset consisted of 230 sample plots (2582 trees) from the Third Spanish National Forest Inventory (SNFI) randomly split into training data (152 plots) and validation data (78 plots). Two different methodologies were used for modelling local (Petterson) and generalized height-diameter relationships (Cañadas I): GWR, with different bandwidths, and linear mixed models. Finally, the quality of the estimated models was compared throughout statistical analysis.

Main results: In general, both LMM and GWR provide better prediction capability when applied to a generalized height-diameter function than when applied to a local one, with R^2 values increasing from around 0.6 to 0.7 in the model validation. Bias and RMSE were also lower for the generalized function. However, error analysis showed that there were no large differences between these two methodologies, evidencing that GWR provides results which are as good as the more frequently used LMM methodology, at least when no additional measurements are available for calibrating.

Research highlights: GWR is a type of spatial analysis for exploring spatially heterogeneous processes. GWR can model spatial variation in tree height-diameter relationship and its regression quality is comparable to LMM. The advantage of GWR over LMM is the possibility to determine the spatial location of every parameter without additional measurements.

Keywords: Spatial analysis; *Pinus halepensis* Mill; forest modelling.

Abbreviations used: GWR (Geographically Weighted Regression); LMM (Linear Mixed Model); SNFI (Spanish National Forest Inventory).

Funding: This research was supported by the Internal Grant Agency of the Faculty of Forestry and Wood Technology, Mendel University in Brno within the development project IGA LDF_VP_2015009.

Competing interests: The authors have declared that no competing interests exist.

Correspondence should be addressed to María Quirós-Segovia: xquiross@mendelu.cz

Introduction

In forestry, spatial heterogeneity is theorized as one of the major drivers of biological diversity (Wiens, 1976). Spatial heterogeneity results from the spatial interactions between a number of biotic and abiotic factors and the differential responses of organisms to these factors (Milne, 1991). It may have significant influences on many ecosystem processes at multiple spatial scales (Turner, 1989). The spatial heterogeneity of vegetation patterns (i.e., landscape heterogeneity) is a structural property of landscapes (Li & Reynolds, 1994) that can be defined by the complexity and variability of ecological system properties in space. The spatial heterogeneity of a tree variable (height, diameter, etc.) in a forest stand results from the complex historical and environmental mosaic imposed by competition and systematic environmental heterogeneity (Zhang *et al.*, 2004); which implies a spatio-temporal heterogeneity. Forest researchers have realized that the spatial pattern of tree locations strongly affects (1) competition among neighbouring trees, (2) size variability and distribution, (3) growth and mortality and (4) crown structure (Zhang *et al.*, 2004). Ignoring spatial heterogeneity in forest modelling causes biased parameter estimates, misleading significance tests, and sub-optimal prediction (Anselin & Griffith, 1988).

Global and local regression models have traditionally been used for studying spatially heterogeneous data (Fotheringham *et al.,* 2002). There are important differences between these models, for example, global regression models summarize the characteristics of the spatial pattern over the whole study area and attempt to identify homogeneity (i.e. Ordinary Least Squares). On the other hand, local regression models make explicit the differences in the pattern observed among parts in the study area (i.e. Geographically Weighted Regression) (Fortin & Dale, 2005). This aspect makes global models easy to compute. However, these models are not realistic and the results obtained may not be correct. Local models attempt to identify exceptions and are variable throughout the space. Hence, these models are suitable for specific sample plots but are of limited use for large populations (Fotheringham *et al.,* 2002).

There are several methods for solving the heterogeneity problem, for example, models based on heteroscedasticity specification, generalized additive models, classification and regression trees, linear mixed models (LMM) and geographically weighted regression (GWR). In this document the latter two methods were analysed.

LMM has been widely used for height-diameter modelling (Lappi, 1997; Mehtätalo, 2004; Calama & Montero, 2004). This method estimates fixed and random parameters simultaneously for the same model. The introduction of random parameters into the model, specific to each sampling unit, enables us to model the variability detected for a given phenomenon among different locations, after defining a common fixed functional structure (Lindstrom & Bates, 1990). Mixed models give an unbiased and efficient estimation of the fixed parameters of the model. Furthermore, to correctly apply mixed models in unsampled areas a calibration is required, which is of critical importance when making local predictions (Meng & Huang, 2009).

The second methodology analysed, GWR, is used to explore spatially heterogeneous processes (Brunsdon *et al.*, 1996). The underlying idea of GWR is that parameters may be estimated anywhere in the study area given a dependent variable and a set of one or more independent variables which have been measured at known locations (Charlton & Fotheringham, 2009). Finally, no additional measurement is needed to predict the value of parameters for unsampled locations. GWR extends ordinary least squares regression models by accounting for spatial structure and estimates a separate model for each geographic location within the studied area (Matthews & Yang, 2012). It can be said that GWR is similar to a '*spatial microscope*' in reference to its ability to measure and visualize variations in relationships, unobservable for non-spatial global models (Matthews & Yang, 2012). GWR was originally used in relation to the economy and civil engineering. In the field of ecology, GWR has been applied to remotely sensed data counts with some interesting results (Wang *et al.*, 2005; Salas *et al.*, 2010; Chen *et al.*, 2012). However, in forest management, the GWR method has predominantly been used for small study areas although it was developed for use in large scale studies.

Tree height is an important variable for forest management in several ways: (1) total height together with diameter at breast height allows us to estimate the total volume of the tree, (2) in regular stands, dominant height together with age are used to estimate the site index and (3) the distribution of tree heights in the stand is very important to describe the vertical structure.

Nevertheless, height measurement can be time consuming and consequently expensive, therefore it is common to use height-diameter models instead. There are two kinds of height-diameter models: local models and generalized models.

Local models are designed specifically for forest stands and are the most precise curves for a particular forest stand (Diéguez-Aranda *et al.*, 2009). However, they have several disadvantages, for example: (1) they require the heights of many trees in the stand to be measured, therefore are very expensive and (2) due to forest heterogeneity (age, quality, density) and the silvicultural state of stands within a given forest, a local function cannot be adjusted to all situations (Diéguez-Aranda *et al.*, 2009). Generalized height-diameter models could improve the results in such forests. These models include stand variables for obtaining precise and unbiased estimations (Gaffrey, 1988). Dominant height is usually included because it is easy to measure and is independent of silvicultural treatments. Other variables such as basal area, density or quadratic mean diameter could also be included (Sloboda *et al.*, 1993; Páscoa, 1987).

The main aim of this paper is to analyse the suitability of GWR for modelling height-diameter relationships in a large area and to compare it with LMM, which is one of the most frequently used methods for including spatial heterogeneity in these models.

To test whether GWR provides as good results as the more frequently used LMM, we applied both methodologies to two different functions: the Petterson (1955) local function and the Cañadas *et al.*, (1999) generalized function, and compared their ability to predict heights when applied to an independent dataset.

Material and methods

Study area

To study GWR, we selected mono-specific stands of Aleppo pine (*Pinus halepensis* Mill.) in the Region of Murcia, in South-East Spain (between latitude 37° 47' 36" N and 38° 6' 11" N, and longitude 1° 53' 48" W and 1° 26' 17" W). The study area covered around 1 300 km² in Espuña, Cambrón, Burete, Lavia and Quipar mountain ranges (Fig. 1). In this region, Aleppo pine was growing in low-density stands with relatively low dominant heights, probably reflecting medium to poor site qualities (Condés & Sterba, 2008).

Data were obtained from 230 sample plots with 2582 trees belonging to the Third Spanish National Forest Inventory (SNFI). SNFI design is based on permanent sample plots located at the nodes of one kilometre square grid, re-measured in an inventory cycle of ten years. The sample plots consisted of four concentric circles of 5, 10, 15 and 25 meter radius, in which the diameters and heights of all trees with a breast height diameter over 7.5, 12.5, 22.5 and 42.5 cm respectively were measured. Furthermore, plot centre coordinates, together with the polar coordinates of each measured tree, i.e. distance and azimuth form the plot centre, were recorded (MAGRAMA, 2015).

Data in the studied area were randomly split into two subsets. The first consisted of 152 plots with 1697 trees and was used to develop the model (training data). The second was used to validate the model (validation data) and consisted of 78 plots with 885 trees (Fig. 1). Table 1 shows the main stand variables for both subsets estimated by weighting individual tree data according to the area of the concentric SNFI subplots.

Modelling height-diameter relationships

Two different height-diameter functions were selected for the study. The first of these was the local height-diameter Petterson's function (Petterson, 1955), (eq. 1), which has been successfully used in similar, previously published studies (e.g. Juárez de Galíndez *et al.*, 2007; Drápela, 2011; Adamec, 2014; Adamec & Drápela, 2015)

$$h = 1.3 + \frac{1}{\left(a_0 + \frac{a_1}{d}\right)^3} \tag{1}$$

where h is the total height (m), d is the diameter at breast height (cm) and a_0 and a_1 are the parameters to be estimated.

The second selected function was the Cañadas I generalized height-diameter function (Cañadas *et al.*, 1999), (eq. 2). This function was selected because it offered the possibility to perform the computation with GWR together with the fact that it had previously provided good results for Aleppo pine forests (Cabanillas, 2010).

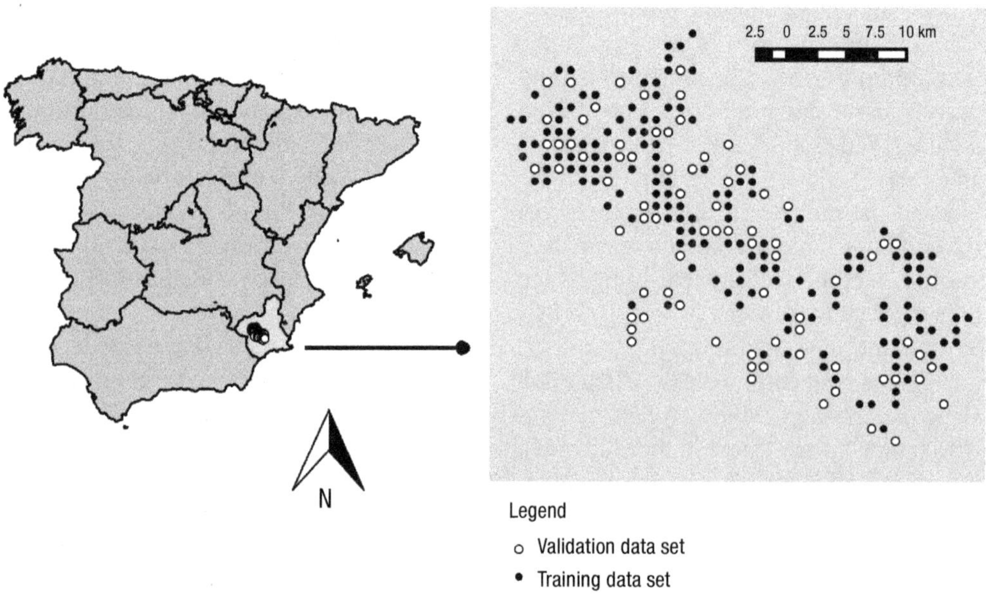

Figure 1. Study area & subsamples.

Table 1. Statistical analysis of data

	Training dataset 152 plots / 1697 trees						
	d	h	N	G	dg	H_o	D_o
Min	7.5	2.0	14	0.6	10.2	4.0	11.1
Max	55.7	21.5	1351	27.3	38.3	16.3	39.0
Mean	21.4	8.6	361	9.1	20.2	8.8	24.3
Standard Deviation	2.6	8.3	283	5.6	6.3	2.4	5.9
	Validation dataset 78 plots / 885 trees						
Min	17.8	2.4	19	1.3	9.2	3.7	12.5
Max	53.1	15.5	1489	32.4	40.4	12.7	41.0
Mean	20.0	8.0	420	9.3	18.9	8.3	23.4
Standard Deviation	7.4	2.1	334	6.2	6.0	1.9	5.4

d - Diameter at breast height (cm); h - Total height (meters); N - Number of stems per hectare (stems/ha); G - Plot basal area (m2/ha); d_g - Quadratic mean diameter (cm); H_o - Dominant height (m); D_o - Dominant diameter (cm).

$$h = 1.3 + (H_0 - 1.3) \cdot \left(\frac{d}{D_0} \right)^{b_0} \quad (2)$$

where h is total height (m), d is diameter at breast height (cm), H_o is dominant height (m) obtained according to the Hart criteria (Hart, 1928), D_0 is dominant diameter, calculated as the average of the 100 thickest trees per hectare, and b_0 is the parameter to be estimated.

As GWR is basically a technique developed for linear regressions (Fotheringham *et al.*, 2002), to fit these equations it was necessary to linearize them. Consequently, to fit these equations it was necessary to linearize them, which was done as follows in equations 3 and 4:

Petterson: $$\frac{1}{\sqrt[3]{h-1.3}} = a_0 + a_1 \cdot \frac{1}{d} \quad (3)$$

Cañadas I: $$\ln\left(\frac{h-1.3}{H_0-1.3}\right) = b_0 \ln\left(\frac{d}{D_0}\right) \quad (4)$$

These two height-diameter functions were fitted following the two different methodologies detailed below.

Linear Mixed Models

In a first step, linear mixed models were used for fitting the height-diameter equations. As data came from a hierarchical structure, with several trees measured in the same plot, a mixed model was used with the plot as the grouping structure of the random effects. We included random effects in the a_0 and a_1 term of Petterson local model (eq. 3) and in b_0 for the Cañadas I (eq. 4).

The models were fitted using the maximum likelihood method of the lme procedure (R Development Core Team, 2014) to allow the comparison of results. A level of $p = 0.05$ was used for significance testing of the variables in the model, and Akaike's information criterion (AIC) was used to compare results. The significance of random effects was tested by comparing the models with similar ones developed using a constant as grouping variable. Goodness-of-fit measures for linear mixed models were calculated using the lmmR2 procedure (R Development Core Team, 2014).

Geographically Weighted Regression

In a second step, the models were fitted using the GWR with three different bandwidths and an optimal number of trees.

Two of these bandwidths, calculated taking into account the SNFI grid, were 1000 m and 500 m. These two values were used to determine whether the SNFI design influenced the GWR results. The third bandwidth was calculated according to the calibration recommended by Fotheringham *et al.*, (2002), which was based on the optimization of the studied formula according to a kernel and a weighted function. In this case, the use of a fixed kernel plus a gaussian weighted function was recommended. A fixed kernel assumes that the bandwidth at each centre *i* is a constant across the study area. The gaussian weighted function (eq. 5) is a continuous function of the distance between two observation points (Gollini *et al.*, 2015).

Gaussian weighted function:

$$w_{ik} = e^{\left(-\left(d_{ik}/b\right)^2\right)} \quad (5)$$

where w_{ik} is the *k-th* element of the diagonal of the matrix of geographical weights, and d_{ik} is the distance between observations *i* and *k* (in our case observations were trees) and *b* is the bandwidth.

Finally, the optimal number of trees was calculated according to the calibration recommended by Fotheringham *et al.*, (2002). This was based on the optimization of the number of trees, which is to be applied in every height-diameter formula. As in the case of bandwidth, this optimization is based on the type of kernel and weighted function. For this situation, an adaptive kernel together with a bi-square weighted function was recommended. This kernel adapts itself in the size to the data density. The bi-square weighted function (eq. 6), is a discontinuous function, giving null weights to observations with a distance greater than a specific point (Gollini *et al.*, 2015).

Bisquare weighted function:

$$w_{ij} = \left[1-\left(d_{ij}/n\right)^2\right]^2 \quad \text{When } d_{ij} \leq n \qquad (6)$$

$$w_{ij} = 0 \quad \text{When } d_{ij} \leq n$$

where w_{ij} is the *j-th* element of the diagonal of the matrix of geographical weights, and d_{ij} is the distance between observations *i* and *j*, and *n* is the number of neighbourhoods (number of trees).

GWR provides a range of coefficients geographically located in the study area, and a continuous GWR map of coefficients can be obtained by the coefficients interpolation. The Inverse Distance Weighting (Watson & Philip, 1985) was selected as interpolation methodology because this interpolation is widely used in natural sciences (Chen *et al.*, 2012) and is easily applicable to GWR (Tardanico, 2006). See for instance Fotheringham *et al.*, (2002) for further description of GWR.

GWR was fitted using the GWModel package in R language (Binbin *et al.*, 2015; R Development Core Team, 2014). A level of $p = 0.05$ was used for significance testing of the variables in the models, and Akaike's information criterion was used to compare results.

Model validation

Models were validated by applying GWR as well as LMM to the validation dataset, obtaining a set of predicted heights. In the case of GWR, the predicted values were obtained through the GWR map of coefficients.

In order to validate LMM, it was necessary to calibrate the model. Very few measurements can be used in the calibration, even less than the number of random parameters to be predicted (Calama & Montero, 2004). Four trees per plot were used for calibrating the model in this study, this is a very common practice in forest inventories in Spain (Calama & Montero, 2004). The technique of best linear unbiased predictor (BLUP) (Robinson, 1991) was used to estimate of random parts of the parameters.

The validation was performed by comparing observed and predicted heights through a statistical analysis of the residuals. Mean error (ME), mean absolute error (MAE), root mean square error (RMSE) and coefficient of determination (R^2) were calculated according to the formula in Table 2.

Table 2. Residual formulas

Mean error	$ME = \frac{\sum(y_i - \hat{y}_i)}{n}$
Root Mean Square Error	$RMSE = \sqrt{\frac{\sum(y_i - \hat{y}_i)^2}{n}}$
Mean Absolute Error	$MAE = \frac{\sum\lvert y_i - \hat{y}_i\rvert}{n}$
Coefficient of determination	$R^2 = 1 - \frac{\sum_{i=1}^{n}(y_i - \hat{y}_i)^2}{\sum_{i=1}^{n}(y_i - \overline{y}_i)^2}$

Where n is the sample size; y_i and $\hat{y}_i$ are respectively the observed and predicted tree heights and $\overline{y}_i$ the average of individual tree heights.

Results

Table 3 shows the estimated coefficients for equations 3 and 4, firstly for the LMM and GWR when applied to the Petterson height-diameter function, and secondly when they were applied to the Cañadas I height-diameter function. In this table, the minimum, mean, standard deviation and maximum values of the coefficients obtained for the whole study area (considering the map of coefficients for GWR and the fixed and random effects for LMM) were detailed together with R-squared (R^2) and the Akaike information criterion (AIC).

For the Petterson height-diameter function, the coefficient a_0 mean value varied between 0.37 and 0.55 for LMM and also for GWR independently of the bandwidth (between 0.34 and 0.52 for a bandwidth of 1000, between 0.32 and 0.69 for a bandwidth of 500, between 0.34 and 0.71 for a Fixed Kernel bandwidth, and between -0.19 and 0.64 for an adaptive kernel bandwidth). The range of this parameter was lowest when obtained by LMM or by GWR with a bandwidth of 1000 m, but increased for the bandwidth of 500 and the fixed kernel, and reached a maximum for the adaptive kernel. The coefficient a_1, i.e. the slope of the curve, showed higher variation with larger standard deviations and wider ranges than the a_0. So, for LMM the standard deviation of a_1 was 0.48 while for GWR were from 0.65 to 1.39. However, mean values along the study area were similar for all the methodologies, from 1.79 to 2.06. Once again, the largest variations were found for the GWR fitted with fixed or adaptive kernel, proving to be more sensitive to the spatial variability of the data. Moreover, the smallest AIC was obtained with the fixed kernel, representing an optimal bandwidth of 363 m; and with the adaptive kernel, representing an optimal number of trees of 26. However, all GWR methods showed certain incongruent results such as negative or very large parameter values. An example of estimation using the Petterson map of coefficients is presented in Fig. 2, which provides a graphical representation of the previously explained parameter distribution for GWR with a predefined bandwidth of 1000 m.

In the case of the Cañadas-I height-diameter function it can be seen that the coefficient b_0 had slightly larger standard deviations and ranges for GWR than for LMM. Additionally, the mean values were almost the same between both methods. The AIC for this function was smaller when GWR was fitted with a predefined bandwidth of 500 m. For this generalized height-diameter function, the optimal bandwidth calculated by fixed kernel was 755 m, and the optimal number of trees in the adaptive kernel was 65. Again, some incongruent minimum values can be seen for the GWR methodology in Table 3. A representative map of coefficients for the Cañadas I estimation is shown in Fig. 3, and the parameter distribution of GWR is presented with a predefined bandwidth of 1000m.

As response variables in the Petterson and Cañadas-I functions were different, the AIC and R^2 values were presented in Table 3. However, AIC were not comparable between models.

Model validation

Table 4 shows the errors obtained for the validation dataset when different methodologies were used. Predicted values were estimated according different assumptions: (1) LMM with fixed effects, (2) LMM with random effects and (3) GWR, including all the analysed variations.

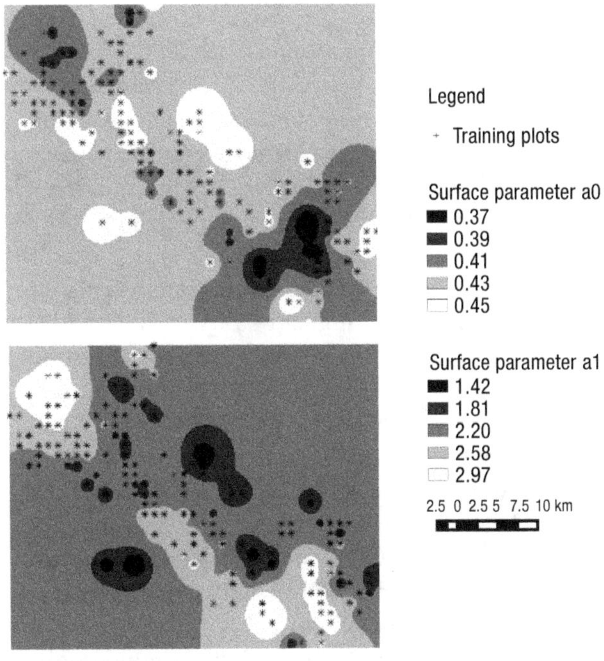

Figure 2. GWR map of regression coefficients. Example of Petterson height-diameter model (1000 m predefined bandwidth), coefficient a_0 and a_1.

The validation revealed that the Petterson height-diameter function had poorer predictive capacity than the Cañadas I generalized function. Consequently, it can be seen that the R^2 for the LMM with only fixed effects increased from 0.27 to 0.71, when calibration the improvement was from 0.63 to 0.72 and for GWR the R^2 ranged between 0.29 and 0.35 for Petterson while for Cañadas was 0.71.

When analysing the errors obtained from Petterson model, it was observed that, independently of bandwidth or kernel, GWR resulted in lower bias than LMM with only fixed effects (ME=0.6) or in similar than the calibrated (ME=0.1). Moreover, MAE and RMSE were very similar for GWR and LMM with only fixed effects (MAE=1.3 and RMSE=1.7 for LMM without calibration; and MAE=1.3 - 1.4 and RMSE=1.7 for GWR). However both MAE and RMSE were smaller when calibrating LMM (MAE= 0.9 and RMSE=1.3). Regarding the Cañadas-I height-diameter function, the error analysis showed similar results for all the GWR and LMM methods with and without calibration (Table 4).

The standardized residual plots showed that residuals are symmetrically distributed in a random pattern around the horizontal axis for any of the functions and fitting methodologies (Figures 4 and 5). Judging from the errors, it can be stated that GWR applied to generalized function works as well as the more frequently used LMM methodology.

Discussion

Tree height is an important variable which is used for estimating stand volume or site quality (Diéguez-Aranda *et al.*, 2009). Forest variables are typically highly heterogeneous and in this paper, two different methods were presented for modelling height-diameter relationships and their spatial heterogeneity. The first method was LMM, in which the height-diameter curves use fixed and random coefficients and which gives very accurate predictions when random effects can be predicted (Nanos *et al.*, 2004). However, to calibrate the LMM prior to use, additional height measurements of trees within the stand are required (Lappi, 1997). The second method was GWR, which is a geostatistical method that can predict height-diameter models without the necessity for additional measurements. This effect is very interesting in case of large inventory areas such as national forest inventories, because the only available data are those collected in the inventory plots.

Table 3. Coefficient distribution of models - In LMM and GWR using linearized equations (eq. 3 and 4)

		Petterson							Cañadas I						
Method		**Coefficient**	**Min**	**Mean**	**SD**	**Max**	**R2**	**AIC**	**Coefficient**	**Min**	**Mean**	**SD**	**Max**	**R2**	**AIC**
LMM		a_0+a_{0j}	0.37	0.44	0.03	0.55	0.82	-6304	b_0+b_{0j}	0.30	0.66	0.14	1.31	0.59	-1161
		a_1+a_{1j}	1.80	1.96	0.48	2.76									
GWR	**1000**	a_0	0.34	0.42	0.03	0.52	0.68	-5923	b_0	0.18	0.66	0.16	1.35	0.57	-1260
		a_1	-0.23	2.06	0.65	4.07									
	500	a_0	0.32	0.43	0.05	0.69	0.81	-6676	b_0	-0.42	0.65	0.21	1.49	0.60	-1319
		a_1	-1.09	1.85	0.81	4.31									
	Fixed Kernel	a_0	0.34	0.44	0.05	0.71	0.83	-6884	b_0	0.01	0.66	0.18	1.37	0.58	-1287
		a_1	-2.46	1.79	0.88	4.76									
	Adaptive Kernel	a_0	-0.19	0.43	0.06	0.64	0.82	-6809	b_0	-0.04	0.66	0.17	1.34	0.57	-1267
		a_1	-0.48	1.87	1.39	21.88									

Where Min is the minimum value; SD is the standard deviation; Max is the maximum value of coefficient; R^2 is the coefficient of determination, in case of LMM is referred to conditional R^2; AIC the Akaike's information criterion; a_0, a_1 and b_0 are the fixed factor in the case of LMM models; a_{0j} and b_{0j} are random factors in the case of LMM. In GWR a_0, a_1 and b_0 are the parameters estimated by the model.

These two methodologies were tested according to a local (Petterson) and a generalized (Cañadas I) height-diameter function as well as with different bandwidths in the case of GWR.

In the case of the Petterson model, it was observed that this method had insufficient predictive quality when GWR or uncalibrated LMM was applied. It is possible that the large heterogeneity of the studied area made it difficult for the Petterson function to reach the necessary quality. This could be due to the fact that the Petterson formula only had diameter at breast height as an independent variable and would appear to be too simple to reproduce the forest heterogeneity. Nevertheless, when Petterson equation was validated with calibrated LMM, a high improvement in the predictive quality was observed. In the case of height-diameter model, random component represents the specific conditions of every stand in the model. It causes better precision of the model for particular location in comparison with the model with fixed parameters. This could be important in a local model such as Petterson, designed specifically for giving precise prediction in individual forest stands (Diéguez-Aranda *et al.*, 2009).

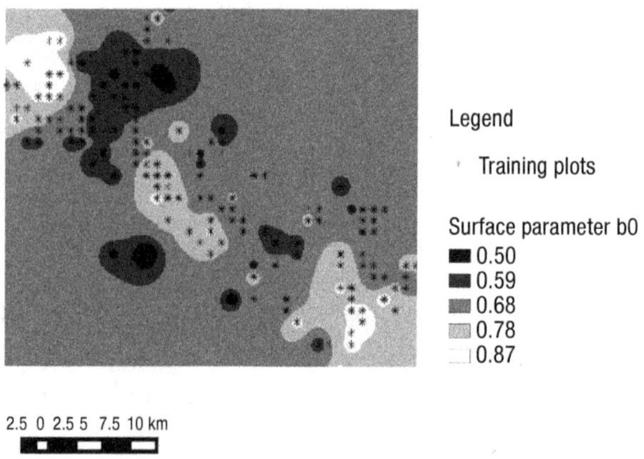

Figure 3. GWR map of regression coefficients. Example of Cañadas I height-diameter model (1000 m predefined bandwidth), coefficient b_0.

In contrast, the Cañadas I model revealed high stability in the model fitting and high predictive quality in the validation in both LMM and GWR. In these cases, R^2 values in the validation improve from local to generalized regressions from 0.63 to 0.72 in calibrated LMM. Similarly, MAE and RMSE were lower for the generalized models. The comparison between local and generalized equations is analogous with previous studies which concluded that the inclusion of stand-density variables such as dominant height or dominant diameter in the base height–diameter function increased the accuracy of prediction (Temesgen & Gadow, 2004; Diéguez-Aranda *et al.*, 2009).

The predictive quality of the model was improved when few additional height measurements were used for estimating the random parameters for the validation dataset. Calibration using small subsamples of trees is one of the main utilities of mixed models in forest applications (Calama & Montero, 2004). However, when additional measurements are not available GWR provides predictions with higher quality than the LMM.

In some of the GWR models the presence of odd, extreme values was observed. With the Petterson equation these values appear in every GWR model and did not correspond to specific plots but rather, were scattered randomly in the study area. This could be another indicator of the weakness of these models. However, in the case of Cañadas I, these odd values only appeared in the models with a bandwidth of 500 m and adaptive kernel. Specifically, this odd value was located in plot number 1473, which contained only 3 trees and was located in a relatively isolated area. Odd values have also been observed in similar height-diam-

Table 4. Comparison of errors obtained for the validation dataset

	VALIDATION	ME	MAE	RMSE	R^2
PETTERSON	LMM fixed effect	0.6	1.4	1.7	0.27
	LMM random effect	0.1	0.9	1.3	0.63
	GWR	0.1	1.3-1.4	1.7	0.29-0.35
CAÑADAS I	LMM fixed effect	-0.1	0.8	1.1	0.71
	LMM random effect	-0.2	0.8	1.1	0.72
	GWR	-0.1	0.8	1.1	0.71

Where ME is the mean error; MAE is the mean absolute error; RMSE is the root mean squared error; R^2 is the coefficient of determination.

eter modelling studies (Zhang *et al.*, 2004; Zhang *et al.*, 2008; Zhang *et al.*, 2009). It is important to be aware of the presence of these odd values in order to select the correct model. The inclusion of these values in the interpolation could lead to erroneous results in the GWR coefficient map.

In our study, the observed GWR models gave the same residuals regardless of the calibration used. However, according to Zhang *et al.*, (2009), the GWR with adaptive kernel was expected to produce more desirable model residuals in terms of spatial autocorrelation and heterogeneity than the GWR with fixed kernel model for the plots. There are a wide range of calibration options so it is possible that future modifications of settings could improve the current results. For instance, it would be possible to test different bandwidths, or to add robustness to GWR (Fotheringham *et al.*, 2002), or even to calculate the optimal bandwidth differently or the optimal number of trees with different kinds of weighted functions (Gollini *et al.*, 2015).

The maps of coefficient in Figures 2 and 3 illustrate the spatial heterogeneity of the height-diameter relationship. In both figures it can be seen that there are greater slopes in both models in the upper-left area and in the bottom-right area. Additionally, more flat areas exist in the centre of the studied areas. In Fig. 2, two areas can be observed where a_0 takes the highest values and a_1 the lowest; therefore in Fig. 3, only the left side is visible. It would be interesting to compare these contours with local terrain layers (slopes, kind of soil, vegetation), in order to determine whether there is a relationship between fittings and local conditions. Since the spatial variation of local conditions and competition were taken into account in the GWR model, it showed better fitting results for the height-diameter regression relationship (Zhang *et al.*, 2004).

When comparing the LMM and GWR models in this study, it was found that GWR estimation produces similar errors to the LMM when generalized equation was applied. Analogous results have been found in other studies, supporting our finding that GWR gives similar quality estimations to the LMM methodology and predicts the response variable accurately (Zhang *et al.*, 2005; Zhang *et al.*, 2008, Zhang *et al.*, 2009). However, GWR showed not so good predictive quality with the Petterson function, showing no important improvement over LMM without calibration. We suppose that GWR quality could be improved by more sophisticated calibration, which needs a deeper investigation.

LMM and GWR provide different perspectives and insights for data analysis and model prediction. In LMM, the fixed component explains the impact of different variables as with the ordinary least squares regression (Yang & Huang, 2011), and the random component explains the heterogeneity and randomness by both known and unknown factors (Vonesh & Chinchilli, 1997). LMM is able to characterize the spatial covariance structures in the data with different geostatistic models (Zhang *et al.*, 2009), and produces accurate predictions for the response variable by accounting for the effects of spatial autocorrelation through the empirical best linear unbiased predictors (Littell *et al.*, 1996; Schabenberger & Pierce, 2002). However, GWR has several very important advantages over LMM, which include the following: (1) It is a useful tool for exploring spatial nonstationarity at different scales (Jetz *et al.*, 2005), which was achieved by changing the bandwidth of study. (2) Its capacity for testing the heterogeneity of the study area (Danlin & Yehua, 2013). The heterogeneity is shown in the coefficient maps (Figures 2 and 3) and can be tested numerically by applying the relative spatial heterogeneity index (Zhang *et al.*, 2009). (3) GWR is able to work with spatial data and every kind of attribute data (Danlin & Yehua, 2013); this is particularly useful in the case of Forest Inventories where attributes of measured trees and their location are known. (4) The residuals for GWR are at least similar as for LMM (Zhang *et al.*, 2004; Zhang & Shi, 2004; Zhang *et al.*, 2009), as confirmed in our research. (5) GWR results, unlike LMM model results, are mappable; these maps of coefficients facilitating the in-

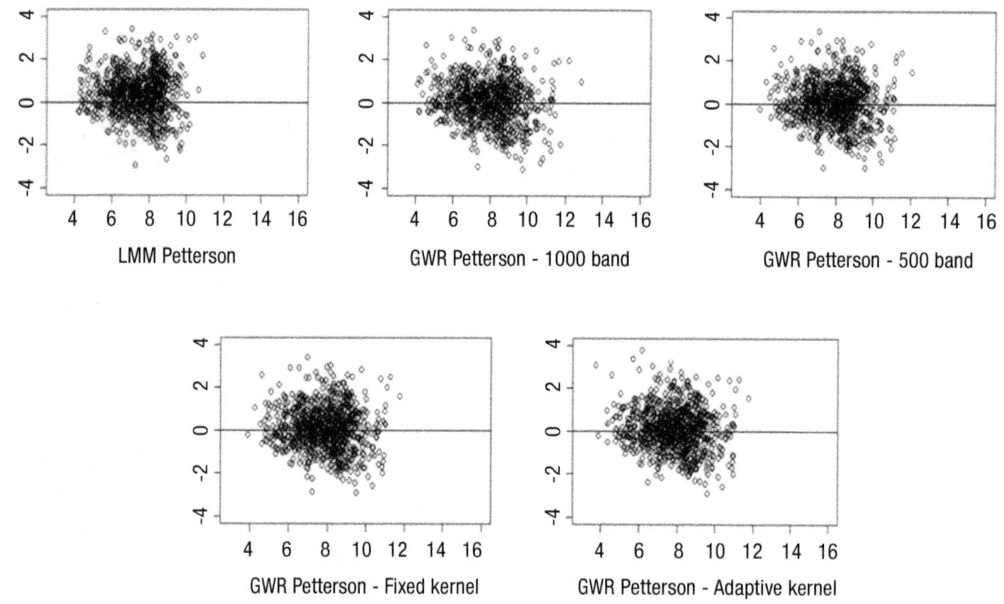

Figure 4. Standardized Residuals versus predicted height for the LMM and the different GWR methodologies when applied to Petterson equation.

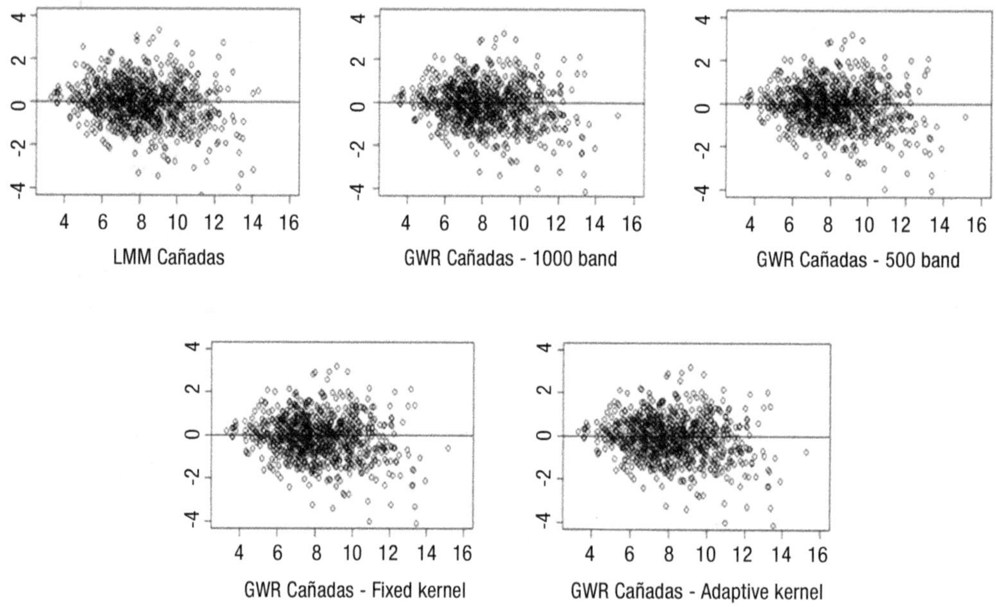

Figure 5. Standardized Residuals versus predicted height for the LMM and the different GWR methodologies when applied to Cañadas I equation.

terpretation based on spatial context and known characteristics of the study area (Goodchild & Janelle, 2004). The maps of coefficients in our study are shown in Figures 2 and 3. (6) In contrast to the LMM method, the GWR regression does not require any additional measurements or calibrations.

Conclusions

Two different methodologies were used to model local and generalized height-diameter relationships: the geostatistical method GWR and LMM methodology. Additionally, two different height-diameter functions were selected to fit these models: a local function and a generalized function. Finally, the quality of the estimated models was compared through statistical analysis.

As expected, it was observed that in general, both LMM and GWR provide better prediction quality when applied to a generalized height-diameter equation. However, error analysis revealed that there were no great differences between these two methodologies, evidencing that GWR applied to a generalized function provides results which are at least as good as those obtained using the calibrated LMM methodology. However, GWR has a number of advantages over LMM. For instance, GWR provides information on a continuous map of regression coefficients within the studied area. Consequently, is possible to calculate tree heights in every single location without the need for additional measurements. Furthermore, the GWR method can help us to understand spatial processes and may have interesting applications in the area of forest yield and production.

Acknowledgements

The authors would like to thank two anonymous reviewers and handling editor for their valuable comments that helped improve the quality of this study.

References

Adamec Z, 2014. Využití moderních regresních metod pro modelování výškové křivky. Doctoral thesis. Mendelova Univerzita v Brně. Czech Republic.

Adamec Z, Drápela K, 2015. Generalized additive models as an alternative approach to the modelling of the tree height-diameter relationship. J For Sci 61(6): 235-243. http://dx.doi.org/10.17221/14/2015-JFS.

Anselin L, Griffith DA, 1988. Do spatial effects really matter in regression analysis. Papers in Regional Science Association 65: 11-34. http://dx.doi.org/10.1111/j.1435-5597.1988.tb01155.x.

Binbin L, Harris P, Charlton M, Brunsdon C, Nakaya T, Gollini I, 2015. R-CRAN Package "GWModel". Available at: http://cran.r-project.org/web/packages/GWmodel/GWmodel.pdf.

Brunsdon C, Fotheringham S, Charlton M, 1996. Geographically Weighted Regression: A method for explore spatial non-stationarity. Geogr Anal 28 (4): 281-298. http://dx.doi.org/10.1111/j.1538-4632.1996.tb00936.x.

Cabanillas A, 2010. Bases para la gestión de masas naturales de (*Pinus halepensis* Mill.) en el Valle del Ebro. Doctoral Thesis. Dept. de Ingeniería Forestal, Univ. Politécnica de Madrid, Spain.

Calama R, Montero G, 2004. Interregional nonlinear height-diameter model with random coefficients for stone pine in Spain. Can J For Res 34:150-163. http://dx.doi.org/10.1139/x03-199.

Cañadas N, García C, Montero G, 1999. Relación altura-diámetro para *Pinus pinea* L. en el Sistema central. Actas del Congreso de Ordenación y Gestión Sostenible de Montes, Santiago de Compostela, Spain. I: 139:153.

Charlton M, Fotheringham S, 2009. Geographically Weighted Regression. White Paper for Science Foundation, Ireland: 1-14. http://www.geos.ed.ac.uk/~gisteac/fspat/gwr/gwr_arcgis/GWR_WhitePaper.pdf.

Chen G, Zhao K, McDermid G, Hay G, 2012. The influence of sampling density on geographically weighted regression: a case study using forest canopy height and optical data. Int J Remote Sens 33 (9): 2909-2924. http://dx.doi.org/10.1080/01431161.2011.624130.

Condés S, Sterba H, 2008. Comparing an individual tree growth model for Pinus halepensis Mill. in the Spanish region of Murcia with yield tables gained from the same area. Eur J Forest Res 127: 253-261. http://dx.doi.org/10.1007/s10342-007-0201-7.

Danlin Y, Yehua DW, 2013. Geographically Weighted Regression. http://slideplayer.com/slide/3428243/.

Diéguez-Aranda U, Rojo Alboreca A, Castedo-Dorado F, Álvarez González JG, Barrio-Anta M, Crecente-Campo F, González González JM, Pérez-Cruzado C, Rodríguez Soalleiro R *et al.*, 2009. Herramientas selvícolas para la gestión forestal sostenible en Galicia. Dirección Xeral de Montes, Consellería do Medio Rural, Xunta de Galicia, Spain. 259 pp.

Drápela K, 2011. Regresní modely a možnosti jejich využití v lesnictví. Habilitační práce. Mendelova univerzita v Brně. Czech Republic.

Fortin MJ, Dale M, 2005. Spatial Analysis: A Guide for ecologist. First ed. Cambridge: Cambridge University Press, UK.

Fotheringham S, Brunsdon C, Charlton M, 2002. Geographically Weighted Regression: The Analysis of Spatially Varying Relationships, John Wiley & Sons, New York, NY, 269 pp.

Gaffrey D, 1988. Forstamts-und bestandesindividuelles Sortimentierungsprogramm als Mittel zur Planung, Aushaltung und Simulation. Diplomarbeit Forscliche Fakultät, Univ. Göttingen. Germany.

Gollini I, Binbin L, Charlton M, Brunsdon C, Harris P, 2015. GWmodel: An R Package for Exploring Spatial Heterogeneity Using Geographically Weighted Models. J Stat Sof 63 (17): 1-50. http://dx.doi.org/10.18637/jss.v063.i17.

Goodchild MF, Janelle DG, 2004. Spatially integrated social science. New York, NY: Oxford University Press.

Hart HMF, 1928. Stamtal en dunning; een orienteerend onderzoek naar de beste plantwijdte en dunning swijze woor den djati. Veenman & Zonen. Wageningen. Netherlands.

Jetz W, Ahbek CR, Ichstein JWL, 2005. Local and global approaches to spatial data analysis in ecology. Global Ecol Biogeogr 14: 97-98. http://dx.doi.org/10.1111/j.1466-822X.2004.00129.x.

Juárez de Galíndez M, Pece MG, Gaillard de Benítez C, Sanguedolce J, Mariot V, Mazzuco R, 2007. Ecuaciones altura-diámetro en Prosopis nigra (Griseb) Hieron (algarrobo negro) en Santiago del Estero, Argentina. Foresta Veracruzana 9 (1): 9-14. Recursos Genéticos Forestales. Xalapa, México.

Lappi J, 1997. A longitudinal analysis of height/diameter curves. Forest Sci 43:555-570.

Li H, Reynolds JF, 1994. A simulation experiment to quantify spatial heterogeneity in categorical maps. Ecology 75: 2446-2455. http://dx.doi.org/10.2307/1940898.

Lindstrom MJ, Bates DM, 1990. Nonlinear mixed effects for repeated measures data. Biometrics 46: 673-687. http://dx.doi.org/10.2307/2532087.

Littell RC, Milliken GA, Stroup WW, Wolfinger RD, 1996. SAS System for Mixed Models. SAS Institute Inc, Cary, NC.

MAGRAMA, 2015. Ministerio de Agricultura, Alimentación y Medio Ambiente. Ministra: Isabel García Tejerina. http://www.magrama.gob.es.

Matthews S, Yang TC, 2012. Mapping the results of local statistics: Using geographically weighted regression. Demographic Res 26(6): 151-166. http://dx.doi.org/10.4054/DemRes.2012.26.6.

Mehtätalo L, 2004. A longitudinal height-diameter model for Norway spruce in Finland. Can J Forest Res 34(1): 131-140. http://dx.doi.org/10.1139/x03-207.

Meng SX, Huang S, 2009. Improved calibration of nonlinear mixed-effects models demonstrated on a height growth function. Forest Sci 55(3): 238-248.

Milne BT, 1991. Heterogeneity as a multi-scale characteristic of landscape. In: J Kolasa and TA Pickett, Ecological Heterogeneity: 69-84. Springer-Verlag, New York.

Nanos N, Calama R, Montero G, Gil L, 2004. Geostatistical prediction of height/diameter models. Forest Ecol Manag 195: 221-235. http://dx.doi.org/10.1016/j.foreco.2004.02.031.

Páscoa F, 1987. Estructura, crescimento e produçao em povoamentos de pinheiro bravo. Um modelo de simulaçao. Tesis doctoral. Universidade Técnica de Lisboa. Portugal.

Petterson H, 1955. Barrskogens volymproduktion. Meddelanden från Statens skogsforskningsinstitut 45 (1A): 1-391.

R Development Core Team, 2014. R: A language and environment for statistical computing. R Foundation for Statistical Computing, Vienna, Austria. https://www.r-project.org/.

Robinson GK, 1991. That BLUP is a Good Thing: The Estimation of Random Effects. Statistical Sci 6(1): 15-32. http://dx.doi.org/10.1214/ss/1177011926.

Salas C, Ene L, Gregoire T.G, Naesset E, Gobakken T, 2010. Modelling tree diameter from airborne laser scanning derived variables: A comparison of spatial statistical models. Remote Sens Environ 144: 1277-1285. http://dx.doi.org/10.1016/j.rse.2010.01.020.

Schabenberger O, Pierce FJ, 2002. Contemporary Statistical Models for the Plant and Soil Sciences. CRC Press, Boca Raton, FL, 738.

Sloboda VB, Gaffrey D, Matsumura N, 1993. Regionale und lokale Systeme von Höhenkurven für gleichaltrige Waldbestände. Allg Forst-u J-Ztg 164: 225-228.

Tardanico R, 2006. GWR&2D Mapping. Florida International University. http://www2.fiu.edu/~tardanic/.

Temesgen H, Gadow K, 2004. Generalized height-diameter models—an application for major tree species in complex stands of interior British Columbia. Eur J For Res 123: 45-51. http://dx.doi.org/10.1007/s10342-004-0020-z.

Turner MG, 1989. Landscape ecology: the effect of pattern on process. Ann Rev Ecol Syst 20: 171-197. http://dx.doi.org/10.1146/annurev.es.20.110189.001131.

Vonesh EF, Chiinchilli VM, 1997: Linear and nonlinear models for the analysis of repeated measurements. Boca Raton, Chapman & Hall/CRC: 560.

Wang Q, Ni J, Tenhunen J, 2005. Application of a geographically weighted regression to estimate net primary production of Chinese forest ecosystem. Global Ecol Biogeogr 14: 379-393. http://dx.doi.org/10.1111/j.1466-822X.2005.00153.x.

Watson DF, Philip GM, 1985. A Refinement of Inverse Distance Weighted Interpolation. Geoprocessing 2:315-327.

Wiens JA, 1976. Population responses to patchy environments. Ann Rev Ecol Syst 7: 81-120. http://dx.doi.org/10.1146/annurev.es.07.110176.000501.

Yang Y, Huang S, 2011: Comparison of different methods for fitting nonlinear mixed forest models and for making predictions. Can J Forest Res 41: 1671-1686. http://dx.doi.org/10.1139/x11-071.

Zhang L, Bi H, Cheng P, Davis CJ, 2004. Modelling spatial variation in tree diameter-height relationships. Forest Ecol Manag 189: 317-329. http://dx.doi.org/10.1016/j.foreco.2003.09.004.

Zhang L, Gove JH, Heath LS, 2005. Spatial residual analysis of six modelling techniques. Ecol Model 186: 154-177. http://dx.doi.org/10.1016/j.ecolmodel.2005.01.007.

Zhang L, Ma Z, Guo L, 2008. Spatially assessing model errors of four regression techniques for three types of forest stands. Oxford Journals 81(2): 209-225. http://dx.doi.org/10.1093/forestry/cpn014.

Zhang L, Ma Z, Guo L, 2009. An evaluation of spatial autocorrelation and heterogeneity in the residuals of six regression models. Forest Sci 55: 533-548.

Zhang L, Shi H, 2004. Local Modelling of Tree Growth by Geographically Weighted Regression. Forest Sci 50 (2): 225-244.

14

Resin tapping activity as a contribution to the management of maritime pine forest

Amélia Palma[1], João Miguel Pereira[2], Paula Soares[2]

[1]*Instituto Nacional de Investigação Agrária e Veterinária. Av. República, Quinta Marquês, 2780-159 Oeiras, Portugal.* [2]*Universidade de Lisboa, Instituto Superior de Agronomia, Centro de Estudos Florestais. Tapada Ajuda, 1349-017 Lisboa, Portugal*

Abstract

Aim of the study: In this work potential resin yield in a region of high forest ability where maritime pine is the main species was estimated in order to understand the viability of promoting resin exploitation.

Area of study: This study was conducted in Castro Daire County in central region of Portugal.

Material and methods: To quantify the resin yield of trees tapped for the first time two plots were installed in a maritime pine stand with average tree age 65 years. Before the beginning of the resin tapping, dendrometric tree variables were measured. Also, in a neighbouring stand, 25 trees were selected to check the relation between tree dbh and resin yield. Gum resin from every tree was weighted during the season. Estimates of potential resin yield in Castro Daire County were made based on data from National Forest Inventory plots, resin tapping legislation and resin yield values obtained in the field. Two scenarios were considered: high and low resin yield. To understand the intentions of forest owners towards restarting resin tapping activity 16 maritime pine forest owners were interviewed.

Main results: The results point out a high yield potential capacity for gum resin production in the County: values between 2,025 and 5,873 tons were obtained.

Research highlights: Results may highlight the important socio-economical role of the resin tapping activity and can be used to support national forest policies to the resin sector and give forest owners motivation to reactivate resin tapping activity.

Keywords: non-wood forest product; resin yield potential; forest owner.

Funding: STAR TREE. FP7-KBBE-2012-6 n°311919.

Competing interests: The authors have declared that no competing interests exist.

Correspondence should be addressed to Paula Soares: paulasoares@isa.ulisboa.pt

Introduction

Portugal holds a strong vocation for forest activities. According to the last National Forest Inventory (NFI), maritime pine forests occupy 23% (714,445 ha) of the territorial area (ICNF, 2013). Throughout the 20th century, maritime pine was the main species in the Portuguese forest but in result of poor management, forest fires and the incidence of forest pests such as the pine wood nematode, it became the third species in area (ICNF, 2013).

In the 20th century maritime pine was the most important species used as a source of gum resin obtained from living trees in Portugal. The country was an important world producer of this raw material, with yields higher than 100,000 tons before 1980 (CESE, 1996). Since then, the activity suffered a strong retraction, mostly due to forest fires and to international market competition, especially from China and Brazil, two countries that became large scale resin producers in that decade (Nieto & Herrero, 2013). In 2005, in Portugal, resin yield was around 6,000 tons and today resin tapping activity is residual (AR, 2015).

In the last few years, with the increase of rural unemployment, the social relevance of this activity in certain areas stood out. Also, resin and resin products are object of high demand worldwide, especially by the European chemical industry. Conjugated efforts and commitment of National and Regional level organizations and resin sector agents in Portugal and Spain, aiming at a sustainable exploitation of this resource, lead to an increased interest in developing the activity in both Iberian countries. Gum resin provides also an additional income, and may result in the improvement of forest management practices, namely by understory cleanings and increased surveillance during the summer period when fire risk is high.

Taking all these reasons into account the objectives of this work are:

(1) to quantify the resin yield in trees tapped for the first time

(2) to check the relation between tree size and resin yield

(3) to determine the potential resin yield in regions of high forest ability and where maritime pine is the main species

(4) to check the intention of forest owners toward restarting or initiating resin tapping activity.

Study area and methods

This study was conducted in a region that fulfils the following requirements: interior rural areas with ageing population; high percentage of forest land use mainly with maritime pine; predomination of small dimension ownership; resin tapping activity in the past. Castro Daire County, with 37,907 ha, where 15,688 ha are occupied by forest and 11,184 ha by maritime pine forest, was selected (AFN, 2010). In this region climate has Atlantic influence, with warm to hot summers and cold winters. Annual precipitation is of around 1,200 mm, and mean annual air temperatures are close to 14 °C (PMDFCI, 2013).

The field work took place during the 2014 resin tapping season in neighbouring counties where we found two forest owners who intended to tap. At Sátão, the resin tapping work was performed by a non-local forestry services company. The area was a pure mature stand with 65 years of age. Two plots, with 2,000 m^2 each, were installed. Dbh, total height and height at the live crown basis were measured and the number of tapped faces was recorded in all the trees in the plots and stand variables were calculated (table 1). The tapping operations lasted from July to the end of October and only two resin harvestings took place. Resin from every tree was weighted just before the harvesting operation. Five woundings with application of acid paste were done, and two of them were coincident with resin harvestings.

At Figueiredo Alva, resin exploitation was undertaken by the forest owner who lived near the forest area. Here the trees were wounded once a week with application of acid paste too. In this case, it was not a typical forest stand, but the trees were distributed along a narrow strip. It was not possible to install plots. So, a set of mature pine trees were selected and tapped for the first time with the objective to check the relation between tree size (dbh) and resin yield and to obtain a reference value to be used in the scenario "high resin yield" assuming a positive relation between resin yield and number of woundings (Palma, 2007) . Dbh was measured in all the trees (table 1). The tapping operations lasted from May to November and six resin harvestings took place. Twenty woundings were done. Only one face was open in each tree.

The potential resin yield of Castro Daire County was estimated using data from 17 inventory plots from National Forest Inventory (NFI5) (AFN, 2010) in pure and mixed dominant maritime pine stands in the County: 14 plots with measured trees, 2 plots classified as "burned" and 1 plot classified as "occupied by young trees". For the calculations, in each one of the trees of these plots, legal restrictions for resin tapping in Portugal were taken into account: (a) trees with dbh < 25.5 cm were not included; (b) trees with $25.5 \leq$ dbh < 35 cm, one tapped face; (c) trees with dbh ≥ 35 cm, two tapped faces. Mean yield values per face obtained at Figueiredo Alva and at Sátão were used in order to define two yield scenarios – respectively high and low resin yield. The resin yield was initially estimated at tree level, then summed over each NFI plot and finally reported to the hectare. Resin yield estimates for the Castro Daire County were based on the mean resin yield per hectare considering the 17 NFI plots mentioned above and reported to the total area of pure

Table 1. Characteristics of the two plots (Sátão) and the 25 tapped trees (Figueiredo Alva).

Site	N (ha^{-1})	G (m^2ha^{-1})	hd (m)	n	dmin (cm)	dg (cm)	dmax (cm)	hmin (m)	hm (m)	hmax (m)
Sátão – Plot 1	240	23.5	23.5	48	26.8	35.3	50.0	19.1	22.2	27.8
Sátão – Plot 2	280	26.3	24.6	56	25.3	34.6	46.3	17.5	23.4	29.6
F. Alva	–	–	–	25	27.7	37.3	47.8	–	–	–

N, number of trees; G, basal area; hd, dominant height; n, number of tapped trees; dmin, minimum diameter; dg, quadratic mean diameter; dmax, maximum diameter; hmin, minimum height; hm, mean height; hmax, maximum height.

and mixed dominant maritime pine stands in the County (AFN, 2010), considering the representativity of each NFI plot to the maritime pine area of the County.

In order to understand the intentions of forest owners towards restarting or initiating resin tapping activity, 16 maritime pine forest owners in Castro Daire County were interviewed. The questions focused on the socio-economic characteristics of the owners, experience with resin activity, and possible motivations to restart or initiate resin tapping.

Results and discussion

The variation in resin yield in each of the sites is great (Figure 1). Table 2 presents the resin yield obtained in Sátão and Figueiredo Alva. At Sátão, and considering both plots, the average yield was 1.91 kg/tree, with a maximum of 4.53 kg/tree and a minimum of 0.66 kg/tree. At Figueiredo Alva an average yield of 3.75 kg/tree was obtained, with a maximum of 7.42 kg/tree and a minimum of 1.95 kg/tree. This difference may be the result of the length of harvesting season, and number of resin wounds (Palma, 2007). Mean resin yield values per tree obtained in both sites are consistent with literature: 2-2.5 kg/face (Oliveira *et al.*, 2000), 2-higher than 4 kg/tree (Nanos *et al.*, 2001), 3.7 kg/tree (Tadesse *et al.*, 2001), 2.39-3.21 kg/tree (Rodríguez-García *et al.*, 2014, 2015). Mean resin yield values per face were used in the definition of the yield scenarios – high (3.75 kg/face) and low (1.31 kg/face) resin yield.

In Sátão, it is not observed a relation between tree size - dbh, total height and crown ratio (ratio between crown length and total height) - and resin yield per face (Figure 2). However a slight relation is detected between dbh and resin yield per tree (Figure 3). The short length of the tapping season, and the small number of wounds (five) may justify not only the low resin yield obtained but also the founded weak correlation. However in Figueiredo Alva trees were resin tapped throughout all the season (May-November) but the relation between dbh and resin yield was not observed (Figure 3). Rodríguez-Garcia *et al.* (2014) found that, on average, high yielder trees were taller and had larger dbh than low yielders with R^2 0.38 (between resin yield and dbh).

Resin yield has been reported to increase, in high yielders, in the years following the first wound opening (Rodriguez-Garcia *et al.*, 2014). Hence, considering that pine trees were tapped for the first time in 2014, it is possible to admit that in next campaigns greater resin yields will be obtained in the studied plots.

The potential resin yield for Castro Daire County, considering the low and high resin yield scenarios, is 2,024.5 tons and 5,872.6 tons, respectively. These values are quite optimistic knowing that the global resin yield in Portugal was slightly above the 6,000 tons in 2011 and 2012 (AR, 2015). These values can be considered the resin yield potential. However, the results from the 16 interviews revealed that local forest owners have a low motivation to restart or to initiate resin tapping activity and the main reasons pointed out are: the non-existence of a resin market, the lack of forest workers, high costs associated to resin activity when the work is sub-contracted, high risk of fire (re)occurency. These reasons are enhanced by the forest characteristics: small property dimension, not easy accessibility, unevenaged structure, natural regenerated with high density after a fire occurrence, and dense understory revealing lack of management.

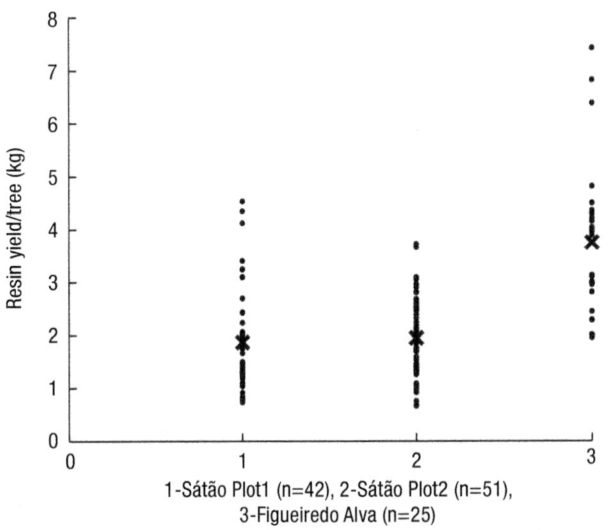

Figure 1. Variation of resin yield per tree in each site. Crosses represent the mean values; n, number of tapped trees.

Table 2. Resin yield in Sátão and Figueiredo Alva.

	Sátão*	**Figueiredo Alva**
	Total yield per face (kg)	
Average	1.31	3.75
Maximum	2.91	7.42
Minimum	0.55	1.95
	Total yield per tree (kg)	
Average	1.91	3.75
Maximum	4.53	7.42
Minimum	0.66	1.95

* values presented are the mean values of Plot 1 and Plot 2.

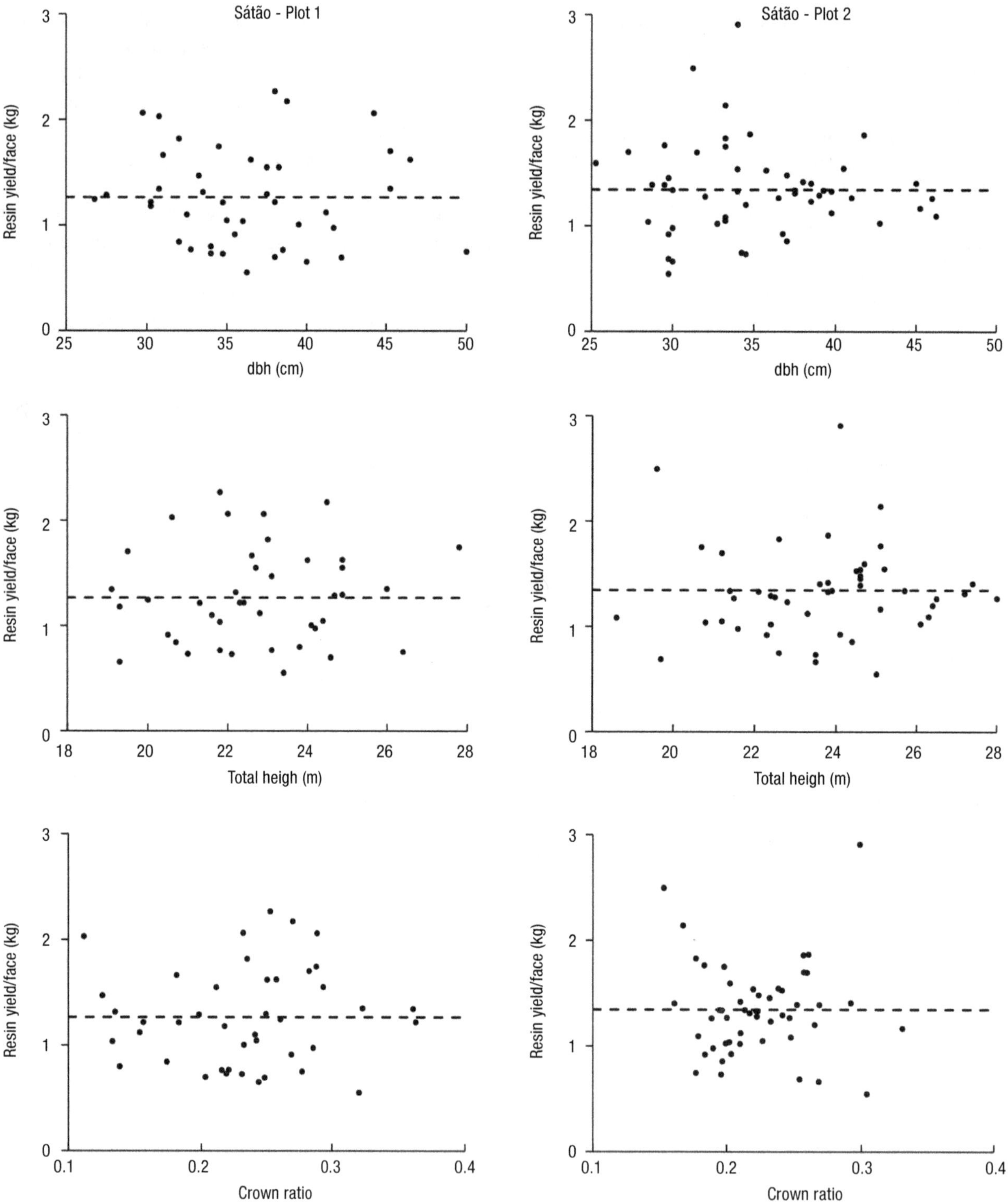

Figure 2. Relation between tree size and resin yield per face in Sátão (Plot 1 and Plot 2). Dotted lines represent mean resin yield.

Conclusion

Under proper management, maritime pine forests can provide multiple economic benefits. Resin tapping activity can be interesting for the maintenance and enhancement of the maritime pine forests, as it increases its internal rate of return and allows the forest owner to obtain annual revenues to practice a more active forest management. However, forest owners reveal low interest to restart or maintain the activity. Hence, promoting tapping as a financially attractive activity is essential to encourage these agents, as resin production depends essentially on forest owners.

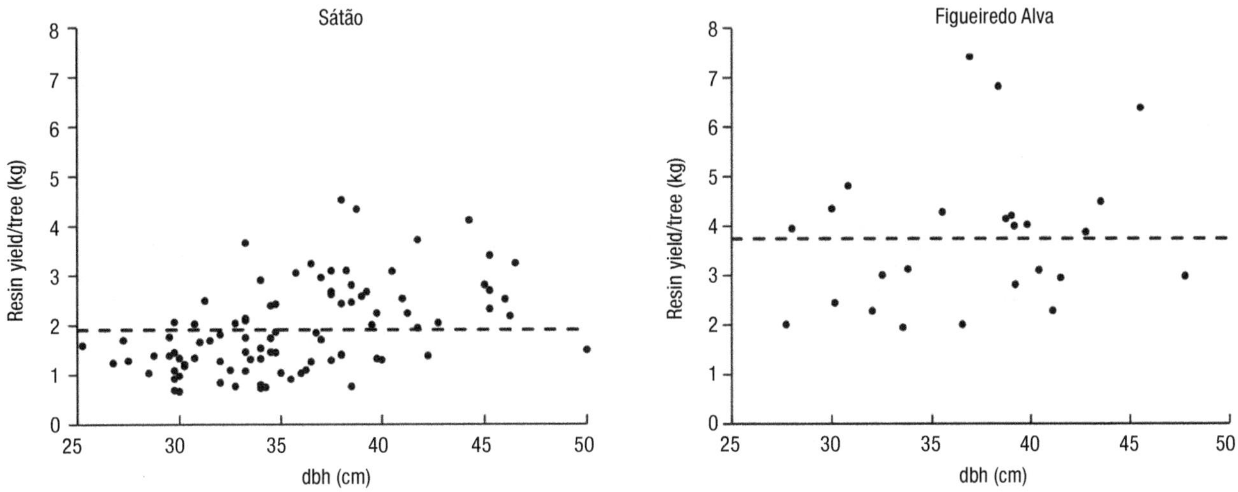

Figure 3. Relation between tree dbh and resin yield per tree in Sátão (Plot 1 + Plot2) and Figueiredo Alva. Dotted lines represent mean resin yield.

References

AFN, 2010. Inventário florestal nacional Portugal Continental IFN5 2005-2006. Lisboa.

CESE, 1996. O Setor florestal português. Conselho Ensino Superior Empresa, Ministério da Educação, Póvoa de Varzim, Portugal.

ICNF, 2013. 6º Inventário Florestal Nacional: áreas dos usos do solo e das espécies florestais de Portugal Continental. Resultados preliminares. Instituto da Conservação da Natureza e das Florestas, Lisboa.

Nanos N, Tadesse W, Montero G, Gil L, Alia R, 2001. Spatial stochastic modelling of resin yield from pine stands. Can J For Res 31: 1140-1147. http://dx.doi.org/10.1139/x01-047.

Nieto AP, Herrero FP, 2013. La resinación en España y en el mundo en 2013: situación y perspectivas. II International Symposium on Natural Resins 2013, Coca-Segovia, Ministerio de Agricultura, Alimentación y Medio Ambiente, Madrid.

Oliveira AC, Pereira, JS, Correia, AV, 2000. A silvicultura do pinheiro bravo. Centro Pinus. Viana do Castelo, Portugal.

Palma A, 2007. Capacidade produtiva de resina de pinheiro bravo. Breve panorâmica do sector resineiro em Portugal. Dissertação para Investigador Auxiliar, INRB, Oeiras.

PMDFCI, 2013. Plano Municipal de Defesa da Floresta Contra Incêndios de Castro Daire. Câmara Municipal de Castro Daire, Portugal.

Rodríguez-García A, Martín JA, López R, Mutke S, Pinilos F, Gil L, 2015. Influence of climate variables on resin yield and secretory structures in tapped *Pinus pinaster* Ait. in central Spain. Agricultural and Forest Meteorology 202: 83-93. http://dx.doi.org/10.1016/j.agrformet.2014.11.023.

Rodríguez-García A, López R, Martín JA, Pinillos F, Gil L, 2014. Resin yield in *Pinus pinaster* is related to tree dendrometry, stand density and tapping-induced systemic changes in xylem anatomy. Forest Ecology and Management 313: 47-54. http://dx.doi.org/10.1016/j.foreco.2013.10.038.

Tadesse W, Nanos N, Au-on FJ, Alía R, Gil L, 2001. Evaluation of high resin yielders of *Pinus pinaster* Ait. Forest Genetics 8(4): 271-278.

15

Asset-building payments for ecosystem services: assessing landowner perceptions of reforestation incentives in Lebanon

Arbi J. Sarkissian[1,2], Robert M. Brook[1], Salma N. Talhouk[2] and Neal Hockley[1]

[1]*Bangor University; College of Natural Sciences; School of Environment, Natural Resources and Geography; Deiniol Road, Bangor, Gwynedd, LL57 2UW. United Kingdom.* [2]*American University of Beirut; Faculty of Agricultural and Food Sciences, Dept. of Landscape Design and Ecosystem Management; P.O. Box 11-0236/ AUB Riad El Solh, Beirut 1107 2020. Lebanon*

Abstract

Aim of study: Incentivising landowners to supply ecosystem services remains challenging, especially when this requires long-term investments such as reforestation. We investigated how landowners perceive, and would respond to, distinct types of incentives for planting diverse native trees on private lands in Lebanon. Our aim was to understand landowners' attitudes towards hypothetical Payments for Ecosystem Services (PES) contracts options; their likely participation; and the potential additionality they would provide.

Area of study: Highland villages situated within eight of Lebanon's 20 Important Plant Areas.

Material and methods: Mixed-methods surveys were conducted with 34 landowners to determine past, present and future land-use strategies. Study participants were presented with three differently structured reforestation contract options (or schemes). The three schemes (results-based loan, action-based grant, and results-based payments) differed in their expected risks and benefits to landowners. Qualitative debriefing questions followed each of the schemes presented.

Main results: Although the results-based loan did deter uptake relative to the lower risk action-based grant, results-based payments did not significantly increase uptake or planting area, suggesting asymmetric attitudes to risk. Qualitative probing revealed economic, social (*e.g.* trust) and institutional factors (*e.g.* legal implications of planting forest trees on private land) that limited willingness to participate in the results-based contract option.

Research highlights: This study demonstrates the importance of combining qualitative and quantitative methods to better understand landowner perceptions of incentives and risks, particularly in challenging socio-political contexts.

Additional keywords: agro-ecosystems; biodiversity; conditionality; displacement; mixed-methods; participation; PES.

Abbreviations used: AFDC (Association for Forests, Development and Conservation), AUB-NCC (American University of Beirut – Nature Conservation Center); ES (ecosystem services); IPA (Important Plant Areas); LRI (Lebanese Reforestation Initiative); MOA (Ministry of Agriculture); MOE (Ministry of Environment); NGO (non-governmental organisations); PES (payments for ecosystem services).

Authors´ contributions Conception and design of research experiment; analyses and interpretation of the data: AJS and NJH. Data acquisition; drafting and editing of the manuscript: AJS. Administrative, technical and fieldwork support: SNT and RMB. Critical revision of the manuscript for intellectual content: NJH, RMB and SNT.

Funding: Khaldoun Barakat Research Fund (to the Nature Conservation Center at the American University of Beirut, AUB-NCC).

Competing interests: The authors have declared that no competing interests exist. The study received ethics approval from both Bangor University and AUB review boards.

Correspondence: should be addressed Arbi J. Sarkissian: arbi.sarkissian@outlook.com

Introduction

Economic theory postulates that many environmental problems exist because markets have not been fully developed for biodiversity or most ecosystem (or environmental) services (Pattanayak *et al.*, 2010). Increasing demand for agricultural commodities has therefore undermined important ecosystem services such as carbon sequestration and watershed protection, making agricultural expansion one of the major drivers of deforestation and biodiversity loss globally (Gibbs *et al.*, 2010). Despite the steady rise in protected areas in the last decade, conserving biodiversity is expected to become more challenging due to climate change and increasing competition for land (Pullin *et al.*, 2013). However, policy instruments such as payments for ecosystem services (PES) and agri-environment schemes are being adopted widely to incentivise landowners to supply off-farm ecosystem services from private lands (Schomers & Matzdorf, 2013).

PES are defined by Wunder (2007) as voluntary and conditional transactions between at least one buyer and one seller for the supply of additional units of a clearly-defined ecosystem service, or land-uses likely to generate those services. PES has become an attractive environmental policy instrument given its voluntary nature, allowing for public and private participation at various scales, and its flexibility in combining economic incentives with existing regulatory policies (Barrett *et al.*, 2013). Yet important challenges in designing PES include the trade-offs between efficiency and social equity, which can influence long-term ecological outcomes (Pascual *et al.*, 2014). From an economic perspective, PES investments often compete with existing land-uses (*e.g.* agriculture) and maintaining lower payments would attract landowners with lowest opportunity costs. This approach may have distributional consequences, since landowners with larger holdings and lower opportunity costs are favoured over those with smaller holdings whose incomes are tied to farming (McDermott *et al.*, 2013). Competitive PES schemes could also displace agriculture or other productive activities leading to land conversion and intensification elsewhere sometimes referred to as 'leakage' (Pattanayak *et al.*, 2010).

Hitherto, PES have largely focused on use-restricting strategies, *e.g.* avoided deforestation, but are increasingly employed to finance reforestation (or afforestation), referred to as asset-building schemes (Wunder, 2008). However, recent studies have criticised carbon-focused PES and Reduced Emissions from Deforestation and Degradation (REDD+) for incentivising monoculture plantations, negatively impacting biodiversity and local livelihoods (*e.g.* Lindenmayer *et al.*, 2012). Locatelli *et al.* (2014) argued that carbon-focussed incentives would not automatically result in bundled co-benefits for biodiversity and local ecosystem services. Designing biodiversity-enhancing reforestation schemes (using even mixes of native species) that are both cost-effective and attract participants remains challenging. We administered a mixed-methods survey to explore the willingness of Lebanese landowners from highland villages to accept incentives for planting diverse native tree species on private lands. Survey participants were presented with three alternative PES contracts schemes (Table S1 [suppl]): Scheme 1, a results-based loan (involving repayments conditional on seedling survival: negative conditionality); Scheme 2, an action-based grant (conditional on planting only); and Scheme 3, results-based payments (conditional on seedling survival: positive conditionality). Our aim was to understand landowners' attitudes towards these three differently structured hypothetical PES contracts options; factors influencing decisions to participate; and the likely displacement that could result.

Material and methods

Reforestation in Lebanon

While reforestation efforts were traditionally conducted by the Ministry of Agriculture (MOA), in 1998 the Lebanese parliament transferred funds instead to the Ministry of Environment (MOE) to develop a National Reforestation Plan (NRP) to increase forest cover from 13% to 20% (Regato & Asmar, 2011). Early phases of the NRP suffered high seedling mortality, partly due to a lack of funds for maintenance (*e.g.* irrigation and protection from grazing). Therefore, the last phase of the NRP (c. 2009-2012) was developed as a quasi-PES scheme, where selected municipalities were paid at different stages based on area planted and survival outcomes (MOE/UNDP/GEF, 2014). The Lebanese government has also shown interest in using the Kyoto Protocol's Clean Development Mechanism (CDM) to fund re/afforestation projects to reduce its net greenhouse gas emissions (MOE/UNDP/GEF, 2009). More recently, the MOA initiated a campaign to plant 40 million trees through its National Afforestation/Reforestation Program initiated in 2014 and intends to adopt a forest and landscape restoration approach (Mohanna *et al.*, 2017). Non-governmental organizations (NGOs) with interests in using PES have increasingly become active in re/afforestation in recent years (R. Paton, 2012, *pers. comm.*).

Our study therefore assumes that ecosystem services (ES) buyers could be either the public sector (*e.g.* MOA/MOE), NGOs, or both (with funding often provided through partnering international donor agencies). We determined in a previous study that these reforestation stakeholders are interested in increasing forest cover to enhance a broad array of forest ES (including landscape beauty, soil and water conservation, as well as biodiversity) rather than paying for specific ES (*e.g.* carbon sequestration). However, the ES anticipated heavily depends on the kinds and ratios of species being planted and managed. For example, plantations of fast-growing trees (*e.g.* eucalypts) may sequester carbon much more efficiently than most slow-growing natives, yet may also limit certain regulating services (*e.g.* water and nutrient cycling, pollination, disease mitigation). In fact, stakeholders expressed concerns over the lack of species diversity in past reforestation as well as the recent increase in exotics (*e.g. Paulownia* spp.), and have begun addressing the importance of maintaining resilient forest ecosystems through diversifying the planting of native species (R Paton & S Bou Fakhreddine, 2012, *pers. comm.*). Fruit trees are commonly planted in these regions and apples account for a sizeable proportion of crop production. Apple orchards may contribute to certain ecosystem services such as carbon sequestration (Wu *et al.*, 2012); however, commercial orchards often require high inputs (*e.g.* irrigation, pesticides and fertilizers), which can negatively impact biodiversity and other ES (*e.g.* pollination and watershed maintenance).

Recent reforestation efforts in Lebanon have focussed predominantly on municipal lands. Some implementing stakeholders expressed doubts about transacting with private landowners due to uncertainties with long-term tree retention and costs (R Paton, 2012, *pers. comm.*). However, while opportunity costs may sometimes be lower on municipal lands, transaction costs (*e.g.* monitoring to ensure compliance) are often higher compared to private landowners with proper titles (Engel *et al.*, 2008), particularly under reforestation contracts that often extend beyond the political terms of elected mayors. Yet contract attributes that characterise conditionality are what mainly influence transaction costs, thus posing a significant challenge for designing cost-effective PES contracts (Peterson *et al.*, 2015). Risk and uncertainty appear to be ubiquitous in many farming decisions, *e.g.* crop-selection and adopting new technologies, where decisions are made based on both attitudes towards risks and subjective beliefs (Menapace *et al.*, 2013). These factors, along with opportunity costs and the institutional context where transactions occur, can influence landowner decisions to participate in PES schemes. Understanding Lebanese landowners' perceptions of asset-building PES is therefore critical for informing future reforestation policy.

Factors affecting PES uptake

While participation in PES schemes often depends on landowners' opportunity costs (Chen *et al.*, 2010), the literature has identified other factors that affect participation in asset-building PES schemes, including contract design and social-institutional factors.

PES must be conditional on verified actions (*e.g.* planting trees) or results (*e.g.* carbon sequestration), requiring monitoring of sellers to ensure compliance (Honey-Rosés *et al.*, 2009). In asset-building programmes like reforestation, with high short-term costs and delayed benefits, a fundamental issue of concern to PES buyers is ensuring long term delivery of ecosystem services (Pattanayak *et al.*, 2010). For PES buyers, contract designs often involve trade-offs between supplier uptake, transaction costs, and expected outcomes (Engel *et al.*, 2008). Contracts that are highly bureaucratic or involve excessive conditionality are perceived as being too onerous or risky, reducing landowner uptake (Hudson & Lusk, 2004). In contrast, lack of conditionality or monitoring could result in non-compliance (*e.g.* hidden action) by sellers (Wunder *et al.*, 2014). The choice of payment by actions or results, together with the optimal level of conditionality and monitoring, will depend on the context: the strength of the connection between actions and results, the ease of monitoring each, and the level of risk aversion of sellers and buyers (Gibbons *et al.*, 2011). Asset-building PES may therefore require a mixture of results- and action-based payments over time to cover high initial costs whilst ensuring tree retention (Wunder *et al.*, 2014). Payments are often frontloaded and gradually decreased once private benefits from planted trees were available to participants, but this is best suited to productive species (Hegde *et al.*, 2014). Setting conditions for ensuring mixed native species are planted and retained is more challenging (Montagnini & Finney, 2011).

Understanding farmers' identities and how they perceive risks or uncertainties towards livelihood changes is also important (Duesberg *et al.*, 2013). Social-institutional factors such as trust in (or experience with) incentive-based schemes, local norms and values, dependence on farm-based activities, as well as age and level of education also influence landowners' decisions to join PES schemes (Chen *et al.*, 2009; Fisher, 2012). Participants in asset-building PES tend to have relatively large landholdings, with enough land unsuitable for agriculture, and whose incomes are largely off-farm (Cole, 2010). The context under which the farming system is structured, along with secure tenure and technical or financial know-how may also determine uptake (Kosoy *et al.*, 2008). Factors such as commitment period and required percentage of landholdings allocated have also been found to affect farmer uptake into PES schemes (Kisaka & Obi, 2015). Building trust in the institutions responsible for ensuring payments often takes time, and poorer more risk-averse landowners may be less willing to participate (Fisher, 2012). These issues are particularly critical in cases where governments are buyers or intermediaries, yet have lost the confidence of farmers through previous policies. Beyond this, PES is even more challenging to implement under circumstances where legal and property institutions are weak, which is common in many developing countries (Matzdorf *et al.*, 2013). Even in developed countries like Germany, land tenure implications and contractual uncertainties were principal reasons behind farmers' reluctance to join PES schemes (Schleyer & Plieninger, 2011).

Study area

Lebanon is a small (10,452 km^2), predominantly mountainous country located in the eastern Mediterranean basin and recognised as a centre for plant diversity (Davis *et al.*, 1994). Recognising threats to plant diversity in the eastern Mediterranean, a small team of scientists from the American University of Beirut's Nature Conservation Center (AUB-NCC) started a project to define Important Plant Areas (IPAs) in Lebanon (Yazbek *et al.*, 2010). Designated IPAs are also shown to represent the major ecosystems and unique habitats of Lebanon. Our study area comprised the western slopes of Mount Lebanon where eight of Lebanon's 20 newly designated IPAs are located (Radford *et al.*, 2011). These areas are characteristic of eu-mediterranean (> 1,000 m) to oro-mediterranean (2,200 - 2,800 m) bioclimatic zones, averaging between 1000-1200 mm/yr, precipitation mostly occurring between November and March. While this region has been characterised as being predominantly semi-arid, many microclimates can exist between and even within some IPAs selected for our study (Yazbek *et al.*, 2010). The vegetation types are typical of Mediterranean forest, woodland and scrub communities containing coniferous, deciduous and mixed forest/woodlands, interspersed with semi-natural agro-ecosystems (Makhzoumi *et al.*, 2012). The region produces a variety of tree crops, predominantly apples and stone fruits (Salibi, 2007). The main rainfed crops are cherries, often planted at much higher elevations. Irrigation is a major limiting factor for farmers in these steep and rocky landscapes, requiring extensive terracing to preserve soils and enable irrigation using canals. Irrigation comes from the numerous springs that form the tributaries of seasonal rivers and streams that flow along the western flank of Mt. Lebanon into the sea. Habitats are increasingly threatened by land-uses that include intensive agriculture, overgrazing, urbanisation and quarrying, as well as fires (Sattout & Abboud, 2007).

Sampling

We focused on IPAs with reforestation potential but were unable to conduct our research in high-risk parts of the country, *i.e.* the Bekaa Valley, South Lebanon and near the Syrian border. Eight of the 20 IPAs located along the west-facing slopes of the Mt Lebanon were selected for this study (Fig. 1). Due to security concerns, many villages in the Akkar district (LB07) near the Syrian border were also excluded. A total of 248 villages were identified using Google Earth images embedded with IPA layers that were copied and transposed over administrative maps showing all village/municipal boundaries. Villages were stratified according to IPA, estimated geographic size, population, rurality and elevation. A stratified random sample of 18 villages within these IPAs were selected (see Table S2 [suppl]). Security concerns also necessitated obtaining landowner contact details from mayors and other key informants from sampled villages who acted as our gatekeepers and facilitated our research. We obtained contact details for 52 landowners who were sole proprietors of their holdings, who were then telephoned. After at least two attempts we spoke to 46 landowners, informed them of the study objectives, and asked for their oral consent. Twelve landowners declined to participate because of a lack of land, land tenure issues (*e.g.* inheritance), age or inconvenience. The final survey was conducted with 34 newly recruited participants with their written consent who had not participated in a previous extensive pilot. Our research team was faced with substantial safety risks given the turmoil in Syria, which at the time began showing signs of potentially spilling over into Lebanon. This limited our sample size.

Data acquisition, survey instruments and analyses

The survey (see Table S3 [suppl]) was conducted in Arabic by the first author and a field assistant in the participants' villages, either at their farm, home, workplace, or the municipality office. After obtaining written consent, each participant was given an overview of the study and its objectives. After discussing current and intended land-use, the interviewer introduced the three hypothetical PES schemes in succession (see Table S1 [suppl]) to gauge their acceptability and to stimulate discussion of the key research themes identified above (the schemes were presented to each respondent in the same order for this reason). Entry into any of the schemes only required that they plant a minimum of 1,000 m^2 of contiguous land that they had titles to with the seedlings provided under the programme. Study participants were provided with a list of available native species (see Table S4 [suppl]) that would be used in the PES programme and told that the kinds and quantities of each species would be determined by the programme team. Seedling survival would be estimated by a monitoring team using randomly selected plots (Griscom *et al.*, 2005), assessed on a yearly basis during the five-year period. Follow-up questions were asked after each scheme was presented, which included where they would plant the seedlings and how much area. They were also asked whether the schemes would change

Figure 1. Partial map of Lebanon showing the eight important plant areas (IPAs) in the study area, indicated as shaded polygons (Yazbek *et al.*, 2010). Landowners from 17 villages located within these IPAs were sampled for this study. Source for base mapping: Google Earth.

their intended planting plans (*e.g.* to plant crop trees) for that plot. We did not specify what land-use/land-cover type would be replaced under each scheme, thus study participants were free to decide where to plant trees and how much area this would entail. Follow-up questions determined the extent of agricultural displacement expected (*e.g.* croplands *vs* abandoned/marginal lands). In addition, they were asked open-ended qualitative questions (coded with responses seen only by the interviewer) regarding perceived benefits of the proposed schemes. Respondents who did not wish to participate in any of the schemes were prompted to discuss why they would opt out. These questions were designed to assess the kinds of risks and uncertainties associated with PES schemes of this nature with respect to landowners' perceived benefits in being paid to reforest with diverse native species. The survey concluded by asking what sort of constraints or future land-use changes the participants envisaged, followed by some basic socioeconomic questions.

Quantitative data was analysed using SPSS version 20 (Pallant, 2010) to determine (i) whether there was a significant difference in uptake and area enrolled for reforestation under each consecutive scheme, and (ii) whether land-owner type, age or landholding size influenced participation and land enrolment into corresponding schemes. Qualitative data was transcribed and translated into English by the field assistant. Audio recordings and transcripts were analysed by the first author to identify important themes.

Results

Basic attributes of the sample

All participants in the sample (n=34) were males between the ages of 30 and 81 with a median age of 57. Median household size was five. Over three quarters of the respondents were permanent residents of their villages while the remainder (n=8) spent only summers there. This is likely to be an artefact of sampling but we believe our sample is broadly representative of the relevant population, *i.e.* landowners with some active level of interest in managing the land. Ninety-one per cent indicated that their landholdings were located within villages where they resided. Respondents were generally well educated (Fig. 2a) and included full-time commercial farmers, part-time farmers and hobby or retired farmers (Fig. 2b). Aggregated landholding area of the sample was approximately 227 ha. Parcels ranged from 0.15 to 30 ha (median=3 ha). Nine landowners owned property over 10 ha, consisting of mainly hobby/retired farmers. Landholding size differed weakly between levels of education (Kruskal-Wallis H-test=7.810, p=0.099). Part-time and hobby farmers did not have smaller landholdings than full-time farmers (Kruskal-Wallis H-test=0.258, p=0.879; Fig. 2b).

Past and intended future planting (in the absence of PES)

The sample was highly skewed in terms of landholding size, recent planting area, and number of crop trees planted per respondent. Apples (*Malus domestica* Borkh) were the main commercial crop trees planted, followed by stone fruits (*e.g. Prunus* spp.). A large portion of the commercial tree crops were planted on previously abandoned croplands (Fig. 3). Four respondents indicated that they had planted productive native trees, *e.g.* stone pine (*Pinus pinea* L.), but none had planted other native trees. Nearly 75% had planted over 100 commercial saplings within the last 10 years.

Eighteen respondents intended to plant more trees in the near future. Fifteen hectares was the approximate total area expected to be planted with over 75% taking place on previously abandoned lands. Apples, stone fruit, and nut-bearing trees were the main commercial trees to be planted, with mean anticipated areas of 7.1, 5.5 and 2.1 ha, respectively. None mentioned intentions of planting native forest trees in the future other than stone pine.

Participation and land enrolment in the PES schemes

Twenty-two landowners would be willing to participate in the results-based loan (Scheme 1), offering 21.9 ha of land for reforestation (approximately 10% of total landholding area). Participation increased to 27 farmers with 35.5 ha land enrolled (c16%) for the action-based grant (Scheme 2), but the results-based payments (Scheme 3) did not change the number participating, and only slightly increased the land area to 37.5 ha (17%). A Friedman test indicated a statistically significant difference in land enrolment between schemes (Friedman's ANOVA χ^2 (2)=25.10, p<0.001). *Post hoc* Wilcoxon tests found a significant increase in land enrolment from Scheme 1 to Scheme 2 (T=169, r=−0.62, p<0.001) and Scheme 1 to Scheme 3 (T=198, r=−0.60, p<0.001), but not from Scheme 2 to Scheme 3 (T=77.5, r=−0.27, p=0.116). We tested whether total landholding size (in ha), age, and landowner type (divided between 'full-time farmer' and 'other') influenced participation in each of the three schemes using logistic regression following preliminary analyses to ensure underlying assumptions of models were not violated (Pallant, 2010). Younger landowners and those with larger holdings were more likely to participate in Scheme 1, but these effects disappeared for Schemes 2 and 3 as a greater number of older landowners and

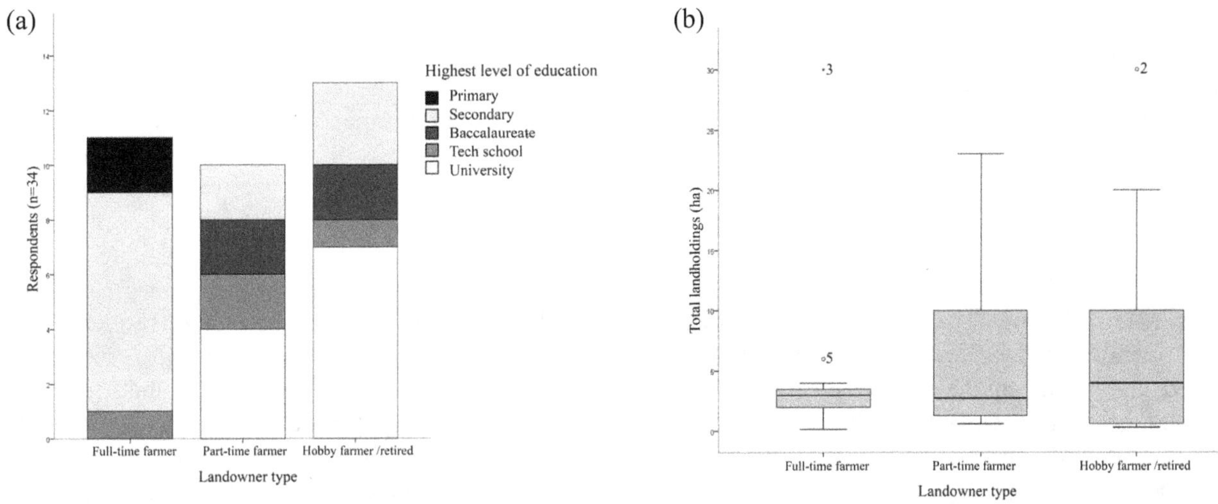

Figure 2. Landowner type subdivided by education (a). Total landholdings by landowner type (b). Landowner type was divided between full-time farmer (most income derived from farming), part-time farmer and hobby / retired farmer.

Table 1. Logistic regression for predicting likelihood of enrolling in PES schemes[a]

		B	SE	Wald	d.f.	Sig.	OR	95% CI for OR	
								Lower	Upper
Scheme 1	(Constant)	3.969	2.235	3.152		0.076	52.911		
	Landowner type	1.182	1.079	1.200	1	0.273	3.260	0.393	27.015
	Age	-0.090	0.043	4.342	1	*0.037	0.914	0.839	0.995
	Landholding size	0.368	0.183	4.044	1	*0.044	1.4450	1.009	2.068
Scheme 2	(Constant)	0.689	1.994	0.119		0.730	1.991		
	Landowner type	-1.808	1.271	2.024	1	0.155	0.164	0.014	1.979
	Age	-0.009	0.036	0.065	1	0.799	0.991	0.923	1.063
	Landholding size	0.863	0.557	2.399	1	0.121	2.370	0.795	7.064
Scheme 3	(Constant)	2.574	2.092	1.514		0.219	13.112		
	Landowner type	-0.072	0.997	0.005	1	0.943	0.931	0.132	6.568
	Age	-0.039	0.036	1.159	1	0.282	0.962	0.896	1.032
	Landholding size	0.319	0.236	1.821	1	0.177	1.375	0.886	2.185

[a]Collinearity diagnostics showed that there was no violation of multicollinearity assumptions with the variables tested (VIF=1.015). Normal probability plots of the regression standardised residuals showed there were no outliers (critical value=13.82; Mahal maximum distance=8.84). *$p<0.05$

landowners with smaller holdings were attracted to the schemes (Table 1).

Agricultural displacement under PES schemes

Seventeen per cent of reforestation would be on cultivated lands (or land in use) under the results-based loan (Scheme 1), 11.5% under the action-based grant (Scheme 2), and 12.4% under the results-based payments (Scheme 3). However, over 65% of respondents that indicated cultivated lands under any of the schemes mentioned they would plant at the margins (*e.g.* borders) of existing cultivation. Eight respondents stated their intended planting plans would change under schemes (*i.e.* native trees would be planted in place of crop trees) of which four mentioned plantings would take place on cultivated lands.

Respondents who declared they would participate in at least one of the three schemes (n=29) were asked if they would foresee any possible land-use changes that may impact the trees in the future. Twelve mentioned no foreseeable changes, ten indicated they may build on those plots, four mentioned passing lands onto children, and three indicated possible agricultural land-use changes. Ten respondents mentioned on-farm benefits of forest trees as possible reasons for maintaining trees beyond the life of the scheme. These included erosion prevention, regulating local climates, filtering the air, and as windbreaks. Four respondents also mentioned increasing landscape beauty as a benefit, related to potential future investments in ecotourism activities. Finally, over half of the participating respondents indicated they would be interested in longer term payments.

Landowner perceptions of PES schemes

The hypothetical schemes were used to initiate a discussion of landowners' perceptions of PES schemes in general, and specific characteristics of the three schemes. Respondents' views of PES varied with a greater portion seeing advantages of providing financial and technical support for farmers. One respondent claimed he would buy more land to enrol if these types of support were genuine and trustworthy. Unsurprisingly, respondents showed a greater keenness towards the action-based grant (Scheme 2) over the results-based loan (Scheme 1) due to relaxed conditions of the latter (*i.e.* lower risk), but the results-based payments (Scheme 3) was no more popular. Some respondents discussed higher payoffs as the main advantage that Scheme 3 had over Scheme 2. One respondent mentioned continuity of payments as a major advantage, and increased the enrolled land by 1 ha from Scheme 2 to Scheme 3:

"The 3rd [scheme] ensures a certain continuity to the [reforestation] plan by [incentivising] the

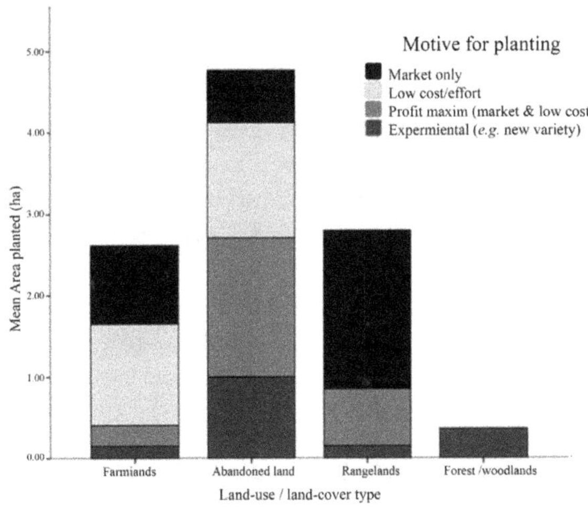

Figure 3. Mean area planted with commercial crops under different land-use/land-cover types subdivided by farmers' motives for planting the crops mentioned. 'Low cost / effort' refers to easy management of the trees, 'Market only' refers to high market value of the crop, and 'Profit maxim' denotes respondents who mentioned both low cost and high market value. *Note:* An outlier was excluded to better present the results in this figure.

farmer to [put] more effort in [ensuring] high rates of [survival] so he can get the highest amount of money" (Resp. #6)

Yet, fifteen respondents enrolled the same amount of land for all three schemes. Availability of land unsuitable for agriculture was the main constraint to participation and land enrolment mentioned. While there was a marginal increase of land enrolled for those who would participate in all three schemes (n=6), one respondent would enrol less land for Scheme 3 than Scheme 2. His reasoning was that since there is no need for maintenance under the action-based grant (Scheme 2), he would plant a much larger yet more remote plot with limited access, whereas he would plant borders of his orchard under the results-based payment (Scheme 3) for ease of care. However, many non-participants simply did not see any benefit of planting native trees regardless of the contract type or money offered. Most of the respondents who opted out of all schemes shared a dislike of non-productive native trees and/or diversified land-uses. For instance, one respondent mentioned he would not even consider diversifying his production, preferring to plant one profitable crop ("nothing beats apples in this region"). If given the option to plant native trees, the most likely candidate would be stone pine for its revenues from pine nuts, but most might still prefer to plant apples:

"Landowners won't grow forest trees on their agricultural lands for the following reasons: Fruit trees are more profitable [in the short-run] because they require less time to produce as opposed to [productive] forest trees; fruit trees can be secured as a source of revenue while the majority of forest [species] don't generate revenues" (Resp. #34)

In general, respondents' comments on PES schemes would suggest that opportunity costs were too high. Yet institutional factors may also influence uptake. For example, at least three of our respondents referred to the legal implications of planting forest trees. Since permits are required for cutting or removal of native conifers even on private lands (Regato & Asmar, 2011), landowners (especially farmers) may be reluctant to plant non-productive species on productive farmlands. And while Lebanon's current forestry policies may have contributed to relative gains in forest cover on abandoned farmlands, they may also have hindered effective forest management and made landowners reluctant to plant more forest trees:

"In the past, forests were well managed and protected by the local people because they were a source of [fodder], wood, and medicinal and aromatic plants... Today, more restrictions have been implemented by the MOA to protect forest areas, but this has actually discouraged people to preserve their forested lands because [these new laws have made forests] 'useless'. Now violations, neglect and forest fires have increased... [because only] when people find a benefit from something, they will work to protect [it]." (Resp. #1)

Lack of experience with incentive-based mechanisms, or attitudes towards government-sponsored agricultural programmes such as the MOA's 'Green Plan' (subsidies aimed at rehabilitating abandoned farmlands), may have also contributed to negative perception of PES schemes in general:

"...the lack of trust in the governmental institutions and the incapacity of the farmer to invest in such [agricultural] projects is the main reason most farmers won't apply [for] the Green Plan." (Resp. #21)

There were other factors respondents mentioned that contributed to lack of uptake and/or land-enrolment besides land availability and the land-use types in question (*e.g.* lack of land unsuitable for agriculture). In addition to negative attitudes towards native species, changing trends in land market prices were also important factors respondents raised that impact participation. Fewer younger landowners are actively managing their holdings than before, hence age will likely be a factor affecting uptake as well. Respondents in our sample were quite aged (median 57), which is reflective of a declining agricultural sector driving many households into cities in search of work. For instance, one respondent chose not to subscribe to any of the schemes both due to his age and the fact that his children no longer live in the village. He gave interesting insights that point to potential constraints in implementing a PES programme with landowners in Lebanon:

"Nowadays, the younger generation is not interested in agriculture and the older generation is no longer able to maintain the land... so [younger landowners] are selling their lands instead of [maintaining] and cultivating them... " (Resp. #17)

This suggests that reforesting private lands may be hindered by a lack of human resources to manage land on some farms and by development on others.

Discussion

Asset-building PES such as reforestation requires long-term maintenance to ensure future additionality of off-farm ES. Though results-based schemes may be more effective in ensuring long-term tree retention than action-based schemes, they depend heavily on landowners' perceptions of the credibility of such long-term payments. Frontloading payments to cover direct costs of planting and maintenance is common in PES using productive trees with private benefits (Hegde *et al.*, 2014), but this is much more challenging under biodiversity-focused PES as in this study. Cost-effective PES aimed at enhancing biodiversity therefore involve trade-offs for both buyers and potential suppliers with respect to risks (Banerjee *et al.*, 2017). For example, buyers would have to weigh trade-offs between efficiency (*e.g.* low payments, transaction costs, and displacement) and effectiveness (*e.g.* supplier uptake, extent of land enrolled, tree retention) when designing contracts while sellers weigh the risks and reward of those contracts (Table 2).

While it plays a key role in asset-building schemes, conditionality tends to limit participation if landowners perceive it as too risky (Chen *et al.*, 2009). The PES schemes in our experiment were designed specifically to investigate how landowners perceive risks related to conditionality. The first two schemes (results-based loan and action-based grant) differed substantially in their level of risk to landowners, and a reduction in risk predictably increased enrolment. However, surprisingly the addition of results-based payments (Scheme 3) did not significantly increase uptake, despite higher payoffs in the long run. This may be due to landowner perceptions of risk and uncertainty in general, particularly for those with incomes tied to farming and forestry (Blennow *et al.*, 2014; Menapace *et al.*, 2013), as well as risks specifically attributed to results-oriented schemes (Burton & Schwarz, 2013). For example, monitoring could be viewed as both an annoyance and loss of autonomy (Hudson & Lusk, 2004). Lack of trust was a key issue raised in the study by a handful of study participants and qualitative probing revealed some interesting responses by landowners with regards to trust in the schemes, as well as with PES in general. PES was as a novel concept to most of the study participants with only few ever having participated in reforestation. It would be expected that participants would feel some degree of uncertainty and distrust in PES until they could see how well it works in reality (*e.g.* experiences shared by neighbouring farmers participating in a PES programme). While compliance to monitoring appeared to have discouraged some respondents, this was not shared by most. There is also the possibility that Scheme 3 was not considered to be credible over the long timescale required to ensure tree retention by landowners; especially in a country which has experienced considerable socio-political turmoil.

In Lebanon, national agricultural policies have overshadowed multifunctional land-use strategies traditionally employed by local communities for managing natural resources (Makhzoumi *et al.*, 2012). This has led not only to poor management of forests (*e.g.* thickets prone to fires), but has also discouraged landowners from planting forest trees. Within this institutional context, landowners may be especially reluctant to plant trees that offer little private benefit in the long-run (as is the case with most native tree species). With respect to asset-building PES such as reforestation, additionality of most ecosystem services occurs over much longer time scales than with use-restricting PES. Perceptions of PES schemes and their subsequent adoption require evaluating long-term uncertainties (*e.g.* tenure, opportunity costs, market or political stability, climate, etc.), often affected by present day conditions of institutions and policies (Zanella *et al.*, 2014). The Lebanese may inherently exhibit more caution and scepticism in making decisions that require long-time commitments due to historically persistent political instability (Makdisi, 2004). Likewise, some participants mentioned having a general distrust in public institutions (*e.g.* MOA) due to previous experiences in subsidized programs like the 'Green Plan'. This raises questions on whether perceptions of PES would change based on who the buyers are, *e.g.* government *vs* NGOs *vs* private sector. Recovery from a 15-year civil war is being hampered by socio-political divisions that continue to paralyze the nation's public institutions. The public sector's inability to regain control of its institutions, due also in part to its lack of transparency, leads to widespread corruption (often referred to as '*wasta*', which appears to have become a social norm). A growing number of NGOs have begun to fill this void and have gained more trust than government institutions (Solberg, 2014) Such factors may have negative implications for PES in Lebanon as government institutions desperately try to re-establish oversight of the forestry sector through consolidating re/afforestation efforts and enforcing policies.

Many studies have found that participation in asset-building PES is contingent upon farm-based incomes (*i.e.* opportunity costs), farming systems (*e.g.* available marginal lands), landholding size, and age (*e.g.* Cole, 2010; Kisaka & Obi, 2015). Recent studies on incentives for re/afforestation have also shown that decisions to participate in such schemes may not be solely based on actual or perceived opportunity costs, but also on non-financial factors related to risks and uncertainties (Duesberg *et al.*, 2013). Our results suggest that consideration of opportunity costs was ubiquitous amongst respondents, especially if they could foresee possibilities of bringing land in disuse back into cultivation, or the prospect of developing their land in the future. Moreover, landowners' opportunity costs could vary from one plot to the next, and perhaps even within the same plot (Wunder, 2007). Respondents in our sample owned modest size holdings, and may have been conservative with how much land they would be willing to enrol. Landowners would have to consider important trade-offs when selecting plots with the lowest opportunity costs, such as direct costs of planting and irrigating seedlings on difficult terrain. This is particularly critical for less experienced tree planters who may underestimate the level of difficulty or work involved, which is especially relevant for those that are quite aged. Reasons why landowners in our study (particularly full-time farmers) would opt out of schemes accord with other studies, particularly if livelihoods would be affected by having native species on farms, such as loss of tenure and negative perceptions of biodiversity (Zubair & Garforth, 2006). More recent studies have also indicated that uptake of asset-building PES initiatives depends more on landowner attitudes and perceptions of how such policies affect future livelihoods (Trevisan *et al.*, 2016).

Table 2. Trade-offs between efficiency and effectiveness of schemes

	Scheme 1	Scheme 2	Scheme 3
Farmer uptake	Medium	High	High
Area enrolled	Low	Medium	Medium
Displacement	Medium	Low	Low
Risks (to farmers)	High	Low	Low/Medium
Transaction costs [a]	High	Low	High
Payment costs [a]	Low	Med	High

[a] These are reasonable estimates of the costs to buyers involved in mounting the schemes

The overall success of an asset-building PES programme in Lebanon requires not only long-term tree retention, but would have to factor in the programme's potential for displacing agriculture. This is important to consider in the context of an agricultural sector that is changing rapidly with emigration of rural households, combined with urbanisation and increasing land prices in some areas. Our results suggest that these schemes would not result in significant displacement, as they are not competitive with agriculture, but may therefore result in small and fragmented reforestation. Of the estimated 9,800 ha of abandoned farmlands in the six districts where our study was conducted, more than half was suitable for agriculture (Salibi, 2007). Many are abandoned due to lack of access to water and roads, in which case road-building and agricultural development projects could potentially increase opportunity costs. If an asset-building PES programme were implemented these infrastructural improvements could be stimulated by the programme itself.

Finally, we acknowledge that a more representative sample size would have helped tremendously in quantitative analysis of each scheme and would have helped draw a more cohesive picture of landowner perceptions to conditionality. However, this study was concerned with developing a more qualitative assessment of PES and the schemes presented, as has been conducted in other case-studies (*e.g.* Zanella *et al.*, 2014). We also acknowledge the limitations of mixed-methods studies in that it would have been difficult to combine qualitative analysis with a much larger sample size. A more representative sample would also have to include absentee landowners, whose incomes are presumably not tied to farming. Therefore, our study focussed on full-time residents, most with vested interests in farming. Despite this, there was considerable heterogeneity amongst farmers and their preferences, yet social and institutional aspects appeared to play an important role in uptake for most. These included issues with credibility and trust in new institutions as well as legal implications of planting native trees on private lands, resulting in high opportunity costs and unforeseeable risks in the future. Future research may want to examine whether absentee landowners, presumably having larger holdings with little or no commercial farming, are less risk averse than those in our sample, and thus may display a greater willingness-to-accept PES and enrolling more land. If PES buyers would prefer targeting absentee landowners with lower opportunity costs over farmers, they must be reminded that long-term ecological outcomes are closely tied to efficiency and equity trade-offs.

Conclusion

This paper examined the potential for PES to incentivize landowners to plant diverse native trees on private property. The objective of this mixed-methods study was to examine how Lebanese landowners perceive PES schemes and how different forms of conditionality might affect participation. Combined qualitative and quantitative methods enabled us to gauge landowners' perceptions towards schemes, helping to identify factors that would influence uptake, land enrolment and establishment of native trees on private property in the long run. Lebanese landowners from montane villages are heterogeneous in their occupations, landholdings, and preferences. Despite this, many appeared willing to participate in asset-building PES aimed at enhancing biodiversity. Qualitative probing revealed some of the constraints and challenges perceived by landowners, which helped strengthen our quantitative results. We found that the addition of results-based payments (Scheme 3) did not increase participation or land enrolment, possibly due to a lack of trust in long-term programmes, especially in a society facing constant turmoil. We also identified the importance of uncertain future opportunity costs in a rapidly changing rural context. This study demonstrates the importance of combining qualitative and quantitative data collection in studies of PES and shows that the potential for tailoring PES schemes to supply off-farm ecosystem services will depend on understanding landowners' perceptions.

Acknowledgements

The authors would like to thank Dr Mariana Yazbek for providing maps and expertise on Important Plant Areas of Lebanon as well as Mr Khaled Sleem and Mrs Dima Ousta for their help with translations of survey instruments and supporting documents. The corresponding author would like to thank Mr Edward Antoun for his transcriptions and translations, and for his invaluable contributions in the field. The authors would also like to thank the two anonymous reviewers for their comments and suggestions, which considerably improved our paper.

References

Banerjee S, Cason TN, de Vries FP, Hanley N, 2017. Transaction costs, communication and spatial coordination in Payment for Ecosystem Services Schemes. J Environ Econ Manage 83: 68-89. https://doi.org/10.1016/j.jeem.2016.12.005

Barrett CB, Bulte EH, Ferraro PJ, Wunder S, 2013. Economic instruments for nature conservation. In: Key Topics in Conservation Biology 2; Macdonald DW & Willis KJ (eds). pp: 59-73. John Wiley & Sons, Oxford. https://doi.org/10.1002/9781118520178.ch4

Blennow K, Persson J, Wallin A, Vareman N, Persson E, 2014. Understanding risk in forest ecosystem services: implications for effective risk management, communication and planning. Forestry 87 (2): 219-228. https://doi.org/10.1093/forestry/cpt032

Burton RJF, Schwarz G, 2013. Result-oriented agri-environmental schemes in Europe and their potential for promoting behavioural change. Land Use Policy 30 (1): 628-641. https://doi.org/10.1016/j.landusepol.2012.05.002

Chen X, Lupi F, He G, Liu J, 2009. Linking social norms to efficient conservation investment in payments for ecosystem services. P Natl Acad Sci USA 106 (28): 11812-11817. https://doi.org/10.1073/pnas.0809980106

Chen X, Lupi F, Viña A, He G, Liu J, 2010. Using cost-effective targeting to enhance the efficiency of conservation investments in payments for ecosystem services. Conserv Biol 24 (6): 1469-1478. https://doi.org/10.1111/j.1523-1739.2010.01551.x

Cole RJ, 2010. Social and environmental impacts of payments for environmental services for agroforestry on small-scale farms in southern Costa Rica. Int J Sust Dev World 17 (3): 208-216. https://doi.org/10.1080/13504501003729085

Davis SD, Heywood VH, Hamilton AC, 1994. Europe, Africa, South West Asia and the Middle East. In: Centres of plant diversity: a guide and strategy for their conservation; Heywood VH & Davis SD (eds). IUCN Publ Unit, pp: 354. Cambridge.

Duesberg S, O'Connor D, Dhubháin AN, 2013. To plant or not to plant—Irish farmers' goals and values with regard to afforestation. Land Use Policy 32 (0): 155-164. https://doi.org/10.1016/j.landusepol.2012.10.021

Engel S, Pagiola S, Wunder S, 2008. Designing payments for environmental services in theory and practice: An overview of the issues. Ecol Econ 65 (4): 663-674. https://doi.org/10.1016/j.ecolecon.2008.03.011

Fisher JA, 2012. No pay, no care? A case study exploring motivations for participation in payments for ecosystem services in Uganda. Oryx 46 (1): 45-54. https://doi.org/10.1017/S0030605311001384

Gibbons JM, Nicholson E, Milner-Gulland EJ, Jones JPG, 2011. Should payments for biodiversity conservation be based on action or results? J Appl Ecol 48 (5): 1218-1226. https://doi.org/10.1111/j.1365-2664.2011.02022.x

Gibbs HK, Ruesch AS, Achard F, Clayton MK, Holmgren P, Ramankutty N, Foley JA, 2010. Tropical forests were the primary sources of new agricultural land in the 1980s and 1990s. P Natl Acad Sci USA 107 (38): 16732-16737. https://doi.org/10.1073/pnas.0910275107

Griscom HP, Ashton MS, Berlyn GP, 2005. Seedling survival and growth of native tree species in pastures: Implications for dry tropical forest rehabilitation in central Panama. For Ecol Manage 218 (1-3): 306-318.

Hegde R, Bull GQ, Wunder S, Kozak R, 2014. Household participation in a payments for environmental services programme: the Nhambita Forest Carbon Project (Mozambique). Environ Dev Econ (0): 1-19.

Honey-Rosés J, López-García J, Rendón-Salinas E, Peralta-Higuera A, Galindo-Leal C, 2009. To pay or not to pay? Monitoring performance and enforcing conditionality when paying for forest conservation in Mexico. Environ Conserv 36 (2): 120-128. https://doi.org/10.1017/S0376892909990063

Hudson D, Lusk J, 2004. Risk and transactions cost in contracting: Results from a choice-based experiment. J Agric Food Ind Organ 2 (1): 1-17. https://doi.org/10.2202/1542-0485.1046

Kisaka L, Obi A, 2015. Farmers' preferences for management options as payment for environmental services scheme. Int Food Agribus Man 18(3): 171-192.

Kosoy N, Corbera E, Brown K, 2008. Participation in payments for ecosystem services: Case studies from the Lacandon rainforest, Mexico. Geoforum 39 (6): 2073-2083. https://doi.org/10.1016/j.geoforum.2008.08.007

Lindenmayer DB, Hulvey KB, Hobbs RJ, Colyvan M, Felton A, Possingham H, Steffen W, Wilson KA, Youngentob K, Gibbons P, 2012. Avoiding bio-perversity from carbon sequestration solutions. Conserv Lett 5 (1): 28-36. https://doi.org/10.1111/j.1755-263X.2011.00213.x

Locatelli B, Imbach P, Wunder S, 2014. Synergies and trade-offs between ecosystem services in Costa Rica. Environ Conserv 41 (1): 27-36. https://doi.org/10.1017/S0376892913000234

Makdisi S, 2004. The lessons of Lebanon: The economics of war and development. I.B. Tauris & Co., Ltd., London. 248 pp.

Makhzoumi JM, Talhouk SN, Zurayk R, Sadek R, 2012. Landscape approach to bio-cultural diversity conservation in rural Lebanon. In: Perspectives on nature conservation - Patterns, pressures and prospects; Tiefenbacher J (ed). pp: 179-200. InTech/Creative Commons. https://doi.org/10.5772/33343

Matzdorf B, Sattler C, Engel S, 2013. Institutional frameworks and governance structures of PES schemes. Forest Policy Econ 37(0): 57-64. https://doi.org/10.1016/j.forpol.2013.10.002

McDermott M, Mahanty S, Schreckenberg K, 2013. Examining equity: A multidimensional framework for assessing equity in payments for ecosystem services. Environ Sci & Policy 33: 416-427. https://doi.org/10.1016/j.envsci.2012.10.006

Menapace L, Colson G, Raffaelli R, 2013. Risk aversion, subjective beliefs, and farmer risk management strategies. Am J Agric Econ 95 (2): 384-389. https://doi.org/10.1093/ajae/aas107

MOE/UNDP/GEF, 2009. Towards 2010 Biodiversity Target: Fourth National Report of Lebanon to the Convention of Biological Diversity. Beirut, Lebanon. 243 pp.

MOE/UNDP/GEF, 2014. Safeguarding and Restoring Lebanon's Woodland Resources Technical Report. Beirut, Lebanon. 96 pp.

Mohanna C, Adada F, Besacier C, 2017. Forest and landscape restoration in Lebanon. http://www.fao.org/in-action/forest-landscape-restoration-mechanism/resources/detail/en/c/412643/ [Dec. 17, 2016].

Montagnini F, Finney C, 2011. Payments for environmental services in Latin America as a tool for restoration and rural development. AMBIO 40 (3): 285-297. https://doi.org/10.1007/s13280-010-0114-4

Pallant J, 2010. SPSS survival manual: a step by step guide to data analysis using SPSS. Open University Press/McGraw-Hill, Maidenhead. 345 pp.

Pascual U, Phelps J, Garmendia E, Brown K, Corbera E, Martin A, Gómez-Baggethun E, Muradian R, 2014. Social equity matters in payments for ecosystem services. Bioscience 64 (11): 1027-1036. https://doi.org/10.1093/biosci/biu146

Pattanayak SK, Wunder S, Ferraro PJ, 2010. Show me the money: Do payments supply environmental services in developing countries? Rev Environ Econ Policy 4 (2): 254-274. https://doi.org/10.1093/reep/req006

Peterson JM, Smith CM, Leatherman JC, Hendricks NP, Fox JA, 2015. Transaction costs in payment for environmental service contracts. Am J Agric Econ 97 (1): 219-238. https://doi.org/10.1093/ajae/aau071

Pullin AS, Bangpan M, Dalrymple S, Dickson K, Haddaway N, Healey JR, *et al.*, 2013. Human well-being impacts of terrestrial protected areas. Environ Evid 2 (1): 2-41. https://doi.org/10.1186/2047-2382-2-19

Radford EA, Catullo G, de Montmollin B, 2011. Important plant areas of the south and east Mediterranean region: priority sites for conservation. IUCN, Gland, Switzerland and Málaga, Spain. 108 pp.

Regato P, Asmar F, 2011. Analysis and evaluation of forestation efforts in Lebanon: Expert Report, Lebanese Ministry of Agriculture and FAO, Beirut, Lebanon. 59 pp.

Salibi A, 2007. Marketing Study for olive, olive oil and apple in Lebanon. Lebanese Ministry of Agriculture. Beirut, Lebanon. 30 pp.

Sattout EJ, Abboud M, 2007. Thematic assessment report on biodiversity. NCSA Project No. 00045426, UNDP/GEF/MOE. Beirut, Lebanon. 96 pp.

Schleyer C, Plieninger T, 2011. Obstacles and options for the design and implementation of payment schemes for ecosystem services provided through farm trees in Saxony, Germany. Environ Conserv 38 (4): 454-463. https://doi.org/10.1017/S0376892911000361

Schomers S, Matzdorf B, 2013. Payments for ecosystem services: A review and comparison of developing and industrialized countries. Ecosyst Serv 6 (0): 16-30. https://doi.org/10.1016/j.ecoser.2013.01.002

Solberg M, 2014. Patronage, contextual flexibility, and organisational innovation in Lebanese protected areas management. Conserv Soc 12 (3): 268-279. https://doi.org/10.4103/0972-4923.145138

Trevisan ACD, Schmitt-Filho AL, Farley J, Fantini AC, Longo C, 2016. Farmer perceptions, policy and reforestation in Santa Catarina, Brazil. Ecol Econ 130: 53-63. https://doi.org/10.1016/j.ecolecon.2016.06.024

Wu T, Wang Y, Yu C, Chiarawipa R, Zhang X, Han Z, Wu L, 2012. Carbon sequestration by fruit trees - Chinese apple orchards as an example. PLOS ONE 7 (6): e38883. https://doi.org/10.1371/journal.pone.0038883

Wunder S, 2007. The efficiency of payments for environmental services in tropical conservation. Conserv Biol 21 (1): 48-58. https://doi.org/10.1111/j.1523-1739.2006.00559.x

Wunder S, 2008. Payments for environmental services and the poor: concepts and preliminary evidence. Environ Dev Econ 13 (3): 279-297. https://doi.org/10.1017/S1355770X08004282

Wunder S, Nelson H, Nikolakis W, 2014. Lessons in the design of payments for environmental services: Theory and experience. In: Forests and Globalization: Challenges and Opportunities for Sustainable Development; Nikolakis W & Innes JL (eds). pp: 202. Routledge, NY.

Yazbek MM, Houri N, El-Zein M, Safi S, Sinno-Seoud N, Talhouk SN, 2010. Important plant areas in Lebanon: A preliminary study based on published literature and consultations with national experts. AUB-NCC, Beirut, Lebanon.

Zanella MA, Schleyer C, Speelman S, 2014. Why do farmers join payments for ecosystem services (PES) schemes? An assessment of PES water scheme participation in Brazil. Ecol Econ 105: 166-176. https://doi.org/10.1016/j.ecolecon.2014.06.004

apple orchards as an example. PLOS ONE 7 (6): e38883. https://doi.org/10.1371/journal.pone.0038883

Wunder S, 2007. The efficiency of payments for environmental services in tropical conservation. Conserv Biol 21 (1): 48-58. https://doi.org/10.1111/j.1523-1739.2006.00559.x

Wunder S, 2008. Payments for environmental services and the poor: concepts and preliminary evidence. Environ Dev Econ 13 (3): 279-297. https://doi.org/10.1017/S1355770X08004282

Wunder S, Nelson H, Nikolakis W, 2014. Lessons in the design of payments for environmental services: Theory and experience. In: Forests and Globalization: Challenges and Opportunities for Sustainable Development; Nikolakis W & Innes JL (eds). pp: 202. Routledge, NY.

Yazbek MM, Houri N, El-Zein M, Safi S, Sinno-Seoud N, Talhouk SN, 2010. Important plant areas in Lebanon: A preliminary study based on published literature and consultations with national experts. AUB-NCC, Beirut, Lebanon.

Zanella MA, Schleyer C, Speelman S, 2014. Why do farmers join payments for ecosystem services (PES) schemes? An assessment of PES water scheme participation in Brazil. Ecol Econ 105: 166-176. https://doi.org/10.1016/j.ecolecon.2014.06.004

16

FlorNExT®, a cloud computing application to estimate growth and yield of maritime pine (*Pinus pinaster* Ait.) stands in Northeastern Portugal

Fernando Pérez-Rodríguez*[1,2], Luís Nunes[1,2], Ângelo Sil[1,2], João C. Azevedo[1,2]

[1]*Escola Superior Agrária, Instituto Politécnico de Bragança, Campus de Santa Apolónia, 5300-253, Bragança, Portugal* [2]*Centro de Investigação de Montanha, Campus de Santa Apolónia, 5300-253, Bragança, Portugal*

Abstract

Aim of the study: To introduce and describe FlorNExT®, a free cloud computing application to estimate growth and yield of maritime pine (*Pinus pinaster* Ait.) even-aged stands in the Northeast of Portugal (NE Portugal).

Area of study: NE Portugal.

Material and methods: FlorNExT® implements a dynamic growth and yield modelling framework which integrates transition functions for dominant height (site index curves) and basal area, as well as output functions for tree and stand volume, biomass, and carbon content.

Main results: FlorNExT® is freely available from any device with an Internet connection at: http://flornext.esa.ipb.pt/.

Research highlights: This application has been designed to make it possible for any stakeholder to easily estimate standing volume, biomass, and carbon content in maritime pine stands from stand data, as well as to estimate growth and yield based on four stand variables: age, density, dominant height, and basal area. FlorNExT® allows planning thinning treatments. FlorNExT® is a fundamental tool to support forest mobilization at local and regional scales in NE Portugal.

Keywords: forest management; maritime pine; forest modelling; knowledge transfer tool.

Funding: This research was supported by the SIMWOOD project (Sustainable Innovative Mobilisation of Wood), EU FP7 Collaborative Project 2013-2017 Grant Agreement No. 613762.

Competing interests: The authors have declared that no competing interests exist.

Correspondence should be addressed to Fernando Pérez-Rodríguez: fernando.perez@ipb.pt

Introduction

The ability to predict growth and yield is a fundamental requirement for forest management and planning, therefore of utmost importance for the forest sector in any country or region. Forest growth is usually approached through mathematical modelling at the individual tree or forest stand level, according to a specific period of time and the silvicultural practices followed (Bravo *et al.*, 2011). The equations used for this purpose are, however, often dependent on a particular environmental context. In addition, forest growth equations are difficult to apply in practice when the end user lacks adequate knowledge or experience in forest modelling. In northeastern Portugal there is a high heterogeneity of biophysical conditions and forest management practices which makes it difficult to estimate standing volume. At the same time, there are no tools to facilitate the use of regionally adjusted forest growth and yield models for managers and stakeholders in general.

There are currently numerous forest growth and yield computational applications (Packalen *et al.*, 2013). Although most of them have been developed as desktop tools, e.g., ModisPinaster for maritime pine in the Tâmega river valley (Fonseca, 2004), GesMO for several species in Galicia (Diéguez-Aranda *et al.*, 2009), there are now platforms that take advantage of the available cloud computing capabilities that enhance data update and potentiate real time interactions with users, such as Web-Globulus 3.0 (Palma, 2016).

As part of an ongoing research effort to increase forest mobilization across Europe we developed a cloud computing application (FlorNExT®) to make forest growth and yield simulations accessible to any forest manager or stakeholder in northeastern Portugal to support management and to stimulate forest supply regionally. FlorNExT® has modelling capabilities for both maritime pine (*Pinus pinaster*) and Pyrenean oak (*Quercus pyrenaica*) although, for convenience, we centred this communication on maritime pine only. For that, we selected and analysed several forest growth models for maritime pine with potential application in the region and tested them with data from the region and against the PBRAVO model (Páscoa, 2001). In the development of FlorNExT® we followed as key principles user friendliness, simplicity, scalability, and self-explainability.

Specifications

FlorNExT® was programmed in Visual Studio Web 2013 Express, using MVC technology. The languages used were C# and XML for models, drivers and configuration and Razor, HTML, JavaScript and CSS for displays. There are Portuguese, Spanish and English versions of the application.

System description

Structure and models

The maritime pine module of the application uses a set of models from different authors (Table 1).

The dynamic forest growth models used in the application were:

— Projection of dominant height: Tomé (2001)
— Projection of basal area: Páscoa (1987)
— Initialization of basal area: Páscoa (1987)
— Mortality: self-thinning line equation of Luis & Fonseca (2004)
— Stand volume: Luis & Fonseca (2004)
— Biomass: Diéguez-Aranda *et al.* (2009)
— Carbon: Diéguez-Aranda *et al.* (2009)

The tree equations used for thinning were:

— Average diameter: quadratic mean diameter relationship of Fonseca (2004)
— Generalized height–diameter relationship: Almeida (1999)
— Tree volume over bark: Nunes *et al.* (2010).

The dynamic growth models project key stand variables over time and when thinning is applied, FlorNExT® computes the respective diameter distribution using a bi-parametric Weibull probability density function whose parameters are related to the projected stand variables. The intensity of thinning is defined using the Wilson Factor, computed as $Fw = 100/\left(\sqrt{N} \cdot H_{dom}\right)$. The volume extracted in a particular thinning is computed using the individual-tree volume and the height-diameter equations.

For the validation of FlorNExT®, total volume estimated by the set of equations in the tool was tested against estimates produced by PBRAVO (Páscoa, 2001) based on 5 cm diameter classes data (Table 2). Differences between volume estimates are small for most stand conditions. For a smaller diameter class amplitude (1 cm), individual-tree volume equation estimates are close to those of the stand volume equation (Table 2).

Also, stand and individual-tree volume and generalized height-diameter equations were evaluated with inventory data. All equations showed consistent behaviour producing adequate estimates when compared with measurements in IFN5 (IFN, 2005) plots (see e.g. height-diameter in Table 3). In addition, for a total of 916 pairs

Table 1. Summary characteristics of data in the models tested

	Stand models				
	Luis & Fonseca (2004)	**Fonseca (2004)**	**Tomé (2001)**	**Diéguez-Aranda *et al.* (2009)**	**Páscoa (1987)**
t (years)	7 – 80	12 – 65	9 – 45	9 – 45	10 – 80
H_{dom} (m)	3.2 – 29.8	5.6 – 25.9	3.5 – 23.0	4.6 – 20.4	8.9 – 23.9
G (m²/ha)	0.3 – 52.9	6.0 – 56.9	—	5.1 – 72.5	<7.5 – >27.5
N (trees/ha)	60 – 8480	200 – 7680	—	654 – 3142	<251 – 3333

	Individual tree models	
	Almeida (1999)	**Nunes *et al.* (2010)**
d (cm)	3.0 – 47.5	5.5 – 77.4
h (m)	2.3 – 25.2	5.6 – 36.0

t – Age; H_{dom} – Dominant height; G –Basal area; N – Density; d – Dbh; h – Total height

Table 2. Comparison of FlorNExT® and PBRAVO volume estimates for variable stand structure conditions (input data)

Input data				Volume estimates			
				FlorNExT® Stand	FlorNExT® Tree (DC 1 cm)	PBRAVO Tree DC 5 cm	FlorNExT® Tree DC 5 cm
t (years)	N (trees/ha)	H_{dom} (m)	G (m^2/ha)	Vob* (m^3/ha)	Vob** (m^3/ha)	Vob** (m^3/ha)	Vob** (m^3/ha)
10	1500	5.46	5	14.79	13.26	10.46	10.2
12	1800	6.18	8.7	24.97	26	20.15	16.87
13	1404	7	9.8	35.51	32.52	26.75	20.91
14	2700	6	12.5	39.49	37.2	27.71	24.64
15	1100	9	14.9	66.27	61.36	54.68	43.27
25	1500	16	35	233.07	228.26	195.39	172.14
29	1000	18	32	236.16	232.95	202.78	181.21
32	900	16	40	277.84	269.54	230.74	220.72
35	550	19	40	320.04	312.34	262.46	262.66
48	510	21	40	346.7	345.9	296.89	295.38

t – Age; N – Density; H_{dom} – Dominant height; G – Basal area; Vob – Volume over bark; * –Stand volume equation estimate; ** – Individual-tree volume equation estimate; DC 1 cm – 1 cm diameter class; DC 5 cm – 5 cm diameter class.

of *h* and *d* measurements made by the authors in plots recently established in the region, the Almeida (1999) equation presented a mean of prediction residuals (Bias) of 0.05 m, a mean of absolute prediction residuals (MAR) of 0.75 m and a modelling efficiency of 0.88.

Using FlorNExT®

FlorNExT® is organized in four sections: i) Home page, ii) Growth simulator, iii) Thinning simulator, and iv) About FlorNExT®. In the "Home page" the user is introduced to the basics of the application and has the possibility of choosing between the "Growth simulator" and the "Thinning simulator" sections. The "Growth simulator" allows the user to obtain stand growth data over time based on a small set of inputs and predefined thinning schemes. This section is suited for users who want to project stand growth and yield in the future considering thinning, either predefined or customized. No thinning is also a possibility. The "Thinning simulator" section, on the other hand, is directed to managers who want to define and apply instantaneously a particular thinning operation to a particular stand, although stand growth can be simulated after thinning.

The "Growth simulation" section (Fig. 1) estimates volume, biomass, carbon content, average and current annual volume increments, and site index based on the following inputs: stand age (years), density (number of trees/ha), dominant height (m), and growth projection (years). Stand basal area (m^2/ha) can be provided by the user but it will be estimated if unknown, although this

Table 3. Residuals of tree height estimates with Almeida (1999) and IFN (2005) equations as compared with measured trees in inventory plots in the Nordeste region. Only estimates for plots with more than five measurements are shown

Stand structure					Bias (m)		MAR (m)	
N (trees/ha)	H_{dom} (m)	Dg (cm)	D_{dom} (cm)	t (years)	(Almeida, 1999)	(IFN, 2005)	(Almeida, 1999)	(IFN, 2005)
840	12.82	20.55	30.51	29.50	0.77	0.03	1.16	1.12
300	12.90	23.20	29.20	31.00	0.47	-0.05	1.44	1.31
900	6.29	10.40	13.60	14.50	-0.02	-0.14	0.28	0.34
160	11.03	24.57	27.00	30.00	0.16	-0.36	0.51	0.48
1199	8.90	13.43	16.90	16.50	-0.23	-0.21	0.46	0.40
3658	11.10	11.42	18.38	30.00	-0.24	-0.23	0.46	0.35
260	8.20	14.20	19.80	18.00	-0.25	-0.25	0.81	0.72
200	6.70	15.70	17.80	14.70	-0.23	-0.54	1.03	1.18
840	5.90	11.10	11.60	13.50	0.18	0.19	0.51	0.52
840	15.50	23.60	27.30	26.50	1.01	1.00	1.43	1.32

N – Density; H_{dom} – Dominant height; Dg – Quadratic mean diameter; D_{dom} – Dominant diameter; t – Age; Bias – mean of prediction residuals; MAR– mean of absolute prediction residuals.

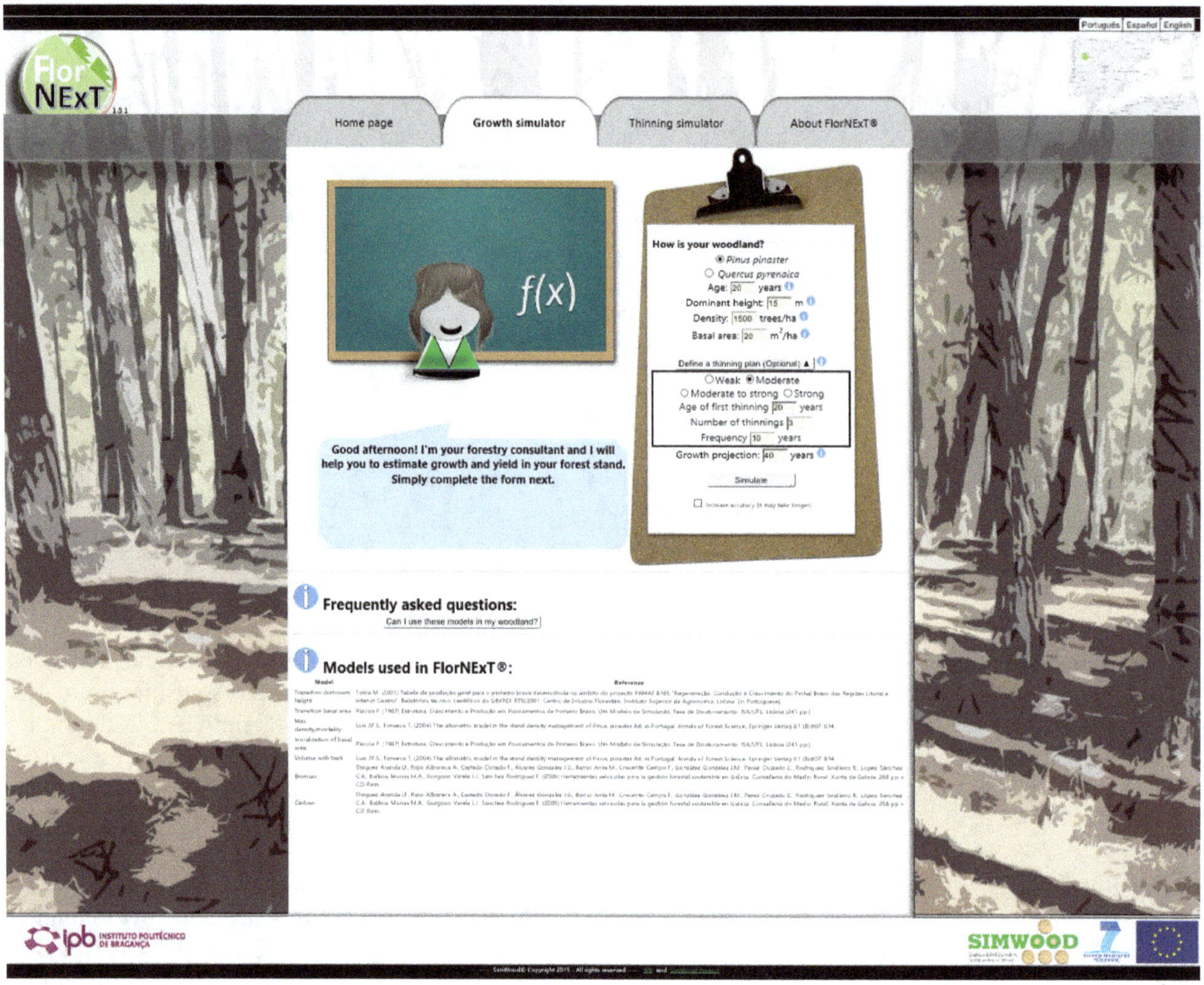

Figure 1. The FlorNExT® "Growth simulator" section.

might result in large errors. It is, therefore, strongly recommended that basal area is inputted in the application. The user is able to choose among thinning intensity levels (weak, moderate, moderate-strong, and strong) based on the Wilson Factor (0.16, 0.20, 0.23 and 0.28, respectively). In addition, the user can choose the age of first thinning, frequency, and number of thinnings. Growth without thinning can also be estimated. The default thinning is moderate from below (first thinning at age 20 years, 10 years frequency, maximum 3 thinnings). The user can also increase the precision level of the estimates using the check box at the end of the inputs form. Volume in this case is calculated yearly based on the stand structure, the diametrical distribution, and the individual tree volume equation (Nunes *et al.*, 2010). When uncheck, the simulation process uses the stand volume equation of Luis & Fonseca (2004). The former process can result in longer processing time.

The output of the growth simulator is a table with stand variables for each year of the simulation period and the result of the application of the prescribed thinning plan. For each thinning in this table a graph and a detailed table of the distribution of standing trees and trees to be removed by thinning are produced.

The "Thinning simulator" section requires as inputs stand age (years), number of trees/ha, dominant height (m) and, optionally, basal area (m^2/ha), such as for the previous section. The user can define the thinning method (from below or systematic) and the number or percentage of trees in the stand to extract. When this number or percentage is not known, the user can check the "I don't know" option for which the system applies the moderate thinning Wilson Factor.

The outputs of the thinning simulators include the volume to extract and the graphical distribution of standing trees per diameter class (d) with indication of trees to be removed by systematic or thinning from below (Fig. 2).

Credits, contact information, and a disclaimer are shown in the section "About FlorNExT®". In addition,

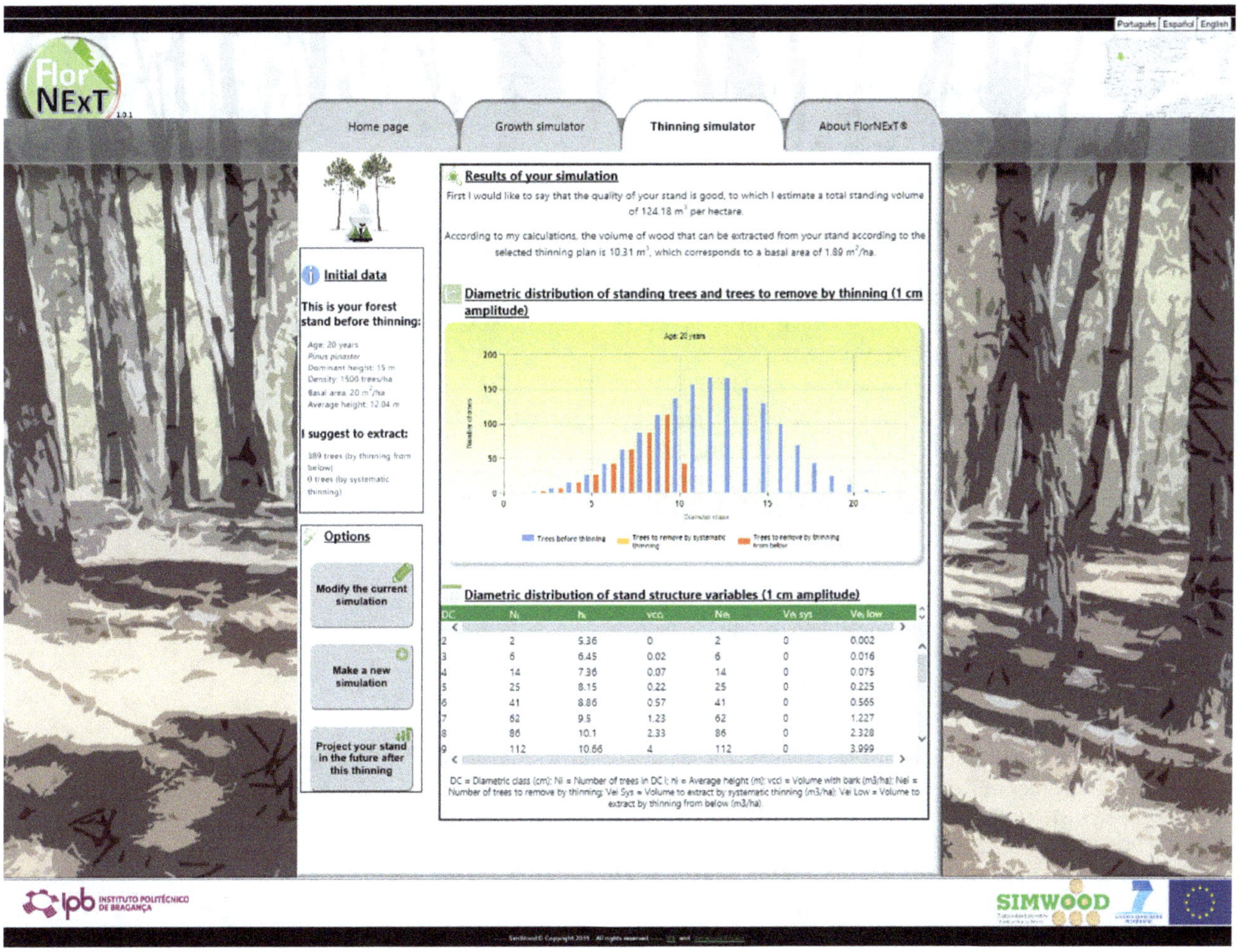

Figure 2. Output of the FlorNExT® "Thinning simulator" section.

User's guides in Portuguese and English are available from the application's website.

Access

FlorNExT® was developed as a public cloud application, i.e., to be accessible via the Internet. Therefore, no specific installation is required and it can be freely accessed from any Internet-connected device (desktop, laptop, mobile phone, smartphone, tablet, etc.) from the address http://flornext.esa.ipb.pt/. FlorNExT® is compatible with the most common browsers. FlorNExT® is hosted in a server owned by the Polytechnic Institute of Bragança (http://www.ipb.pt).

FlorNExT® conditions of use

The application is provided "as is" without any warranty, express or implied. In addition, FlorNExT® developers reserve for themselves the right to make changes and to translate it to other languages without notice.

FlorNExT® use is open to the forestry community but the authors are not responsible for use or interpretation of the results obtained by the application.

Final considerations

The development of tools like FlorNExT® is required to translate complex model equations into simple applications easy to operate by any end user. Cloud computing improves the capability of development of user-friendly tools and simplifies the process of updating them. These kinds of tools are valuable resources for regions where forest mobilization is constrained by lack of usable information and models to support sustainable forest management.

Acknowledgments

We thank an anonymous reviewer for valuable comments and suggestions on the manuscript.

References

Almeida LFR, 1999. Comparação de metodologias para estimação de altura e volume em povoamento s de pinheiro bravo no Vale do Tâmega, Relatório Estágio. UTAD, Vila Real, Portugal.

Bravo F, Álvarez-González JG, del Río M, Barrio M, Bonet JA, Bravo-Oviedo A, Calama R, Castedo-Dorado F, Crecente-Campo F, Condés S, *et al.*, 2011. Growth and yield models in Spain: Historical overview, contemporary examples and perspectives. Forest Syst 20(2): 315-328. http://dx.doi.org/10.5424/fs/2011202-11512

Diéguez-Aranda U, Rojo-Alboreca A, Castedo-Dorado F, Álvarez-González JG, Barrio-Anta M, Crecente-Campo F, González-González JM, Pérez-Cruzado C, Rodríguez-Soalleiro R, López-Sánchez CA, *et al.*, 2009. Herramientas selvícolas para la gestión forestal sostenible en Galicia. Consellería do Medio Rural, Xunta de Galicia. Santiago de Compostela, Spain.

Fonseca T, 2004. Modelação do crescimento, mortalidade e distribuição diamétrica, do pinhal bravo no Vale do Tâmega. Doctoral Thesis, UTAD, Vila Real, Portugal.

IFN, 2005. Inventário Florestal Nacional 5. http://www.icnf.pt/portal/florestas/ifn. Portugal.

Luis JFS, Fonseca T, 2004. The allometric model in the stand density management of *Pinus pinaster* Ait. in Portugal. Ann For Sci 61(8): 807-814. http://dx.doi.org/10.1051/forest:2004077

Nunes L, Tomé J, Tomé M, 2010. A system for compatible prediction of total and merchantable volumes allowing for different definitions of tree volume. Can J For Res 1: 747-760. http://dx.doi.org/10.1139/X10-030

Packalen T, Marques AF, Rasinmäki J, Rosset C, Mounir F, Rodriguez LCE, Nobre SR, 2013. Review. A brief overview of forest management decision support systems (FMDSS) listed in the FORSYS wiki. Forest Syst 22(2): 263-269. http://dx.doi.org/10.5424/fs/2013222-03192

Palma JHN, 2016. Web Globulus 3.0. http://home.isa.utl.pt/~joaopalma/modelos/webglobulus/

Páscoa F, 1987. Estrutura, Crescimento e Produção em Povoamentos de Pinheiro Bravo. Um Modelo de Simulação. Tese de Doutoramento. ISA/UTL. Lisboa, Portugal. 241 pp.

Páscoa F, 2001. Pbravo v.2.0. Modelo de Produção para o Pinheiro Bravo. Federação dos Produtores Florestais de Portugal. Aplicação Informática e Manual do Utilizador. 47 pp.

Tomé M, 2001. Tabela de produção geral para o pinheiro bravo desenvolvida no âmbito do projecto PAMAF 8165. Relatórios GIMREF RT9/2001. CEF, ISA, Lisboa, Portugal.

17

Gmelina arborea "death disease" in fast-growth plantations: Effects of soil and climatic conditions on severity and incidence and its implications for wood quality

Marcela Arguedas, María Rodriguez-Solis, Róger Moya and Alexander Berrocal

Instituto Tecnológico de Costa Rica, Escuela de Ingeniería Forestal, Apartado 159-7050, Cartago, Costa Rica.

Abstract

Aim of study: Plantations are threatened by an emerging disease called "Gmelina death disease". The objective of this study was measured the incidence and severity of this disease and were correlated with the characteristics of the plantations, micro- and macronutrients in the soil and climatic parameters.

Area of study: The present study evaluated 16 symptomatic fast-growth plantations of different age in Costa Rica.

Material and methods: Fungi were identified from xylem of infected trees. Incidence and severity was measured and correlated with the characteristics of the plantations, micro- and macronutrients in the soil and climatic parameters. Root condition and the quality (specify gravity, mechanical and decay resistance and chemical compositions), of the wood of symptomatic and asymptomatic trees were compared.

Main results: Three fungal species *(Chaetomella raphigera, Fusarium solani* and *Rhizomucor variabilis)* were identified from diseased samples. Clay content in the soil profile from 10-20 cm deep explained a significant proportion of the variation in the incidence and severity of the disease, and stand density was related to severity. Although two climatic variables, Holdridge's potential annual evapotranspiration and Thornthwaite's potential evapotranspiration, showed a relationship between the incidence in the trees and symptoms of the disease. Infected wood turned black in symptomatic trees. Specify gravity and mechanical resistance of infected wood decreased, whereas its natural durability was unaffected. Changes were observed in the quantities of Mg, Fe, Ca, K and Zn in infected wood.

Research highlights: Gmelina plantations established in sites with high stand densities and high contents of clay increase susceptibility to this disease.

Additional keywords: death syndrome; soil management; tropical species; pathogen.

Abbreviations used: DBH (diameter at 1.3 m height); h (dominant height); ECEC (effective cation exchange capacity); PDA (potato-dextrose-agar); SG (specific gravity)

Authors´ contributions: Conceived, designed and performed the experiments: MA and AB. Analyzed the data: MRS and RM. Contributed reagents/materials/analysis tools: MA. Wrote the paper: MRS and RM. All authors read and approved the final manuscript.

Funding: Vicerrectoría de Investigación y Extensión del Instituto Tecnológico de Costa Rica (ITCR); Fondo Nacional para el Financiamiento Forestal (FONAFIFO); Oficina Nacional Forestal (ONF).

Competing interests: All authors declared that no competing interests exist.

Correspondence should be addressed to Róger Moya: rmoya@itcr.ac.cr

Introduction

Gmelina arborea Roxb. (Gmelina) is a fast-growing forest species that produces high quality timber used for furniture manufacturing (Vallejos *et al.*, 2015). This species is widely used in commercial programs in tropical countries, for lumber, pulp and energy production (Dvorak, 2004; Tenorio *et al.*, 2016). In Central America (Costa Rica, Nicaragua and Guatemala), approx. 40,000 ha have been planted for commercial purposes (Dvorak, 2004). From those, approx. 18,000 ha correspond to Costa Rica (Moya & Tomazello, 2008). Since the 90's, Costa Rica has begun to develop wood pallet industries for agriculture products and all pallets are made with *G. arborea* wood from commercial plantations (Barrantes & Ugalde, 2017).

A series of problems has been observed in Gmelina plantations related to diseases and insect pests (Wingfield & Robinson, 2004; Umana *et al.*, 2015). In particular, several studies have reported *Gmelina arborea* "death disease". This disease impairs the development of trees, limiting their girth and height growth and thus hampering commercial plantations. The incidence of the disease in commercial plantations can reach up to to 40% (Murillo *et al.*, 2014). Generally, some trunk rot is also observed, called canker. Canker lesions appear with bulging, frequently together with exudation and sprouting in the stem.

Duke (1987) found that seven pathogens were present at the seedling stage. Julian (1982) reported damping-off, mold and stem rot in beds of seedlings caused by several fungi. Wingfield & Robinson (2004) performed an extensive revision and classified the diseases and insect pests of Gmelina according to their association with the foliar system, the trunk or the roots. Maringoni & Furtado (1997) and Umana *et al.* (2015) associated stem gall or stem canker in trees from Gmelina plantations with several fungi and showed canker development with tree death and foliage decrease. Lastly, one of the most severe pathogens attacking Gmelina is *Ceratocystis fimbriata*, which causes stem and branch canker and vascular wilt disease (Muchovej *et al.*, 1978; Ribeiro, 1982; Harrington *et al.*, 2011; Umana *et al.*, 2015).

Furthermore, problems of mortality in gro-ups of trees between 2 and 8 years old in Gmelina plantations has been reported in Costa Rica (Arguedas, 2004; Salas, 2015). Until recently, the death of trees in Gmelina plantations had been known as the "Nectria disease" in Costa Rica and many parts of tropical America (Arguedas, 2004; García-Díaz *et al.*, 2011). The name Nectria comes from the fact that Nectria species' anamorphs include Fusarium species, thus forming complexes of species (Samuels & Nirenberg, 1989; O'Donnell, 2000; Samuels *et al.*, 2009). However, this problem has recently been referred to as "Gmelina death disease" (García-Díaz *et al.*, 2011; Murillo *et al.*, 2014; Chavarría-Vega & Carmona-Solís, 2016).

According to Arguedas (2004), in early stages of the disease, the foliage withers, becomes dry and then falls. Trees often sprout at the bottom of the stem and have sap "weeping holes" at the pruning point, which later become blackish. In general, it is not possible to detect canker clearly during the initial stages of the disease. Rather, it can be observed in cuts made above the sprouts, where depressions of the bark can be detected. The affected cortical tissue shows dark brown necrosis, and thicker trees crack at the canker region. When the disease has progressed, a canker or anthracnose encircles the stem at some point, thus causing death of the apical part.

An important aspect of the disease that causes the greatest loss of trees in Gmelina plantations, with concomitant economic loss, is the scarce amount of knowledge regarding the cause of the damage (Muchovej *et al.*, 1978; Arguedas & Quirós, 1997; Arguedas, 2004; Umana *et al.*, 2015; Chavarría-Vega & Carmona-Solís, 2016). Several researchers have indicated the economic problems caused by Gmelina death disease. However, few studies have focused on evaluating the quality of the wood in stems damaged by the disease.

Recent observations of the disease symptoms show that it is also associated with a widespread black stain of wood (Murillo *et al.*, 2014). Regarding this issue, an extensive study performed by Moya *et al.* (2009) analyzed black stained wood – denominated "wetwood" –, a technical term that refers to wood with high moisture content, darker color after exposure to air, a distinctive odor, low permeability, a higher pH compared to normal wood, slow drying, and the development of drying defects (Ward & Pong, 1980).

In view of the importance of *G. arborea* to the forestry sector in Costa Rica and other countries in tropical America, and considering the lack of knowledge about Gmelina death disease, its causes and implications for the quality of wood, the aim of the present study was to determine the incidence and severity of the disease in juvenile plantations of *Gmelina arborea*. Then, the incidence and severity of the disease were correlated with (i) tree morphology, tree age and stand density (ii) the presence and quantity of micro- and macronutrients in the soil and (iii) the climatic variables of the zones where Gmelina grows. Additionally, fungi were identified from six isolates of the xylem of infected trees from six plantations. Lastly, the root development and wood quality in symptomatic and asymptomatic trees were compared.

Material and methods

Severity and incidence measurements of the disease and stand variables

— **Sampled plantations:** Based on the information provided by the *Oficina Nacional Forestal* (National Forestry Office) and the *Fondo Nacional de Financiamiento Forestal* (National Forestry Financing Fund) of the government of Costa Rica regarding the disease in plantations of *Gmelina arborea*, sampling was conducted in 16 sites between 2 and 6 years old, distributed in various regions of the country (Fig. 1). Two of these plantations showed no symptoms of the disease (sites 2 and 4), whereas diseased trees were evident in the remaining 14 plantations.

— **Sampling within the plantations:** A 14 m radius circular plot (615 m^2) was established. Trees in that area were counted and the projection per hectare was

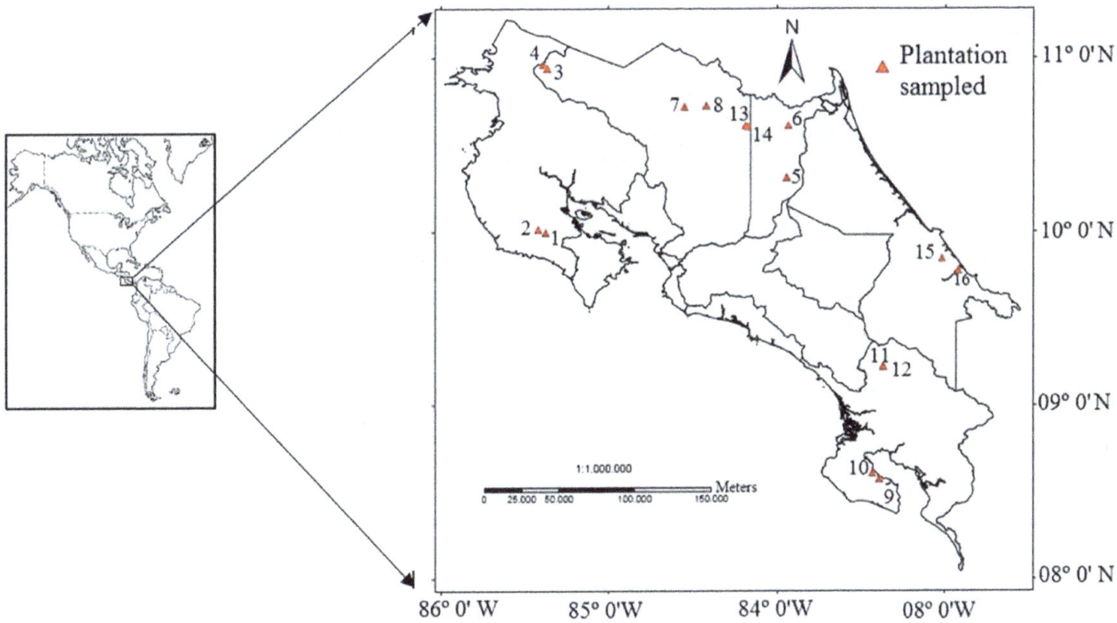

Figure 1. Geographic location of the 16 plantations of *Gmelina arborea* in Costa Rica.

performed (stand density). In addition, the diameter at 1.3 m height (DBH) and the dominant height (h) represented by the three highest trees within the plot, were measured. Healthy trees were identified within the plot, as were trees showing any symptom of the disease, the progress of which was evaluated. According to Arguedas (2004) and Salas (2015) the symptoms of the disease in *G. arborea* are leafless crown or foliar withering (Fig. 2a), canker occurrence (Fig. 2b), sprout occurrence (Fig. 2c) and black exudation in the stem (Fig. 2d). Then, four classes were defined (Table 1).

— **Determination of incidence and severity:** The incidence of the disease and the average index of severity were determined in each plot, following the methodology proposed by Araúz (1998) and Salas (2015).

— **Soil evaluation:** Ten sampling points were randomly selected within each assessment plot. Two soil samples were obtained at each point, employing a manual drill from depths of 0 to 10 cm (profile A) and 10 to 20 cm (profile B). The ten soil portions were mixed, and then the samples per layer were divided into four equal portions and one portion was randomly selected. The soil was characterized according to order and texture. The samples were measured for their pH in water at a 1:2 ratio, and for their proportions of Ca, Mg and K; the effective cation exchange capacity (ECEC), which corresponds to the sum of pH+Ca+Mg+K; the percent base saturation calculated as (pH/ECEC)×100; and the amounts of P, Zn, Cu, Fe and Mn. In addition, the bulk density (g/cm^3), clay (< 2 μm), silt (<2-50 μm) and sand (>50 μm) proportions of the soil in the sampled plantations were determined. For this purpose, soil samples of 80 cm^3 were gathered at two depths: from 5 cm to 10 cm and from 10 cm to 15 cm; these samples were dried at 110 °C for 24 hours to reach dry weight. Then, the density, which corresponds to the ratio of

Figure 2. Main symptoms assessed to classify the effects of the Gmelina death disease in plantations of *Gmelina arborea*: (a) foliar wilting, (b) presence of cankers, (c) sprouting and (d) black exudation in the stem.

Table 1. Classes established to evaluate the degree of severity of the Gmelina death disease in symptomatic trees in forest plantations of *Gmelina arborea*

Class of severity	Symptoms
A	Early foliar wilting; the stem may exhibit small necrotic wounds with mild blackish exudation generally at the pruning points. Sprouting may start. Not all symptoms are present in this class.
B	The tree is visibly affected. The foliage wilts, leaves are smaller and over 50% are lost. Canker lesions may appear (with bulging). Frequent exudation and sprouting in the stem.
C	Tree is almost dead. No foliage, evident bark detachment and loss due to canker, and sprouting.
D	Tree completely dry and rotten.

Source: Salas (2015)

the dry weight (g) and volume of the cylinder (cm^3), was calculated.

— **Climatic characterization:** First, the climate in the regions where plantations are located was characterized following Holdridge's world life zones classification (Holdridge, 1967). Then, the geographic coordinates and elevation (masl) were obtained for each sampled plantation. This was performed by means of the Global Positioning System (Garmin GPS). In parallel, the permanent meteorological stations of the *Instituto Meteorológico Nacional* closest to the sampled plantations were located. Information was gathered about average temperature and monthly precipitation within a period of 10 years (2004-2014).

A vertical temperature gradient of 0.6 °C per 100 m and the difference in elevation between each plantation based on the elevation of each meteorological station were used to adjust the temperature of each plantation. To adjust the precipitation, an interpolation of the data was performed using the QGis® Analysis of Spatial Interpolation.

Later, the hydric balance was calculated for each plot following the Thornthwaite & Mather (1957) method. Data on the annual potential evapotranspiration, actual evapotranspiration and water deficit in the soil were gathered by means of Thornthwaite's table for soil moisture retention, according to the effective depth and soil texture in each plot. Additionally, Holdridge's annual potential evapotranspiration was calculated as (annual biotemperature * 0.58), and the ratio of potential evapotranspiration and average annual temperature equal to Holdridge's annual biotemperature (Holdridge, 1967) were also calculated.

— **Fungal identification:** Wood with symptoms is presents a dark color (Fig. 3a-b), whilst healthy wood does not (Fig. 3c). Trees affected by the Gmelina death disease were cut in six plantations (Fig. 3d). Cross sections were extracted from the stems one meter from the base of the tree (Fig. 3e-g). During cutting, black areas with a strong odor were observed on the stems, perhaps because of a tree pathogen (Fig. 3a-b). A cross section of each tree was cut at breast height for later identification of the pathogen. The cross sections were washed with distilled water and then three small sections from the heartwood were cut from the periphery of the affected areas (black, strong smell), with the help of a sterile scalpel (Fig. 3h-i).

The three samples per tree were placed on separate Petri dishes containing potato-dextrose-agar (PDA; Difco laboratories, Detroit) and incubated for 22 days at 25 °C under continuous darkness. Several colonies developed during that period. Mycelia were collected from the periphery of the colonies and again placed on a Petri dish to grow into PDA. Then, fungal purification was again performed by hyphal tip transfer to Petri dishes with PDA, which were stored in the dark at 25 °C for 22 days. The mycelial samples of each purified culture for phylogenetic analysis were identified by extracting the DNA employing the Murray & Thompson´s method (White *et al.*, 1990). Then, a PCR was conducted and the product was sequenced for one sample for each Petri dish. A total of 18 sequences were obtained (3 trees × 6 plantation or sites × 1 Petri dish). Each sample was added to a 50 μL solution containing 25 pmol of each

Figure 3. Cross sections of *Gmelina arborea* trees showing areas associated to the Gmelina death disease (a and b) and normal cross section without disease (c). Fungal insulation: tree with disease (d), cross-section extractions (e), cross-section infected (f), and process of insolation (g-j).

primer. The primers used were ITS4 and ITS5. PCR products were sequenced by Macrogen. To analyze the obtained sequences, the NCBI Blast tool was employed for comparison with the GenBank database (http://www.ncbi.nlm.nih.gov/genbank/) to identify the isolated fungi.

Comparative study of healthy and infected trees

— **Root morphology evaluation:** Two trees were chosen in each of the plantations where the disease was present, one visibly healthy and the other infected. For each selected tree, two points were located for sampling at the edge of the dripping area (the edge of the crown). At these points, 20 cm^3 of soil with roots was extracted at ground level. The samples were washed with distilled water to eliminate the organic matter and then placed in glass containers with ethanol (70%) to prevent deterioration. Then, two representatives 5 cm long subsamples were extracted from the terminal parts of the absorption roots. Using an Epson Expression 1680 scanner, a digital image was captured of the roots and was processed afterwards with WinRhizo Basic V 2005-b® software. The following variables resulted from the image processing: root total length, root average diameter, number of tips, number of bifurcations and number of crossings.

— **Evaluation of the wood properties:** For the present study, we followed the methods proposed by Jeremic *et al.* (2004) and Moya *et al.* (2009), where 9 trees with symptoms of the disease were selected from a 3-year-old plantation in Penshurst, Limón, Costa Rica. The average annual temperature at this location is 26 °C and the annual precipitation is 3078 mm. The average DBH of the trees was 21 cm and the average total height was 19 m. One-meter-long logs were cut from each selected tree, until reaching the commercial height (13 cm diameter). A cross section approximately 4 cm thick was extracted at the end of each log. Five one-meter-long logs with stained and unstained regions were chosen. Then, the samples for determining wood properties were extracted first from zones affected by the disease, corresponding to wood infected by Gmelina death disease. Later, samples of unaffected wood were extracted from the same log or cross section, at the same distance from the heartwood. The properties evaluated were as follows:

a) Physical properties: Green wood density (green weight/green volume), specific gravity according to ASTM D-143 (ASTM, 2003a) and moisture content according to ASTM D-4442 (ASTM, 2003b). A total of 30 samples of dimensions 2.5 cm × 2.5 cm × 2.5 cm were extracted from the discs of infected and non-infected wood. The infected and non-infected wood samples were collected at the same distance from the heartwood. Each sample was measured for its green weight and green volume (their condition at the moment of extraction of the sample) and then placed in an oven for 24 hours at 105 °C to determine its final weight. This information was used to determine the physical parameters.

b) Mechanical properties: From the 2.7 cm thick boards obtained from the 1 m long logs, approx. 30 samples of dimensions 2.7 cm × 2.7 cm × 1 m were obtained and dried at 22 °C and 66% relative humidity to reach an equilibrium moisture content of 12%. The mechanical properties evaluated were maximum stress in compression, modulus of rupture and modulus of elasticity in static bending. All tests were performed following the ASTM D-143 standard (ASTM, 2003a).

c) Resistance to fungal decay: 240 wood samples measuring 2.5 cm × 2.5 cm × 2.5 cm were cut and divided into three groups: 80 with infected sapwood, 80 with unstained sapwood and 80 with infected heartwood. The samples were inoculated by two strains of the fungi, *Trametes versicolor* (white decay) and *Lenzites acuta* (brown decay), following the procedure set in the ASTM-D 2017-81 standard (ASTM, 2003c). This test yields the percentage of weight loss of the samples after 16 weeks of exposure to fungal attack.

d) Inorganic ash composition of wood: Chips of wood were collected from healthy and infected samples from the cross sections of the sampled trees and then ground and sieved through 40- and 60-sized meshes. Ash analysis was performed using the methods of Sparks (1996) and included determination of the mass of N, P, Ca, Mg, K and S, whose concentrations were reported as percentages, and Fe, Mn, Cu, Zn and B, whose concentrations were reported as milligrams of inorganic element per gram of ash.

e) Wood color: Color was measured in the samples to measure the mechanical properties of the infected and uninfected wood. These samples were conditioned to 12% moisture content and were evaluated in the tangential face. To measure color, the standardized color measuring system CIE L*a*b* was used with a miniScan XE Plus spectrophotometer, following the procedure described by Moya *et al.* (2012).

Statistical analysis

A Pearson correlation matrix was used for determining the most correlated incidence and severity for further analysis. Next, incidence and severity were correlated with the tree morphology (DBH and h), stand density, tree age, soil and climatic characteristics. A forward stepwise analysis was applied to establish the effect that tree morphology, tree age, climate characteristics and soil characteristics in the profile from 10-20 cm had on the percentage of incidence and severity of the disease and on the various damage classes. The analysis of variance, ANOVA, was also performed to determine the level of significance of symptomatic and asymptomatic wood; it was applied to characteristics of the roots and the physical and mechanical characteristics, content of macro- and microelements, loss of weight during fungal attack and different color parameters. Afterwards, the presence of significant differences among the media of these parameters of normal wood and symptomatic wood was verified by means of the Tukey test ($p<0.05$). The statistical software SAS 8.0 was used to perform the statistical analysis of the data.

Results

Incidence and severity of the disease in the plantations

The incidence of the disease varied from 0% to 58% of the total number of trees in the plantation. When sorted the 16 sampled plantations from highest to lowest incidence and severity, three different inflection points could be observed. According to this, four groups were established: (i) plantations where incidence was equal to 0 (plantations 2 and 4); (ii) a second group with values of incidence from 3% to 4% (plantations 13, 7 and 9); (iii) a third group with values between 17% and 35% (plantations 1, 8, 6, 15, 12, 16, 10 and 11) and (iv) a fourth group with values of incidence between 48% and 58% (plantations 3, 5 and 14) (Fig. 4a). For severity, it ranged between 0% and 28% (Fig. 2b) and four groups were determined as well: (i) the first group includes healthy plantations (plantations 2 and 4), (ii) the second group showed severities between 1% and 4% (plantations 13, 7, 9, 8 and 6), (iii) the third group showed severities between 10% and 13% (plantations 1, 12, 16, 3, 15 and 10) and (iv) the last group showed values between 21% and 28% (plantations 5, 11 and 14) (Fig. 4b).

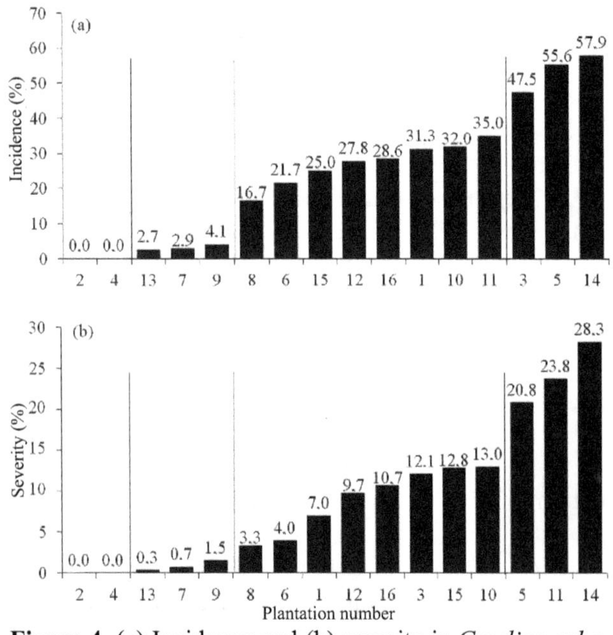

Figure 4. (a) Incidence and (b) severity in *Gmelina arborea* plantations infected with the Gmelina death disease in Costa Rica.

The percentage of the damage class A (wilting, exudation and small sprouting) was higher in plantations 1, 3, 6, 8, 12 and 16, while the percentage of damage class B (wilting, small leaves, 50% loss of foliage, canker or swelling, exudation and sprouting) was higher in plantations 5 and 10. For damage class C, the percentage was higher in plantations 14 and 15. Finally, plantation 11 showed the highest number of trees with damage class D (Table 2).

Fungal identification

The DNA sequence analysis of the six types of wood from samples collected from symptomatic trees revealed the presence of three species of fungi: *Fusarium solani* was present in 3 sites (2, 3 and 4) showing a BLAST match of 99% in the GenBank database (AF178407.1, KF255997.1 and AF178407.1, respectively). *Rhizomucor variabilis* was present in sites 5 and 6 with a BLAST match of 99% with HM623314.1 of GenBank database. Finally, *Chaetomella raphigera*, with a 98% BLAST match with AY487076.1 of the GenBank database, was present only in the xylem of the tree in site 1.

Effect of the parameters of the plantations, environmental conditions and soil fertility.

The amount of clay in the soil layer between 10 cm and 20 cm accounted for 66% of the variation of incidence of the disease, followed by K in the same layer explaining 22% of the variation (Table 3). The amount of clay was positively correlated with the incidence of the disease (Fig. 5a). In the case of K, contrariwise, the correlation with the incidence was negative (Fig. 5b). However, this relation was influenced by two points, 0.5 and 0.6 in K content. If these two points were removed, then the relationship would become non-significant (R^2=0.02). Altogether, clay and K content explained 87% of the total variation of the incidence (Table 3).

With respect to severity, it was first related to stand density, which explained 33% of the variation, after which the amount of clay in segment 10-20 cm explained 27% of the variation (Table 3). Once more, K appeared in the 10-20 cm profile and explained 21% of the variation.

Table 2. Percentage of trees in the different classes of damage severity in *Gmelina arborea* plantations in Costa Rica.

Plantation No.	Density (N/ha)	Age (years)	Healthy trees (%)	Damage class (%)			
				A	B	C	D
1	520	3	68.75	21.88	6.25	3.13	0.00
2	585	2	100.00	0.00	0.00	0.00	0.00
3	991	2	52.46	34.43	3.28	8.20	1.64
4	1104	2	100.00	0.00	0.00	0.00	0.00
5	585	3	44.44	19.44	25.00	2.78	8.33
6	974	2	78.33	20.00	0.00	0.00	1.67
7	1104	2	97.06	1.47	1.47	0.00	0.00
8	974	2	83.33	13.33	1.67	1.67	0.00
9	796	4	95.92	2.04	0.00	2.04	0.00
10	406	4	68.00	8.00	16.00	4.00	4.00
11	650	4	65.00	5.00	2.50	7.50	20.00
12	585	6	72.22	16.67	2.78	2.78	5.56
13	601	5	97.30	2.70	0.00	0.00	0.00
14	309	2	42.11	15.79	10.53	21.05	10.53
15	585	3	75.00	5.56	0.00	19.44	0.00
16	682	2	71.43	11.90	7.14	7.14	2.38

All these variables explained 81% of the variation of severity (Table 3). The stand density was negatively correlated with severity (Fig. 5c); the amount of clay in 10-20 cm profile was again positively correlated with severity (Fig. 5d).

Root morphology

The average number of tips, forks and crossings and the root length was statistically lower in infected trees than in healthy trees. Meanwhile, no significant difference was observed between both conditions of sampled trees regarding root diameter (Fig. 6).

Table 3. Multiple regression analysis for the relationship between incidence, severity scores and classes established for severity grades for 16 plantations of *Gmelina arborea* infected with Gmelina death disease in Costa Rica.

Incidence (%)	Severity (%)
Clay between 10-20 cm deep** (R_p^2=0.66; R_T^2=0.66)	Stand density** (R_p^2=0.33; R_T^2=0.33)
K between 10-20 cm deep** (R_p^2=0.22; R_T^2=0.87)	Clay between 10-20 cm deep * (R_p^2=0.27; R_T^2=0.60)
	K between 10-20 cm deep * (R_p^2=0.21; R_T^2=0.81)

R_p^2: coefficient of partial correlation. R_T^2: coefficient of total correlation. **Statistically significant at 99% confidence, * statistically significant at 95% confidence.

Wood quality

— **Appearance:** symptoms of the disease in *Gme-lina arborea* trees were evident in both the sapwood and the heartwood (Fig. 3). The affected region was differentiated from the surrounding tissue by its darker color and occasional sap secretion (Fig. 3). Fermentation was another characteristic of these dark regions as it causes a rancid to fetid rumen-like odor. Affected wood showed light (L*) tonality, statistically lower than unaffected wood, in both green and dry wood (Fig. 7a). A higher tonality was also observed in redness (a*), although this was statistically higher in the dry condition (Fig. 7b) compared to the wet condition. Finally, less yellowish tonality (b*) was observed in green wood in relation to normal wood (Fig. 7c).

— **Wood properties:** it was observed that affected wood presented a statistically higher value of density in green condition, higher moisture content, less resistance to compression stress, less modulus of elasticity and module rupture in the bending test compared to unaffected wood

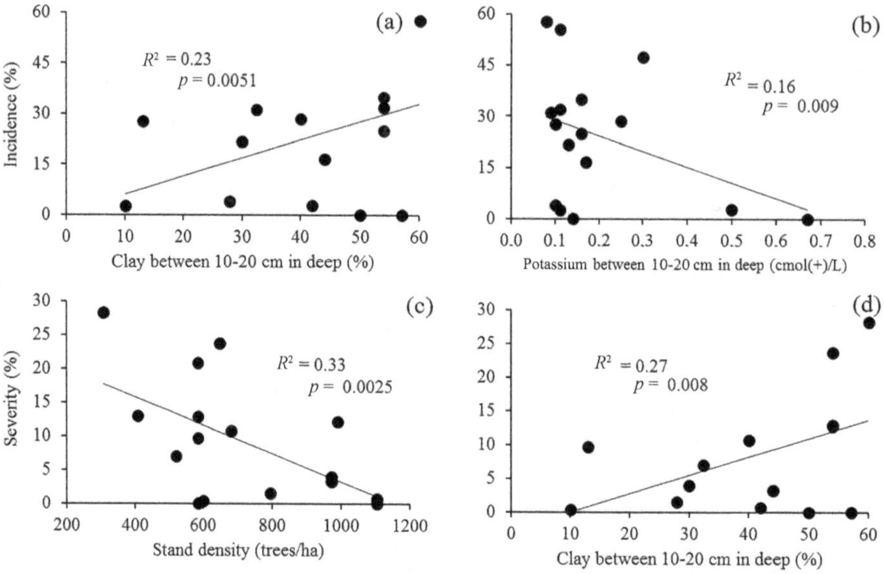

Figure 5. Correlation between incidence of Gmelina death disease and quantity of clay and potassium in the second profile 10-20 cm deep (a-b); correlation between severity and stand density (c) and quantity of clay in the second profile 10-20 cm deep (d).

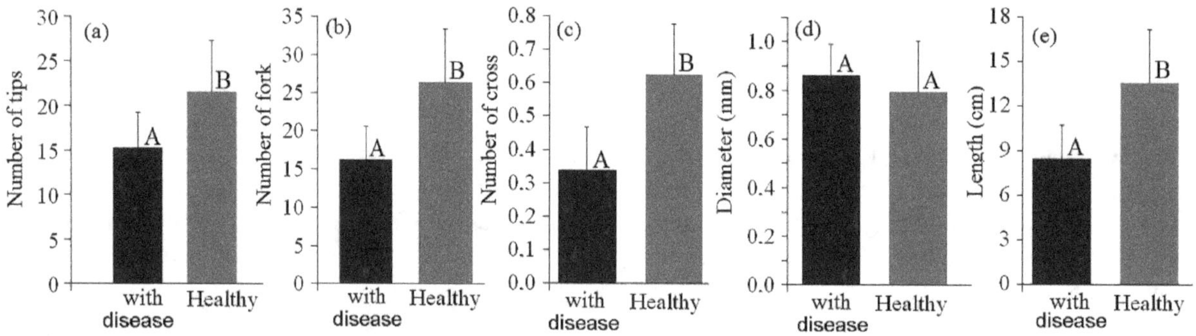

Figure 6. Number of tips (a), forks (b) and crossings (c), diameter (d) and length (e) of roots in trees infected with Gmelina death disease in plantations of *Gmelina arborea* in Costa Rica. Different capital letters indicate a statistical significance of 99% and bars indicate confidence intervals. The averages were derived from root samples in 16 sampled plantations.

(Table 4). The specify gravity was not statically different in affected and unaffected wood (Table 4). Regarding the durability of the wood to white-decay (*T. versicolor*) and brown-decay (*L. acuta*) fungal attack, there was no statistical evidence that symptomatic wood was less durable than normal wood (Fig. 7d-e).

On the other hand, the evaluation of micro- and macronutrients of wood ash showed that the amount of Mg and Fe is statistically higher and quantities of Ca, K and Zn are statistically lower in symptomatic wood compared to asymptomatic wood (Table 4). The amounts of remaining nutrients were not statistically different between symptomatic and asymptomatic wood. For other chemical characteristics, the pH and electrical conductivity were statistically lower in wood from trees with "Gmelina death disease" than in normal wood (Table 4).

Discussion

The incidence (0 to 58%) and severity (0 to 28%) found in this study show the variation in the presence of the Gmelina death disease in *G. arborea* plantations in Costa Rica (Fig. 4). Currently, it is possible to observe juvenile plantations with such variations throughout the country. These values are higher than those from the study by Murillo *et al.* (2014) of three-year plantations in a wet tropical climate region also in Costa Rica, which found incidence rates of 23.5 to 26.2% and a severity of 8.54 to 17.55%. Meanwhile, Singh *et al.* (2003) and Soni *et al.* (2005) recorded 8 to 15% incidence of wilt disease in *Gmelina arborea* and other species such as *Dalbergia sissoo, Tectona grandis* and *Buchnania lanzan*, but their measurements were performed during the seedling stage.

Trees in C and D classes could pose pro-blems regarding their use because of their darker tonality

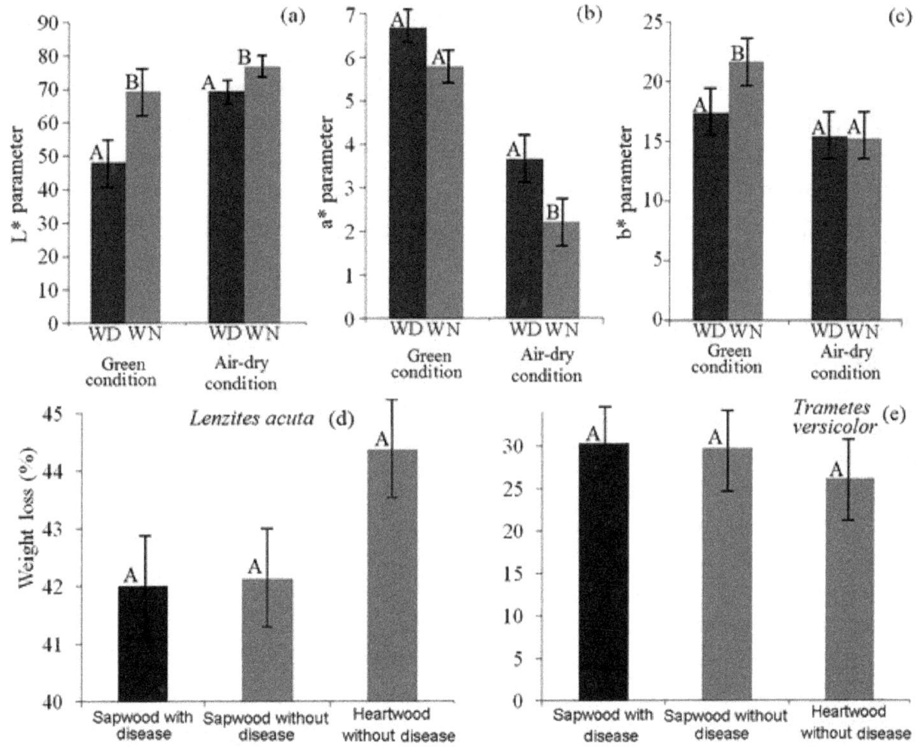

Figure 7. Colour parameters in green and air-dry condition in wood with diseased (WD) and without diseased or normal wood (WN) (a-c), and weight loss in wood for *Lenzites acuta* and *Trametes versicolor* (d) in normal wood and wood from trees with Gmelina death disease growing in *G. arborea* plantations. Different capital letters indicate a statistical significance of 99%

Table 4. Physical and mechanical properties, and micro- and macronutrients in normal wood and in wood showing symptoms of Gmelina death disease in *Gmelina arborea* plantations in Costa Rica.

Property	Variable	Normal wood	Symptomatic wood	Percentage of difference
Physical	Density (g/cm^3)	1.11^{A}	1.02^{B}	-8.82
	Specific gravity (g/cm^3)	0.46^{A}	0.44^{A}	4.35
	Moisture content (%)	123.7^{A}	154.7^{B}	-25.06
Mechanical	Compression stress (MPa)	0.24^{A}	0.13^{B}	45.83
	Compression density (g/cm^3)	0.48^{A}	0.48^{A}	0.00
	Flexural modulus of elasticity (MPa)	11166^{A}	8792^{B}	21.26
	Flexural modulus of rupture (MPa)	63.84^{A}	52.76^{B}	17.36
Ash content	Nitrogen (%)	0.23^{A}	0.25^{A}	-8.70
	Phosphorus (%)	0.02^{A}	0.02^{A}	0.00
	Calcium (%)	0.19^{B}	0.14^{A}	26.32
	Magnesium (%)	0.04^{B}	0.03^{A}	25.00
	Potassium (%)	0.54^{B}	0.30^{A}	44.44
	Sulphur (%)	0.01^{A}	0.01^{A}	0.00
	Iron (mg/kg)	27.17^{B}	63.00^{A}	-131.87
	Copper (mg/kg)	4.00^{B}	3.17^{A}	20.75
	Zinc (mg/kg)	13.33^{B}	11.83^{A}	11.25
	Manganese (mg/kg)	0.00^{A}	0.00^{A}	0.00
	Boron (mg/kg)	3.67^{A}	3.33^{A}	9.26
Other	pH in water	6.47^{B}	5.80^{A}	10.36
	Electrical conductivity (mS/cm)	1.93^{B}	1.03^{A}	46.63
	Carbon (%)	46.90^{A}	46.42^{A}	1.02
	Ratio C/N	211.24^{A}	187.83^{A}	11.08

compared to normal wood (Fig. 3), their higher moisture content and their lower resistance in some mechanical properties (Table 4). Although the percentages of incidence and severity (58% and 28%) can be considered high, the highest severity classes (C and D) did not appear in all plantations. Approximately 20% of damage class C appeared in sites 14 and 15 and damage class D in site 11 (Table 2). These results can be somewhat confounded by tree age, as plantations varied in age (2-6 years old range) and each one was taken as a single point in the analysis, *i.e.* within-plantation variation was not considered. However, age was weakly correlated with damage class, as the oldest plantations, 6 and 5 years (plantations 12 and 13, respectively) old, had low severity damage (Table 2).

Although three species of fungi were found in stained wood (*Fusarium solani, Chaetomella raphigera* and *Rhizomucor variabilis*), only *F. solani* has been reported by other researchers as being present in infected trees of *G. arborea* (García-Díaz *et al.*, 2011; Murillo *et al.*, 2014; Umana *et al.*, 2015; Chavarría-Vega & Carmona-Solís, 2016). On the other hand, *Chaetomella raphigera* is reported to occur together with fruit rot pathogens, and can be linked to fungi of the *Nectria* and *Fusarium* species (Gajbhiye *et al.*, 2016). However, it is necessary to know the actual association of *C. raphigera* and Fusarium species in Gmelina trees in a plantation. Finally, *Rhizomucor variabilis* was renamed after *Mucor irregularis* (Voigt *et al.*, 1999) and many reports indicate that this pathogen can cause plant disease.

On the other hand, the *Fusarium solani* species complex (FSSC) may embrace approx. 50 phylogenetic species (O'Donnell, 2000; O'Donnell *et al.*, 2008; Coleman, 2016) that dwell in soil, plants and animals (Nalim *et al.*, 2011). In trees, this fungus is considered a problem since it reportedly affects timber production in certain species, such as *Dalbergia sissoo* (Basak & Basak, 2011) and *Terminalia nigrovenulosa* (Seo *et al.*, 2013). The presence the *Fusarium solani*, the anamorph of *Nectria*—now Neocosmospora—*haematococca* (Nalim *et al.*, 2011; Coleman, 2016), agreed with several studies that identified the same tree fungi as "Gmelina death disease" (García-Díaz *et al.*, 2011; Murillo *et al.*, 2014; Chavarría-Vega & Carmona-Solís, 2016). However, we only isolated strains from 6 sites and other important tests must be performed. Therefore, we cannot infer that this fungus or FSSC is a causal agent of "Gmelina death disease".

An additional important effect relative to the development of root rot disease in other agricultural crops occurs in root development (Román-Avilés *et al.*, 2004). It was demonstrated that symptomatic Gmelina trees exhibit different root characteristics (Fig. 6), that is, the number of tips, forks and crossings and root length were reduced relative to asymptomatic trees. This is because pathogen infection reduces root density by killing roots and may thus attenuate the functional efficiency of the remaining roots (Román-Avilés *et al.*, 2004). Root disease becomes more severe when roots are unable to escape the pathogen due to edaphic factors, such as high clay content (Barros *et al.*, 2014). In fact, clay content was positively correlated with incidence and severity of *Gmelina* death disease (Table 4). Soils with a high clay content tend to affect the roots of many agricultural crops (Leslie & Summerell, 2006) and probably the xylem of woody plants too because high clay content implies a high moisture storage capacity and poor drainage of the soil (Belete *et al.*, 2013).

Stand density, on the contrary, did not contribute to the development of disease symptoms, but high stand density decreased the severity in Gmelina trees (Table 3). This result disagreed with the findings of Arguedas (2004), who mentioned that trees growing in high density are unable to develop appropriately and create mechanisms of protection against pathogens. Most likely, if high density is present in a juvenile plantation, then either trees do not develop the disease, or two to six years is not enough time for juvenile trees develop the disease. On the contrary, the low stand density present in adult plantations was associated with high severity levels and some trees probably died (Fig. 5c). A positive correlation ($p<0.0001$) between severity and tree age confirmed that mature plantations present trees with high severity, due to adult trees that had developed the disease.

In relation to the negative relationship between K and the incidence of Gmelina death disease, no clear conclusions could be drawn. The reason is that even though a statistically significant relationship was found (Fig. 5b), this was strongly influenced by two observations. Therefore, more research is necessary on this point.

As noted, wood quality was impaired (Table 4), color being the most negatively affected parameter (Fig. 7) in regard to appearance. Color change is likely produced by changes in the content of extractives in the wood, which give color to Gmelina (Moya *et al.*, 2009; Moya *et al.*, 2012). Other affected characteristics (Table 4) relate to wood structure, lower density and mechanical resistance. As explained by Xu *et al.* (2001), the lower specific gravity (SG) of infected wood in *Quercus rubra* and *Q. falcata* is possibly due to wood mass loss resulting from degradation of infected wood associated with bacteria and wood decaying fungi. They also stated that volatile constituents such as methane, acetic acid, propionic acid and butyric acid present in infected wood are lost during oven drying, resulting in a lower SG and reduction of wood resistance.

However, a positive aspect of symptomatic infected wood is that the wood's resistance to fungi is not impaired. Symptomatic wood is characterized by brown stains in heartwood (or sapwood) that are susceptible to decay. The decreased pH of infected wood in some species may make it less susceptible to acid-forming brown-rot fungi, but offers no protection from white-rot fungi (Moya *et al.*, 2009).

Inorganic elements are essential for tree growth and are measured as micro- and macronutrients in ash (Xu *et al.* 2001). An increasing amount of inorganic elements is associated with the presence of infected wood (Jeremic *et al.*, 2004). The present study showed that the amounts of Ca, Mn, Fe, Cu, K and Zn differ between infected wood and normal wood. Murdoch *et al.* (1987) suggested that elevated levels of inorganic elements in infected wood resulted from the selective degradation of cell membranes by bacteria, leading to the accumulation of unused micronutrients.

Gmelina death disease management

Gmelina plantations have been established in Costa Rica in sites with high stand densities and high contents of clay in the profile 10 to 20 cm deep, which are inappropriate conditions that increase susceptibility to Gmelina death disease. Different solutions may be implemented for this problem. One solution is related to the selection of suitable sites where the amount of clay is below 40%, to allow the soil to conveniently drain the levels of rainfall characteristic of reforestation sites in Costa Rica. The second solution refers to managing infected trees at early ages to grow healthy trees capable of combating pathogens.

In summary, the quantity of clay present in the 10-20 cm layer offers conditions for the development of the Gmelina death disease. Lack of management of plantations also promotes the development of the disease (high severity) though increased stand density. The characteristics of climate had no effect on the disease. Although two climatic variables, Holdridge's potential annual evapotranspiration (PAEH) and Thornthwaite's potential evapotranspiration (PET), showed a relationship with the incidence in the trees and symptoms of the disease, it was not statistically significant. The visual quality of the wood was affected by Gmelina death disease. The affected area turned black, and mechanical resistance was also reduced. The natural durability of the wood, however, was not affected. There were changes in ash composition, specifically in the levels of Mg, Fe, Ca, K and Zn. An appropriate selection of sites, with the amount of clay below 40%, allows adequate growing of trees in commercial plantations in Costa Rica, and adequate management at early ages allows to grow healthy trees capable of combating pathogens.

References

Araúz F, 1998. Fitopatología: un análisis agroecológico. San José, Costa Rica, EUCR. 469 pp.

Arguedas M, 2004. Problemas fitosanitarios de la melina (*Gmelina arborea* (Roxb)) en Costa Rica. Kurú Rev For 1 (2): 1-7.

Arguedas M, Quirós L, 1997. Experiencias y perspectivas del manejo de plagas forestales en Costa Rica. Manejo Integrado de Plagas 45: 34-42.

ASTM, 2003a. Standard test methods for small clear specimens of timber. Annual Book of American Society for Testing and Materials Standards, Vol 04.10, Philadelphia, USA. (ASTM D 143-94, Reapproved 2000), 25 pp.

ASTM, 2003b. Standard test methods for direct moisture content measurement of wood and wood-base materials. Annual Book of American Society for Testing and Materials Standards, Vol 04.10, Philadelphia, USA (ASTM D 4442-92), 8 pp.

ASTM, 2003c. Standard method of accelerated laboratory test of natural decay resistance of woods. Annual Book of American Society for Testing and Materials Standards, Vol. 04.10, Philadelphia, USA (ASTM D 2017-81, Reapproved 1994), 15 pp.

Barrantes A, Ugalde S, 2017. Usos y aportes de la madera. Estadísticas 2016. Oficina Nacional Forestal, San José, Costa Rica. 40 p.

Barros JA, Medeiros EV, Notaro KA, Moraes WS, Silva JEM, Nascimento TC, Moreira KA, 2014. Different cover promote sandy soil suppressiveness to root rot disease of cassava caused by *Fusarium solani*. Afr J Microbiol Res 8 (10): 967-973. https://doi.org/10.5897/AJMR2014.6607

Basak AC, Basak SR, 2011. Biological control of *Fusarium solani* sp. dalbergiae, the wilt pathogen of *Dalbergia sissoo*, by *Trichoderma viride* and *T. harzianum*. J Trop For Sci 23 (4): 460-466.

Belete E, Ayalew A, Ahmed S, 2013. Associations of bi-ophysical factors with faba bean root rot (*Fusarium so-lani*) epidemics in the northeastern highlands of Ethi-opia. Crop Prot 52: 39-46. https://doi.org/10.1016/j.cropro.2013.05.003

Chavarría-Vega M, Carmona-Solís R, 2016. Efecto de micro-organismos antagonistas en el control de la enfermedad denominada "Nectria" en la melina *Gmelina arborea* Roxb. Rev For Mesoam Kurú 13 (32): 21-29. https://doi.org/10.18845/rfmk.v0i0.2549

Coleman JJ, 2016. The *Fusarium solani* species complex: ubiquitous pathogens of agricultural importance. Mol Plant Path 17 (2): 146-158. https://doi.org/10.1111/mpp.12289

Duke JA, 1987. *Gmelina arborea* Roxb. Purdue University, Centre for New Crops and Plant Products. West Lafayette, IN, USA. 25 pp.

Dvorak WS, 2004. World view of *Gmelina arborea*: opportunities and challenges. New Forest 28: 111-126. https://doi.org/10.1023/B:NEFO.0000040940.32574.22

Gajbhiye M, Sathe S, Shinde V, Kapadnis B, 2016. Morphological and molecular characterization of pomegranate fruit rot pathogen, *Chaetomella raphigera*, and its virulence factors. Ind J Microbiol 56 (1): 99-102. https://doi.org/10.1007/s12088-015-0554-4

García-Díaz SE, Cibrián-Tovar D, Pérez-Vera OA, 2011. *Haematonectria haematococca* (Berk. & Broome) Samuels & Nirenberg y su anamorfo (*Fusarium solani* (Mart.) Sacc.) asociado a muerte descendente de teca y melina. XVI Simposio Nacional de Parasitología Forestal, Cuernavaca, Morelos, 26-28 Oct 2011, Mexico. pp: 184-186.

Harrington TC, Thorpe DJ, Alfenas AC, 2011. Genetic variation and variation in aggressiveness to native and exotic hosts among Brazilian populations of *Ceratocystis fimbriata*. Phytopathology 101 (5): 555-566. https://doi.org/10.1094/PHYTO-08-10-0228

Holdridge L, 1967. Life zone ecology. Tropical Science Center. San Jose, Costa Rica. 206 p.

Jeremic D, Cooper P, Srinivasan U, 2004. Comparative analysis of balsam fir wetwood, heartwood, and sapwood properties. Can J For Res 34: 1241-1250. https://doi.org/ 10.1139/x03-287

Julian E. 1982. Plantation forestry in the tropics. Oxford Univ Press, NY. 50 pp.

Leslie JF, Summerell BA, 2006. The Fusarium Laboratory Manual. Black-Well Publ, 388 pp. https://doi.org/10.1002/9780470278376

Maringoni AC, Furtado EL, 1997. Pathogenicity of *Agrobacterium tumefaciens* on Gmelina. Summa Phytopatol 23: 248-251.

Moya R, Tomazello M, 2008. Variation in the wood anatomical structure of *Gmelina arborea* (Verbenaceae) trees at different ecological conditions in Costa Rica. Rev Biol Trop 56 (2): 689-704.

Moya R, Muñoz F, Dragica J, Berrocal A, 2009. Visual identification, physical properties, ash composition and water diffusion of wetwood in *Gmelina arborea*. Can J For Res 39 (3): 537-545. https://doi.org/10.1139/X08-193

Moya R, Soto R, Jiménez P, Tenorio C, 2012. Relationship between wood colour parameters measured by the CIELab system and extractive and phenol content in *Acacia mangium* and *Vochysia guatemalensis* from fast-growth plantations. Molecules 17 (4): 3639-3652. https://doi.org/10.3390/molecules17043639

Muchovej JJ, Albuquerque FC, Ribeiro GT, 1978. *Gmelina arborea* - A new host of *Ceratocystis fimbriata*. Plant Dis Rep 62 (8): 717-719.

Murdoch RJ, Campana J, Biermann CJ, 1987. Physical and chemical properties of wetwood in American elm (*Ulmus americana*). Can J Pl Path 9: 20-23. https://doi.org/10.1080/07060668709501906

Murillo O, Badilla Y, Rojas F, Mata X, 2014. Uso de biocontroladores y materiales tolerantes a los patógenos asociados al síndrome de la muerte descendente de la teca (*Tectona grandis*) y cancro Nectria de la melina (*Gmelina arborea*). Informe Final de proyecto de Investigación. Instituto Tecnológico de Costa Rica, Cartago, Costa Rica, 51 pp.

Nalim FA, Samuels GJ, Wijesundera RL, Geiser DM, 2011. New species from the *Fusarium solani* species complex derived from perithecia and soil in the Old World tropics. Mycologia 103 (6): 1302-1330. https://doi.org/10.3852/10-307

O'Donnell K, 2000. Molecular phylogeny of the *Nectria haematococca-Fusarium solani* species complex. Mycologia 92 (5): 919-938. https://doi.org/10.2307/3761588

O'Donnell K, Sutton DA, Fothergill A, McCarthy D, Rinaldi MG, Brandt ME, Zhang N, Geiser DM, 2008. Molecular phylogenetic diversity; multilocus haplotype nomenclature, and in vitro antifungal resistance within the *Fusarium solani* species complex. J Clin Microbiol 46: 2477-2490. https://doi.org/10.1128/JCM.02371-07

Ribeiro GT, 1982. Avaliação preliminar da resistência de árvores de *Gmelina arborea* Lineaus, mediante inoculações do fungo *Ceratocystis fimbriata* El. & Halst., causador do cancro em gmelina. Fitopatol Bras 7: 517-524.

Román-Avilés B, Snapp SS, Kelly JD, 2004. Assessing root traits associated with root rot resistance in common bean. Field Crops Res 86 (2): 147-156. https://doi.org/10.1016/j.fcr.2003.08.001

Salas A, 2015. Determinación de la incidencia y severidad de la pudrición del tronco de genotipos de melina (*Gmelina arborea* Roxb.) en el Pacífico Sur de Costa Rica. Univ Nac de Costa Rica, Fac Cienc Tierra y Mar, Esc Ciencias Ambientales. Tesis de Licenciatura en Ciencias Forestales. Heredia, Costa Rica. 50 pp.

Samuels GJ, Lu BS, Chaverri P, Candoussau F, Fournier J, Rossman AY, 2009. *Cyanonectria*, a new genus for *Nectria cyanostoma* and its *Fusarium* anamorph. Mycol Prog 8: 49-58. https://doi.org/10.1007/s11557-008-0577-x

Samuels J, Nirenberg H, 1989. Nectria and Fusarium. 1. *Nectria setofusariae* and its anamorph *Fusarium setosum*. Can J Bot 67 (11): 3372-3377. https://doi.org/10.1139/b89-409

Seo DJ, Lee HB, Kim IS, Kim KY, Park RD, Jung WJ, 2013. Antifungal activity of gallic acid purified from *Terminalia nigrovenulosa* bark against *Fusarium solani*. Microb Path 56: 8-15. https://doi.org/10.1016/j.micpath.2013.01.001

Singh Y, Verma RK, Jamaluddin J, 2003. Combination of biocontrol agents, organic matter and biofertilizer to suppress Fusarium wilt and improve growth of *Gmelina arborea* seedlings. Ind J Trop Biodiv 11: 74-84.

Soni KK, Asaiya AJK, Nandeshwar DL, Jamaluddin J, 2005. Fusarium wilt of *Buchnania lanzan* Spreng - A new disease record from India. Ind J Trop Biodiv 13: 111-112.

Sparks D, 1996. Methods of soil analysis. Part 3, Chemical methods. Soil Sci Soc Am, Inc., Madison, WI, USA.

Tenorio C, Moya R, Arias-Aguilar D, Briceño-Elizondo E, 2016. Biomass yield and energy potential of short-rotation energy plantations of *Gmelina arborea* one year old in Costa Rica. Ind Crop Prod 83: 63-73. https://doi.org/10.1016/j.indcrop.2015.12.005

Thornthwaite CW, Mather JR, 1957. Instructions and tables for computing potential evapotranspiration and the water balance. Drexel Inst Techn Climat 10 (3): 185-311.

Vallejos J, Moya R, Serrano R, 2015. Effects of thinning on diameter, heartwood, density and drying defects of *Gmelina arborea*. Maderas. Cienc Tecnol 17 (2): 365-372. https://doi.org/10.4067/S0718-221X2015005000034

Umana EJ, Akwaji PI, Markson AA, Udo SE, 2015. *Gmelina arborea* Roxb: associated mycoflora and diseases in Cross River State, Nigeria. Global J Sci Front Res: C Biol Sci 15 (4): 1-15.

Voigt K, Cigelnik E, O'donnell K, 1999. Phylogeny and PCR identification of clinically important Zygomycetes based on nuclear ribosomal-DNA sequence data. J Clin Microbiol 37: 3957-3964.

White TJ, Bruns T, Lee WT, Taylor JW, 1990. Amplification and direct sequencing of fungal ribosomal RNA genes for phylogenetics. In: PCR protocols: A guide to methods and applications; Innis MA *et al.* (eds). pp: 315-322. https://doi.org/10.1016/B978-0-12-372180-8.50042-1

Xu Z, Leininger TD, Lee AWC, Tainter FH, 2001. Chemical properties associated with bacterial wetwood in red oaks. Wood Fiber Sci 33: 76-83.

Ward JC, Pong WY, 1980. Wetwood in trees: a timber resource problem. United USDA, Pacific NW For Range Exp, General Tech. Rep. PNW-112, 45 pp.

Wingfield MJ, Robinson DJ, 2004. Diseases and insect pests of *Gmelina arborea*: real threats and real opportunities. New Forest 28: 227-243. https://doi.org/10.1023/B:NEFO.0000040950.01256.ed

18

National assessment of throughfall sensitivity to changes in storm magnitude for the forests of Iran

Pedram Attarod[1,2], Qiuhong Tang[2], John T. Van Stan II[3] and Xingcai Liu[2]

[1]*Forestry and Forest Economics Department, Faculty of Natural Resources, University of Tehran, Iran.* [2]*Key Laboratory of Water Cycle and Related Land Surface Processes, Institute of Geographic Sciences and Natural Resources Research, Chinese Academy of Sciences, China.* [3]*Department of Geology and Geography, Georgia Southern University, USA.*

Abstract

Aim of study: To understand throughfall (TF) sensitivity to variability in rainfall amount (P_g) for typical forest sites across the main climate types of Iran.

Area of study: Nine forest stands of several common native and introduced tree species situated in all common Iranian climate types, but located primarily in northern Iran.

Material and methods: A nondimensional relative sensitivity coefficient was employed to predict responses of TF to P_g changes. Projected P_g changes over the measurement sites for the period 2020-50 were estimated using one of the Coupled Model Intercomparison Project phase 5 (CMIP5) known as HadGEM2-ES under low and high emission scenarios (RCP 2.6 and 8.5).

Main results: TF displayed strong positive linear relationships with P_g at all sites [TF=0.66 P_g -0.16; R^2=0.91]. The sensitivity coefficient ranged from 0.96-2.35 across the nine forest sites and large sensitivity coefficient differences were found between small (< mean annual P_g) and large (> mean annual P_g) storms for arid and Mediterranean plantations. Shifts in P_g and increased small storm frequency are predicted for these regions (2020-50) under low and high emission scenarios.

Research highlights: TF sensitivity may be a useful variable when selecting tree species for afforestation to buffer expected shifts in P_g due to climate change.

Additional keywords: climate change; forest ecosystems; precipitation projection; throughfall sensitivity.

Authors´ contributions: Pedram Attarod: Design and conception; Qiuhong Tang: Financial support; John T. Van Stan II: Language editor and scientific comments, and Xingcai Liu: Data preparation.

Funding: This research was supported by the University of Tehran (UT), Iran and by the International Partnership Program of Chinese Academy of Sciences (Grant No. 131A11KYSB20170113) and the National Natural Science Foundation of China (Grant No. 41790424).

Competing interests: The authors have declared that no competing interests exist.

Correspondence should be addressed to Pedram Attarod: attarod@ut.ac.ir

Introduction

A significant portion of rainfall is intercepted on forest canopy surfaces, where it is either returned to the atmosphere as interception or redistributed to the ground as throughfall (TF) and stemflow. This rainfall partitioning by forest cover significantly alters hydrologic cycling along the soil-forest-atmosphere continuum (Miralles *et al.*, 2010) and can impact feedbacks between the hydrologic cycle and global climate (Davies-Barnard *et al.*, 2014). Stemflow is the proportion of rainfall that drains to the ground along the stem, usually accounting for <2% of annual rainfall in most forests (Van Stan & Gordon, 2018). The remaining majority (60-90%) of rainfall passes through canopy gaps or drips from the vegetation, reaching the surface as TF (Levia *et al.*, 2011) where it can influence soil moisture (Raat *et al.*, 2002), physicochemistry (Rosier *et al.*, 2015), fine root distribution (Ford & Deans, 1978), and microbial processes (Moore *et al.*, 2016). As these ecohydrological interactions play key roles in ecosystem functioning, understanding TF's response to natural and anthropogenic variability is critical to forest management.

Studies across forest ecosystems agree that rainfall amount (P_g) is the principal variable driving stand-scale TF amount (Levia & Frost 2006; Levia *et al.*, 2011)—often explaining >90% of inter-storm TF variability—and that TF-P_g relationships are shaped by canopy structures, like leaf area index, crown depth, etc. (Staelens *et al.*, 2008; Toba & Ohta 2005). However, the authors are unaware of previous work quantifying TF sensitivity to changes in P_g—a commonly anticipated climate response to hydrologic intensification (Huntington, 2006). This is surprising since hydrologic intensification has been linked to increased extreme P_g in many regions (*c.f.*, Tollefson, 2016). Moreover, afforestation, reforestation, urban forestry and other "greening" initiatives have grown in popularity and, therefore, increased the forest cover in most developed regions (*i.e.*, McGovern & Pasher, 2016). In Iran, for example, the restoration of semi-arid and arid ecosystems through planting of low-demand and drought tolerant species has become a critical element of national ecosystem management plans (Attarod *et al.*, 2015b).

Iran has invested in vast tree-plantings throughout its major cities for urban greening and air pollution mitigation, including the Chitgar and Lavizan Forest Parks (Sadeghi *et al.*, 2016) and the ongoing "Jam Afforestation Project" which is tasked with large-scale afforestation and reforestation in the Zagros region (FRWO, 2012). Restoration of the natural Caspian deciduous forests (that extend from the Alborz Mountains to the southern coast of the Caspian Sea) has also resulted in significant reforestation projects since the 1960s (Abbasian *et al.*, 2015). Concerns have risen over the impact of these greening initiatives on the hydrological cycle (Sun *et al.*, 2006; Wang *et al.*, 2011), particularly for the 90% of Iran classified as arid or semi-arid (Ul Hassan *et al.*, 2007). The first process in the rainfall-to-runoff pathway is the partitioning of P_g by forest canopies (Savenije, 2004) and, accordingly, an improved understanding of TF sensitivity to climate change and P_g is essential for addressing these concerns and quantifying the impacts of Iran's (and other nations') large-scale afforestation and reforestation efforts.

A sensitivity analysis is a technique used to determine how different values of an independent variable impact a particular dependent variable under a given set of assumptions. A method for estimating the "sensitivity coefficient" of a dependent variable (in this case, TF) on the relative changes of an independent variable (in this case, P_g) exists (McCuen, 1974) and has been rigorously applied to evapotranspiration and its principal meteorological drivers (Hupet & Vanclooster, 2001). However, the authors are unaware of its application to assess sensitivity of TF to its principal meteorological driver, P_g, for multiple species of contrasting canopy structure. Thus, the objectives of this study are to (1) collect rainfall and TF across common forest species of differing canopy structures and climates in Iran, then (2) quantify and compare their TF sensitivity coefficients. Accomplishing these aims will provide novel information to complement existing data used by forest managers during species selection for restoration and afforestation activities.

Materials and Methods

Sites description

Data were collected in 9 forest stands of several common native and introduced tree species (Table 1) situated in all common Iranian climate types (Table 2), but located primarily in northern Iran (Fig. 1). The selected species represent a diversity of forest canopy architectures, ranging from the smooth-barked, broad-leaved canopy of *Fagus orientalis* (FO) to the rough-barked, needle-leaved canopy of *Pinus eldarica* (PE). Leaf phenology also differs among the selected species as, for example FO is deciduous and PE is evergreen. The PE and *Cupressus arizonica* (CA) throughfall sites are in southern Alborz Mountain Range near the city of Tehran, while the remaining sites—FO, *Quercus castaneifolia* (QC_1-QC_3), *Acer velutinum* (AV), *Pinus brutia* (PB), *Cupressus sempervirens* var. *horizontalis* (CS)—are in northern Alborz Mountain Range and Southern coasts of the Caspian Sea (Fig. 1). Stand structural (and TF) measurements for each forest stand were performed in 0.5 ha plots. Stand density ranged from 112 trees ha^{-1} in the FO forest to 1,600 trees ha^{-1} in the CS plantation (Table 1). Diameter at breast height (1.3 m, dbh) varied greatly among the measured forests, with the smallest values in CS (12 cm) and the largest values in QC_3 (65 cm) (Table 1). Mean canopy coverage also exhibited a wide range across sites, from 45% for CS and 95% for FO (Table 1).

Meteorological data were obtained from the nearest synoptic meteorological stations recording reliable long-term meteorological data (Fig. 1). The range in meteorological data records is from 1951 to 2015 (Table 2). There exists relatively long distances between the Mehr-Abad, Sari, and Gorgan meteorological stations (Fig. 1), but there is no significant topography between the measurement sites and the meteorological stations, and no closer meteorological stations exist in the region. Due to lack of meteorological station inside the Caspian forest, we used meteorological data recorded by Nou-Shahr meteorological station regardless of elevation difference. The ranges of annual precipitation (P) and temperature (T) in the weather station sites are 16.4-17.4 °C and 230-1291 mm, respectively (Table 2). The "De Martonne" climate classification, *i.e.*, De Martonne aridity index (I_{DM}), as described by Baltas (2007), ranged roughly from 49 to 8.5, so that the study sites

Table 1. Forest sites where throughfall (TF) data were collected, including their climate, geographical coordinates, elevation, and stand characteristics. "G" and "D" refer to growing and dormant seasons, respectively, and dbh refers to diameter at breast height.

Species	Code	Climate^	Elevation (m)	Longitude	Latitude	Mean dbh (cm)	Mean height (m)	Closure (%)	Density (trees ha^{-1})	Study period (year)	Storms (n)
Fagus orientalis	*FO*	VH	1410	51°37'	36°35'	50	32	95	112	08, 09 (G)	53
Quercus castaneifolia	*QC_1*	VH	1550	51°37'	36°35'	36	28	80	175	09, 10 (G)	28
Quercus castaneifolia	*QC_2*	SH	360	52°14'	36°28'	21	18	70	864	12 (G)	20
Acer velutinum	*AV*	SH	360	52°14'	36°28'	20	19	80	292	12 (G)	20
*Pinus brutia**	*PB*	SH	360	52°14'	36°28'	14	12	85	541	12 (G)	20
Quercus castaneifolia	*QC_3*	M	250	55°14'	37°15'	65	21	55	198	13 (G)	24
Cupressus sempervirens var. *horizontalis*	*CS*	M	250	55°14'	37°15'	12	7	70	1600	13 (G)	24
*Pinus eldarica**	*PE*	A	1220	51°08'	35°42'	22	9	60	1020	11-14 (G, D)	165
*Cupressus arizonica**	CA	A	1220	51°08'	35°42'	20	8	45	960	11-14 (G, D)	165

*Non-native. ^VH: Very humid, SH: Semi-humid, M: Mediterranean, A: Arid. See text for climate classification method.

Table 2. Details of weather stations included in this study, their climate types according to the De Martonne aridity index (I_{DM}: Baltas, 2007), and the duration of their measurement record.

Station	Dates of record	Elevation (m asl)	Longitude	Latitude	Annual temperature (°C)	Annual precipitation (mm)	Climate (I_{DM} value)
Nou-Shahr	1977-2015	-21	51°30'	36°39'	16.4	1291	Very humid (49)
					16.4	1291	Very humid (49)
Sari	1985-2015	23	53°00'	36°33'	16.9	742	Semi-humid (27)
					16.9	742	Semi-humid (27)
					16.9	742	Semi-humid (27)
Gorgan	1955-2015	0	54°24'	36°54'	17.8	572	Mediterranean (21)
					17.8	572	Mediterranean (21)
Mehr-Abad	1951-2015	1191	51°19'	35°41'	17.4	230	Arid (8.5)
					17.4	230	Arid (8.5)

were grouped into very humid (FO, QC_1), semi-humid (QC_2, AV, PB), Mediterranean (QC_3, CS), and arid (PE, CA) climates (Table 1).

Field measurements

Field measurements were performed from July-2008 to March-2014 during growing and dormant seasons (Table 1). A discrete rain event was defined as a period with >0.1 mm of rainfall. The minimum inter-event dry period between discrete storms was 4-10 h, depending on the site. The effect of pre-storm canopy wetness was assumed negligible as the canopy is assumed to be dry after the minimum inter-event time, which is a common assumption among rainfall partitioning studies (Carlyle-Moses & Gash, 2011). Snowfall was ignored. P_g at each forest site was measured by 3-10 funnel-type plastic collectors with 10 cm funnel diameter and 20-30 cm heights, placed in the nearest open area away from the forest stands. The average of water amount measured in all rain-gauges at a site was used to estimate P_g for each site. Quantities of water in the collectors were measured manually using a graduated cylinder. Storms that reached only the weather stations but not the forest sites, or vice versa, were ignored. P_g volumes were measured at the same time as TF volumes at each site,

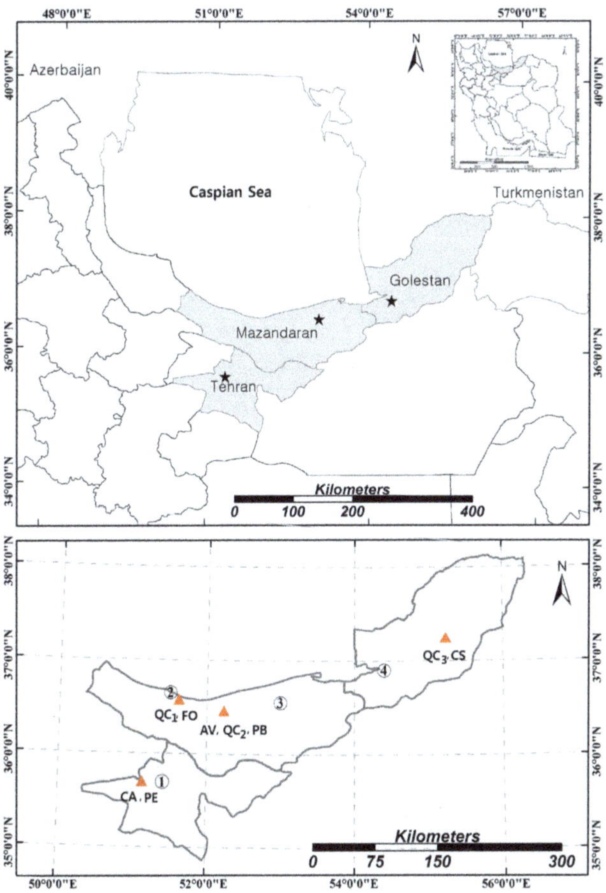

Figure 1. Locations of the measurement sites (triangles) and synoptic meteorological stations (numbers) in the northern provinces of Iran. See Tables 1 and 2 for species abbreviations. Meteorological stations are numbered as: (1) Mehr-Abad; (2) Nou-Shahr; (3) Sari; and (4) Gorgan.

either immediately after a storm or at sunrise following a night storm.

TF was measured using 20-50 rain collectors of the same design as those used to quantify P_g. TF collectors were randomly distributed beneath the canopy and fabric covered the neck of the collectors to avoid litter, needles, and debris from entering (Abbasian *et al.*, 2015). TF data based on multiple field campaigns that used a different number of collectors is a common, often necessary, issue in national-to-international scale studies (*i.e.*, Wallace *et al.*, 2013; Návar, 2017); however, it is important to note that this methodological variability may introduce uncertainty to the stand-scale TF estimates (Vose *et al.*, 2016). Variability about mean P_g and TF will be expressed in standard error (SE) throughout.

Throughfall sensitivity coefficients

A practical method of presenting a sensitivity analysis is to plot relative changes of a dependent variable (in this case, TF) against relative change of an independent variable (*i.e.*, P_g) as a curve (*e.g.*, Singh & Xu, 1997; Goyal, 2004, Attarod *et al.*, 2015a). This is different from a standard regression of TF versus P_g (Fig. 2a) as the mean sensitivity coefficient is calculated as the slope of a correlation between the percent changes in P_g against percent changes in TF (Fig. 2b). The sensitivity coefficient represents the fraction of change in P_g transmitted to the change of TF, *i.e.* a sensitivity value of 0.1 would suggest that a 10% increase in P_g may be predictable to increase TF by 1%. Negative coefficients would indicate that a decrease in TF would result from an increase of P_g, which is not expected due to the universally reported positive relationship between P_g and TF (Friesen *et al.*, 2015). In the present study, the sensitivity coefficient of TF was determined in climate classifications only in response to changes in P_g on an event-basis.

Projected P_g changes over the measurement sites

General Circulation Models (GCMs) are numerical models representing physical processes in the atmosphere, ocean, cryosphere, and land surface. These models are the most advanced tools currently available for simulating the response of the global climate system to increasing greenhouse gas concentrations. GCM simulations for the fifth Assessment Report (AR5) of the Intergovernmental Panel on Climate Change (IPCC) have become available (Taylor *et al.*, 2012; Miao *et al.*, 2014). Comparing to the IPCC AR4, the GCMs in AR5 include a more varied set of model types (*i.e.*, climate/earth system models with more interactive components such as atmospheric chemistry, aerosols, dynamic vegetation, ice sheets and carbon cycle) (Liu *et al.*, 2013). Several improvements in the physics, numerical algorithms and configurations are implemented in the IPCC AR5 models with a new set of scenarios called Representative Concentration Pathways (RCPs) for energy and industry CO_2 emissions (Moss *et al.*, 2010). The RCPs span a large range of stabilization, mitigation and non-mitigation pathways. Phase 5 of the Coupled Model Intercomparison Project (CMIP5) is a standard experimental protocol for studying the output of GCMs which provides a community-based infrastructure in support of climate model diagnosis, validation, intercomparison, documentation and data access (Jones *et al.*, 2011). Longer time-scale ("centennial") experiments have been performed at the Met Office Hadley Centre with the HadGEM2-ES Earth System model—one of the CMIP5 climate models (Collins *et al.*, 2011; Jones *et al.*, 2011; Miao *et al.*, 2014). The HadGEM2-ES model used here is one of the state-of-the-art GCMs and involves many typical and advanced

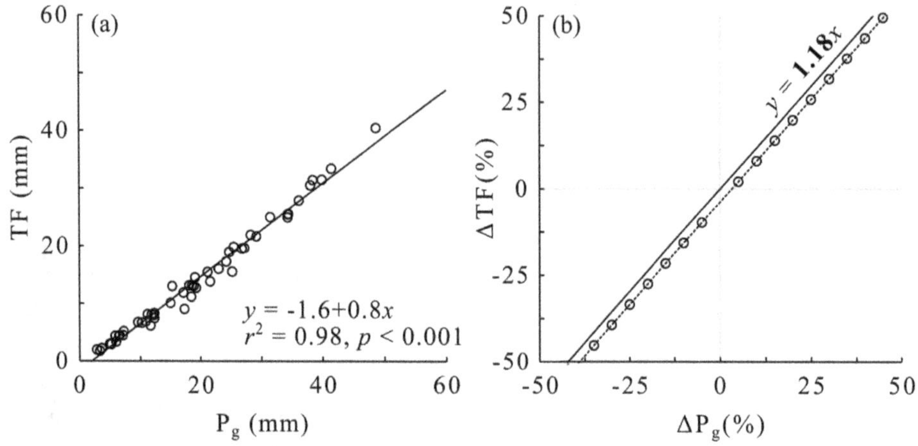

Figure 2. Example regression of throughfall (TF) against storm magnitude (P_g) for data collected at the *Fagus orientalis* site. Through shifting P_g by any percentage (ΔP_g), the regression equation (in panel a) can be used to calculate a corresponding percent change in TF (ΔTF). These shifts are (b) plotted and a regression calculated where the slope is a nondimensional coefficient of "sensitivity."

representations of land and ocean processes (Jones *et al.*, 2011). The HadGEM2-ES climate data has been widely used for climate studies (Huntingford *et al.*, 2013). The *r1i1p1* ensemble of HadGEM2-ES was used in this study. The *r1i1p1* ensemble is the most accessible ensemble in the CMIP5 archive.

To understand P_g variations under changing climate over the measurement sites, we focused on the precipitation projection for two scenarios: RCP 2.6 and RCP 8.5. RCP 2.6 represents a "low" emissions scenario featured by the radiative forcings of 2.6 Wm^{-2} and atmospheric CO_2 concentration of 421 ppm by 2100. RCP 8.5 represents a "high" emission scenario with the radiation forcing of 8.5 Wm^{-2} and CO_2 concentration of 936 ppm by 2100. The time resolution of projected storm magnitude (*i.e.*, P_g) is daily. Future climate data in grids with 0.5°×0.5° horizontal resolution across measurement sites were obtained from the HadGEM2-ES model projections (Hempel *et al.*, 2013).

Delta change method

Because of the limitations of coarse-resolution GCM climate data (used in these predictions), the delta change (DC) method is generally used to derive scenarios of future climate (*e.g.*, Fischer *et al.*, 2007; Shahid, 2011; Chung & Nkomozepi, 2012). The method consists of simply scaling the observed climate data using monthly change factors calculated from the differences in climatology predicted by GCMs for the current and future periods. In this way, P_g for future time periods was derived by scaling the historically observed climate data (OBS) by the GCM-computed change. This results in a new P_g time series scaled according to the GCMs, but based on historical observations. In this study, the OBS from 1974 to 2004 (reference period) were scaled to derive the 30-yr future climate scenario for the period 2020-2050. Relative change factors (ΔVar) were then applied to the observed flux variable to calculate future P_g. Following Leng & Tang (2014), the DC method are formulated as Eq., (1):

$$\mathrm{Var}_\Delta(\mathrm{i,j}) = [\mathrm{Var}_{\mathrm{OBS}}(\mathrm{i,j})][\Delta\mathrm{Var}(\mathrm{j})]; \quad \mathrm{i} = 1,2,\ldots\ldots 31; \quad \mathrm{j} = 1,2,\ldots\ldots 12, \tag{1}$$

where Var_Δ is the scaled flux variable using the DC method and $\mathrm{Var}_{\mathrm{OBS}}$ is the observed flux variable in the historic period. The suffixes i and j stand for the day and the month, respectively, and ΔVar is the monthly DC factor, which is calculated as follows:

$$\Delta\mathrm{Var}(\mathrm{j}) = \frac{\overline{\mathrm{Var}}_{\mathrm{future}}(\mathrm{j})}{\overline{\mathrm{Var}}_{\mathrm{current}}(\mathrm{j})}; \quad \mathrm{j} = 1,2,\ldots\ldots,12 \tag{2}$$

where $\overline{\mathrm{Var}}_{\mathrm{future}}$ (j) and $\overline{\mathrm{Var}}_{\mathrm{current}}$ (j) are the mean values of the time series for month j for the future and current time periods by GCMs.

Results

Overview of rainfall and throughfall events

There were 290 P_g events recorded from 2008-2014 across all measurement sites and P_g ranged from 0.5-54.7 mm. Mean P_g was 10.8±0.7 mm across all forest sites, but site-specific mean P_g ranged from 17-20 mm for forests in the very humid, semi-humid, and Mediterranean climate types (Table 3). Mean P_g from the arid forest sites was 4.4 mm (Table 3). At arid sites,

Table 3. Characteristics of recorded rainfall events (P_g) in the measurement sites with respect to the different climate types. n refers to the number of recorded events.

Climate type	n	Mean (mm)	Max (mm)	Min (mm)	SE (mm)	%< Mean	%< 5 mm	%>15 mm
Very Humid	81	20.0	50.9	2.8	1.4	58	13	85
Semi-humid	20	17.3	35.0	0.5	2.4	45	25	65
Mediterranean	24	19.4	54.7	0.5	3.0	58	29	62
Arid	165	4.4	19.0	0.5	0.3	67	70	5

67% of collected storms were < mean P_g; however, for other climate types, this fraction was generally around 50% (Table 3). "Large" storms (P_g > 15 mm) were frequent at the very humid forest sites (85%), yet only 5% of storms exceeded 15 mm at arid sites (Table 3). "Small" storms (P_g < 5 mm) were the norm for arid forests in this study (70%: Table 3). The respective percentages of P_g < 5 mm or P_g > 15 mm were similar for the semi-humid and Mediterranean forests (Table 3).

Strong positive linear relationships between event mean TF and P_g were observed across all sites [TF=0.66 (P_g) - 0.16; R^2 = 0.91] and at each individual site (Table 4). Slopes of the regression lines between TF and P_g (often assumed to be linked to varying canopy structures) were wide-ranging: 0.39 for CS to 0.81 for FO (Table 4). Relative TF (TF:P_g) varied roughly from 40% (Mediterranean CS) to 75% (very humid QC_1) (Table 4). In arid and Mediterranean climates, minimum TF:P_g were measured at zero, however, in other climates the minimum was approximately 47% (Table 4). TF:P_g was, surprisingly, nearly the same when averaged for all the needle-leaved forests (55.2%) and broad-leaved forests (50%) across the climate zones (Table 4). Marked differences in TF:P_g on average, however, were observed for plantations versus natural forests, where plantations were found to produce 53.4% of TF compared to 71.9% for natural forests. For all sites, QC_1 produced the highest TF:P_g in very humid conditions (QC_1: 74.5%) and the lowest TF:P_g under Mediterranean conditions (CS: 39.3%) (Table 4). Generally higher standard errors were observed for TF:P_g in the Mediterranean (3.25%) and arid sites (3.35%) compared to forests in other climates (1.34%) (Table 4).

Historical annual storm characteristics

Historical records at the meteorological stations show that mean annual precipitation decreased from very humid (1291 mm y^{-1}) to arid climate (230 mm y^{-1}) (Table 5). Compared with other climates, the arid climate had relatively constant annual precipitation, varying only by 9 mm y^{-1} between 1951-2015 (Table 5). The ratio of maximum P_g to mean annual precipitation was respectively 11%, 16%, 18%, and 22% for the semi-humid, very humid, Mediterranean, and arid climates (Table 5). Although mean P_g had a descending trend from very humid (10.5 mm) to arid (3.9 mm), the proportion of mean P_g to maximum P_g in different climates ranged from 8.5% in the semi humid climate to 5.0% in the very humid climate (Table 5).

Throughfall sensitivity to changes in storm magnitude

TF exhibited varying degrees of sensitivity to P_g, showing large fluctuations (*i.e.*, doubling in sensitivity)

Table 4. Relationships between throughfall (TF) and rain event magnitude (P_g), percent of event based average relative throughfall (TF: P_g), and related statistics. See Table 1 for the tree species represented by the location codes. n refers to the number of recorded events.

Tree species	n	Mean TF: P_g (%)	Max TF: P_g (%)	Min TF: P_g (%)	SE TF: P_g (%)	Regression	r^2
FO	53	69.4	85.0	52.1	1.3	TF=0.81 P_g -1.60	0.98
QC_1	28	74.5	87.8	60.1	1.1	TF=0.80 P_g -0.92	0.99
QC_2	20	66.4	80.0	45.4	1.3	TF=0.70 P_g -0.17	0.97
AV	20	44.2	59.2	30.9	1.3	TF=0.54 P_g -1.03	0.96
PB	20	59.0	78.6	43.7	1.7	TF=0.62 P_g -0.55	0.91
QC_3	24	58.4	88.3	0	3.1	TF=0.57 P_g -0.88	0.87
CS	24	39.3	76.0	0	3.4	TF=0.39 P_g -0.64	0.80
PE	165	50.7	86.8	0	3.9	TF=0.79 P_g -0.50	0.98
CA	165	55.8	89.2	6.0	2.8	TF=0.73 P_g -0.34	0.97

Table 5. Historical characteristics of annual precipitation, number of storms, and other related rain event magnitude (P_g) characteristics recorded in the meteorological stations nearest to the measurement sites.

Station Climate (Record)	Annual precipitation (mm)			P_g (mm)	
	Mean ± SE	Max.	Min.	Mean	Max.
Nou-Shahr Very humid (1977-2015)	1291±30	1628	921	10.5	208
Sari* Semi-humid (1985-2015)	742±28	1007	442	7.2	85
Gorgan Mediterranean (1955-2015)	577±17	962	314	6.0	105
Mehr-Abad Arid (1951-2015)	230±9	403	100	3.9	50

*event number is for the period 2000-2015 (1666 events); In the very humid, Mediterranean, and arid climates, 4819, 5762, and 3772 events were recorded, respectively.

for species planted in different climates (Fig. 3). TF from the arid PE forest was most sensitive to fluctuations in P_g (= 2.35) and TF from the Mediterranean QC_3 forest was least sensitive (= 0.96) (Fig. 3). Needle-leaved plantations (*i.e.*, PE, CA) generally had higher sensitivity (> 1.5) to changing P_g compared to broadleaved forests (~ 1.0) (Fig. 3). For broadleaved forests, QC_1 was less sensitive than FO in a very humid climate (Fig. 3). QC_1, QC_2, and QC_3 had similar sensitivity coefficients regardless of climate: very humid (=1.09) to semi-humid (=1.07) and Mediterranean climates (= 0.96) (Fig. 3). TF sensitivity was tested for larger and smaller storm sizes than the mean P_g in each climate (Table 6). Excluding AV, PB, and QC_1, TF was found to be more sensitive to small storms (Table 6). Plantations in arid or Mediterranean climates had large differences in sensitivity coefficient between small and large events, *e.g.* 2.95 *vs.* 1.14 for PE and 1.78 *vs.* 0.74 for CS plantation (Table 6). Interestingly, the FO forest had nearly identical sensitivity values for storms smaller (1.06) and larger (1.05) than mean P_g (Table 6). Sensitivity of the very humid QC_1 natural forest was greater for large events (1.07 against 1.24) in comparison with the semi-humid QC_2 and Mediterranean QC_3 forests (Table 6).

Projected changes in annual precipitation and storm magnitude

During the 2020-2050 period, the low emissions (RCP 2.6) scenario predicted increased annual precipitation on average in all climate types, excluding the semi-humid climate, yet the high emissions (RCP 8.5) scenario predicted decreased annual precipitation on average in all climates excluding the arid climate (Table 7). Projected changes to mean P_g from the low emission scenario was +16.0%, +3.2%, 0%, and -1.8% for Mediterranean, very humid, arid, and semi-humid forests, respectively (Tables 5 and 7). The high emissions scenario predicts that mean P_g will change by -5% in very humid forests and -10% in arid forests, but mean P_g will not change for forests in the Mediterranean and semi-humid climates (Tables 5 and 7).

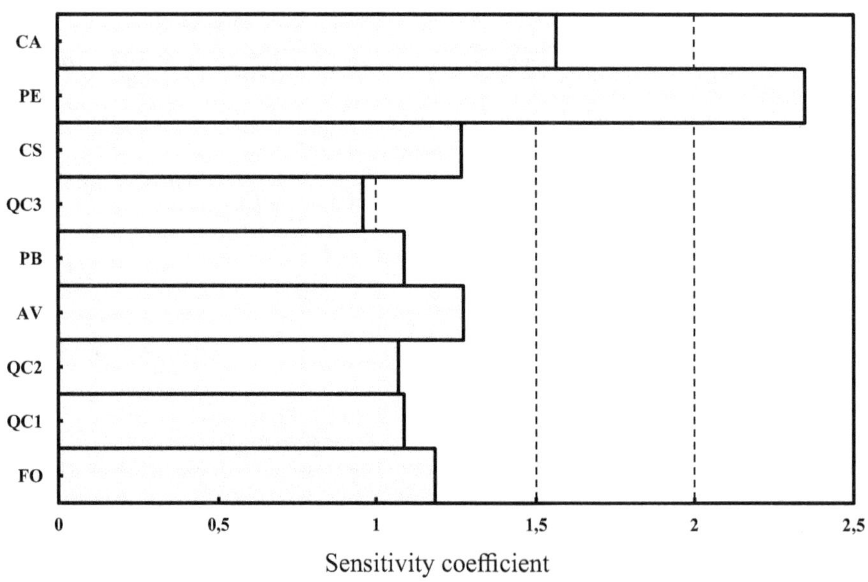

Figure 3. Mean throughfall (TF) sensitivity coefficients for the measurement sites. See Table 1 for the tree species represented by the location codes.

Table 6. Throughfall (TF) sensitivity coefficients classified for storm magnitudes (P_g) smaller and larger than mean P_g recorded in the measurement sites. See Table 1 for the tree species represented by the location codes.

Tree species	Mean P_g (mm)	Mean sensitivity for < mean P_g	Mean sensitivity for > mean P_g
FO	18.9	1.06	1.05
QC_1	23.2	1.07	1.24
QC_2	17.3	1.08	0.81
AV	17.3	1.14	1.34
PB	17.3	0.96	1.45
QC_3	19.4	1.20	0.80
CS	19.4	1.78	0.74
PE	4.4	2.95	1.14
CA	4.4	1.57	1.21

Table 7. Projections of annual precipitation, and mean storm magnitude (P_g) during the period of 2020-2050 predicted by the GCM under RCP 2.6 (low emission), and RCP 8.5 (high emission) scenarios.

Climate	Annual precipitation (mm)			P_g (mm)
	Mean±SE	Max.	Min.	Mean
Very humid				
RCP 2.6	1306±34	1587	871	10.8
RCP 8.5	1207±31	1489	845	10.0
Semi-humid				
RCP 2.6	736±28	989	563	7.0
RCP 8.5	670±24	903	515	7.2
Mediterranean				
RCP 2.6	598±18	867	405	7.0
RCP 8.5	518±16	743	357	6.0
Arid				
RCP 2.6	271±13	448	144	3.9
RCP 8.5	247±11	391	139	3.5

Discussion

Storm magnitude and throughfall

We collected a wide range of storms across sites hosting common forests in all of Iran's climate types (Tables 1-2; Fig. 1). Stands for which comparable TF data exist (*e.g.*, FO and QC_1) are comparable to those reported elsewhere, for example: FO versus *Fagus sylvatica* (e.g., Staelens *et al.*, 2008) and QC versus *Quercus serrata* and *Quercus acutissima* (*e.g.*, Toba & Ohta, 2005). Relationships between P_g and TF for individual forest often indicate differences in the canopy-meteorological interactions that control TF production by plants (Zhang *et at.*, 2016). In this study, as in others, P_g accounted for most (nearly all) of TF variability regardless the species, yet the slope and intercept of the P_g-TF correlation varied across the forests of differing species-climate combination (Table 4). Relative TF:P_g in our study, 39.3-74.5%, appears to be strongly tied to forest structure (*e.g.*, density, seasonal change, vegetation area index, gap fraction, and canopy storage capacity), and climate conditions (P_g and, perhaps storm intensity and wind conditions) as shown by many others (Crockford & Richardson, 2000; 2011; Staelens *et al.*, 2008; Muzylo *et al.*, 2012; Motahari *et al.*, 2013). Results indicated that, when the same species experiences different climate conditions, they can differentially partition P_g into TF—something particularly evident QC in three different climates (Table 4).

Throughfall sensitivity to storm amount

Large fluctuations in TF sensitivity were observed between different species, climate types, and storm sizes (greater or less than mean annual P_g) common to Iran (Table 6). For forest managers, it is apparent that different species selected for planting in different climates will exhibit differing TF sensitivities to shifting P_g (Fig. 3). Practically, managers can use these sensitivity coefficients (Fig. 3 or Table 6) and projected changes to P_g (Tables 5 versus 7) and estimate the potential shift in TF supply to the forest surface. For example, PE's sensitivity coefficient (2.35) indicates that a 10% decrease in P_g could approximately reduce TF by 23%. Although we identified little change in the total yearly precipitation in the arid climate for the most recent decade (where these PE plantations are generally situated), an approximate estimation of 8% decrease in the storm size observed in the recent decade can induce an 18% decrease in TF in these plantations. Greater sensitivity coefficients for PE (and CA) plantations may be a result of (1) these species having larger storage capacities than other tree species in this study (Sadeghi *et al.*, 2015), (2) their arid climate allowing their canopies to dry more efficiently between storms, and (3) most-to-nearly all the storms being small (Table 6). The higher sensitivity of these species to P_g has implications for forest managers dealing with climate change, since both are the most widely-used species for afforestation in the arid and semiarid regions of Asia (*e.g.*, Iran, Lebanon, Syria, Pakistan, Iraq, and Afghanistan) and are frequently selected due to their greater tolerance to drought, high/low temperature extremes, and being faster-growing than native broad- or needle-leaved tree species (Jazirei, 2009). Under a changing climate, water resource management in arid regions is therefore complicated by afforestation initiatives and forest managers may be able to use TF sensitivity as a variable in their decision to choose a species with the least sensitivity to expected shifts in P_g.

The remaining species located in the Caspian forests of northern Iran with very humid, semi-humid and Mediterranean climates exhibited roughly the same sensitivity (mean = 1.13; Fig. 3). However, according to our results, the replacement of QC_3 natural forest with CS man-made in the Mediterranean areas of the eastern Caspian region will likely increase TF sensitivity by 0.31 (Fig. 3). This increased TF sensitivity with converting Mediterranean QC forests to CS may be a product of greater tree density of young CS (Table 1) or, more theoretically, greater roughness lengths and zero-plane displacement heights (Rutter *et al.*, 1975; Valente *et al.*, 1997). Moreover, CS's scale-like needle-leaves have different leaf shedding habits compared to QC_3's broadleaves, which will differentially interact with P_g to alter TF and its sensitivity (Pypker *et al.*, 2011).

Natural broad-leaved forests in the Caspian region (FO and QC_1) showed roughly the same sensitivity (Fig. 3). The primary function of the Caspian forests, other than wood production, is conservation of soil and water resources (Sagheb Talebi *et al.*, 2014). Thus, it is expected that the low TF sensitivity of FO and QC_1 to shifts in P_g will help buffer the region against climate changes. But, restoration of the Caspian deciduous forest of northern Iran by planting species, like CS (which had a greater TF sensitivity coefficient), were extensively performed by Iran's Forest, Rangeland, and Watershed organization, and may have considerable effects on ecosystem ecohydrology through altered TF supply. In contrast, introduction of PB in the semi humid climate of the central Caspian region for restoration of degraded forests showed roughly the same, even lower, sensitivity compared to native species (AV and QC_2) (Fig. 3). Thus, understanding the relationship between hydrologic cycling and the impact of afforestation projects on these variables can be useful for forest management and selection of suitable species for reforestation of degraded forest ecosystems.

Although the consistently-observed strong correlation between stand-scale TF and storm amount across forest types and climates confirms that storm amount is the primary factor affecting TF generation after canopy structures are saturated (Levia *et al.*, 2011), the degree to which stand-scale TF responds to storm amount has been linked to vegetation structure (leaf area index, crown depth, etc.) (Staelens *et al.*, 2008; Toba & Ohta 2005). Consequently, it is advisable to incorporate vegetation characteristics with TF sensitivity in response to changes in P_g responses in future investigations.

Combining throughfall sensitivity and climate projections

Changes in P_g to an area are expected to be compounded by the TF sensitivity of each forest type. An example "rough" estimation using the 16% increase in P_g predicted by RCP 2.6 scenario in the Mediterranean climate results in a 20% and 15% increase in TF generation in CS and QC_3 forests, respectively. The projected 3.2% increase (by RCP 2.6) and 5% decrease (by RCP 8.5) in mean P_g, however, in the very humid climate of Iran where the Caspian forests are located (Table 7) is not anticipated to significantly alter TF supply to the forest floor due to the tree species' low sensitivity (Fig. 3). Despite the highest sensitivity of plantations in the arid climate (~2), very slight changes in mean P_g are predicted by the RCP 2.6 scenario which may not influence TF receipt at the surface. However, under the RCP 8.5 scenario, a 10% decrease in mean P_g in the arid climate could decrease TF by 23% and 15% per event beneath PE and CA plantations, respectively. Fewer impacts may be realized in the semi-humid climate where the dominant tree species showed low TF sensitivity and not significant changes in mean event size are predicted by both RCPs. Clearly, these climate change scenarios (and others) can be a reference for setting minimum and maximum configurations of forest cover in water management and planning associated with adaptation. There will be large uncertainty among GCM models, most GCM models may have different performances across regions, and, as a consequence of lacking large scale observations, downscaling from GCM grid data to local areas is difficult. Moreover, the delta change factor strategy used in this study does not adjust climate projections, but assumes that the signal or changes are reasonably projected by climate models, even though the models are biased (Leng & Tang, 2014). A key feature of this approach is that, because the method uses historical precipitation as its basis and GCM data only to change the magnitude of the historical precipitation, it fails to account for changes in P variability predicted by different GCM models (Leng & Tang, 2014). Still, the projections and downscaling of changes in P_g under the most conservative (low emission) and the highest greenhouse gas emissions pathway (high emission) scenarios (Table 7) underscores the necessity of work to understand TF sensitivity of common forest types in a region.

Conclusion

This national assessment of throughfall beneath common forest types in Iran showed large fluctuations in throughfall sensitivity for species planted in different climates. Arid needle-leaved plantations exhibited generally higher sensitivity responses to changing storm amount, while very humid, natural broad-leaved forests had low sensitivity responses. Projections of mean storm magnitude under a low emission scenario indicated that, compared to historical data, modest changes would be expected for most climates (excluding arid) on average during 2020-2050. However, under a high emission scenario, larger alterations are expected in all climate types except the arid climate. Results indicate that any projected change in storm size will not simply be translated directly to a change in throughfall. Rather, the increase or decrease in throughfall amount may be better predicted as a product of the projected change in storm magnitude and the forest-specific throughfall sensitivity. These findings have implications for selection of the most appropriate and adapted species for afforestation under climate change.

References

Abbasian P, Attarod P, Sadeghi SMM, Van Stan II JT, Hodjati SM, 2015. Throughfall nutrients in a degraded indigenous Fagus orientalis forest and a Picea abies plantation in North of Iran. Forest Syst 24(3) e035. https://doi.org/10.5424/fs/2015243-06764

Attarod P, Kheirkhah F, Khalighi Sigaroodi Sh, Sadeghi SMM, 2015a. Sensitivity of reference evapotranspiration to global warming in the Caspian region, North of Iran. J Agr Sci Tech-Iran 17: 869-883.

Attarod P, Sadeghi SMM, Pypker TG, Bagheri H, Bagheri M, Bayramzadeh V, 2015b. Needle-leaved trees impacts on rainfall interception and canopy storage capacity in an arid environment. New Forest 46: 339-355. https://doi.org/10.1007/s11056-014-9464-2

Baltas E, 2007. Spatial distribution of climate indices in northern Greece. Meteorol Appl 14: 69–78. https://doi.org/10.1002/met.7

Carlyle-Moses DE, Gash JHC, 2011. Precipitation interception loss by forest canopies. In: Levia DF, Carlyle-Moses D, Tanaka T (eds) Forest hydrology and biogeochemistry: synthesis of past research and future directions. Ecolog studies 216: 407–424.

Chung SO, Nkomozepi T, 2012. Uncertainty of paddy irrigation requirement estimated from climate change projections in the Geumho River basin, Korea. Paddy Water Environ 10: 175–185. https://doi.org/10.1007/s10333-011-0305-z

Collins WJ, Bellouin N, Doutriaux-Boucher M, Gedney N, Halloran P, Hinton T, Hughes J, Jones CD, Joshi M, Liddicoat S, *et al.*, 2011. Development and evaluation of an Earth-system model HadGEM2. Geosci Model Dev 4: 997–1062. https://doi.org/10.5194/gmdd-4-997-2011

Crockford RH, Richardson DP, 2000. Partitioning of rainfall into throughfall, stem-flow and interception: effect of forest type, ground cover and climate. Hydrol Process 14: 2903–2920. 3.0.CO;2-6" target="_blank">https://doi.org/10.1002/1099-1085(200011/12)14:16/17<2903::AID-HYP126>3.0.CO;2-6

Davies-Barnard T, Valdes PJ, Jones CD, Singarayer JS, 2014. Sensitivity of a coupled climate model to canopy interception capacity. Clim Dynam 42(7-8): 1715-1732. https://doi.org/10.1007/s00382-014-2100-1

Fischer G, Tubiello FN, van Velthuizen H, Wiberg DA, 2007. Climate change impacts on irrigation water requirements: Effects of mitigation, 1990–2080. Technol Forecast Soc 74: 1083–1107. https://doi.org/10.1016/j.techfore.2006.05.021

Ford ED, Deans JD, 1978. The effects of canopy structure on stemflow, throughfall and interception loss in a young sitka spruce plantation. J Appl Ecol 15(3): 905-917. https://doi.org/10.2307/2402786

FRWO (Forest, Range, and Watershed Organization of Iran), 2012. Project of Participatory Afforestation in the Zagros Region.

Friesen J, Lundquist J, Van Stan II JT, 2015. Evolution of forest precipitation water storage measurement methods. Hydrol Process 29: 2504–2520. https://doi.org/10.1002/hyp.10376

Goyal RK, 2004. Sensitivity of evapotranspiration to global warming: a case study of arid zone of Rajasthan (India). Agr Water Manage 69: 1–11. https://doi.org/10.1016/j.agwat.2004.03.014

Huntingford C, Zelazowski P, Galbraith D, Mercado LM, Sitch S, Fisher R, Lomas M, Walker A, Jones C, Booth, B, *et al.*, 2013. Simulated resilience of tropical rainforests to CO2-induced climate change. Nature Geoscience 6: 268–273. https://doi.org/10.1038/ngeo1741

Hempel S, Frieler K, Warszawski L, Schewe J, Piontek F, 2013. A trend-preserving bias correction - the ISI-MIP approach. Earth Syst Dynam 4(2): 219-236. https://doi.org/10.5194/esd-4-219-2013

Huntington TG, 2006. Evidence for intensification of the global water cycle: review and synthesis. J Hydrol 319: 83-95. https://doi.org/10.1016/j.jhydrol.2005.07.003

Hupet F, Vanclooster M, 2001. Effect of the sampling frequency of meteorological variables on the estimation of the reference evapotranspiration. J Hydrol 243: 192–204. https://doi.org/10.1016/S0022-1694(00)00413-3

Jazirei MH, 2009. Dryland Afforestation. University of Tehran Press, Iran. pp 560. ISBN 978-964-03-6122-1.

Jones CD, Hughes JK, Bellouin N, Hardiman SC, Jones GS, Knight J, Liddicoat S, O'Connorl FM, Andres RJ, Bell C, *et al.*, 2011. The HadGEM2-ES implementation of CMIP5 centennial simulations. Geoscie Model Dev 4(3): 543-570. https://doi.org/10.5194/gmd-4-543-2011

Leng G, Tang Q, 2014. Modeling the impacts of future climate change on irrigation over China: sensitivity to adjusted projections. J Hydrometeorol 15(5): 2085-2103. https://doi.org/10.1175/JHM-D-13-0182.1

Levia DF, Frost EE, 2006. Variability of throughfall volume and solute inputs in wooded ecosystems. Prog Phys Geog 30: 605–632. https://doi.org/10.1177/0309133306071145

Levia DF, Keim RF, Carlyle-Moses DE, Frost EE, 2011. Throughfall and stemflow in wooded ecosystems, in Forest Hydrology and Biogeochemistry: Synthesis of Past Research and Future Directions, Ecol Studr Ser 216: 425–443.

Liu J, Song M, Horton RM, Hu Y, 2013. Reducing spread in climate model projections of a September ice-free arctic. P.12571-6 in Proc. of Conf. of National Academy of Sciences of the United States USA 110 (31).

McCuen RH, 1974. A sensitivity and error analysis of procedures used for estimating evaporation. Water Resour Bull 10 (3): 486–498. https://doi.org/10.1111/j.1752-1688.1974.tb00590.x

McGovern M, Pasher J, 2016. Canadian urban tree canopy cover and carbon sequestration status and change 1990–2012. Urban For Urban Gree 20: 227-232. https://doi.org/10.1016/j.ufug.2016.09.002

Miao C, Duan Q, Sun Q, Huang Y, Kong D, Yang T, Ye A, Di Zh, Gong W, 2014. Assessment of CMIP5 climate models and projected temperature changes over Northern Eurasia. Environ Res Lett. https://doi.org/10.1088/1748-9326/9/5/055007

Miralles DG, Gash JH, Holmes TRH, de Jeu RAM, Dolman AJ, 2010. Global canopy interception from satellite observations. J Geophys Res 115. https://doi.org/10.1029/2009JD013530

Moore LD, Van Stan II JT, Gay TE, Rosier C, Wu T, 2016. Alteration of soil chitinolytic bacterial and ammonia oxidizing archaeal community diversity by rainwater redistribution in an epiphyte-laden Quercus virginiana canopy. Soil Biol Biochem 100: 33-41. https://doi.org/10.1016/j.soilbio.2016.05.016

Moss RH, 2010. The next generation of scenarios for climate change research and assessment. Nat 463: 747–56. https://doi.org/10.1038/nature08823

Motahari M, Attarod P, Pypker TG, Etemad V, Shirvany A, 2013. Rainfall interception and canopy storage capacity of a Pinus eldarica plantation in a semi-arid climate zone: an application of the Gash model. J Agr Sci Tech-Iran 15: 981–994.

Muzylo A, Llorens P, Domingo F, 2012. Precipitation partitioning in a deciduous forest plot in leafed and leafless periods. Ecohydrology 5: 759–767. https://doi.org/10.1002/eco.266

Návar J, 2017. Fitting rainfall interception models to forest ecosystems of Mexico. J Hydrol 548: 458–470. https://doi.org/10.1016/j.jhydrol.2017.03.025

Pypker TG, Levia DF, Staelens J, Van Stan II JT, 2011. Canopy structure in relation to hydrological and biogeochemical fluxes. XVII. In: Levia, D.F., Carlyle-Moses, D.E., Tanaka, T. (Eds.), Forest Hydrology and Biogeochemistry: Synthesis of Past Research and Future Directions. Ecological Studies Series, Springer-Verlag, Heidelberg 216: 371–388. https://doi.org/10.1007/978-94-007-1363-5_18

Raat KJ, Draaijers GPJ, Schaap MG, Tietema A, Verstraten, JM, 2002. Spatial variability of throughfall water and chemistry and forest floor water content in a Douglas fir forest stand. Hydrol Earth Syst Sci 6: 363–374. https://doi.org/10.5194/hess-6-363-2002

Rosier CL, Van Stan II JT, Moore LD, Schrom JOS, Wu T, Reichard JS, Kan J, 2015. Forest canopy structural controls over throughfall affect soil microbial community structure in an epiphyte-laden maritime oak stand. Ecohydrology. https://doi.org/10.1002/eco.1595

Rutter AJ, Morton AJ, Robins PC, 1975. A predictive model of rainfall interception in forests II. Generalization of the model and comparison with observations in some coniferous and hardwood stands. J Appl Ecol 12: 367–380. https://doi.org/10.2307/2401739

Sadeghi SMM, Attarod P, Van Stan II JT, Pypker TG, Dunkerley D, 2015. Efficiency of the reformulated Gash's interception model in semiarid afforestations. Agr Forest Meteorol 201: 76-85. https://doi.org/10.1016/j.agrformet.2014.10.006

Sadeghi SMM, Attarod P, Van Stan II JT, Pypker TG, 2016. The importance of considering rainfall partitioning in afforestation initiatives in semiarid climates: A comparison of common planted tree species in Tehran, Iran. Sci Total Environ 568: 845-855. https://doi.org/10.1016/j.scitotenv.2016.06.048

Sagheb Talebi K, Sajedi T, Pourhashemi M, 2014. Forests of Iran, A treasure from the past, a hope for the future. Plant Vegetation 10: 149.

Shahid S, 2011. Impacts of climate change on irrigation water demand in northwestern Bangladesh. Climatic Change 105: 433–453. https://doi.org/10.1007/s10584-010-9895-5

Savenije HHG, 2004. The importance of interception and why we should delete the term evapotranspiration from our vocabulary. Hydrol Process 18(8): 1507-1511. https://doi.org/10.1002/hyp.5563

Singh VP, Xu C-Y, 1997. Sensitivity of mass transfer-based evaporation equations to errors in daily and monthly input data. Hydrol Process 11: 1465–1473. 3.0.CO;2-X" target="_blank">https://doi.org/10.1002/(SICI)1099-1085(199709)11:11<1465::AID-HYP452>3.0.CO;2-X

Staelens J, Schrijver AD, Verheyen K, Verhoest NEC, 2008. Precipitation partitioning into throughfall, stemflow, and interception within a single beech (Fagus sylvatica L.) canopy: influence of foliation, rain event characteristics, and meteorology. Hydrol Process 2: 33-45. https://doi.org/10.1002/hyp.6610

Sun G, Zhou GY, Zhang ZQ, Wei XH, McNulty SG, Vose JM, 2006. Potential water yield reduction due to forestation across China. J Hydrol 328: 548–558. https://doi.org/10.1016/j.jhydrol.2005.12.013

Taylor KE, Stouffer RJ, Meehl GA, 2012. An overview of Cmip5 and the experiment design B. B Am Meteorol Soc 93: 485–98.

Toba T, Ohta T, 2005. An observational study of the factors that influence interception loss in boreal and temperate forests. J Hydrol 313: 208–220. https://doi.org/10.1016/j.jhydrol.2005.03.003

Tollefson J, 2016. Global warming already driving increases in precipitation extremes. Nature. https://doi.org/10.1038/nature.2016.19508

Ul Hassan M, Qureshi AS, Heydari N, 2007. A proposed framework for irrigation management transfer in Iran: Lessons from Asia and Iran. Colombo, Sri Lanka: International Water Management Institute. 37 pp. (IWMI Working Paper 118)

Valente F, David JS, Gash JHC, 1997. Modelling interception loss for two sparse eucalypt and pine forests in central Portugal using reformulated Rutter and Gash analytical models. J Hydrol 190: 141–162. https://doi.org/10.1016/S0022-1694(96)03066-1

Van Stan II JT, Godon DA, 2018. Mini-Review: Stemflow as a Resource Limitation to Near-Stem Soils. Front Plant Sci 9(248): 1-9. https://doi.org/10.3389/fpls.2018.00248

Vose S, Zimmermann B, Zimmermann A, 2016. Detecting spatial structures in throughfall data: The effect of extent, sample size, sampling design, and variogram estimation method. J Hydrol 540: 527-537. https://doi.org/10.1016/j.jhydrol.2016.06.042

Wallace J, Macfarlane C, McJannet D, Ellis T, Grigg A, Van Djik A, 2013. Evaluation of forest interception estimation in the continental scale Australian water resources assessment-landscape (AWRA-L) model. J Hydrol 499: 210–223. https://doi.org/10.1016/j.jhydrol.2013.06.036

Wang YH, Yu P, Feger KH, Wei X, Sun G, Bonell M, Xiong, W, Zhang S, Xu L, 2011. Annual runoff and evapotranspiration of forestlands and non-forestlands in selected basins of the Loess Plateau of China. Ecohydrology 4: 277–287. https://doi.org/10.1002/eco.215

Zhang Z, Zhao Y, Li X, Huang L, Tan H, 2016. Gross rainfall amount and maximum rainfall intensity in 60-minute influence on interception loss of shrubs: a 10-year observation in the Tengger Desert. Sci Rep-UK 6: 26030. https://doi.org/10.1038/srep26030

19

Sharing the gains in forest management: Insights from the application of principles of collective action

Maria Alló[1] and Maria L. Loureiro[2]

[1]*Universidad de A Coruña. Facultad de Economía y Empresa. Dept. Economía. Campus de Elviña, s/n, 15071 A Coruña. Spain.* [2]*Universidad de Santiago de Compostela. Facultad de C. Económicas. Dept. Fundamentos da Análise Económica. Avda do Burgo das Nacións, s/n, 15782 Santiago de Compostela. Spain.*

Abstract

Aim of study: To analyze the factors that motivate the sharing of a contribution to a common pool resource (CPR).

Area of study: We obtained data from Galicia communal forests during 2013-2014.

Material and methods: A survey among forest owners in which questions about decisions, mimicking those in public good games, were included. In addition, the compliance with the principles of collective action (PCA), and their implications in the management of CPR were tested.

Main results: PCA are not functioning perfectly in our sample of communal forest owners. In line with previous literature, results suggest that individuals are willing to share an important amount of endowment in a CPR. Examining the role that PCA play in individuals' decisions, it was observable that when owners face the sharing of an endowment, the *monitoring, conflict resolution mechanism* and *minimal recognition of rights* imply more cooperative results. Current communal forests should promote a better application of these PCA in order to obtain a more cooperative behavior from their members.

Research highlights: Communal forest owners are quite generous according to the results obtained. In addition, it has also been found that the endowment is an important factor to consider, while social aspects represented through the PCA also matter when explaining sharing decisions. The present study may be useful in order to promote stronger cooperation in local communities.

Additional keywords: public good game; endowment; communal forest; survey.

Abbreviations used: CPR (Common Pool Resource); OLS (Ordinary Least Squares); PCA (Principles of Collective Action); PGG (Public Good Game); RE (Random Effects).

Authors´ contributions: Design of the survey and game: MA and MLL. Drafting of the manuscript: MA. Critical revision and editing: MLL.

Funding: Xunta de Galicia (Grant 2017 GPC GI-2037 Gru ED431B 2017/14).

Competing interests: The authors have declared that no competing interests exist.

Correspondence should be addressed to Maria Alló: maria.allo.pazos@udc.es

Introduction

Cooperation in groups at times is challenging to achieve. Specifically, with regards to the management of natural resources, Cardenas & Ostrom (2004) emphasized the need of resolution of collective action problems. They indicated the importance of understanding how groups make decisions about the use of the resources and how self-governed policies can promote the sustainable use. When explaining human behaviour, the traditional economic theory argues that individuals are rational and selfish. In terms of environmental resource markets, the most typical result will find the existence of non-cooperative and free-riding behavior; thus, zero cooperation is the predicted result for most environmental resource extraction games (Hardin, 1968). However, several studies have shown that deviations from the output of traditional economic theory may take place (Gächter & Herrmann, 2006). Specifically, Ostrom (1990) pointed out that communities can govern their own natural resources without overharvesting. Some of the most used models in economic literature to test the hypothesis of rational and selfish individuals are the Public Good Games (PGG).

In this paper, the aim is focused on common pool resources (CPR), specifically in communal forests. At this regard, Ostrom *et al.* (1992) argued that individuals may reach successful agreements to manage resources efficiently in CPR. In this sense, provision of information, communication, and the possibility of sanctioning are important factors that can help management strategies. Other studies analyze relevant aspects to sustain cooperation. Fehr & Fischbacher (2003) stated that individual heterogeneity and the interaction between altruistic and selfish individuals are crucial in cooperation, as well as the environment in which these individuals act. According to their results, more altruistic individuals can influence selfish ones. More recently, Rustagi *et al.* (2010) pointed out that structural factors, such as the characteristics of the resource, the size of the group and socioeconomic differences, as well as other social behaviors, such as conditional cooperation or costly enforcement of the norm can explain these deviations. Cardenas (2011) emphasized that it is necessary to mediate in the conflict between self-interest and group interest to avoid overexploitation through incentives, controls, sanctions or community-based recommendations, among others. In this respect, literature has focused significantly on the role that institutions, regulations and economic incentives may play. Nevertheless, studies such as Cardenas (2011) concluded that policymakers should also take into account that their regulations can alter the normative behavior of users, as they combine the valuation of the regulation with their subjective thoughts regarding this mechanism. Janssen *et al.* (2013) also emphasized that it is important to allow for the enforcement of rules and social norms as they can act as complements or substitutes. Therefore, aspects such as communication, information, sanctioning, social behavior and other additional factors, including the characteristics of resources, or of the environment, matter.

This empirical study is conducted in the North West of Spain, Galicia. This is one of the most important forest areas of Europe. This forest surface is about 1.4 million hectares, which represents 48% of the total of the Galician territory. Furthermore, more than 120,000 ha correspond to certified forest surface, which places Galicia at the front run of Spain in terms of sustainable forest management (Xunta de Galicia, 2014). This paper focusses on communal forests[1], which have existed for centuries and have played an important role in agriculture (Caballero, 2014). They were regulated under a common law until the year 1968 (BOE, 1968), when an institutional framework for communal forests was established, without individual quotas of property. These communities are also a signal of identity, of local culture and are economic and productive resources, due to the creation of jobs and wealth (Xunta de Galicia, 2014). To be a member of the communal forests, the individual has to reside in the local community and attend meetings and assemblies, where one member of each household or family is represented and decisions about the CPR are taken collectively.

Conflicts among communal forests may arise. Thus, internal conflicts highlight disagreements among commoners, whereas external conflicts demonstrate issues involving various parties: between neighbor communities and government organizations (based on different perspectives between owners and government); with enterprises (cession rights of resources and their uses); but also external socio-economic and environmental conflicts with non-governmental organizations (Gómez-Vázquez *et al.*, 2009). The objective of this study is to analyze the sharing rules in collective forest managements (CPR) through a modified public good game and assess the interplay of social factors. Two aspects of this paper can be considered a novelty: the first is the fact that the study is conducted with actual forest owners, instead of students or other subjects; and the second contribution is the analysis of the Principles of Collective Action (PCA) proposed by Ostrom (1990) used to understand their implications in terms of sharing.

Material and methods

Public good games

To analyze individuals' decisions, a survey was used employing questions which reflected the choices to be made in a PGG. This game is usually played in a laboratory by n individuals. This game has been simulated in the context of structured surveys with actual CPR owners. In a lab setting, all individuals i receive an endowment e and simultaneously and independently decide whether to keep this endowment for themselves or invest a given amount $g_i\ \varepsilon[0,e]$ in the CPR in question. The total amount contributed by all n participants together equals $g = \Sigma g_i$, where $0 \leq gi \leq e$ is multiplied by a number m (marginal per capita return), with $0 < m < 1 < nm$. Therefore, when the PGG is played once, and considering that players are rational and selfish in maximizing utility, contributing 0 is a

[1]With regards to the management of Galician forests 98% of the territory is managed by private owners and 2% is owned by the public sector (Gómez-Vázquez *et al.*, 2009). The private forests are managed both by single owners, as well as collective owners known as communal forests (Comunidade de Montes Veciñais en Man Común, CMVMC). There are around 2,800 communal forests which occupy about 700,000 ha (Xunta de Galicia, 2014). They represent around 33% of the total forest surface. The BOE (2012) defines them as "*private forests, with independence of: origin, productive capacity, current utilization and agrarian vocation, are of the neighbors' communities. In addition, communal forests are exploited in a community regimen, without allocation of quotas among neighbors. Furthermore, these forests are indivisible, inalienable, imprescriptible and indefeasible goods.*"

dominant strategy. However, the joint group outcome is maximized when everybody contributes with the full endowment. The payoff function of users can be expressed as:

$$p_i(g) = e - g_i + m\sum_{j=1}^{n} g_j \quad if\ g_i > 0 \qquad (1)$$

A paper by Chaudhuri (2011) conducted a survey about the PGG literature. The main conclusions of this study are that the most typical issues analyzed in PGG are conditional cooperation, costly punishments and other means to sustain cooperation. In addition, this author also highlighted that the main conclusions obtained with respect to the investment in public goods are that in one-shot experiments the contributions are about 40% and 60% of the optimal level with large variations from 100% to 0%. In addition, when individuals play the game repeatedly, contributions decline over time and more participants choose to "free-ride" (Ledyard, 1995). Fehr & Gintis (2007) found two types of behavior: free-rider or conditional cooperators. The first type corresponds to those who never contribute to a public good, and the second to those who increase their contribution when they expect that others will also contribute. Fischbacher *et al.* (2001) conducted a PGG, finding that around 50% of the subjects can be classified as conditional cooperators. Furthermore, the key factors that favor the cooperation among individuals are communication, the establishment of a threshold and a higher marginal per capita return. Nevertheless, there are other factors that can also influence choices; for example: gender, the size of the group, and reciprocal motivations, among others. Ostrom *et al.* (1992) stated that information, communication and sanctions can help to reach agreements efficiently. Andreoni (1995) concluded that studies of preferences for cooperation should also be considered in the analysis.

In this paper, a survey employing questions reflecting a one shot PGG (individuals only decide one time and in an individual basis[2]) where members of communal forests have to make a decision about how to share an endowment was conducted. First, this PGG is modified due to the fact single users are told to consider the rest of the members (50 members[3]) in their decisions (*n=50*). Previous studies usually employ groups of four people. In our specific case, the group under consideration will be composed by all users of the community in order to represent a similar situation to the reality. A second important aspect to take into account is that the contributions to the fund will allow different management efforts in the communal forests to be carried out, from which the community will benefit, specified with a percentage over the initial amount invested (m). Thus, in order to keep a perfect similarity with the reality, m represents a future re-investment which is not specified, depending on each specific community and it is based on previous rents obtained and shared. At this respect, our PGG is described as a real situation that collective owners can face frequently as owners of the communal forest (Rodríguez *et al.*, 2017). It is important to highlight that the game is conducted through an interview and no real money is being played. Bethwaite & Tompkinson (1996) also carried out an ultimatum game under the same conditions. These practical approaches should be taken into consideration when comparing these results with those obtained by other studies. Finally, it is important to note that each participant was presented with three different sizes of endowment, in order to assess whether the size of the economic incentives matter in sharing decisions. A similar exercise employing the ultimatum game was conducted by Andersen *et al.* (2011) in poor villages of Northeast India, finding that the size of the endowment matters.

Data description

A survey was carried out in North-west Spain, in the region of Galicia, from winter 2013 until summer 2014, interviewing a sample of 75 forest owners. In order to facilitate the first contact with forest owners, information from the websites of communal forests was obtained. After an initial phone call or contact e-mail, we visited the place where they live (a total of 29 different places) conducting face to face surveys at their homes; or after attending their annual association meetings. Some participants preferred to answer via online rather than face to face, petition that was accommodated. Information from 29 communal forests belonging to 31 different councils was collected. The four provinces of Galicia are represented in this small sample; with 55.71% of the observations from the

[2]In this paper, it is used a one-shot game and individuals play individually. Nevertheless, respondents belong to actual forest communities. Thus, the sharing decisions (though simulated) are framed in real communal forests and although the game is not played in group, a concrete group of reference does exist in the mind of the respondents (the communal forests to which they belong), with which they have had repeated interactions. Therefore, provided some real conditions in the community (communication level, overall cooperation/free-riding level, concrete social norms, past experiences, etc.) and individual preferences, the respondent indicates a sharing decision. We thank a reviewer for suggesting this reflection.
Therefore, provided some real conditions in the community (communication level, overall cooperation/free-riding level, concrete social norms, past experiences, etc.) and individual preferences, the respondent indicates a sharing decision. We thank a reviewer for suggesting this reflection.
[3]An average size of 50 members was chosen, given communities contacted had a great variety in size. Therefore, that the fact that this is a hypothetical situation was explained. According to Balboa *et al.* (2006) the average number of owners by communal forest is about 54.

province of A Coruña, 22.86% from Lugo, 17.14% from Pontevedra and 4.29% from Ourense. The survey requested information about the characteristics of communal forests, from which they are members, with the objective to reflect members' knowledge. In addition, questions related to their opinion about public authorities were also included, and the way in which decisions related to forest are made. Furthermore, the survey also included questions to test whether PCA were functioning. PCA related questions can be seen in Table 1 and a full description is added in the Table S1 [suppl.]. The PGG was presented in another section of the questionnaire.

Table 2 shows an example of the game presented to participants. Basically, forest owners had to decide about the sharing of endowment that the community receives as a consequence of the actions carried out in order to keep forests in good conditions. As it can be

Table 1. Questions to test whether the PCA are functioning and percentage obtained.

		Degree of compliance (%)
Clearly defined boundaries	Do you think that in your community all community members know each other?	72.39
	Do you think that the residents who are not community members are clearly not part of the community?	49.47
	Do you think that in practice the neighbors respect community members?	68.75
Congruence between appropriation and provision rules and local conditions	Do you think that the rules of forestry, that have to be complied with, are appropriate to conserve the resource?	53.64
	Do you think they are effective?	50.00
	Do you believe these rules respect the local traditions and beliefs?	59.37
Collective choice arrangements	In terms of decisions around forestry issues, are these taken collectively, together and equally represented?	85.41
Monitoring	Are there controls to ensure compliance by the community?	50.52
	Are there controls to ensure compliance by neighbors who are not members of the community?	44.27
Graduated sanctions	Are there penalties for owners who do not comply with the obligations laid down in the law?	51.04
	If there are sanctions, are they variable depending on the offense committed?	35.41
Conflict resolution mechanisms	Are there quick mechanisms to solve the conflicts that can arise?	22.39
Minimal recognition of rights	Do you think that your rights and decisions are respected by others, who are not members of the community and the administration?	46.35

Table 2. Example of the PGG presented to participants.

Imagine that you are a member of a communal forest that is formed by 50 neighbors. Your community has received a reward by the "Law against fires" as a consequence of the effort that this communal forest organization has made to fight against forest fires and the good condition of the forest. These funds can be used freely by collective owners. Therefore, you can collect your proportional amount or reinvest it in the community fund. The amount that you deposit in the fund will be used to finalize works in the forest from which the community will obtain some benefits equivalent to 40% of the initial amount invested. Using the following table, please indicate your preference:

Reward	When splitting the reward among 50 members, you have the right to:	I take:	I deposit in the fund:
If the reward is €500	€10		
If the reward is €5,000	€100		
If the reward is €50,000	€1,000		

seen, each common owner has to decide the sharing of three different endowments. In this way, we can test whether the size of the endowment is important as has been found in other previous literature. In addition, the survey also included some questions to understand the degree of effort made by members to maintain a forest in good condition, and therefore assesses their opinion of other forest owners' efforts (these questions were asked before participating in the PGG). It is important to highlight that the PGG was conducted individually. The same strategy was employed by Braaten (2014) with the goal of promoting comprehension amongst participants. Finally, socio-demographic characteristics were elicited in the last part of the survey.

Empirical approach

Individuals' allocations made in the stated preference exercise were modelled as a function of the size of the endowment (reward), the forest characteristics of individuals and the most relevant social and socio-economic characteristics, among other variables. Furthermore, the PCAs were also incorporated into the regression model in order to test the effect of other perceptions towards the collective management.

A first Ordinary Least Squares (OLS) regression was estimated. Using the Breusch-Pagan (1979) test, it was assessed the presence of heteroskedasticity, finding that the test carries a value of 24.93 for a Chi-squared with 1 degree of freedom and an associated p-value of 0.00. Therefore, we reject homoscedasticity. To deal with this problem, an OLS model clustered by individual was estimated[4]. In addition, and in order to take into account the panel nature of our data, based on the fact that three different responses from each individual were collected, the between and within variability was also analyzed using a Random Effects (RE) model[5]. The dependent variable in this regression equation is the vector of the total distributions (the amount of the endowment shared with regards to the total amount of endowment available), labeled as (***Y***). This vector represents the percent of endowment that individuals decide to send to the public fund. The vector of explanatory variables (X_i) is grouped into four categories that include: the size of the endowment to be shared X_s, the forest characteristics X_f, the socio-economic characteristics X_{se}, while X_{pca} represents the PCA.

The estimated model corresponds to the estimation of the following equation:

$$Y = \beta_0 + \beta_s X_s + \beta_f X_f + \beta_{se} X_{se} + \beta_{pca} X_{pca} + \varepsilon \quad (2)$$

where Y represents the cooperation of the individuals within their groups when they face the economic incentive; β_0 is the usual constant term, the corresponding β are the coefficients associated with the respective explanatory variables to be estimated. Thus, β_s represents the amount of endowment to be shared; the β_f contains the characteristics of the forest where individuals are; β_{se} includes the socio-economic characteristics of the participants while the vector β_{pca} represents the seventh PCA proposed by Ostrom (1990), and ε is a vector of the error term, independently and identically distributed.

Research hypotheses

One of the main aims of this paper is to analyze the impact of PCA in sharing decisions. It is important to note that with the aim of including the effect of PCA several questions were included in the survey (see Table 1); and in order to measure the impact of PCA, we constructed indicators to measure the existence of these principles. Table 1 provides information about the type of questions used to estimate each individual PCA. Questions were answered with a "yes" or "no", and recoded as 1 and 0, respectively. Thus, in order to create the indicators, the affirmative responses were added up.

It is expected that PCA may have a significant impact on explaining the sharing of the endowment. The aim was to investigate whether individuals feel that they are involved in the decision-making process; whether they are aware that their rights are respected; and whether there are controls to assure them, among others factors. Thus, the hypothesis proposed is to test whether the fulfillment of PCA enumerated by Ostrom (1990) influence sharing decisions in the community.

$$\begin{aligned} H_0 &: \beta_{pca} = 0 \\ H_1 &: \beta_{pca} \neq 0 \end{aligned} \quad (3)$$

The second question of analysis is whether the size of the endowment matters in sharing decisions. To study this hypothesis, a variable which represents the size of each individual endowment has been included. In this sense, Anderson *et al.* (1998) found that contributions increase with the endowment; while Andersen *et al.* (2011) concluded that stakes are important but in the case of ultimatum games. Therefore, this is a novelty aspect in the literature of PGG, due to the fact that no

[4]In our PGG each individual faces three decisions, this means, the sharing of three different amounts of endowment to be shared. Therefore, clustering by individual this aspect has been taken into consideration.

[5]A Hausman test was conducted in order to test whether a Fixed Effects model performed better than a RE model. The test conducted carried out a value of $\chi^2_1 = 0.080$ with a p-value of 0.771. Therefore, the RE model is selected as the model that best fits the data.

previous studies were found that analyzed the effect of the size of endowment in this type public good game. Specifically, the hypothesis to be tested is:

$$H_0 : \beta_{se} = 0$$
$$H_1 : \beta_{se} \neq 0 \quad (4)$$

Results

Data results

In terms of the dataset composition, 64% of the respondents were men, with an average *age* of 55 years and the average number of household members was around three persons. With respect to the level of studies, 52% of the respondents have a basic level of studies or no formal studies (*basic studies*), and with regards to the income-related questions, about 24% of participants earn less than €1,000/monthly (*lower income*).

With regards to the characteristics of forests, about 44% of them had more than 100 communal members (*over 100 members*). It was also detected that in terms of location, 58% of the communities in this sample were close to the sea (*near to the coast*) (as opposed of being distant rural communities). This geographical indicator is important, since it provides information about the degree of economic dynamism of the area. Previous studies have shown that communities located near the coast are more active and dynamic, mirroring the economic local conditions (Alló & Loureiro, 2016). With the goal to know more about their engagement in forest management, participants were asked about their level of effort compared with the rest of the owners; and 21% stated that their own effort is higher than the rest of the members (*greater effort*).

With the aim of obtaining more information about their social characteristics, additional questions related to whether they were born in locations near to forests were included, with 75% answering in a positive way (*place born*). About 52% responded that their relationships with the rest of owners were very or quite good (*good relationship*).

Finally, it was also taken into account the survey mode: whether the surveys have been completed via online instead of face to face, with 33% preferring the online mode (*online*). Table 3 shows the summary statistics for the variables analyzed.

Public good game

The distribution of the endowment is presented in Figure 1. It is important to note that each individual had to make three allocation decisions due to the three different sizes of the endowment. The amount of endowment shared can be considered as a proxy of

Table 3. Summary statistics for the variables analyzed.

Variable	Description	Mean	Std. Dev
Fund	Percentage of endowment allocated to the fund	83.698	29.309
Size of endowment	10, if the size of the endowment is €10; 100 if the size is €100 and 1000 if the size is €1000	370.000	448.158
Clearly defined boundaries		0.406	0.492
Congruence between appropriation and provision rules and local conditions		0.406	0.492
Collective choice arrangements		0.854	0.354
Monitoring		0.411	0.493
Graduated sanctions		0.865	0.911
Conflict resolutions mechanisms		0.224	0.418
Minimal recognition of rights		0.464	0.500
Age	Age of respondents	54.578	15.174
Born place	1, if respondents were born in the place where forests are; 0 otherwise	0.750	0.434
Basic studies	1, if the responds have a basic level of studies; 0 otherwise	0.505	0.501
Lower income	1, if respondents earn less than €1,000 per month; o otherwise	0.240	0.428
Near to the coast	1, if the communal forest is located close to the coast; 0 otherwise	0.578	0.495
Over 100 members	1, if the collective forest has more than 100 members; 0 otherwise	0.443	0.498
Greater effort	1, if owners think that their own effort is greater than the rest of the owners; 0 otherwise	0.214	0.411
Good relationship	1, if owners respond that their relationships with the rest of owners are very or quite good; 0 otherwise	0.521	0.501
Online	1, if the respondent answered the survey via online; 0 otherwise	0.333	0.473

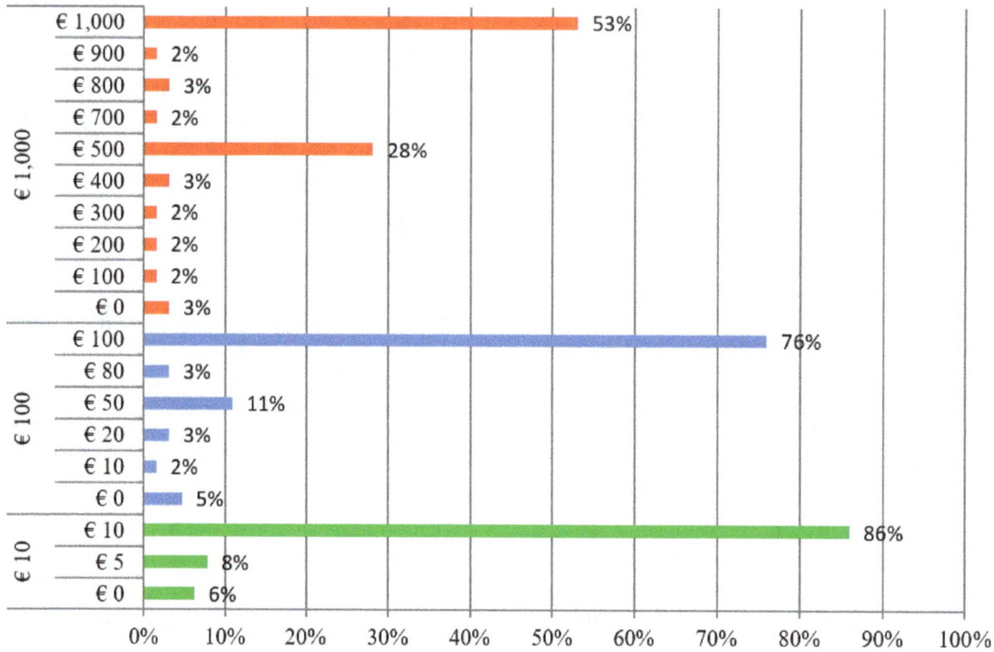

Figure 1. Allocations of the endowment (%) in a fund account for different endowment sizes (€10, €100, €1000).

cooperation (Stoop *et al.*, 2012). Assessing the results, it was observable that when the endowment to share was €10, around 86% decided to allocate the full amount to the common fund, and around 8% decided to invest half of the endowment in the public fund, while the rest (6%) decided to keep the endowment for themselves. Thus, on average the mean contribution was about €9. However, when the endowment was higher, the allocation changed. Specifically, for an endowment of €100, 5% decided not to share this endowment, also another 5% decided to send a positive value below the half to the public fund; while 11% decided to split the endowment at 50%. Moreover, around 3% sent to the public fund more than half, but less than 100% of the endowment, and 76% allocated the total €100 to the fund. On average, the contribution to the fund was about €85. This decision changed slightly when the endowment was even higher. For an endowment of €1000, 3% kept the full amount, around 9% decided to share less than half with the fund, while 28% decided to share half with the public fund. In addition, 7% of the owners share more than half with the public fund, and finally, 53% decided to send the total amount to the fund. On average, the mean contribution to the fund for this endowment was about €759. Therefore, the higher the size of the endowment, the higher the amount of endowment that collective owners keep for themselves. Nevertheless, it is important to highlight that in spite of that, they were quite generous, given that the percentages of endowment kept by themselves was quite small (around €1 on average for an endowment of €10, around €13 for an endowment of €100, and about €240 for an endowment of €1000). In the survey of literature conducted by Chaudhuri (2011), it was found that average contributions were about 40% to 60% of the optimal level. This is an interesting result in terms of cooperation, particularly in the absence of mechanisms to encourage it (such as penalties, or communication strategies, among others). Stoop *et al.* (2012) concluded that without any kind of contact among players, cooperation is not guaranteed.

The role of PCA

The survey results related to the PCA are presented in Table 1. With regards to the principle of *clearly defined boundaries*, around 72% of respondents believed that in their communities all members knew each other; 49% thought that residents who were not members of the communal forest knew this fact, while the 69% considered that neighbors respected community members in practice.

When examining *congruence between appropriation and provision rules and local conditions*, it is important to note that only 53% of commoners thought that the rules that they had to follow were appropriate for resource conservation. With regards to the regulation's effectiveness; only 50% believed that it was *effective*. Moreover, 59% thought that the *existing rules respect local traditions*. Analyzing the results for the principle of *collective choice arrangements*, 85% confirmed that forest decisions were taken collectively with individuals equally represented. In addition, around 50% of participants stated that there were controls to ensure compliance by the forest community, and 44% stated that there were also controls for outside members (*principle of monitoring*). With regards to the principle of *graduated sanctions*, 51% of commoners affirmed that there were penalties for people who did not comply with the requirements, and 35% stated that these penalties varied depending on the degree of infraction in the case of the endowment. Only 22% expressed that there was *a quick mechanism to resolve conflicts* and 46% believed that their *rights and decisions are respected* by non-members and the administration for the endowment treatment.

OLS and RE results

In order to understand the factors that drive the allocations of endowment to the CPR, we estimated a robust OLS and a RE models. The main results are summarized in Table 4. First, we examined and discussed the estimates with regards to the size of the endowment for both models. We observed that the coefficient obtained for *size of endowment* was negative and statistically significant. Therefore, when the endowment to be shared was larger, less proportion was shared with the public fund. This means that with lower endowments, individuals are willing to share proportionately a greater amount of endowment with the public fund.

We also considered the effect of different characteristics of owners to understand their decisions. Thus, Bechtel & Scheve (2014) concluded that socio-demographic characteristics play an important role in social dilemmas. Specifically, we obtained that the *lower income* indicator was negative and statistically significant. Moreover, those individuals who thought that they made a *greater effort* than the rest, were more likely to allocate the endowment to the public

Table 4. Results obtained from the OLS and RE models.

	OLS			Random effects		
	Coef.	**Robust SE**	**P>|t|**	**Coef.**	**SE**	**P>|t|**
Size of endowment	-0.012	0.004	0.001	-0.012	0.003	0.000
Near to the coast	-1.848	5.801	0.751	-2.565	7.724	0.740
Members over 100	8.358	6.453	0.200	7.752	7.874	0.325
Age	0.203	0.344	0.557	0.178	0.294	0.546
Born place	2.705	6.010	0.654	1.752	8.013	0.827
Basic studies	4.889	8.314	0.559	3.909	8.142	0.631
Lower income	-15.172	6.598	0.025	-13.556	7.526	0.072
Greater effort	23.481	5.218	0.000	23.332	7.996	0.004
Good relationship	3.239	7.195	0.654	3.855	6.225	0.536
Online	-2.536	8.283	0.760	-4.198	9.687	0.665
Clearly defined boundaries	0.966	6.624	0.885	0.958	8.349	0.909
Congruence between appropriation and provision rules and local conditions	-15.587	5.746	0.009	-16.000	7.390	0.030
Collective choice arrangements	-4.921	7.397	0.508	-5.875	9.006	0.514
Monitoring	19.649	6.443	0.003	20.764	9.211	0.024
Graduated sanctions	-7.208	2.385	0.004	-6.829	3.717	0.066
Conflict resolutions mechanisms	15.854	5.501	0.005	15.624	8.180	0.056
Minimal recognition of rights	16.944	5.561	0.003	17.307	9.260	0.062
Constant	64.552	20.428	0.002	67.974	20.763	0.001
Sigma_u				18.325		
Sigma _e				19.500		
Rho				0.469		
Wald χ^2 (17)				48.120		
F statistic	8.480					
p-value	0.000			0.000		
R^2	0.319					
Root mean squared error	25.347					
N	192			192		

fund. These may be more altruistic or more committed commoners.

With regards to the effect of PCA, we observed that the principles related with the *congruence between appropriation and provision rules and local conditions* and *graduated sanctions* showed negative and statistically significant coefficients. Therefore, individuals who believed that these principles were being fulfilled are less willing to share the endowment. Furthermore, we also observed that three PCA had a positive impact promoting more cooperative decisions, specifically, the *monitoring*, the *conflict resolution mechanisms* and the *minimal recognition of rights*. These results may be implying that when individuals feel that they are involved in the management process, are more willing to share the endowment towards the CPR. Therefore, results show that PCA fulfilment may encourage more generous decisions, but not always (as in the case of *congruence between appropriation and provision rules and local conditions* and *graduated sanctions*).

Discussion

The aim of this paper was to analyze, in the context of a CPR, how common owners made decisions. Through a survey using questions mimicking the decisions of a PGG, the question of how landowners shared out an endowment was analyzed. It is important to highlight that responses obtained from the survey were assessed as a proxy of the real contributions. In addition, we would like to emphasize that we were focusing on a sample of forest owners who were very active in CPR management. As an example, around 79.2% affirmed that they attended the meetings organized by the communal forest, and also 56% participated in the activities proposed. We acknowledge that this fact may bias some of the results.

Therefore, and spite of having a sample of very active members, and as a conclusion, we observed that PCA were not functioning perfectly in our sample of communal forest owners. Thus, the adoption of these PCA could be improved as a way to deal with the current management problems faced by these CPR. Regression models were estimated in order to assess PCA effects in sharing decisions. In addition, the size of the endowment was also considered in order to understand whether the sharing of an endowment is dependent on the size of this.

Evaluating how endowments were shared, obtained results were in line with previous studies, finding that on average common owners made higher contributions to a CPR than suggested by traditional economic theory. Specifically, it was found that communal forest owners were quite generous. It was also found that the endowment size was an important factor to consider. In this sense, Andersen *et al.* (2011) concluded that the size of the stake matters in ultimatum games.

With regards to the PCA, we have included a number of questions in the survey that were used through the construction of indicators. Examining the role each play in individuals' decisions, it was observable that when owners face the sharing of an endowment, the *monitoring*, *conflict resolution mechanism* and *minimal recognition of rights* imply more cooperative results. Therefore, current communal forests should promote a better application of these PCA in order to obtain a more cooperative behavior from their members. The present results may be useful in order to promote stronger cooperation in local communities.

Acknowledgements

The authors wish to thank the editor, Mario Soliño, and the reviewers for helpful comments.

References

Alló M, Loureiro ML, 2016. Evaluating the fulfilment of the principles of collective action in practice: A case study from Galicia (NW Spain). Forest Policy Econ 73: 1-9. https://doi.org/10.1016/j.forpol.2016.08.002

Andersen S, Ertac S, Gneezy U, Hoffman M, List JA, 2011. Stakes matter in ultimatum games. Am Econ Rev 101: 3427-3439. https://doi.org/10.1257/aer.101.7.3427

Anderson SP, Goeree JK, Holt CA, 1998. A theoretical analysis of altruism and decision error in public good games. J Public Econ 70: 297-323. https://doi.org/10.1016/S0047-2727(98)00035-8

Andreoni J, 1995. Cooperation in public goods experiments: kindness or confusion? Am Econ Rev 85(4): 891-904.

Balboa X, Besteiro B, Fernández X, Fernández L, Jordán M, López E, Soto D, Viso, P, 2006. Os montes veciñais en man común: o patrimonio silente. Natureza, economía, identidade e democracia na Galicia rural. Vigo: Edicións Xerais de Galicia.

Bechtel MM, Scheve KF, 2014. Public goods, reciprocity and the causal effect of expected cooperation in representative samples. https://pdfs.semanticscholar.org/9406/d8993f6bf764b91fa138726786e8a0c26276.pdf?_ga=2.132334474.528702125.1531734437-374087474.1531734437 [16 July 2018]

Bethwaite J, Tompkinson P, 1996. The ultimatum game and non-selfish utility functions. J Econ Psychol 17: 259-271. https://doi.org/10.1016/0167-4870(96)00006-2

BOE, 1968. Law 52/1968, of 29th July of Communal forests. Boletin Oficial del Estado (Spain) No. 181, 27/07/1968.

BOE, 2012. Law 7/2012, of 28th June of Galician forests. Boletin Oficial del Estado (Spain) No. 217, 08/09/2012.

Braaten RH, 2014. Land rights and community cooperation public goods experiment from Peru. World Dev 61: 127-141. https://doi.org/10.1016/j.worlddev.2014.04.002

Breusch T, Pagan A, 1979. A simple test for heteroskedasticity and random coefficient variation. Econometrica 47: 1287-1294. https://doi.org/10.2307/1911963

Caballero G, 2014. Community based forest management institutions in the Galician communal forests: a new institutional approach. Forest Policy Econ 50: 347-356. https://doi.org/10.1016/j.forpol.2014.07.013

Cardenas JC, Ostrom E, 2004. What do people bring into the game? Experiments in the field about cooperation in the commons. Agricultural Systems 82(3): 307-326. https://doi.org/10.1016/j.agsy.2004.07.008

Cardenas JC, 2011. Social norms and behavior in the local commons as seen through the lens of field experiments. Environ Resour Econ 48: 451-485. https://doi.org/10.1007/s10640-010-9452-8

Chaudhuri A, 2011. Sustaining cooperation in laboratory public goods experiments: a selective survey of the literature. Exp Econ 14: 47-83. https://doi.org/10.1007/s10683-010-9257-1

Fehr E, Fischbacher U, 2003. The nature of human altruism. Nature 425:785-791. https://doi.org/10.1038/nature02043

Fehr E, Gintis H, 2007. Human motivation and social cooperation: experimental and analytical foundations. Annu Rev Sociol 33: 43-64. https://doi.org/10.1146/annurev.soc.33.040406.131812

Fischbacher U, Gächter S, Fehr E, 2001. Are people conditionally cooperative? Evidence from a public goods experiment. Econ Lett 71: 397-404. https://doi.org/10.1016/S0165-1765(01)00394-9

Gächter S, Herrmann B, 2006. Human cooperation from an economic perspective. Cooperation in Primates and Humans: 279-301.

Gómez-Vázquez I, Álvarez-Álvarez P, Marey-Pérez MF, 2009. Conflicts as enhancers or barriers to the manageme gnt of privately owned common land: a method to analyze the role of conflicts on a regional basis. Forest Policy Econ 11: 617-627. https://doi.org/10.1016/j.forpol.2009.09.001

Hardin G, 1968. The tragedy of the commons. Science 162: 1243-1248. https://doi.org/10.1126/science.162.3859.1243

Janssen MA, Bousquet F, Cardenas JC, Castillo D, Worrapimphon K, 2013. Breaking the elected rules in a field experiment on forestry resources. Ecol Econ 90, 132-139. https://doi.org/10.1016/j.ecolecon.2013.03.012

Ledyard JO, 1995. Public goods: a survey of experimental research. In: Handbook of experimental economics; Kagel J, Roth AE (eds). pp: 111-194. New York: Princeton University Press.

Ostrom E, 1990. Governing the commons. Cambridge: Cambridge University Press. 280 pp. https://doi.org/10.1017/CBO9780511807763

Ostrom E, Walker J, Gardner R, 1992. Covenants with and without a sword: self-governance is possible. Am Polit Sci Rev 86(2): 404-417. https://doi.org/10.2307/1964229

Rodríguez-Cebreiro D, Aboal Viñas J, Picos Martín J, 2017. Análisis de los ingresos de montes vecinales en mano común de Galicia y su reinversión. Proc VII Congreso Forestal Español, Cáceres (Spain), June 26-30. pp: 1-10.

Rustagi D, Engel S, Kosfeld M, 2010. Conditional cooperation and costly monitoring explain success in forest common management. Science 330, 961-965. https://doi.org/10.1126/science.1193649

Stoop J, Noussair CN, van Soest D, 2012. From the lab to the field: cooperation among fishermen. J Polit Econ 120(6): 1027-1056. https://doi.org/10.1086/669253

Xunta de Galicia, 2014. Consellería do Medio Rural e do Mar. Sector Forestal. http://www.medioruralemar.xunta.es/es/areas/forestal/presentacion/ [16 July 2018]

20

Accounting for windthrow risk in forest management planning: a Romanian tailor-made solution

Marian Drăgoi and Ionut Barnoaiea

University of Suceava, Suceava, 13, Universitatii str., 720229, Romania.

Abstract

Aim of study: To better estimate the annual allowable cut reserve (AACR), taking into consideration the endemic windthrows (EW), we combined a series of existing algorithms into a coherent methodology to use the data available at district level, without any additional fieldworks.

Area of study: The algorithm was tested on the EW occurred in the last 20 years in Brosteni FD (Eastern Carpathians, Romania) that covers 21,013 ha and we found that every year from an AAC of 37,000 m^3 no more than 2,700 m^3 shall be spared for EW that might occur next year.

Material and methods: We considered three EW enabling factors (stand slenderness, location on pits and mounds, and the vicinity of canopy gap) and three contingency tables of the EW produced between 1999 and 2008, one for each 40-year age group. Then we calculated a Bayesian model for all six permutations of enabling factors, each of them being tested on the data referring to 2008-2017 period.

Results: Plugging the posterior EW likelihoods into a Markov chains (MC) model, we produced a formula that enables a better estimation of the optimal AACR that could be replaced with salvage cuttings every next year. Other options of using the EW likelihoods are also presented at length, such as the type of age-class structure that requires no AACR, that is a "U" shape age structure, as well as a rough assessment of the additional demand for seedlings needed to re-plant the stands affected by EW. The relatively short period of time the input data refer to, which is ten years, equals the time window of the forest planning and this parity allows a ten-year forecast period, enough for modeling the stationary age-structure of even-aged forests.

Research highlights: A new model for optimizing the annual allowable cut (AAC) in even-age forests in the context of endemic windthrows (EW) scenario has been developed and evaluated.

Additional keywords: Bayes' rule, forest management planning, endemic windthrows.

Authors´ contributions: MD conceived and designed the work, compiling literature review and writing the text; IB carried out GIS data processing including Excel tabulations.

Funding: The authors did not receive specific funds for this work.

Competing interests: The authors have declared that no competing interests exist.

Correspondence should be addressed to Marian Drăgoi: marian.dragoi@usm.ro

Introduction

Windthrows are natural hazards and a significant body of literature has been published on this particular issue (Ulanova, 2000; Klaus *et al.*, 2011; Nilsson *et al.*, 2004; Schliemann & Bockheim, 2011; Boon, 2012). Nowadays, windthrows are included into the large concept of natural disturbances (Popa, 2008), along with wildfires, droughts, and insects' attacks.

One important outcome and symptom of the climate change is the frequency and geographical distribution of endemic windthrows (EW), which are affecting larger and larger areas (Res *et al.*, 2006; Senf & Seidl, 2017). EW were defined by Lanquaye-Opoku & Mitchell, (2005) as disturbances produced by peak winds, with return intervals shorter than 5 years. The amount of wood blown down may vary from tens of cubic meters to hundreds and this wide interval of variation is caused by the 'domino effect', quite difficult to take into consideration (Nolet, 2012; Nolet & Béland, 2017). Lohmander & Helles, (1987) summarized a series of theories on the causes of windthrows while (Mitchell, 1998) came up with a methodological framework based on the so-called windthrow triangle, which makes use of data concerning the stand condition, soil, and exposure.

Due to the large variety in geographic distribution and economic loss, studying natural disturbances under climate change is very provocative in statistical terms.

Yousefpour *et al.*, (2012) ranked the methodological approaches and found out that the most used models are geometric Brownian motion and probability distribution. The authors also produced some interesting statistics of the articles published on risk-related issues on forestry and concluded that the trend of articles published each year is exponential. Markov Chains and Bayesian models, we have used in our exercise, were among the least used approaches.

Long-term sustained yield has long been modeled as MC (Kouba, 2002) and Buongiorno & Zhou, (2011) demonstrated that even the Faustmann's formula is a particular case of MC, assuming that all transitions from one phase to the next one are certain, meaning that all stands will reach the prescribed maturity age.

Holecy & Hanewinkel (2006) produced a forest insurance model based on probabilities given by Weibull functions plugged into a MC model while Strigul *et al.*, (2012) developed a MC model to study the dynamic mosaic of forest stands; they concluded that age-class structures of the forest are skewed to the right, which is consistent with a MC stationary distribution.

The new paradigm of adaptive forest management stems from a mixture of extensive use of recent knowledge, risk management and a series of measures able to maintain the ecosystem resilience (Bolte *et al.*, 2009). In line with this new paradigm, the forest management planning shall come up with a more flexible approach of forest sustainability and, in particular, a deeper understanding of the sustained yield principle.

An extensive literature review on assessing the natural hazard risks in forestry has been published by Hanewinkel *et al.* (2010) who came up with a flowchart about the risks pertaining to different geographical scales of the decision-making process. However, the first step of the procedure, that is the framework analysis, takes into consideration just the regional climate models and the general circulation model, which are relevant for extreme events only (Popa, 2008).

The sustained yield principle (SYP) has long been adopted by Romanian forest policy makers and, it has been implemented into the algorithm of calculating the AAC too. As anyone would expect, SYP was an important pillar of forest management under the command and control economy, and it made sense for many reasons. During the transition to market economy SYP has been justified by the social concern that freeing too much the forest planning will encourage the forest owners to use any legal loophole to make the most of their own forests (Bouriand, 2005). Obviously, many illegal loggings reported by the media and the Court of Accounts in 2013 (Anonymous, 2013) gave to politicians a convincing reason for keeping a tight control on the annual harvest, despite that, after 2008, the allowable cut adopted by the forest management planning had been rigorously checked by the public authority.

However, on a long run, pursuing only SYP could be ineffective and somehow dangerous for the forest economy and forest ecosystems, for two reasons at least: 1) SYP cannot go along with any ecological leeway simply because the planning methods based on SYP do not account for any uncertainty; and 2) pursuing SYP per se has created a positive feed-back loop because more salvage fellings in stands older than 60 years old have replaced mature stands initially included into the cutting budget and the harvesting plan.

Having salvage fellings instead of regular harvesting operations is even more convenient from a sheer economic standing point: according to the Forest Act, the regeneration fund, which is a sort of insurance deposit for having enough money to regenerate a given stand after final felling, is fed by 10% of the income provided by the main yield only. Such a 'tax relief' for salvage cuttings makes sense for catastrophic events, like the windthrow that took place in 1995, when more than 8 million m^3 have fallen in one night (Popa, 2005). At that time, the broken trees were left un-hauled in remote areas for more than three years, due to extremely difficult harvesting conditions. Noteworthy, the Forest Act was adopted in 1996 when the consequences of that catastrophic event were apparent. On the contrary, when it comes to EW such a legal provision turns into a loophole, because more salvage fellings instead of regular harvest allow forest owners to avoid feeding this regeneration fund.

The aforementioned feedback loop has been implemented into an official regulation which states that every year the annual harvest is made of 80% of the current AAC, plus 20% of AAC of the previous year unless salvage products are to be harvested from stands older than 60 years.

If salvage fellings occur in stands older than 60 years, due to whatever reason, their total volume is deducted from the previous year's left AACR. If the salvage volume is higher than the previous year reserve, the difference will be deducted from the next year 20% AAC, and so forth. Eventually, having salvage cuttings each year, by the end of the planning period (every ten years) at least 20% of the harvestable volume will have been left uncut. This 20% AACR has neither economic nor technical support, being based on a rough assessment of the contribution of the catastrophic windthrows to the national cutting budget since 1995.

From the standpoint of mid and long-term planning, sooner or later the mature stands, postponed from harvesting, will eventually pile up into the last age class and the AAC stops being a means to normalizing the forest age structure. This particular situation shall also be addressed, in order to find out how the process of getting the mature stands older and older can be stopped.

Often, due to logging settings, salvage cuttings brought about by the windthrow are not applied only to the broken trees but also to some healthy trees that have to be removed in order to make broken trees accessible, which makes sense from the technological point of view (Waldron *et al.*, 2013). Therefore, the scope of any perturbation is always larger than the forest area covered by the damaged trees.

When it comes to windthrows, the gap between case-studies based on different methodologies and effective headways towards a better forest planning is still large. Therefore, this study aims to producing an algorithm that utilizes the data referring to EW occurred in the past to improve the decisions regarding the annual cutting budget every next year. We developed a cascading Bayes model (Keith *et al.*, 2003) to assess the EW likelihoods considering all possible permutations of three EW contributing factors (proximity of a prior EW, terrain conditions and slenderness). The EW likelihoods are further plugged into a MC model that shows the tendency of having a slant age class structure, regardless the normal age pursued by the forest planning. Finally, the share of each age class produced by MC is plugged into an intuitive formula that helps the forester decide the share of each annual allowable cut (AAC) that must be put off for the coming year in order to compensate for the EW volume that might occur in the current year.

Material and Methods

Pilot area

The proposed algorithm in this study uses data collected from Brosteni FD located in Eastern Carpathians, in Romania (Figure 1). The training dataset refers to the EW occurred between 1999 and 2009, while the testing data refer to the EW occurred after 2009. The main features of the two data sets are summarized in Table 1.

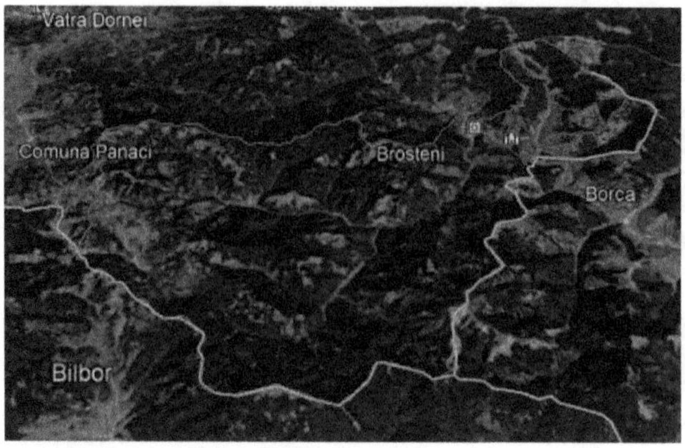

Figure 1. Location of Brosteni FD.

According to the current management plan Brosteni FD covers 21,013 hectares of full-stocked forest, with elevations ranging between 137 and 1650 m a.s.l. Part of the forest (7341.9 hectares) is strictly protected for different reasons (steep slopes with outcrops, patches of old growth forest, biodiversity and conservation). The main species is Norway spruce (*Picea abies* L.,) encompassing 88% of the area, followed by silver fir (*Abies alba*) with 4%, beech (*Fagus sylvatica*, 3%, and some other less important broadleaf species such as birch (*Betula alba*). The main yield allowable cut for 1999 forest management plan was 55,081 m^3 per year; the total length of the forest roads network is 137 km (an average 6.5 m/ha). The forest area restituted to individual forest owners is labeled on the map as 'no data available' (Figure 2). Prior to forest restitution, which started in 2000, the total forest area was 30,957 ha.

Eight years later, when the management plan was updated, the area shrank to 21,032.70 ha and the allowable cut decreased to 36,626 m^3 per year. The total current growth of commercial stands, which may be harvested if the age structure is normalized, is 60,352 m^3, meaning that the age structure is unbalanced. The total area of commercial forests is 12,941 ha (61% of the total forest area).

Bayesian assessment of EW likelihoods

We actually adopted the windthrow triangle defined by Mitchell (1995), without hypothesizing the relative

Table 1. Main features of training and testing datasets (a triple column for each 40 yr. age group of stands).

Data set	No. of cases	Total area (ha)	Avera area per stand (ha)	No. of cases	Total area (ha)	Avera area per stand (ha)	No. of cases	Total area (ha)	Avera area per stand (ha)
Training	1920	13006.9	6.77	966	9895.1	10.24	763	7563	9.91
Testing	1312	6466.5	4.93	837	8275.8	9.89	744	6290.4	8.45
Training/Testing	1.46	2.01	1.37	1.15	1.19	1.04	0.97	1.2	1.17

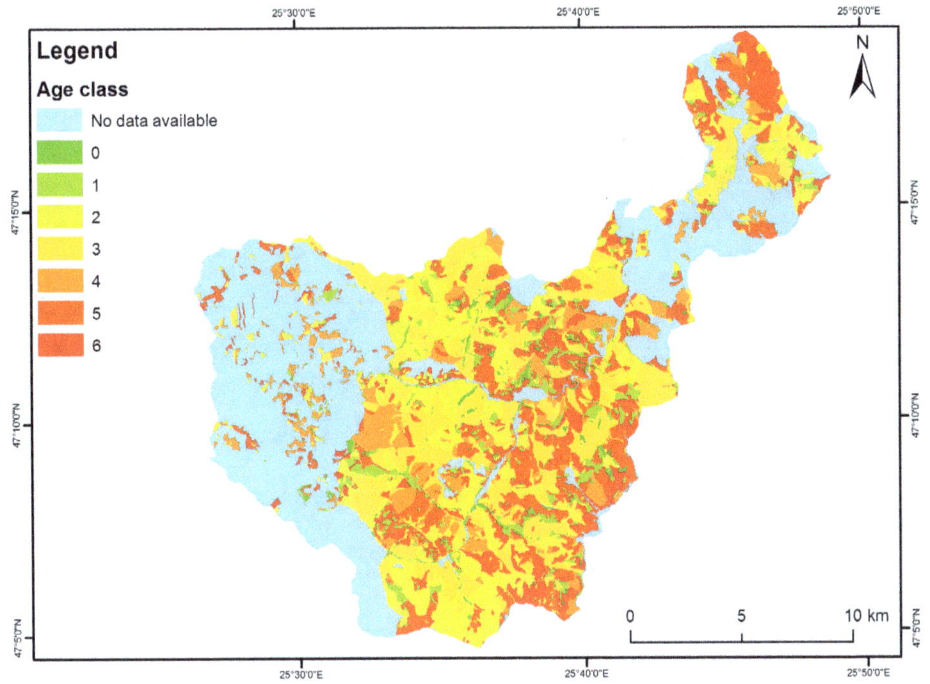

Figure 2. Age class distribution of Brosteni FD.

importance of any enabling factor. Based on literature (Scott & Mitchell 2005, Panferov & Sogachev 2008) and data-screening, we considered that EW depends on the following three factors: the stem taper, or slenderness (Bošela *et al.*, 2014) (symbolized by 'S'), 'pits and mounds' terrain condition (symbolized by 'T'), and the vicinity to a gap in tree canopy ('V'), which might be a previous EW. The existence of a canopy gap makes an EW more likely, due to the margin effect (Gadow, 2000) and fragmentation (Holeksa *et al.*, 2017). We calculated stem tapper as ratio between the height, in meters, and the diameter at breast height in centimeters, according to Olofsson & Blennow (2005).

All three factors were conveyed in binomial variables for all stands either affected by EW or not affected at all. Having the contingency table of EW occurred between 1999 and 2008, we calculated the likelihood of having an EW in any compartment, after 2008, in six different ways, considering all permutations of the three enabling factors as follows: STV, SVT, TSV, TVS, VTS, and VST, where each capital letter signifies one of the three enabling factors.

The Bayesian cascade works as follows: the posterior likelihood given by the interaction between wind and the first factor turns into the prior of the interaction of wind with the second factor, and with the third factor respectively. For example, equations 1-3 show how we estimated the EW posterior probability, given the terrain conditions, gap vicinity and slenderness. More precisely, equation (1) gives the posterior of having an EW on pits and mounds (terrain conditions), equation (2) gives the posterior of having EW on pits and mounds and near an existing canopy gap, having the prior given by equation (2), and equation (3) gives the final EW posterior, given terrain condition, vicinity, and slenderness (stem tapper higher than one). The posterior likelihood given by the interaction between wind and the first factor turns into the prior of the interaction of wind with the second factor, and with the third factor respectively.

$$p(W|T) = \frac{p(T|W) \cdot p(W)}{p(T)} \quad (1)$$

$$p(W|T \cap V) = \frac{p(T \cap V|W) \cdot p(W|T)}{p(T \cap V)} \quad (2)$$

$$p(W|T \cap V \cap S) = \frac{p(\mathrm{T} \cap \mathrm{V} \cap \mathrm{S}|W) \cdot (W|\mathrm{T} \cap \mathrm{V})}{p(\mathrm{T} \cap \mathrm{V} \cap \mathrm{S})} \quad (3)$$

The symbols have the following meanings: P(W|T) – the probability of EW, given the stand location on pits and mounds (location sensitivity); P(T|W) – conditional probability the stand is located on pits and mounds, given EW; P(W) - windthrow prevalence, or the share of the stands blown down in the last 10 years irrespective to their location, stem tapper or vicinity; P(T) – likelihood that a stand is located on pits and mounds, P(V) – likelihood of being adjacent to a canopy gap (vicinity), P(S) – probability of stem taper higher than one.

The affected and not affected areas were broken down in a contingency table (Table 2), given the EW recorded by FD staff after 1999. The EW taken into consideration occurred on commercial or noncommercial forests

Table 2. Contingency tables of EW occurred in Brosteni FD between 1999 and 2008.

	Without EW (ha)	With EW (ha)	Reference area (ha)
1st age group (1-40 years)			
Slenderness ≥ 1.00	3404	289.7	
Slenderness < 1.00	8872	441.2	13006.9
Adjacent gap	21.5	112.2	
Adjacent close canopy	12254.5	618.7	13006.9
Pits and mounds terrain	321.6	41	
Flat terrain	11954.4	689.9	13006.9
2nd age group (41-80 years)			
Slenderness ≥ 1.00	4895.5	589.6	
Slenderness < 1.00	4219.7	190.3	9895.1
Adjacent gap	4.3	130.5	
Adjacent close canopy	9110.9	649.4	9895.1
Pits and mounds terrain	158	7.1	
Flat terrain	8957.2	772.8	9895.1
3rd age group (older than 80 year)			
Slenderness ≥ 1.00	2511.7	455	
Slenderness < 1.00	3909.8	686.5	7563
Adjacent gap	12.2	121.3	
Adjacent close canopy	6409.3	1020.2	7563
Pits and mounds terrain	269.9	144.6	
Flat terrain	6151.6	996.9	7563

(i.e. forests managed for wood production, or for conservation purposes, respectively).

The locations of stands with EW produced in the last decade are presented in Figure 3. The vicinity of a new stand with a prior EW was established in a buffer of 30 m, as it approximates the mean height of a mature Norway spruce stand. The GIS buffers were applied to both 1998 and 2008 GIS maps. Further, the two buffers were intersected, resulting an area that contains the attributes of the two adjacent sub-compartments, with new EW next to prior EW within a maximum distance of 60 m. A screenshot of the GIS buffers is presented in Figure 4. The location of stands with stem tapper greater than one is shown in Figure 5 while the stands located on pits and mounds are shown in Figure 6.

For each sub-compartment, the volume affected by EW between 1999 and 2008 was summed up. The corresponding area was estimated by diving the EW volume to the average growing stock per hectare of each sub-compartment. Even though the patches of EW couldn't be precisely located, their areas were better estimated. If two or more consecutive windthrows are reported in the same compartment, the area of that compartment goes entirely into the corresponding cell of the contingency table shown in Table 2.

This sketchy manner to operate with data is precise enough to address EW, without any account on catastrophic events. After having calculated six arrays of EW likelihoods, based on the training dataset of decade 1999-2008, we tested the likelihoods on the next decade dataset (Figure 7).

For both periods we split the data into three 40-year subsets to take into consideration the age of each stand. Having the contingency table of EW occurred in each age group, and considering all six permutations of enabling factors, we calculated 48 likelihoods of having EW, considering all possible combination of the binary values taken by the enabling factors. In so doing we have computed six different likelihoods of EW for any stand, (one likelihood for each permutation of factors) according to the values taken by the three enabling factors.

Markov Chains

We used the stochastic age class model produced for the first time by Kouba (2002) described as follows:

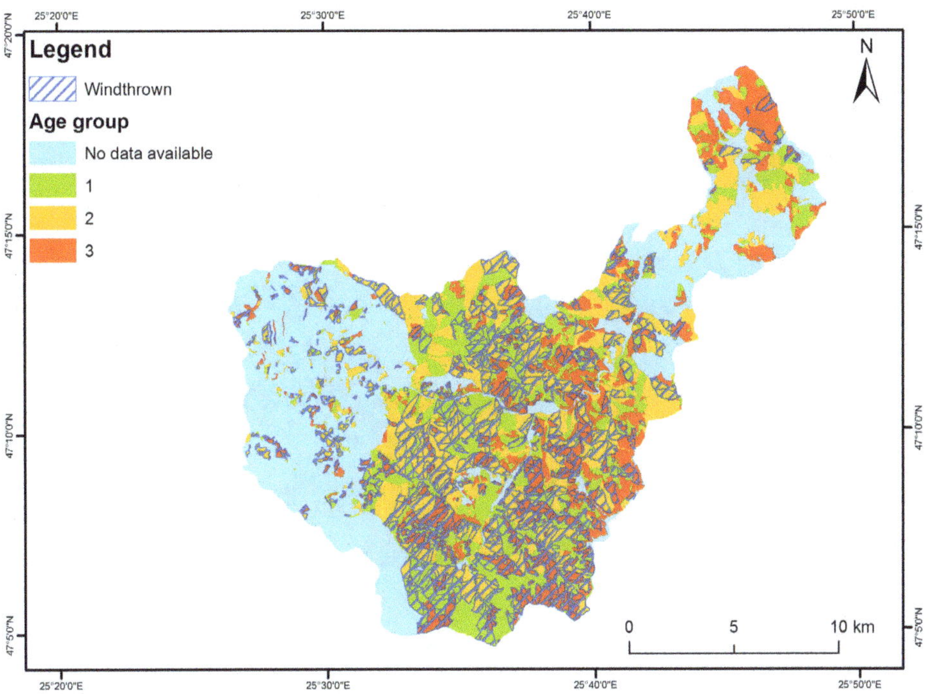

Figure 3. Spatial distribution of EW against age groups.

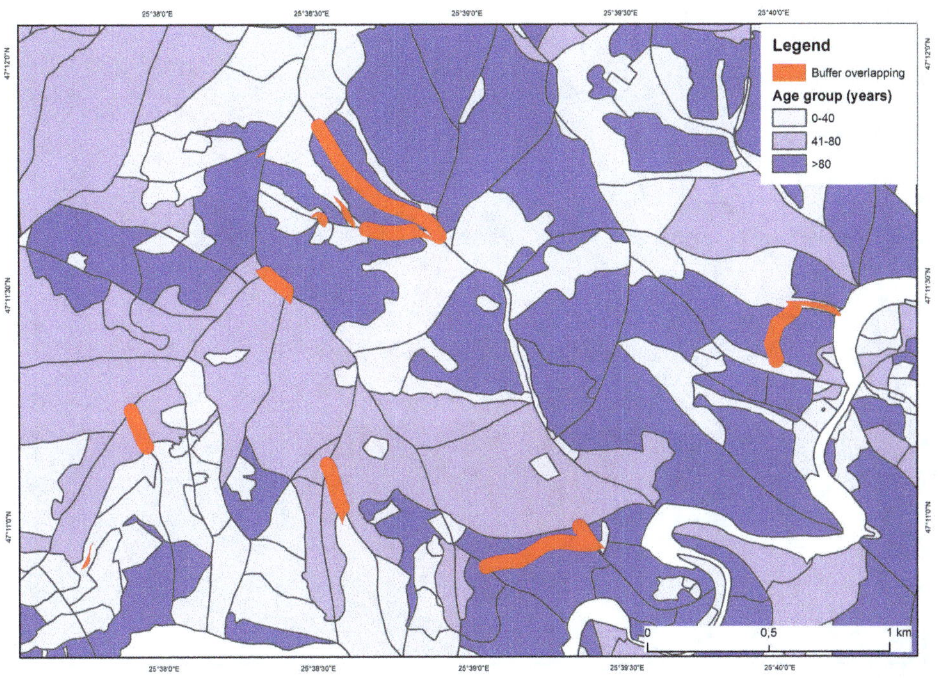

Figure 4. Buffers used to tag EW vicinity.

$$a_i = \frac{\prod_{i=1}^{6} p_{i,i+1}^{max}}{\sum_{k=1}^{6} \prod_{i=1}^{k-1} p_{i,i+1}} \quad (4)$$

where a_i is the share of the ith age class into the whole forest, and $p_{i,i+1}$ is the probability of having the ith age class passed into the following age class (transition probability).

When all transition probabilities are equal to one the transition process is fully deterministic and $a_i=1/n$, for all $i=1,n$ where n is the number of age classes. If all or some transition likelihoods are smaller than one, then the age structure is skewed and the sum of the a_i is smaller than one, meaning that at any moment a certain portion of the forest is waiting for being replanted after a disturbance (in our case, an EW).

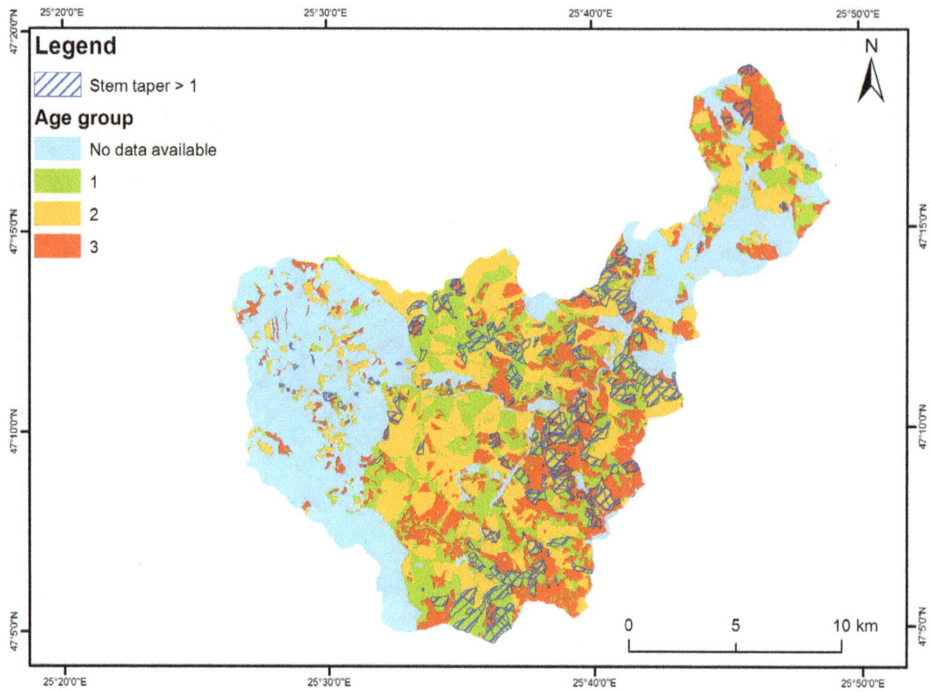

Figure 5. Spatial distribution of stands with stem tapper higher than one.

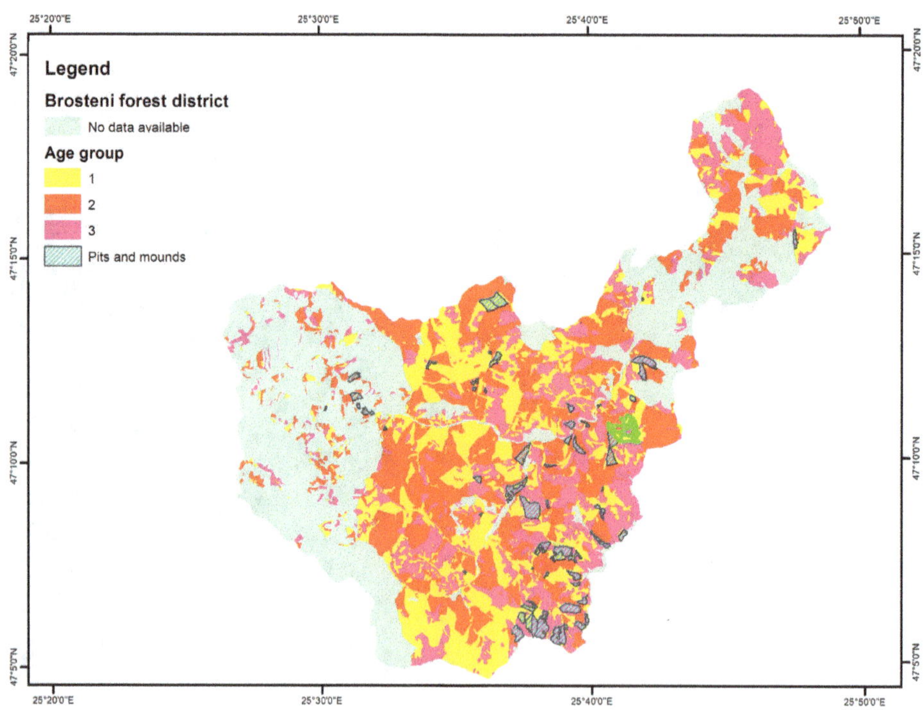

Figure 6. Pits and mounds distribution.

Adjusting the transition probabilities of age classes to forecast accuracy

Appraising the transition likelihood for each age class was a real challenge, for the many reasons that have been already discussed in the introductory section. A Bayesian method was more tempting because it allows a step-by-step assessment of the windthrow likelihood, based on the prior events produced in the last decade. However, the forecast accuracy shall be considered, in order to avoid any overestimation of EW likelihoods. Therefore, for each permutation of enabling factors, within each age group, we calculated the Receiver Operating Characteristic (ROC) and the corresponding

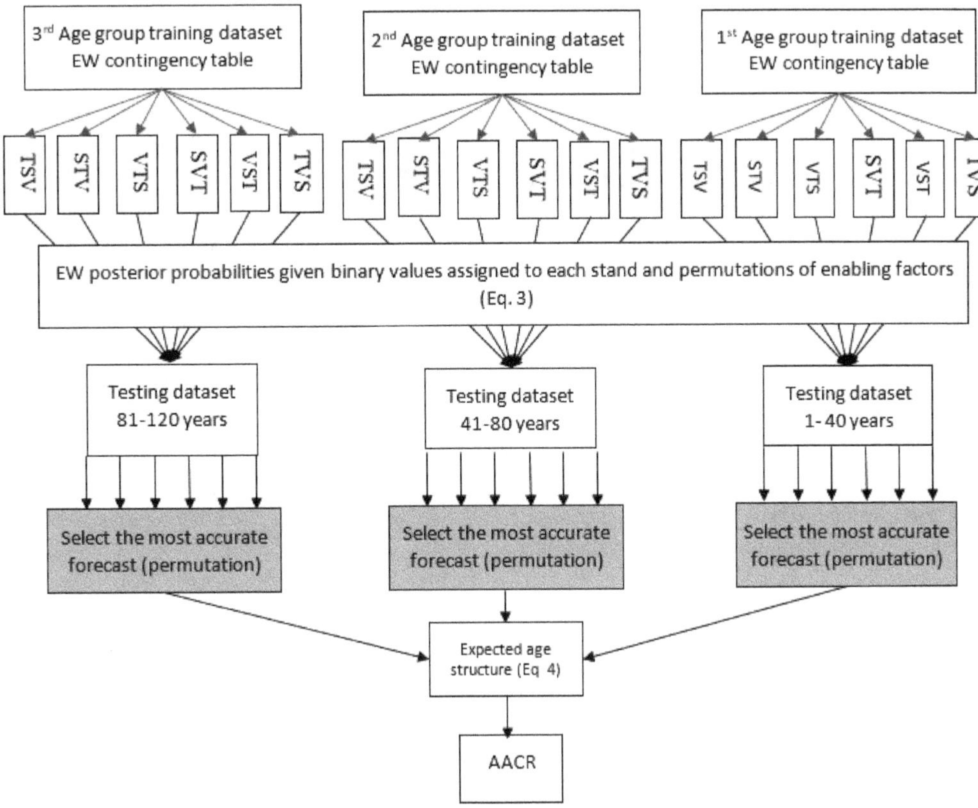

Figure 7. Flowchart of calculating AACR.

Area Under Curve (AUC), to adjust the likelihoods produced by Bayesian inferences, as equation (5) shows:

$$p_{i,i+1}^{max} = p_{i,i+1}^{Pr^{max}} \cdot AUC^{max} \cdot 2 \quad (5)$$

The ROC curve is the geometrical position of the cumulative frequencies of true positives (TP) and false positives (FP), after having the posterior likelihoods of EW ranked in descending order. Based on the coordinates of ROC points, we further calculated the AUC having *true positives* (TP) and *false positives* (FP) estimated by equations (5) and (6).

$$TP = \sum_{i=1}^{n} S_i(p_i - 0.5) \quad for\ all\ i \begin{cases} f_i = 1 \\ p_i > 0.5 \end{cases} \quad (6)$$

$$FP = \sum_{i=1}^{n} S_i\, p_i \quad for\ all\ i \begin{cases} f_i = 0 \\ p_i > 0.5 \end{cases} \quad (7)$$

Where S_i is the area of the ith stand, p_i is the posterior likelihood of EW, and f_i is the real state of that stand (1 =affected by EW in the last decade; 0 –not affected by EW).

The annual AACR vs. the cutting budget to EW occurrence

Having an AAC given by the forest plan and the expected returns to the first age class from older stands we can estimate the AACR lagged for the next year (R) by equation (8).

$$R = \sum_{i=1}^{6} s_i \cdot g_i \cdot \left|\frac{1}{6} - a_i\right| \quad (8)$$

where s_i is the area of the ith age class (i=1,..,n), and g_i is the current growth per year and hectare for the ith age class. When all a_i equal 1/n, meaning that no EW had occurred, the AACR is zero, and the whole AAC can be entirely harvested in the current year.

As already mentioned, the SYP has been adopted long ago and it means more than equal yields year after year. Therefore, two different methods are used to calculate AAC and both methods take into account any likely short supply of wood that may occur in the next 10, 20, 40 and 60 years. By "short supply" we mean the lowest ratio between the total volume that will be harvestable in a given period of time and the length, in years, of that period (next 10, 20, 40 or 60 years).

The whole algorithm has been described at length in Dragoi (2010). The indicative growth method we have referred to, makes use of a ratio labeled with q, which makes the difference between two different situations: too much yield in the next 20 years, due to numerous mature and over-mature stands ($q>1$) or too few mature stands ($q<1$). For each situation different algorithms are being used. For Brosteni FD the second situation applies (deficit of mature stands) and there is no risk of triggering the positive feed-back we highlighted in the introduction.

If the age class distribution were imbalanced and both first and third age groups exceed the second one (the age-class structure of commercial forests resembles a "U" shape), it means that many mature stands were put off over and over after each salvage (i.e. EW) cutting. In such circumstances, indicating a positive feed-back loop, the AAC shall not be diminished with the AACR, because the AAC has long stopped steering the forest to a normal structure, where all age classes are, more or less, even.

When a higher AACR is the norm, not the exception, the "deterministic" age structure will never be balanced but skewed, and it makes no sense to pursue a normal age-class structure, without any account of EW occurrence.

Adjusting the demand for seedlings

Assuming a deterministic model, and no shelterwood system (i.e. no natural regeneration), every ten years the demand of seedlings shall cover the corresponding area of mature stands clear-felled before. In case of EW, the total area which shall be regenerated within each decade is larger, because not only mature stands were harvested but also the damaged trees and the damaged portions of the younger stands. If all $p_{i,i+1}$ are smaller than one, the sum of all a_i given by equation (4) is also smaller than one, and the additional area that shall be replanted is given by equation (9):

$$P = F \cdot \Sigma_i^6 \left(\frac{1}{6} - a_i\right) \quad (9)$$

Where P is the additional area to be planted in the next 20 years, F is the total area of commercial forests.

Results

Adapting cutting budget to EW occurrence

The contingency table used to calculate priors for every permutation of factors is presented in Table 2. Table 3 shows the posterior likelihoods considering all permutations of enabling factors (slenderness, gap vicinity, and terrain conditions) and all possible combinations of values taken by the three enabling factors. Having these probabilities stored in a matrix, we assigned to each sub-compartment included into the testing dataset six probabilities, one for each permutation of enabling factors, considering the age group wherein each sub-compartment falls. The AUC for each permutation and age groups are presented in Table 4 while the EW likelihoods, weighted on area for each age class are shown in Table 5. The AUC for all permutations within the third age group are zero, meaning that posterior likelihoods of having EW are not reliable at all for further calculations. All input variable plugged into equation 8 are shown in Table 6 where the AACR is shown in the last row.

As aforementioned the total AAC of Brosteni FD is 36,626 m^3 and according to the current regulations, the formal reserve for every next year is 7325 m^3. According to the Markovian model, an AACR of 2,696 m^3 is enough to compensate for the effect of EW on sustained yields.

Assessing the extra-demand of seedlings

The gaps left into the forest after EW must be regenerated within a two year period after having hauled all the affected trees. The actual management planning system has no allowance for replanting the gaps produced by EW. For our pilot areas, having the EW likelihoods, the additional area that shall be replanted every ten years is 336 ha, which means an average 33.6 ha/year. Having a norm of 5000 seedlings/ha, the additional number of seedlings per year reaches 168500 plants.

Discussion

The key issue discussed here is whether or not the sheer Bayes' rule is a good option for making statistical inferences. The Bayes' rule assumes interchangeable roles between hypotheses and evidences (Nöth & Weber, 2003). This assumption is counterintuitive because most of the people are inclined rather to think in terms of cause and effect; once this epistemological difficulty is overcome things become clearer if we associate the hypothesis with a causal chain strengthen by more evidence, which turns into a new hypothesis, and so forth (Wintle & Lindenmayer, 2008).

Based on the old concept of windthrow triangle (Mitchell, 1995) we considered just three enabling factors, in all possible orders conveyed into six permutations. Having one more attribute, say the position on the slope, given that in many cases the stands located on foothills are blown down by the wind coming from the opposite direction, the number of all possible permutations (i.e. hierarchies of factors to be tested) rises to 24. From the computational point of view this is not big challenge because all calculations can be plugged into spreadsheet and the input data

Table 3. Transition likelihoods for all permutations of factors and possible combinations of values taken by enabling factors.

Slenderness (S)	Yes	Yes	Yes	Yes	No	No	No	No
Vicinity (V)	Yes	Yes	No	No	Yes	No	No	Yes
Pits and mounds (T)	Yes	No	Yes	No	Yes	Yes	No	No
1st age group 1(-40 years)								
TVS	0.941	0.903	0.134	0.083	0.879	0.065	0.039	0.809
VST	0.941	0.879	0.904	0.809	0.141	0.074	0.039	0.065
SVT	0.941	0.879	0.141	0.065	0.904	0.074	0.039	0.809
VTS	0.941	0.903	0.879	0.809	0.134	0.065	0.039	0.083
STV	0.941	0.134	0.879	0.065	0.903	0.809	0.039	0.083
TSV	0.941	0.134	0.903	0.083	0.879	0.809	0.039	0.065
2nd age group (41-80 years)								
TVS	0.803	0.803	0.803	0.803	0.803	0.803	0.037	0.942
VST	0.958	0.977	0.898	0.942	0.060	0.000	0.037	0.092
SVT	0.958	0.977	0.060	0.092	0.898	0.000	0.037	0.942
VTS	0.957	0.894	0.977	0.942	0.050	0.092	0.037	0.019
STV	0.957	0.050	0.977	0.092	0.894	0.942	0.037	0.019
TSV	0.957	0.050	0.894	0.019	0.977	0.942	0.037	0.092
3rd age group (older than 80 years)								
TVS	0.968	0.967	0.328	0.322	0.902	0.129	0.125	0.900
VST	0.968	0.902	0.967	0.900	0.332	0.319	0.125	0.129
SVT	0.968	0.902	0.332	0.129	0.967	0.319	0.125	0.319
VTS	0.968	0.967	0.902	0.900	0.328	0.129	0.125	0.322
STV	0.968	0.328	0.902	0.129	0.967	0.900	0.125	0.322
TSV	0.968	0.328	0.967	0.322	0.902	0.900	0.125	0.129

are produced by a simple cross-tabulation, provided that all important attributes had been included into the GIS table of attributes (deFillipi *et al.*, 2004).

A 30 m wide buffer may indicate as adjacent two stands separated by a 59.99 m wide strip of trees which were not yet blown down, if the two stands affected by EW are separated by a stand of trees narrower than 60 m. Such situations are represented by very thin buffers, some of them being screen-captured in Figure 4 and marked with yellow circles. Hence, prior EW vicinities might have been overestimated in some cases, but such situations can instantly be spotted.

For the first age group of stands the best predictability corresponds to a seemingly unexpected permutation of factors: vicinity-terrain-slenderness. It is quite surprising because most of the young stands are not actually thinned, and slender. Even more surprisingly, slenderness plays the most important role for the EW produced in middle-aged stands (Konopka *et al.*, 1987). For the last age group none of the permutations

Table 4. AUC on age groups and permutations of enabling factors.

Permutations of enabling factors					
TVS	VST	SVT	VTS	STV	TSV
1st age group					
0.581	0.955	0.580	0.964	0.918	0.909
2nd age group					
0.00	0.741	0.566	0.00	0.829	0.206
3rd age group					
0.00	0.00	0.00	0.00	0.00	0.00

Table 5. Rough Bayesian estimations of EW.

Age class	1	2	3	4	5	6
EW likelihood given by Bayesian inference averaged against area	0.01	0.09	0.01	0.03	0.16	0.27
Appropriate permutation of factors	VTS	VTS	STV	STV	N-A	N-A

Table 6. Estimation of the annual reserve.

Variables of eq. 16	Age class					
	1	2	3	4	5	6
Adjusted EW likelihoods	0.01	0.18	0.05	0.02	0.00	0.00
Actual area (ha)	1,066.00	3,469.40	4,130.70	1,632.40	1,599.70	1,042.80
Area according to normal age structure (ha)	2,156.83	2,156.83	2,156.83	2,156.83	2,156.83	2,156.83
Transition likelihoods	0.99	0.82	0.95	0.98	1.00	1.00
a_i coefficients	0.172	0.142	0.135	0.132	0.132	0.132
stochastic normal age (ha)	2,221.93	1,831.61	1,747.58	1,709.19	1,709.19	1,709.19
Age class shares according to deterministic age class structure (1/6)	0.167	0.167	0.167	0.167	0.167	0.167
Differences	0.01	-0.03	-0.03	-0.03	-0.03	-0.03
Growth per age class m^3/ha/year	3.70	9.39	9.27	5.93	4.18	2.02
Reserve per age class (m^3)	20	819	1211	335	231	73
Total reserve (m^3)						2689

of factors proved to be a serious threat, considering the combinations of EW and values taken by the three enabling factors.

We tested the accuracy of predictions with the ROC-AUC procedures. The main disadvantage of gauging any prediction made by discrete-output procedures – and Bayes' rule turns discrete likelihoods – is the ROC shape (Schindler *et al.*, 2009). All discrete values taken by EW likelihoods are shown in Table 3 but, associated to the stands, they appear in lesser combinations, depending on how numerous are the stands that fall into one of the three age groups. Yet the most complex situations are characterized by eight likelihoods, only the ones with likelihoods grated than 0.5 are being considered, as shown in Figure 8. The first one (a) was the most complex (with four discrete values) but most of other ROC look like (b), with three discrete probabilities and (c), with only two discrete values.

The fourth situation (d) typifies all ROC curves for the last age group. In Figure 8 d) we have only two true positives situations, without any false positives. Because the cumulative frequencies are plotted against

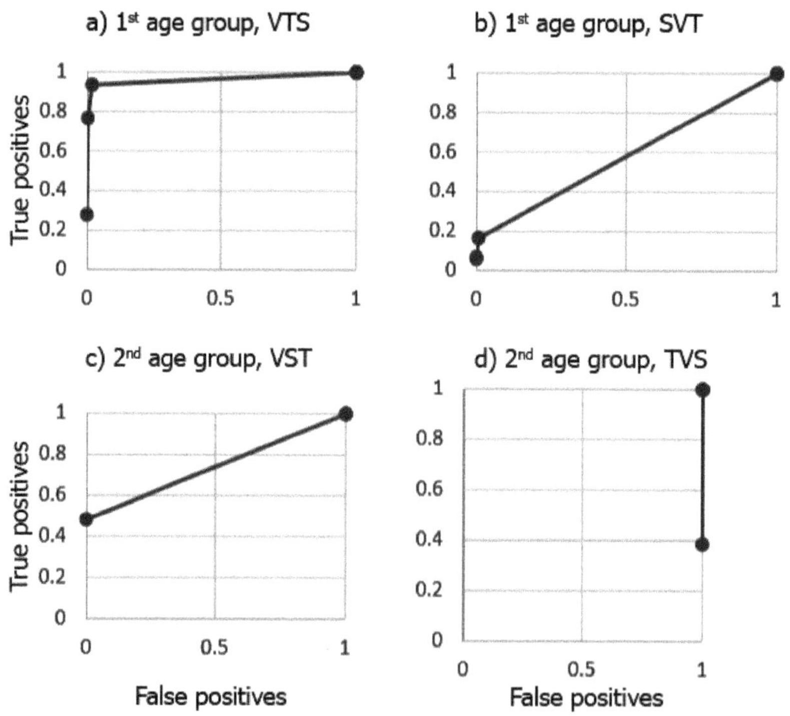

Figure 8. Four typical ROC curve for discrete EW likelihoods.

the two axis, summed up from the lowest to the highest likelihood, the two pints are vertically aligned to the right, having no area below. However, the loss of information is not important because the EW likelihood of the last age class must be forced to one, because all mature stands are supposed to be harvested within the next two decades.

If the young stands had higher growths, the AACR would have been higher: an average of 3.7 m^3/year/ha, as Table 6 shows, is a hint of having unstocked stands in the first two age classes.

By cascading the Bayes tree probabilities for each development stage, we assumed that it must be an order in which all three enabling factors weaken most of the trees, although the order is yet unknown. Hence, we had to conceive all possible orders of factors and see which ones behave better in terms of accuracy of prediction.

Our case study showed that what really matters is not the predefined set of enabling factors, but the number of cases analyzed in order to train the model. Having more cases in the first age group, four ROC, out of six, are meaningful (greater than 0.5), while the second age group has just two significant ROC (for STV, and VST respectively, see Table 4).

We have confined our study to two periods of ten years each to match the data required by the model with the data supplied at no additional cost by the Romanian forest planning system. Worth noting, for Brosteni FD, as well as for any public FDs in Romania, the last and the last but one forest management plans have been produced in GIS technology, allowing cheap data processing; consequently, we couldn't go back too much in the past. In other conditions, having a greater amount of data and more GIS maps for the same area.

Conclusions

This study is the first attempt in calculating AACR, which shall be put off for the next year, given a certain pattern of EW (or whatever disturbance) for Romanian forest planning. The study demonstrated that 10% AACR is enough in mountainous area, where EW are quite frequent. The 10% or 20% of the AAC left unharvested from the previous year, and the same amount of wood left uncut for the coming year is not, allegedly, a substantial improvement of forest management. However, a slimmer AACR is advantageous for logging companies who are bidding for harvesting contracts every year, because they can make more combinations with the tracts worth bidding for, simply because the share of the main yield auctioned every year by the forest managers is 10 % greater.

If the SYP fails to normalize the age-class structure of the forest due to the positive feed-back loop previously described, the model allows a better control on the total yield, and eventually can be used to improve the calculation of AAC, considering a slant age-class structure, not an even one. Having better predictions of EW, knowing their future locations (elevation, slope and aspect) and the hydrological effects of a clear-felling, the forest planners will be able to integrate this information into more complex decision-making models, able to better balance the wood production with water runoff management.

We have considered only three enabling factors, but the list is open: besides slenderness, terrain conditions and EW vicinity, other factors can be added, such as vertical structure of the stands, presence of wind-resisting species of trees, or their location (uphill or downhill). The hints for taking into consideration a new factor are the values stored into the contingency table; and the more permutation of enabling factors are at hand, the more curved ROC will be, allowing better predictions of further EW.

Acknowledgements

We are deeply grateful to the two peer reviewers for their supportive suggestions. We also acknowledge the staff of Brosteni FD who provided us the input data.

References

Anonymous, 2013: Summary of the Audit Report concerning "the patrimonial state of the Romanian Forest Fund between 1990 and 2012. Court of Accounts of Romania.

Bolte A, Ammer C, Löf M, Madsen P, Nabuurs G-J, Schall P, Spathelf P, Rock J, 2009. Adaptive forest management in central Europe: Climate change impacts, strategies and integrative concept. Scand J Forest Res 24(6): 473–482. https://doi.org/10.1080/02827580903418224

Boon S, 2012. Snow accumulation following forest disturbance. Ecohydrology 5 (3): 279–285. https://doi.org/10.1002/eco.212

Bošela M, Konôpka B, Šebeň V, Vladovič J, Tobin B, 2014. Modelling height to diameter ratio – an opportunity to increase Norway spruce stand stability in the Western Carpathians. Forestry J 60: 71. https://doi.org/10.2478/forj-2014-0007

Buongiorno J, Zhou M, 2011. Further generalization of Faustmann's formula for stochastic interest rates. J Forest Econ 17(3): 248–257. https://doi.org/10.1016/j.jfe.2011.03.002

Bouriaud L, 2005. Causes of illegal logging in Central and Eastern Europe. Small-Scale For 4: 269–291.

de Filippi R, Palma J, Reisner Y, Herzog F, 2004. Spatial database for GIS for scaling up. Research Report. Swiss Federal Research Station for Agroecology and Agriculture. 9 pp.

Drăgoi M, 2010. Compensating the opportunity cost of forest functional zoning - two alternative options for the Romanian forest policy. Ann Forest Res 52(1): 81-92.

Gadow K Von, 2000. Evaluating Risk in Forest Planning Models. Silva Fenn 34: 181–191.

Hanewinkel M, Hummel S, Albrecht A, 2010. Assessing natural hazards in forestry for risk management: a review. Eur J Forest Res 130(3): 329–351. https://doi.org/10.1007/s10342-010-0392-1

Holecy J, Hanewinkel M, 2006. A forest management risk insurance model and its application to coniferous stands in southwest Germany. Forest Policy Econ 8(2): 161–174. https://doi.org/10.1016/j.forpol.2004.05.009

Holeksa J, Jaloviar P, Kucbel S, Saniga M, Szewczyk J, Szwagrzyk J, Zielonka T, Żywiec M, 2017, Models of disturbance driven dynamics in the West Carpathian spruce forests. Forest Ecol Manage 388: 79–89. https://doi.org/10.1016/j.foreco.2016.08.026

Keith M, Adams P, Bryant D, Mitchelson K, Cochran D, Lala G, 2003 Inferring an Original Sequence from Erroneous Copies: Two Approaches. J. Bioinform Comput Biol 6(3): 107-114.

Klaus M, Holsten A, Hostert P, Kropp J, 2011. Integrated methodology to assess windthrow impacts on forest stands under climate change. Forest Ecol Manage 261(11): 1799–1810. https://doi.org/10.1016/j.foreco.2011.02.002

Konopka J, Petras R, Toma R, 1987. Slenderness coefficient of the major tree species and its importance for static stability of stands. Lesnictvi 33: 887-904.

Kouba J, 2002. Das Leben des Waldes und seine Lebensunsicherheit, Forstwiss Centralbl (Hamb) 121(4): 211–228. https://doi.org/10.1046/j.1439-0337.2002.02026.x

Lanquaye-Opoku N, Mitchell SJ, 2005. Portability of stand-level empirical windthrow risk models. Forest Ecol Manage 216(1–3): 134–148. https://doi.org/10.1016/j.foreco.2005.05.032

Lohmander P, Helles F, 1987. Windthrow probability as a function of stand characteristics and shelter. Scand J Forest Res 2 (1–4): 227–238. https://doi.org/10.1080/02827588709382460

Mitchell SJ, 1995. The windthrow triangle : A relative windthrow hazard assessment procedure for forest managers windthrow Triangle Soils. Forest Chron 71(4): 446–450. https://doi.org/10.5558/tfc71446-4

Mitchell SJ, 1998. A diagnostic framework for windthrow risk estimation. Forest Chron 74(1): 100–105. https://doi.org/10.5558/tfc74100-1

Nilsson C, Stjernquist I, Bärring L, Schlyter P, Jönsson, AM, Samuelsson H, 2004. Recorded storm damage in Swedish forests 1901-2000. Forest Ecol Manage 199(1): 165–173. https://doi.org/10.1016/j.foreco.2004.07.031

Nolet P, 2012. Predicting Stem Windthrow Probability in a Northern Hardwood Forest Using a Wind Intensity Bio-Indicator Approach. Open J Forest 2(2): 77–87. https://doi.org/10.4236/ojf.2012.22011

Nolet P, Béland M, 2017. Long-term susceptibility of even- and uneven-aged northern hardwood stands to partial windthrow, Forests 8(4). https://doi.org/10.3390/f8040128

Nöth M, Weber M, (2003). Information aggregation with random ordering: Cascades and overconfidence. Economic J 113(484): 166–189. https://doi.org/10.1111/1468-0297.00091

Olofsson E, Blennow K, 2005. Decision support for identifying spruce forest stand edges with high probability of wind damage. Forest Ecol Manag 207: 87–98. https://doi.org/10.1016/j.foreco.2004.10.019

Panferov O Sogachev A, 2008. Influence of gap size on wind damage variables in a forest. Agric Forest Meteorol 148(11): 1869-1881. https://doi.org/10.1016/j.agrformet.2008.06.012

Popa I, 2005. Doborâturi produse de vânt-factor de risc în ecosistemele forestiere montane. (Windthrows – risk factors if mountainous forest ecosystems). Anale ICAS 48(4).

Popa I, 2008. Windthrow risk management: results from Romanian forests. Disturbi in foresta ed effetti sullo stock di carbonio: il problema della non permanenzai, Pubblicazione del Corso di Cultura in Ecologia, Atti del 44 corso, 77-88.

Schlyter P, Stjernquist I, Bärring L, Jönsson AM, Nilsson C, 2006. Assessment of the impacts of climate change and weather extremes on boreal forests in northern Europe, focusing on Norway spruce. Clim Res 31(1): 75–84. https://doi.org/10.3354/cr031075

Schindler D, Grebhan K, Albrecht A, Schönborn, J, 2009. Modelling the wind damage probability in forests in Southwestern Germany for the 1999 winter storm 'Lothar'. Int J Biometeorol 53(6): 543-554. https://doi.org/10.1007/s00484-009-0242-3

Schliemann SA, Bockheim JG, 2011. Methods for studying treefall gaps: A review. For Ecol Manage 261(7): 1143–1151.

Scott R.E, Mitchell SJ, 2005. Empirical modelling of windthrow risk in partially harvested stands using tree, neighbourhood, and stand attributes. Forest Ecol Manag 218(1-3): 193-209. https://doi.org/10.1016/j.foreco.2005.07.012

Senf C, Seidl R, 2017. Natural disturbances are spatially diverse but temporally synchronized across temperate forest landscapes in Europe. Glob Change Biol 3218–3221.

Strigul N, Florescu I, Welden AR, Michalczewski F, 2012. Modelling of forest stand dynamics using Markov chains. Environ Model Softw 31: 64–75. https://doi.org/10.1016/j.envsoft.2011.12.004

Ulanova NG, 2000. The effects of windthrow on forest at different spatial scales: a review. Forest Ecol

Manag 135: 155–167. https://doi.org/10.1016/S0378-1127(00)00307-8

Waldron K, Ruel JC Gauthier S, 2013. Forest structural attributes after windthrow and consequences of salvage logging. Forest Ecol Manag 289: 28–37. https://doi.org/10.1016/j.foreco.2012.10.006

Wintle BA, Lindenmayer DB, 2008. Adaptive risk management for certifiably sustainable forestry. Forest Ecol Manage 256(6): 1311–1319. https://doi.org/10.1016/j.foreco.2008.06.042

Yousefpour R, Jacobsen JB, Thorsen BJ, Meilby H, Hanewinkel M, Oehler K, 2012. A review of decision-making approaches to handle uncertainty and risk in adaptive forest management under climate change. Ann Forest Sci 69(1): 1–15. https://doi.org/10.1007/s13595-011-0153-4

21

Combining innovative mulches and soil conditioners in mountain afforestation with ash (*Fraxinus excelsior* L.) in the Pyrenees (NE Spain)

Jaime Coello[1]*, Míriam Piqué[1], Pere Rovira[1], Carla Fuentes[1] and Aitor Ameztegui[1,2]

[1]*Forest Science and Technology Center of Catalonia (CTFC), Carretera de Sant Llorenç de Morunys, km 2; ES-25280 Solsona, Spain.* [2]*Department of Agriculture and Forest Engineering (EAGROF), University of Lleida, Lleida, Spain.*

Abstract

Aim of study: To assess the effectiveness for improving early seedling performance of the individual and combined application of (i) various doses of an innovative soil conditioner including polyacrylamide-free super-absorbent polymers, fertilizers, root precursors and humic acids; and (ii) innovative mulches based on renewable-biodegradable or recycled raw materials. The assessment was carried out in comparison with reference (commercial) soil conditioners and mulches.

Area of study: Upper montane afforestation site located at 1,430 m altitude in the southern Pyrenees (NE Spain).

Material and methods: We studied the effect of 15 treatments (various combinations of soil conditioners and mulches) on mountain ash (*Fraxinus excelsior* L.), testing survival, diameter and height growth and water and nutrient status during two growing seasons (2014-2015). We also assessed mulch durability during 2014-2016.

Main results: The innovative soil conditioner improved diameter and height seedling growth (92% and 72% respectively) and water and nutrient status. The 40 g/seedling dosage was more cost-effective than the 20 and 80 g/seedling doses. The new formulation performed better in general than the commercial formulation. Mulches led to slight gains compared to control seedlings, and there were no major differences between the mulch models. The combined application of soil conditioners and mulches was not of particular interest.

Research highlights: Soil conditioners consisting of synergic mixtures of water super-absorbent polymers, fertilizers, root growth precursors and humic acids can improve early seedling performance in coarse-textured, stony soils in montane conditions. Small mulches may be only of limited interest as long as weed competitiveness is poor.

Additional keywords: ecotechnology; groundcovers; reforestation; seedling performance; restoration; water super-absorbent polymer; weed.

Abbreviations used: GSn, growing season number 'n'; leaf water potential, LWP; water super-absorbent polymers, SAP; soil conditioner with water super-absorbent polymers, SCwSAP; Soil Plant Analysis Development, SPAD.

Authors´ contributions: Conceived and designed the experiments: JC, PR and MP; data acquisition and management: CF and JC; critical revision of the manuscript for important intellectual content: JC, PR, AA and MP; coordinating the research project: MP.

Funding: The research leading to these results has received funding from the European Union's Seventh Framework Programme managed by REA (Research Executive Agency), FP7/2007 2013, under grant agreement n° 606554 – Sustaffor project: Bridging effectiveness and sustainability in afforestation / reforestation in a climate change context: new technologies for improving soil features and plant performance. AA was supported by the Spanish Government through the 'Juan de la Cierva' fellowship program (IJCI-2016-30049).

Competing interests: The authors have declared that no competing interests exist.

Correspondence should be addressed to Jaime Coello: jaime.coello@ctfc.es

---

Introduction

The history of land use in the Pyrenees, as in other mountain areas of south Europe, spreads over millennia. In the last century the modernization of farming in lowlands resulted in the abandonment of many traditional agricultural and grassland practices (MacDonald *et al.*, 2000), particularly in the difficult to access, small-sized fields predominant in mountain areas. Consequently, forest has expanded significantly, especially pine forests (Ameztegui *et al.*, 2010). This has homogenized the landscape both in terms of composition and structure. Afforestation with broadleaf trees helps to decrease the negative consequences of excessive landscape homogenization, such as an increase in fire risk (Palmero-Iniesta *et al.*, 2017) and negative effects on biodiversity (Ameztegui *et al.*, 2018), among others.

In the Pyrenees, European ash (*Fraxinus excelsior* L.) is a native broadleaf tree with many uses, including timber for tool-making and turnery, fodder (green branches and litter), charcoal and fuelwood (Marie-Pierre *et al.*, 2006; Mottet *et al.*, 2007). Despite these uses, many ash forests were cut down and turned into grasslands in the past (Vigo *et al.*, 2005). However, more recently, its ecological plasticity together with its fast growth and valuable timber has turned ash into a good candidate for afforestation programmes in Europe (Fraxigen, 2005; Weber-Blaschke *et al.*, 2008). The main limiting factor for ash in the southern Pyrenees is its sensitivity to water shortage (Gonin *et al.*, 2013), which is particularly critical during the first years when the root system is not yet well developed (Vallejo *et al.*, 2012). South European montane conditions are characterized by a dry season that coincides with the highest temperatures, and by shallow, coarse-textured soils with high stoniness and steep slopes, resulting in a poor water retention capacity. Soil water content can be increased by support irrigations, which are expensive and often inapplicable in these conditions (Carminati *et al.*, 2010). An alternative option is to mix water super-absorbent polymers (SAPs or hydrogels) with the soil of the planting pit. These polymers can absorb and store up to 400 times their weight in water (Bouranis *et al.*, 1995), which is then available to the plants over an extended time period (Hüttermann *et al.*, 2009). SAPs can increase soil water content by reducing evaporation and percolation losses, which is of particular interest in soils with a poor water retention capacity, i.e. coarse-textured soils (Bhardwaj *et al.*, 2007; Koupai *et al.*, 2008; Del Campo *et al.*, 2011). SAPs have been reported to alleviate soil and plant water potential, and increase plant water use efficiency and the time to reach permanent wilting point, ultimately enhancing plant survival and growth (Sivapalan, 2001; Hüttermann *et al.*, 2009; Del Campo *et al.*, 2011). SAPs are commercialized alone or as synergistic mixtures with other ingredients, especially fertilizers and various organic compounds, with the intention of improving both the physical and chemical soil properties and not only the water-related parameters. Moreover, SCwSAPs aim to prevent some of the limitations of pure SAPs, such as the reduction of NO_3^- and NH_4^+ availability and the risk of being washed away (Rowe *et al.*, 2005). These mixtures are regarded as soil conditioners with SAP (SCwSAP), and have been reported to enhance soil nutrient levels and seedling performance (Machado *et al.*, 2016; Coello *et al.*, 2018). However, the application of SAPs and SCwSAPs at afforestation sites has obtained contrasting results depending on their composition, application method, dosage and soil features (Navarro *et al.*, 2005; Del Campo *et al.*, 2011; Coello *et al.*, 2018).

Despite their interest, there are some constraints involved in using SAP or SCwSAP, as most formulations are based on cross-linked polyacrylamide, which causes social concern due to the residual presence of non-polymerized acrylamide. Although the concentrations are within the legal limits (Rowe *et al.*, 2005) and acrylamide has a short half-life in the soil (Lande *et al.*, 1979), SAP manufacturers are currently developing polyacrylamide-free versions that still need to be tested in the field and compared to current commercial formulations. Moreover, a limiting factor of SAPs is that the improvement in conditions at micro-site level is positive for the seedling but also for the competing vegetation, which is a major threat to young afforestations (Willoughby *et al.*, 2009). Extant vegetation can outcompete newly established seedlings in the struggle for water, light and nutrients (Navarro-Cerrillo *et al.*, 2005). The most widespread weeding techniques are to apply herbicide, which is cost-effective but is of growing social concern (Willoughby *et al.*, 2009), and mechanical weeding, which is expensive and can only be applied at sites that can be accessed easily.

An alternative option is to use mulches, also known as groundcovers or weeding mats. A mulch is an opaque layer covering the soil around the seedling, impeding weeds from germinating in its vicinity (Maggard *et al.*, 2012). Previous works have demonstrated that mulching has positive effects on seedling survival and growth (Van Sambeek, 2010; Maggard *et al.*, 2012, Coello *et al.*, 2017), increases soil water content because it reduces evaporation and transpiration by weeds (Hueso-González *et al.*, 2015), improves nutrient uptake (Van Sambeek & Garrett, 2004) and buffers soil temperatures (Cregg *et al.*, 2009; Coello *et al.*, 2017). Although plastic mulches are the most widespread (Arentoft *et al.*, 2013), several new environmentally friendly mulch materials are being developed, including biofilms, i.e. plastic-like materials made from renewable sources (Kapanen *et al.*, 2008). Other similar materials include composites made from paper residues (Shogren & Rousseau, 2005) and jute tissues (Debnath, 2014).

Despite the interest in these new materials and techniques for afforestation, the combined application of soil conditioners and groundcovers has rarely been tested in field conditions (Navarro *et al.*, 2005; Coello *et al.*, 2018). This study aims to assess the effectiveness on early seedling performance (survival, growth and water and nutrient status) of the individual and combined application of (a) new mulches based on renewable or recycled raw materials, either biodegradable or reusable; and (b) polyacrylamide-free SCwSAP, in south European montane conditions. We also studied mulch durability. We compared the performance of these new techniques with commercially available mulches and SCwSAP.

Our working hypotheses were: i) both the use of mulches and SCwSAP will have a positive effect on all seedling parameters compared to their respective controls, while the combined use of the two techniques will lead to a synergistic increase in performance; (ii) the performance of SCwSAP will be proportional to the application dosage; and (iii) the commercial (including polyacrylamide) and the new SCwSAP (polyacrylamide-free) will perform similarly when applied at the same dosage.

Materials and Methods

Description of the study area

The experiment was conducted in Fontanals de Cerdanya, in the Pyrenean mountains of Catalonia, NE Spain (42°23'9.11N; 1°55'53.90E). The plot is located at a mean altitude of 1,430 m a.s.l, on a north aspect with an average slope inclination of 30%. The mean annual temperature is 7.5°C while the mean annual precipitation is 887 mm, 272 mm of which occur in summer. According to the Köppen classification, the climate is between Cfc (Temperate) and Dfb (Continental). The soil has a loamy-sandy texture (22% clay, 21% silt, 57% sand), the pH is 5.2 and organic matter content is 2.56%. Most soils in this area are Humic Dystrudepts (USDA, 1999) with a mesic temperature regime (Poch & Boixadera, 2008).

This plot was used for cattle grazing until it was abandoned in 2013. The main woody vegetation species, with a very low density, are the trees ash (*Fraxinus excelsior*) and wild pear (*Pyrus communis* L.), and the shrubs *Rosa canina* L. and *Rubus idaeus* L.

The annual summer precipitation was monitored continually with a weather station (Davis Instruments, USA). The summer of the first growing season (2014, GS1 hereinafter) was rather wet (404 mm) compared to the historical reference (272 mm, Ninyerola *et al.*, 2005), while the summer of the second growing season (2015, GS2 hereinafter) was drier (220 mm).

Experimental design and treatments

We planted 450 seedlings following a randomized incomplete block design with six blocks. In each block we planted 75 seedlings that were randomly assigned, in groups of 5 seedlings, to the 15 experimental treatments or combinations of various mulching and soil conditioning techniques. In total each treatment was applied to 30 seedlings. Table 1 shows the details of the techniques applied (mulch and soil conditioner).

Table 1. Description of the experimental techniques.

Technique type	Technique code	Description
Mulch	Control	No mulch applied.
	Com_Plastic	Commercial black polyethylene film, anti-UV treated, 80 μm thick.
	Com_Biofilm	Ökolys®, a commercial green biodegradable woven mat.
	New_Biofilm	Prototype of black biodegradable frame (biopolymer), fused to a black commercial biodegradable film based on PHA (polyhydroxyalkanoate), 80 μm thick, manufactured by Groencreatie and DTC. The frame is to make installation easier.
	New_Jute	Prototype of biodegradable woven jute mat treated with furan bio-based resin for increased durability, manufactured by La Zeloise NV.
	New_Rubber	Prototype of black layer made of recycled rubber, anti-UV treated, 1.5 mm thick to make fixation unnecessary, manufactured by EcoRub BVBA.
Soil conditioner	SC–	No soil conditioner applied.
	New_SC20; New_SC40; New_SC80	TerraCottem Arbor®, at the prototype stage when tested. Product developed for tree and shrub planting. Its formulation includes a new generation of polyacrylamide-free water super-absorbent polymers (36.25% of total weight), fertilizers (14.5%; NPK 3-1-7), humic acids (0.75%), growth precursors (0.25%) and volcanic rock (48.25%). The numbers 20, 40 and 80 indicate the dosage (g/seedling).
	Com_SC40	TerraCottem Universal®, a commercially available soil conditioner with cross-linked polyacrylamide and polyacrylic acid polymers (39.50%), fertilizers (10.50%; NPK 9-2-11), growth precursors (0.25%) and volcanic rock (49.75%). The dosage was 40 g/seedling.

The 15 treatments were organized into three sub-experiments:

(i) Sub-experiment 1: a full factorial design combining the 6 different mulch treatments with a new SCwSAP applied at a dose of 40 g/seedling (New_SC40) as well as without it (SC-). Overall there were 12 treatments with 30 seedlings per treatment, thus 360 seedlings in total. The soil conditioner dose of 40 g/seedling corresponds to the manufacturer's recommendation for the most similar commercial product available.

(ii) Sub-experiment 2: a study of the effect of four different doses of New_SC (0, 20, 40 and 80 g/seedling), combined with a reference mulch (Com_Plastic) in all cases. Each treatment comprised 30 seedlings, with a total of 120 seedlings in this sub-experiment.

(iii) Sub-experiment 3: a study comparing a commercial and a new SCwSAP (Com_SC vs. New_SC), both applied at a dosage of 40 g/seedling and combined with a reference mulch (Com_Plastic). Each treatment comprised 30 seedlings, with a total of 60 seedlings.

The combination Com_Plastic x New_SC40 was present in all three sub-experiments, while the combination Com_Plastic x SC- was present in sub-experiments 1 and 2.

Field trial establishment

We planted the seedlings in late March 2014, during vegetative dormancy. Soil preparation consisted in mechanical soil digging (40 x 40 x 40 cm) with a backhoe spider excavator, which was used to make micro-basins for runoff collection. The plantation frame was 3 x 3 m, for a density of 1,100 seedlings·ha^{-1}. The tree species chosen was mountain ash (*Fraxinus excelsior* L.) from the local provenance *Central Pyrenees*. The seedlings were one-year old and provided in 300 cm^3 containers. They were 15-20 cm high and met the general seedling quality criteria (Cortina *et al.*, 2006). We applied the soil conditioner right before planting following the manufacturer's indications: we dug a sub-pit sized 30 x 30 x 30 cm, put half of the dosage at the bottom of the pit and mixed the other half with the soil used for filling up the pit when the seedlings were planted. We installed the mulches manually right after planting. We chose a small mulch size (40 x 40 cm, similar to the area of the planting pits), to limit costs and because we predicted that there would be low to intermediate weed proliferation during the first years. To protect the afforestation from browsing damage by roe deer (*Capreolus capreolus* L.) we installed a 1.3 m high perimeter fence consisting of four lines of barbed wire.

Assessment of seedling survival and growth

We assessed all seedlings at the time of planting (March 2014) and at the end of the first two growing seasons (October 2014 and 2015) to determine their survivorship, diameter and height. We conducted an additional visual assessment of survival eight weeks after the seedlings had been planted to remove any seedlings from the study that had died soon after planting as a result of poor seedling quality or careless planting (two seedlings in total). We measured seedling diameter at a painted, constant point, 4-5 cm above the ground, with a precision of 0.1 mm using a digital calliper. Seedling height was measured from the ground up to the highest bud, with a cm precision, using a measuring tape. Annual growth rates of alive seedlings were calculated as the difference between the measurement at the end of each growing season and the previous measurement.

Assessment of physiological traits

We measured two traits related to seedling performance: midday leaf water potential (LWP, hereinafter), which is a proxy of seedling water status; and leaf SPAD (Soil Plant Analysis Development), which is a proxy of seedling nutrient status (Djumaeva *et al.*, 2012). We measured LWP using a pressure chamber (Solfranc Technologies, Spain) with N_2 as the pressure gas. We measured the pressure (bars) at which the water within the leaf was ejected through the petiole. In both 2014 and 2015 we conducted 4 fortnightly measurements in July and August. On each of these 8 dates we sampled one leaf from one randomly chosen seedling per treatment and block (n = 6; 90 measures in total). These measurements were taken between 10:00 and 14:00 solar time, i.e., at the highest sun angle. We used a Minolta SPAD-502 instrument (Minolta Camera Co, Japan) to measure SPAD. The LWP and SPAD were measured on the same dates, except for the two August measurements in GS1 when SPAD could not be measured. In each SPAD sampling we measured one randomly chosen alive seedling per treatment and block (n = 6; 72 measurements in total). We calculated the average SPAD value of three leaves from each sampled seedling. Both LWP and SPAD values were obtained from sun-exposed, fully elongated and healthy leaves located in the upper third of the sampled seedling.

Mulch durability

The physical integrity of mulches was assessed visually in October 2016 after 30 months in the field. The mulches were assigned to a damage category depending on the proportion of their surface that was damaged, either physically (torn) or due to weeds growing through and on the mulch layer: (i) intact (no damage); (ii) slight damage (1-25% damage); (iii) intermediate damage (26-50%); severe damage (>50%).

Statistical analyses

We analysed the data independently for each sub-experiment. We considered treatment as a fixed factor and block as a random factor. The LWP and SPAD data collected on different measuring dates were combined to build a more robust dataset.

Seedling survival and mulch durability were analysed with descriptive statistics. Seedling annual diameter and height growth, LWP and SPAD were assessed with an ANOVA, which was two-way in sub-experiment 1 (factors: mulch, soil conditioner, and their interaction) and one-way in sub-experiments 2 and 3. We used a significance level of 0.05 and assessed pairwise differences between treatments with the post-hoc Duncan's multiple range test. When necessary (seedling growth and LWP), values were log or root transformed to meet the ANOVA assumptions of normality and homoscedasticity. Tables and figures show untransformed data. The ANOVAs were run with SPSS v19.0 software.

Results

Seedling survival

The overall survival rate at the end of the first growing season (GS1) was 99%, dropping to 93% at the end of the second growing season (GS2). In sub-experiment 1 the lowest survival rate after two growing seasons corresponded to New_Rubber (82%) and the Control (90%), while the rest of the mulches showed survival rates above 93%. For the soil conditioner, New_SC40 showed a similar survival rate at the end of GS2 as SC- (93% and 90%, respectively). In sub-experiment 2 the treatments leading to the lowest survival rates after two growing seasons were SC- (90%) and New_SC80 (93%), while New_SC20 and New_SC40 resulted in 97% and 100%, respectively. In sub-experiment 3, both Com_SC40 and New_SC40 led to high survival rates at the end of GS2 (97% and 100%, respectively). Table S1 [suppl.] provides the annual survival of all treatments and sub-experiments.

Seedling growth

Table 2 shows the outcomes of the ANOVAs conducted in each sub-experiment. In sub-experiment 1 the mulches did not significantly affect annual diameter growth in GS1 or GS2; however, for height growth

Table 2. Summary of outcomes of the ANOVAs of annual diameter (DG) and height (HG) growth during the first (GS1, 2014) and the second (GS2, 2015) growing seasons; seedling water status (midday Leaf Water Potential, LWP) and seedling nutrient status (Soil Plant Analysis Development, SPAD).

	DG GS1	DG GS2	HG GS1	HG GS2	LWP	SPAD
Sub-experiment 1						
			Mulch (dF=5)			
F	1.02	1.13	2.07	2.92	1.25	1.46
p-value	0.409	0.344	0.072	0.020	0.284	0.203
			Soil conditioner (dF=1)			
F	201.41	8.88	30.592	1.29	6.10	4.49
p-value	<0.001	0.003	<0.001	0.261	0.014	0.035
			Interaction mulch * soil conditioner (dF=5)			
F	2.04	1.78	0.247	1.12	1.66	2.51
p-value	0.072	0.117	0.941	0.361	0.142	0.030
Sub-experiment 2						
			Soil conditioner dose (dF=3)			
F	28.67	4.79	7.73	1.76	0.06	3.71
p-value	<0.001	0.004	<0.001	0.185	0.979	0.013
Sub-experiment 3						
			Soil conditioner formulation (dF=1)			
F	6.85	8.63	3.02	0.07	0.85	8.05
p-value	0.011	0.005	0.092	0.799	0.360	0.006

New_Biofilm induced significantly higher growth rates than the Control in GS2 (Figure 1A). The use of soil conditioner resulted in a higher diameter and height growth than in the Control, although differences were greater for GS1 than for GS2, where differences between treatments were significant for diameter growth but not for height growth (Figure 1B). No significant effect of the interaction between mulch and soil conditioner on diameter growth or height growth was found (Table 2).

In sub-experiment 2 we observed that the diameter growth during GS1 was larger with increasing doses of soil conditioner, although the two highest doses (40 and 80 g/seedling) were not significantly different (Table 2). However, although the addition of soil conditioner also increased diameter growth during GS2, we did not observe significant differences between the tested doses (Figure 2A). In the case of height growth the pattern was unclear in GS1, with New_SC40 leading to higher values than New_SC20, which in turn was not significantly different to New_SC80. In GS2 there were no significant differences between treatments for height growth.

In sub-experiment 3 we observed significantly higher diameter growth when New_SC40 was used compared to Com_SC40 in both GS1 and GS2, while height growth rates were not significantly different (Figure 2B). Table S2 [suppl.] provides the annual growth values for each treatment.

Physiological traits

In sub-experiment 1 neither mulch nor mulch x soil conditioner interaction had a significant effect on LWP (Table 2). However, the addition of soil conditioner (New_SC40) significantly improved seedling water status compared to SC-. In contrast, we observed that SC dosage (sub-experiment 2) and SC formulation (sub-experiment 3) had no significant effect on LWP.

SPAD values were not affected by the different mulches. However, the use of New_SC40 resulted in a higher SPAD than SC-. Mulch x soil conditioner interaction was also significant. Exploring this interaction further we found that, in the presence of soil conditioner, all mulches enhanced the SPAD value compared to the Control, while in the absence of soil conditioner no mulch had any effect on SPAD.

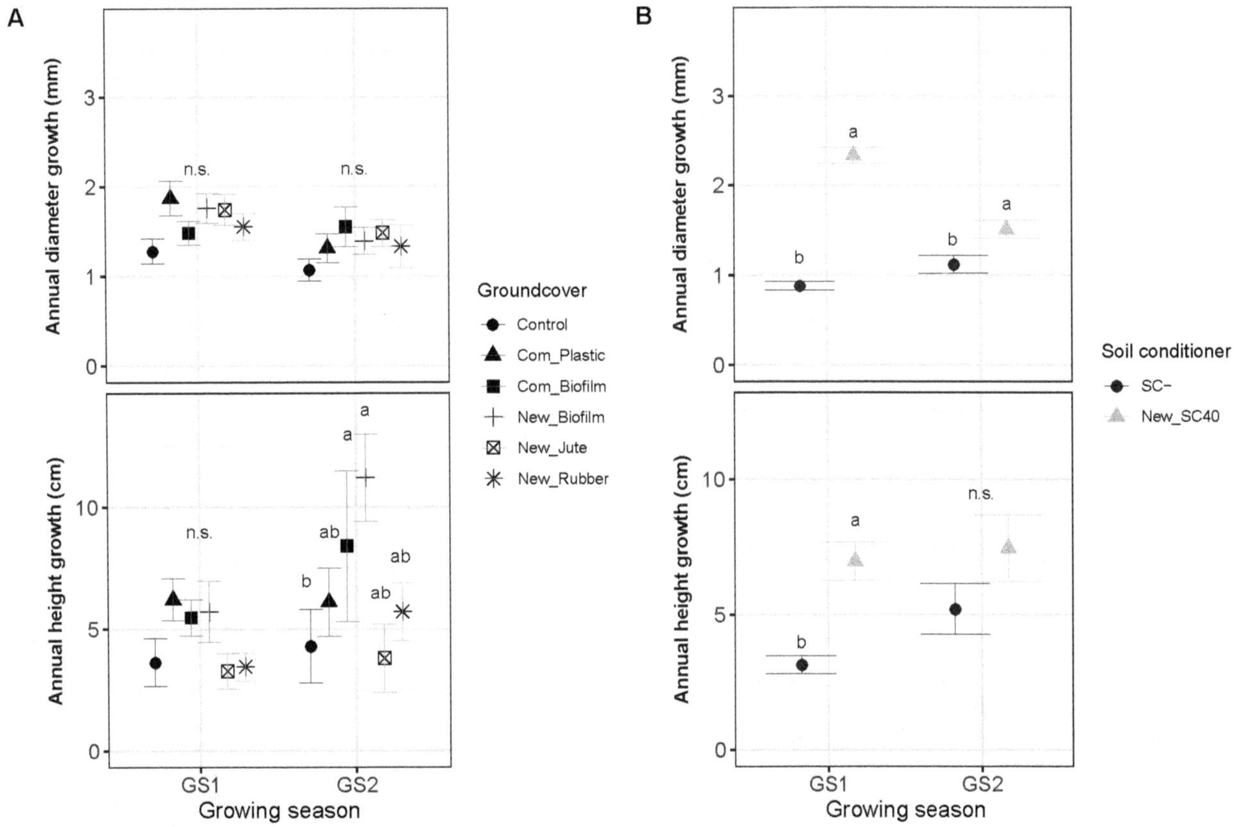

Figure 1. Annual diameter and height growth during the first two growing seasons (GS1-GS2) for sub-experiment 1. A: mulch treatments; B: presence of soil conditioner. For each variable and year, significant differences ($p < 0.05$) between treatments are indicated by different letters (Duncan test grouping), while "n.s." indicates not significant. Whiskers indicate standard error of the mean.

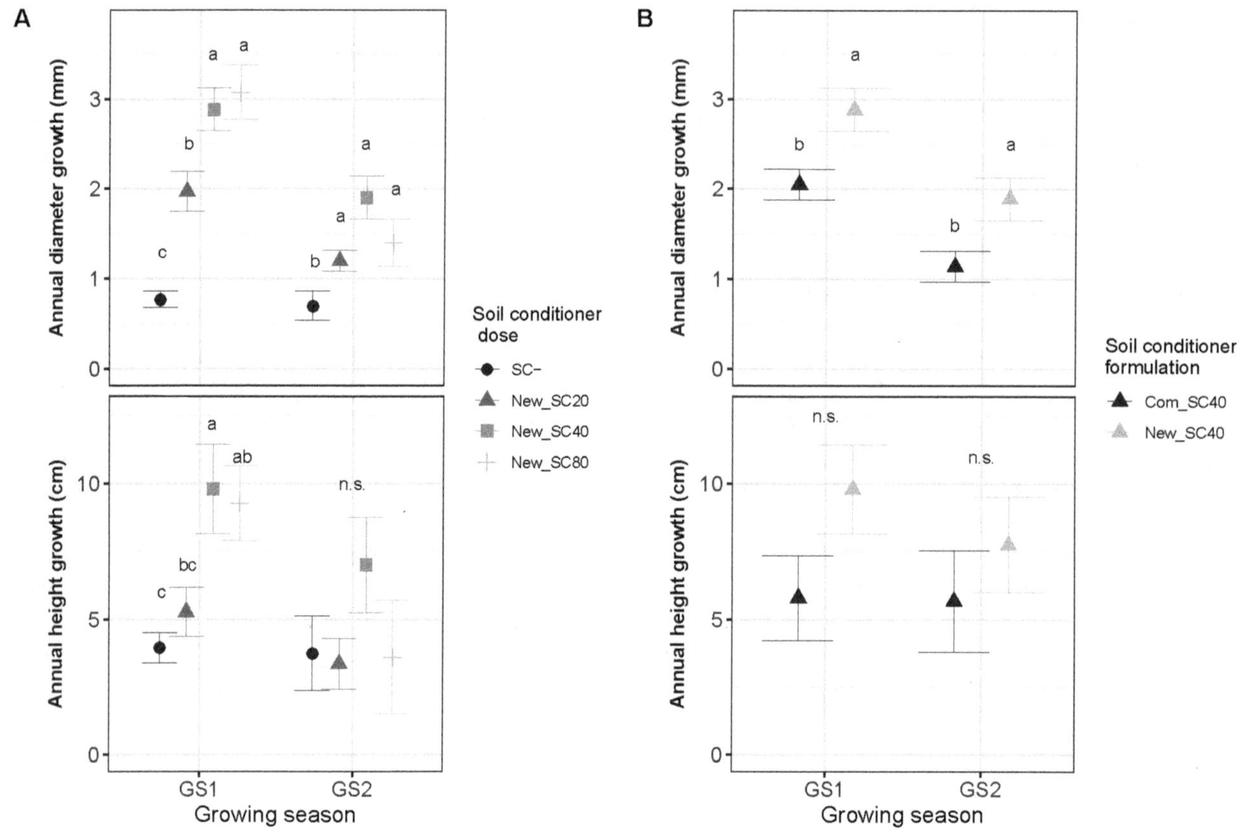

Figure 2. Annual diameter and height growth during the first two growing seasons (GS1-GS2) for (A) sub-experiment 2, and (B) sub-experiment 3. For each variable and year, significant differences ($p < 0.05$) between treatments are indicated by different letters (Duncan test grouping), while "n.s." indicates not significant. Whiskers indicate standard error of the mean.

Increasing the dose of New_SC (sub-experiment 2) to 40 or 80 g/seedling also led to enhanced SPAD compared to SC-. In sub-experiment 3 the new soil conditioner also resulted in significantly higher SPAD than the commercial one.

Table 3 shows the LWP and SPAD values of each treatment.

Mulch durability

The different mulches showed contrasting levels of damage after three growing seasons (Figure 3). The models with the highest integrity were New_Rubber (97% of units intact or with only slight damage) and Com_Plastic (91%). However, the biodegradable mulches had 20%, 8% and 14% (Com_Biofilm, New_Biofilm and New_Jute respectively) of the units severely damaged, while about half of the units were intact.

Discussion

Our study showed the importance that planting techniques can have on improving the early performance of broadleaf seedlings in south European mountain afforestation. The hypotheses of sub-experiment 1 were only partially corroborated: the addition of soil conditioner had a positive effect on all seedling performance parameters, but mulching and the interaction between mulch and soil conditioner (which was additive rather than synergistic) led to benefits below our expectations. In sub-experiment 2, the hypothesis was corroborated because the performance of the soil conditioner was better when the dose was increased from 20 to 40 g/seedling; however, there was a saturating effect at the 80 g/seedling dosage. Finally, the hypothesis of sub-experiment 3 was only partially corroborated, as we found similar height growth and seedling water statuses for the new soil conditioner and the commercial formulation, as initially foreseen; however, the new soil conditioner resulted in unexpected gains in diameter growth and seedling nutrient status.

Nevertheless, our results should be verified with further experiments involving a wider variety of tree species, a longer time span, and whenever possible a thorough study of the changes in the soil water status

Table 3. Midday Leaf Water Potential (LWP) and SPAD of the measurements taken in the summers of 2014 and 2015. Values are expressed as mean ± standard error. Significant differences ($p < 0.05$) between treatments are indicated by different letters, according to Duncan's post-hoc test grouping.

	LWP (bar)	SPAD
Sub-experiment 1 (Mulch and soil conditioner)		
	Mulch	
Control	-22.0 ± 0.3 a	29.1 ± 0.7 a
Com_Plastic	-21.8 ± 0.5 a	30.5 ± 0.7 a
Com_Biofilm	-21.2 ± 0.4 a	31.1 ± 0.7 a
New_Biofilm	-22.2 ± 0.4 a	29.6 ± 0.7 a
New_Jute	-21.5 ± 0.5 a	31.1 ± 0.7 a
New_Rubber	-21.0 ± 0.4 a	30.7 ± 0.7 a
	Soil conditioner	
SC-	-22.1 ± 0.3 b	29.8 ± 0.4 b
New_SC40	-21.2 ± 0.3 a	30.9 ± 0.4 a
Sub-experiment 2 (soil conditioner dose)		
SC-	-21.9 ± 0.7 a	28.8 ± 0.9 b
New_SC20	-22.0 ± 0.5 a	30.7 ± 0.9 ab
New_SC40	-21.7 ± 0.6 a	32.3 ± 0.8 a
New_SC80	-21.8 ± 0.5 a	33.1 ± 1.2 a
Sub-experiment 3 (soil conditioner formulation)		
Com_SC40	-20.9 ± 0.5 a	28.9 ± 0.8 b
New_SC40	-21.7 ± 0.6 a	32.3 ± 0.8 a

throughout the seasonal cycle, and how it is affected by the presence of the mulches and/or soil conditioners. Our results may also have been affected by the exceptionally wet summer (50% higher than the reference value) during the first growing season, which could have contributed to the very high survival rate (99%). If the first summer had been drier, the effect of the soil conditioners and especially the mulches may have been more evident because the seedlings would have been exposed to a more intense water stress that would have been alleviated by the two planting techniques.

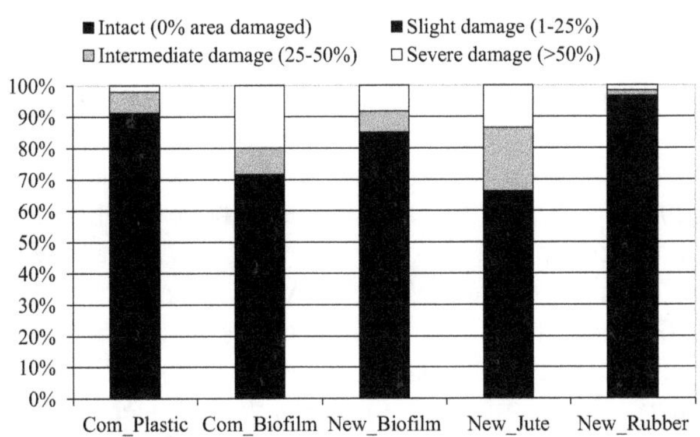

Figure 3. Mulch durability: proportion of the different mulch models in each level of physical damage, after 30 months in the field.

Overall mulch performance

In general, mulching had no significant effect on seedling performance with the only exception of New_Biofilm, which resulted in larger height growth compared to the Control in the second growing season. Although the other variables did not result in significant differences between mulches, the Control was the treatment with the lowest diameter growth and SPAD, and the second lowest height growth and LWP. This seems to imply that mulching has a slightly positive effect. Based on over 110 reports, Van Sambeek (2010) analysed the relative productivity response of different weeding techniques on various tree species, finding a mulch:control response ratio for *Fraxinus* spp. of 222, meaning that mulched seedlings yielded on average 122% more than those without mulch. In our study, however, the mean mulch:control response ratio was 132 for diameter growth and 148 for height growth. Dostálek *et al.* (2007) also found a negligible height growth response in *Fraxinus excelsior* subject to 40 cm wide textile mulching compared to the control, after 5 growing seasons. However, the same authors found a positive effect of fresh bark mulch with similar dimensions as it increased (2-fold) the overall height growth rate compared to the control. The mulches tested here were similar to those evaluated in a previous study (Coello *et al.*, 2018), in which we found no significant differences between them with regard to seedling water status. In contrast, Paris *et al.* (1998) found that 2 m wide plastic mulching had a positive effect on walnut LWP. However, similarly to our study, Cregg *et al.* (2009) and Coello *et al.* (2017) observed that mulching had no effect on seedling early nutrient status. These results question the interest of mulching in the study conditions. That the mulch performance was below the initial expectations could be due to three factors:

(i) the competitiveness of spontaneous vegetation during the first two growing seasons was poor because (a) the soil preparation left 40 x 40 cm of bare soil prior to planting, and (b) the site quality was low due to low temperatures and a stony, coarse-textured soil. Indeed, most of the unmulched planting pits did not show relevant proliferation of extant vegetation, and therefore the expected gains induced by the weeding effect of mulching could have been masked.

(ii) the mulches in our study were rather small (40 x 40 cm) compared to most previous works. Previous studies with small mulches (60 x 60 cm or less) suggest that there are lower gains in seedling performance (Navarro *et al.*, 2005; Dostálek *et al.*, 2007; Valdecantos *et al.*, 2009, 2014; Coello *et al.*, 2018) compared to the more evident benefits of mulches sized 80 x 80 cm or larger (Jiménez *et al.*, 2014, 2016; Vitone *et al.*, 2016; Coello *et al.*, 2017). Davies (1988) found that the mulches with sides sized 120 cm or larger were more effective than those with sides sized 60 cm or less.

(iii) the summer rainfall during the first growing season was 50% higher than the reference value, which may have masked the positive effect of mulching on soil water retention (Barajas-Guzmán & Barradas 2011, McConkey *et al.*, 2012).

Mulch models

There were only minor differences between the mulch models, in line with Maggard *et al.* (2012) and Coello *et al.* (2017, 2018). Only New_Rubber resulted in a much lower survival rate (82%) than the other mulches (95% in average). Overall, and despite the general lack of significant differences between them, the best performance could be attributed to New_Biofilm, which ranked first in overall seedling growth, and to Com_Biofilm, which ranked second in both LWP and SPAD. The sound performance of these models, together with their degradability and renewable origin, make them, along with New_Jute, suitable alternatives to plastic mulching with added benefits from the technical (do not need to be removed) and environmental points of view.

The durability of the mulches was assessed after 30 months, a time span long enough to provide an initial idea of their service life. The plastic mulch in the study performed in line with the literature: Haywood (2000) and Coello *et al.* (2017) found respectively 70% and 66% of plastic mulches with 20% or less damaged area after five years, similarly in our case 91% of units had only 20% of damage or less. The biodegradable mulches also performed in line with previous studies: a biofilm similar to New_Biofilm tested by Coello *et al.* (2017) showed 33% of units severely damaged after 40 months of study, while in our case the values ranged between 8% (New_Biofilm) and 20% (Com_Biofilm). We can therefore initially conclude that the biodegradable mulches are expected to meet the desirable service life of 4-5 years during which more than half of the units should remain intact or be only slightly damaged (Coello & Piqué, 2016).

Overall performance of soil conditioners

The use of SCwSAPs was positive for early seedling performance, with an overall improvement in all assessed parameters, in line with a previous study with the same products evaluated at an afforestation site with *Pinus halepensis* in a coarse-textured soil in semiarid conditions (Coello *et al.*, 2018). However, these results contrast with most previous afforestation studies using SCwSAPs, which found negligible or even negative effects when the SCwSAPs were applied inadequately (1.5 years after planting – Bulíř, 2006; superficial application – Hueso-González *et al.*, 2016), in a low or inaccurate dosage (15 g/seedling – Navarro *et al.*, 2005; unknown dose – Frigola & Nadal, 2013) or in fine-textured, clayish soils (Bulíř, 2005; Navarro *et al.*, 2005; Del Campo *et al.*, 2011; Frigola & Nadal, 2013).

After two growing seasons the average survival for seedlings including any type or dose of soil conditioner (New_SC20, New_SC40, New_SC80 and Com_SC40) was 95%, slightly higher than the 90% of SC- seedlings. Hüttermann *et al.* (2009) also reported a positive effect of soil conditioners on the survivorship of various species and Chirino *et al.* (2011) found a positive effect for *Quercus ilex* L.

With regard to annual diameter and height growth, New_SC40 seedlings grew 92% and 72% more than those of the control treatment, similarly to our previous study (Coello *et al.*, 2018). The growth gains induced by pure SAP (not including fertilizers and other components present in the SCwSAPs tested) in previous afforestation studies with broadleaved species are much less evident than in the present study (Rowe *et al.*, 2005; with *Alnus cordata* (Loisel.) Dub., *Alnus glutinosa* (L.) Gaertn and *Salix x reichardtii*) or even negligible (Clemente *et al.*, 2004 with *Ceratonia siliqua* L., *Olea europaea* L. and *Pistacia lentiscus* L.; Chirino *et al.*, 2011 with *Quercus suber* L.).

The positive effect of SCwSAP or pure SAP on seedling water status has been reported in nursery conditions for other broadleaved species: *Citrus sp* (Arbona *et al.*, 2005); *Populus euphratica* Oliv. (Luo *et al.*, 2009); *Quercus suber* (Chirino *et al.*, 2011) and *Fagus sylvatica* L. (Beniwal *et al.*, 2011; Jamnicka *et al.*, 2013). The improved water status induced by the SCwSAP suggests that the seedlings use the water stored by the polymer (Del Campo *et al.*, 2011; Jamnicka *et al.*, 2013). The improvement in nutrient status induced by New_SC40 was also described by Machado *et al.* (2016), who reported an increase in the level of macro and micronutrients in soils when a SCwSAP was used. In contrast, the use of pure SAP by Clemente *et al.* (2004) did not increase seedling chlorophyll or nitrogen content. This suggests that some of the ingredients of New_SC other than SAP (probably the fertilizers, humic acids, and/or the root growth precursors) help improve tree nutrient status, which is a relevant advantage in soils with a poor nutrient retention capacity, i.e. coarse textured and/or stony soils.

The apparent reduction over time in the soil conditioner effect, which was detected in this study in the second growing season, was also reported by Chirino *et al.* (2011), Oliveira *et al.* (2011) and Coello *et al.* (2018), and could be because (i) the nutrients added via the soil conditioner are progressively exhausted, and/or (ii) the seedling develops new roots out of the planting pit, thus accessing unconditioned soil volume.

Soil conditioner dosage and formulation

Increasing soil conditioner dosage (sub-experiment 2) was generally positive, in line with Al-Humaid & Moftah (2007) and Hüttermann *et al.* (2009). The most cost-effective dose was 40 g/seedling, coinciding with the manufacturer's recommendation for 30 x 30 x 30 cm soil mixing volume. This finding is also in line with our previous study (Coello *et al.*, 2018), in which we found that the 20 g/seedling dose led to better results than the control less often than the 40 g/seedling dose, which provided outcomes that were similar to the highest dose (80 g/seedling) for all variables. Future cost-effectiveness studies considering various dosages, as well as the application of this product at an operational scale, could be used to fine tune the viability and the most recommendable dosage for different site conditions.

In sub-experiment 3 New_SC40 led to better results than Com_SC40 for diameter growth (both growing seasons) and SPAD. These findings contrast with our previous study in semiarid conditions (Coello *et al.*, 2018), in which we generally found similar results for both products. This suggests that the ingredients contained in the new formulation and not in the commercial one (humic acids and the type of root growth precursors and polymers) may be particularly beneficial in cold, montane conditions or for nutrient-demanding species such as *Fraxinus excelsior*. Another strength of the new formulation is that it is expected to have higher social acceptability than most commercially available SAPs or SCwSAPs that contain polyacrylamide (DRI, 2008), like the commercial model in our study. Future experiments with these products will make it possible to ascertain the site conditions and species for which each formulation is particularly efficient.

Interactions between mulch and soil conditioner

As mulching had predominantly no or only minor effects, applying this technique combined with the soil conditioner seems to be of little interest in these conditions. The interaction between the two techniques was only significant in the case of SPAD, where the soil conditioner combined with any mulch type resulted in better seedling nutrient status than when applied alone (unmulched seedlings), while this did not occur in the absence of soil conditioner. This suggests that, for this particular variable (SPAD), the positive effects of mulching may be enhanced when fertilizers are added to the plantation pit, in line with Vincent & Davies (2003).

Implications for management

The soil conditioner with water super-absorbent polymers in a synergistic mixture with other ingredients (fertilizers, humic acids and root growth precursors) is a cheap and easily applied technique that does not require maintenance, and which had a positive effect on all seedling performance indicators in our study. The tested polyacrylamide-free prototype increased both seedling water status, probably due to the polymer; and nutrient status, probably due to the fertilizers and humic acids. This technique ultimately increased seedling survival and growth. The most cost-effective soil conditioner dose was the one recommended by the manufacturer, 40 g/seedling, which often led to better results than the lower dose (20 g/seedling) but never worse than the higher one (80 g/seedling). The new prototype tested seems a suitable alternative to the commercially available version, with technical and social advantages: higher growth rates and better seedling nutrient status, while also being polyacrylamide-free.

On the other hand, the use of small mulches (40 x 40 cm) had a slightly positive effect compared to untreated seedlings, which was usually not statistically significant. Therefore, this technique does not seem a first priority option for enhancing the early establishment of seedlings at poor-quality, montane afforestation sites that have low competitiveness of extant vegetation. Among the different mulch models, the biodegradable ones (prototypes based on biofilm and woven jute, as well as the commercial biofilm) performed similarly to the plastic mulch, and therefore show added value from the technical (do not require removal) and social (come from renewable sources) perspectives.

Acknowledgements

We are grateful to S Martínez, S Navarro, ML Enríquez, M Iacono, C Bellera, A Cunill, E García, A Bothy, P Lumbreras, A Borque, S Busquet, I Krahl, G Martí and A Sala for their support during the establishment and monitoring of the field trial. Special thanks to Francesc Cano and his team at the Catalan Department of Agriculture for essential support in finding a suitable plot for establishing the experiment, installing the perimeter fence and logistics. The study site was kindly provided by the municipality of Fontanals de Cerdanya. Finally, we are thankful to two anonymous reviewers for their constructive comments that significantly contributed to improve the manuscript.

References

Al-Humaid AI, Moftah AE, 2007. Effects of hydrophilic polymer on the survival of buttonwood seedlings grown under drought stress. J Plant Nutr 30: 53–66. https://doi.org/10.1080/01904160601054973

Ameztegui A, Brotons L, Coll L, 2010. Land-use changes as major drivers of mountain pine (Pinus uncinata Ram.) expansion in the Pyrenees. Glob Ecol Biogeogr 19: 632–641. https://doi.org/10.1111/j.1466-8238.2010.00550.x

Ameztegui A, Gil-Tena A, Faus J, Piqué M, Brotons L, Camprodon J. 2018. Bird community response in mountain pine forests of the Pyrenees managed under a shelterwood system. For Ecol Manage 407: 95-105.

Arbona V, Iglesias DJ, Jacas J, Primo-Millo E, Talon M, Gómez-Cadenas A, 2005. Hydrogel substrate amendment alleviates drought effects on young citrus plants. Plant Soil 270: 73–82. https://doi.org/10.1007/s11104-004-1160-0

Arentoft BW, Ali A, Streibig JC, Andreasen C, 2013. A new method to evaluate the weed suppressing effect of mulches: a comparison between spruce bark and cocoa husk mulches. Weed Res 53: 169 175. https://doi.org/10.1111/wre.12011

Barajas-Guzmán MG, Barradas VL, 2011. Microclimate and sapling survival under organic and polyethylene mulch in a tropical dry deciduous forest. Bol Soc Bot Méx 88: 27-34. https://doi.org/10.17129/botsci.303

Beniwal RS, Hooda MS, Polle A, 2011. Amelioration of planting stress by soil amendment with a hydrogel–mycorrhiza mixture for early establishment of beech (Fagus sylvatica L.) seedlings. Ann For Sci 68: 803–810. https://doi.org/10.1007/s13595-011-0077-z

Bhardwaj AK, Shainberg I, Goldstein D, Warrington DN, Levy GJ, 2007. Water retention and hydraulic conductivity of cross-linked polyacrylamides in sandy soils. Soil Sci Soc Am J. 71: 406-412. https://doi.org/10.2136/sssaj2006.0138

Bouranis, DL, Theodoropoulos AG, Drossopoulos JB, 1995. Designing Synthetic Polymers as Soil Conditioners. Comm Soil Sci Plant Anal 26: 1455-1480. https://doi.org/10.1080/00103629509369384

Bulíř, P, 2005 Impact of soil conditioners on the growth of European ash (Fraxinus excelsior L.) on dumps. J For Sci 51: 392–402. https://doi.org/10.17221/4574-JFS

Bulíř, P, 2006. Growth of Austrian pine (Pinus nigra Arnold) treated by soil conditioners on Loket spoil bank. J Forest Sci 52: 556–564. https://doi.org/10.17221/4536-JFS

Carminati A, Moradi BA, Vetterlein D, Vontobel P, Lehmann E, Weller U, Vogel H, Oswald SE, 2010. Dynamics of soil water in the rhizosphere. Plant Soil 332: 163 176. https://doi.org/10.1007/s11104-010-0283-8

Chirino E, Vilagrosa A, Vallejo VR, 2011. Using hydrogel and clay to improve the water status of seedlings for dryland restoration. Plant Soil 344: 99–110. https://doi.org/10.1007/s11104-011-0730-1

Clemente AS, Werner C, Máguas C, Cabral MS, Martins-Louçao MA, Correia O, 2004. Restoration of a limestone quarry: effect of soil amendments on the establishment of native Mediterranean sclerophyllous shrubs. Restoration Ecol 12: 20–28. https://doi.org/10.1111/j.1061-2971.2004.00256.x

Coello J, Piqué M, 2016. Soil conditioners and groundcovers for sustainable and cost-efficient tree planting in Europe and the Mediterranean - Technical guide. Centre Tecnològic Forestal de Catalunya. Solsona. 60 pp.

Coello J, Ameztegui A, Rovira P, Fuentes C, Piqué M, 2018. Innovative soil conditioners and mulches for forest restoration in semiarid conditions in northeast Spain. Ecol Eng 118: 52-65. https://doi.org/10.1016/j.ecoleng.2018.04.015

Coello J, Coll L, Piqué M, 2017. Can bioplastic or woodchip groundcover replace herbicides or plastic mulching for valuable broadleaf plantations in Mediterranean areas? New For 48(3): 415-429. https://doi.org/10.1007/s11056-017-9567-7

Cortina J, Peñuelas JL, Puértolas J, Savé R, Vilagrosa A (coord.), 2006. Calidad de planta forestal para la restauración en ambientes mediterráneos - Estado actual de conocimientos. Organismo Autónomo Parques Nacionales, Ministerio de Medio Ambiente. Madrid. 192 pp.

Cregg BM, Nzokou P, Goldy R, 2009. Growth and Physiology of Newly Planted Fraser Fir (Abies fraseri) and Colorado Blue Spruce (Picea pungens) Christmas Trees in Response to Mulch and Irrigation. Hortscience 44(3): 660-665.

Davies RJ, 1988. Sheet mulching as an aid to broadleaved tree establishment II. Comparison of various sizes of black polythene mulch and herbicide treated spot. Forestry 61(2): 107-124 https://www.dri.edu/images/stories/research/programs/pam/pdf/2008_PAM_Workshop.pdf

Debnath S, 2014. Jute-based sustainable agrotextiles, their properties and case studies. In: Roadmap to Sustainable Textiles and Clothing. Textile Science and Clothing Technology; Muthu SS (ed). Springer Science + Business Media Singapore: 327-355.

Del Campo AD, Hermoso J, Flors J, Lidón A, Navarro-Cerrillo RM, 2011. Nursery location and potassium enrichment in Aleppo pine stock 2. Performance under real and hydrogel-mediated drought conditions. Forestry 84(3): 235-245. https://doi.org/10.1093/forestry/cpr009

Djumaeva D, Lamers JPA, Martius C, Vlek PLG, 2012. Chlorophyll meters for monitoring foliar nitrogen in three tree species from arid Central Asia. J Arid Environ 85: 41-45. https://doi.org/10.1016/j.jaridenv.2012.03.008

Dostálek J, Weber M, Matula S, Frantík T, 2007. Forest stand restoration in the agricultural landscape: the effect of different methods of planting establishment. Ecol. Engn. 29: 77–86. https://doi.org/10.1016/j.ecoleng.2006.07.016

DRI, Desert Research Institute, 2008. Polyacrylamide (PAM) and PAM alternatives workshop. Proc. https://www.dri.edu/images/stories/research/programs/pam/pdf/2008_PAM_Workshop.pdf. [10 May 2018].

Fraxigen, 2005. Ash species in Europe: biological characteristics and practical guidelines for sustainable use. Oxford Forestry Institute, University of Oxford: UK. 128 pp.

Frigola P, Nadal N, 2013. Control de calidad en la ejecución de repoblaciones en la montaña de Portbou (CUP 72/ Elenco 1.007) en el Alt Empordà (Girona). Actas 6º Congreso Forestal Español, Vitoria-Gasteiz. Sociedad Española de Ciencias Forestales.

Gonin P. Larrieu L, Coello J, Marty P, Lestrade M, Becquey J, Claessens H, 2013. Autecology of broadleaved species. Institut pour le Développement Forestier. Paris. 64 pp.

Haywood JD, 2000. Mulch and hexazinona herbicide shorten the time longlife pine seedlings are in the grass stage and increase height growth. New For 19: 279–290. https://doi.org/10.1023/A:1006673509218

Hueso-González P, Martínez-Murillo JF, Ruiz-Sinoga JD, 2016. Effects of topsoil treatments on afforestation in a dry Mediterranean climate (southern Spain). Solid Earth 7: 1479–1489. https://doi.org/10.5194/se-7-1479-2016

Hueso-González P, Ruiz-Sinoga JD, Martínez-Murillo JF, Lavee H, 2015. Overland flow generation mechanisms affected by topsoil treatment: Application to soil conservation. Geomorphology 228: 796-804. https://doi.org/10.1016/j.geomorph.2014.10.033

Hüttermann A, Orikiriza LJB, Agaba H, 2009. Application of superabsorbent polymers for improving the ecological chemistry of degraded or polluted lands. Clean Soil Air Water 37: 517–526. https://doi.org/10.1002/clen.200900048

Jamnická G, Ditmarová L, Kurjak D, Kmeť J, Pšidová E, Macková M, Gömöry D, Střelcová K, 2013. The soil hydrogel improved photosynthetic performance of beech seedlings treated under drought. Plant Soil Environ 59(10): 446–451. https://doi.org/10.17221/170/2013-PSE

Jiménez MN, Fernández-Ondoño E, Ripoll MA, Castro-Rodríguez J, Huntsinger L, Navarro FB, 2016. Stones and organic mulches improve the Quercus ilex L. afforestation success under Mediterranean climatic conditions. Land Degrad Dev 27: 357–365. https://doi.org/10.1002/ldr.2250

Jiménez MN, Pinto JR, Ripoll MA, Sánchez-Miranda A, Navarro FB, 2014. Restoring silvopastures with oak saplings: effects of mulch and diameter class on survival, growth, and annual leaf-nutrient patterns. Agrofor Syst 88: 935–946. https://doi.org/10.1007/s10457-014-9737-y

Kapanen A, Schettini E, Vox G, Itävaara M, 2008. Performance and Environmental Impact of Biodegradable Films in Agriculture: A Field Study on Protected Cultivation. J Polym Environ (2008) 16: 109–122. https://doi.org/10.1007/s10924-008-0091-x

Koupai JA, Eslamian SS, Kazemi JA, 2008. Enhancing the available water content in unsaturated soil zone using hydrogel to improve plant growth indices. Ecohydrol Hydrobiol 11: 67–75. https://doi.org/10.2478/v10104-009-0005-0

Lande SS, Bosch SJ, Howard PH, 1979. Degradation and leaching of acrylamide in soil. J Environ Qual 8: 133–137. https://doi.org/10.2134/jeq1979.00472425000800010029x

Luo ZB, Li K, Jiang X, Polle A, 2009. Ectomycorrhizal fungus (Paxillus involutus) and hydrogels affect performance of Populus euphratica exposed to drought stress. Ann For Sci 66: 106. https://doi.org/10.1051/forest:2008073

MacDonald D, Crabtree JR, Wiegsinger G, Dax T, Stamou N, Fleury P, Gutiérrez Lazpita J, Gibon A, 2000. Agricultural abandonment in mountain areas of Europe: Environmental consequences and policy response. J Environ Manage 59: 47–69. https://doi.org/10.1006/jema.1999.0335

Machado W, Figueiredo A, Guimarães MF, 2016. Initial development of seedlings of macauba palm (Acrocomia aculeata). Ind Crops Prod 87: 14–19. https://doi.org/10.1016/j.indcrop.2016.04.022

Maggard AO, Will RE, Hennessey TC, McKinley CR, Cole JC, 2012. Tree-based mulches influence soil properties and plant growth. HortTechnology 22(3): 353-361.

Marie-Pierre J, Didier A, Gérard B, 2006. Patterns of ash (Fraxinus excelsior L.) colonization in mountain grasslands: the importance of management practices. Plant Ecol (2006) 183: 177-189. https://doi.org/10.1007/s11258-005-9019-x

McConkey T, Bulmer C, Sanborn P, 2012. Effectiveness of five soil reclamation and reforestation techniques on oil and gas well sites in northeastern British Columbia. Can J Soil Sci 92(1): 165-177. https://doi.org/10.4141/cjss2010-019

Mottet A, Julien MP, Balent G, Gibon A, 2007. Agricultural land-use change and ash (Fraxinus excelsior L.) colonization in Pyrenean landscapes: an interdisciplinary case study. Environ Model Assess 12: 293–302. https://doi.org/10.1007/s10666-006-9064-4

Navarro-Cerrillo RM, Fragueiro B, Ceaceros C, del Campo A, de Prado R, 2005. Establishment of Quercus ilex L. subsp. ballota [Desf.] Samp. Using different weed control strategies in southern Spain. Ecol Engn 25: 332-342. https://doi.org/10.1016/j.ecoleng.2005.06.002

Navarro RM, Moreno J, Parra MA, Guzmán JR, 2005. Utilización de tubos invernaderos, mulch plástico y polímeros en el establecimiento de encina y acebuche en el semiárido almeriense. ITEA 101(2): 129-144.

Ninyerola M, Pons X, Roure JM, 2005. Atlas Climático Digital de la Península Ibérica. Metodología y aplicaciones en bioclimatología y geobotánica. ISBN 932860-8-7. Universidad Autónoma de Barcelona, Bellaterra. 45 pp.

Oliveira G, Nunes A, Clemente A, Correia O, 2011. Effect of substrate treatments on survival and growth of Mediterranean shrubs in a revegetated quarry: An eight-year study. Ecol Engn 37: 255–259. https://doi.org/10.1016/j.ecoleng.2010.11.015

Palmero-Iniesta M, Doménech R, Molina-Terrén D, Espelta JM, 2017. Fire behavior in Pinus halepensis thickets: Effects of thinning and woody debris decomposition in two rainfall scenarios. For Ecol Manage 404: 230-240.

Paris P, Olimpieri G, Todaro L, Pisanelli A, Cannata F, 1998. Leaf-water potential and soil-water depletion of walnut mulched with polyethylene and intercropped with alfalfa in central Italy. Agrofor Syst 40: 69–81. https://doi.org/10.1023/A:1006079215567

Poch RM, Boixadera J, (eds) 2008. Sòls de La Cerdanya: Guia de camp. Departament de Medi Ambient i Ciències del Sòl (UdL), Secció d'Avaluació de Recursos Agraris (Generalitat de Catalunya). Lleida. 189 pp.

Rowe EC, Williamson JC, Jones DL, Holliman P, Healey JR, 2005. Initial tree establishment on blocky quarry waste ameliorated with hydrogel or slate processing fines. J Environ Qual 34: 994–1003. https://doi.org/10.2134/jeq2004.0287

Shogren RL, Rousseau RJ, 2005. Field testing of paper/polymerized vegetable oil mulches for enhancing growth of eastern cottonwood trees for pulp. For Ecol Manage 208: 115-122.

Sivapalan S, 2001. Effect of polymer on soil water holding capacity and plant water use efficiency. In: 10th Australian Agronomy Conference; Mendham M (ed.). Australian Society of Agronomy: 1-4.

USDA, 1999. Soil Taxonomy: A basic system of soil classification for making and interpreting soil surveys. Second edition. 886 pp.

Valdecantos A, Baeza MJ, Vallejo VR, 2009. Vegetation management for promoting ecosystem resilience in fire-prone Mediterranean shrublands. Restoration Ecol 17(3): 414–421. https://doi.org/10.1111/j.1526-100X.2008.00401.x

Valdecantos A, Fuentes D, Smanis A, Llovet J, Morcillo L, Bautista S, 2014. Effectiveness of low-cost planting techniques for improving water availability to Olea europaea seedlings in degraded drylands. Restoration Ecol. 22(3): 327–335. https://doi.org/10.1111/rec.12076

Vallejo R, Smanis A, Chirino E, Fuentes D, Valdecantos A, Vilagrosa A, 2012. Perspectives in dryland restoration: approaches for climate change adaptation. New For 43: 561 579. https://doi.org/10.1007/s11056-012-9325-9

Van Sambeek JW, 2010. Database for estimating tree responses of walnut and other hardwoods to ground cover management practices. Proc VI International Walnut Symposium. McNeil DL (ed.): 245-252.

Van Sambeek JW, Garrett HE, 2004. Ground cover management in walnut and other hardwood plantings. Proc 6th Walnut Council Research Symposium; Gen. Tech. Rep. NC-243. St. Paul, MN: U.S. Department of Agriculture, Forest Service, North Central Research Station. 85-100.

Vigo J, Carreras J, Ferré, A, 2005. Manual dels hàbitats de Catalunya. Departament de Medi Ambient i Habitatge. Generalitat de Catalunya.

Vincent A, Davies SJ, 2003. Effects of nutrient addition, mulching and planting-hole size on early performance of Dryobalanops aromatica and Shorea parvifolia planted in secondary forest in Sarawak, Malaysia. For Ecol Manage 180: 261-271.

Vitone A, Coello J, Piqué M, Rovira P, 2016. Use of innovative groundcovers in Mediterranean afforestations: aerial and belowground effects in hybrid walnut. Ann Silv Res 40(2): 134-147.

Weber-Blaschke G, Heitz R, Blaschke M, Ammer C, 2008. Growth and nutrition of young European ash (Fraxinus excelsior L.) and sycamore maple (Acer pseudoplatanus L.) on sites with different nutrient and water statuses. Eur J Forest Res 127: 465–479. https://doi.org/10.1007/s10342-008-0230-x

Willoughby I, Balandier P, Bentsen NS, McCarthy N, Claridge J (eds), 2009. Forest vegetation management in Europe: current practice and future requirements. COST Office, Brussels. 156 pp.

PERMISSIONS

All chapters in this book were first published in FS, by National Institute for Agricultural and Food Research and Technology; hereby published with permission under the Creative Commons Attribution License or equivalent. Every chapter published in this book has been scrutinized by our experts. Their significance has been extensively debated. The topics covered herein carry significant findings which will fuel the growth of the discipline. They may even be implemented as practical applications or may be referred to as a beginning point for another development.

The contributors of this book come from diverse backgrounds, making this book a truly international effort. This book will bring forth new frontiers with its revolutionizing research information and detailed analysis of the nascent developments around the world.

We would like to thank all the contributing authors for lending their expertise to make the book truly unique. They have played a crucial role in the development of this book. Without their invaluable contributions this book wouldn't have been possible. They have made vital efforts to compile up to date information on the varied aspects of this subject to make this book a valuable addition to the collection of many professionals and students.

This book was conceptualized with the vision of imparting up-to-date information and advanced data in this field. To ensure the same, a matchless editorial board was set up. Every individual on the board went through rigorous rounds of assessment to prove their worth. After which they invested a large part of their time researching and compiling the most relevant data for our readers.

The editorial board has been involved in producing this book since its inception. They have spent rigorous hours researching and exploring the diverse topics which have resulted in the successful publishing of this book. They have passed on their knowledge of decades through this book. To expedite this challenging task, the publisher supported the team at every step. A small team of assistant editors was also appointed to further simplify the editing procedure and attain best results for the readers.

Apart from the editorial board, the designing team has also invested a significant amount of their time in understanding the subject and creating the most relevant covers. They scrutinized every image to scout for the most suitable representation of the subject and create an appropriate cover for the book.

The publishing team has been an ardent support to the editorial, designing and production team. Their endless efforts to recruit the best for this project, has resulted in the accomplishment of this book. They are a veteran in the field of academics and their pool of knowledge is as vast as their experience in printing. Their expertise and guidance has proved useful at every step. Their uncompromising quality standards have made this book an exceptional effort. Their encouragement from time to time has been an inspiration for everyone.

The publisher and the editorial board hope that this book will prove to be a valuable piece of knowledge for researchers, students, practitioners and scholars across the globe.

LIST OF CONTRIBUTORS

Aidin Parsakhoo
Assistant Professor, Department of Forestry, Faculty of Forest Sciences, Gorgan University of Agricultural Sciences and Natural Resources, Gorgan, Iran

Kyle Eyvindson
University of Jyvaskyla, Dept. of Biol. Environ. Sci., 40014 University of Jyvaskyla, Finland

Rami Saad and Ljusk Ola Eriksson
Swedish University of Agricultural Sciences, Dept. of Forest Resour. Manage., Skogsmarksgränd, SE-901 83, Umeå, Sweden

Filipe Martins
Simbiente Açores - Engenharia e Gestão Ambiental, Lda., Azores Parque, 9500-794, Ponta Delgada, Portugal

Cristina Alegria
CERNAS - Centro de Estudos de Recursos Naturais, Ambiente e Sociedade, Instituto Politécnico de Castelo Branco, Escola Superior Agrária, Department of Natural Resources and Sustainable Development, Quinta da Senhora de Mércules, Apartado 119, 6001-909 Castelo Branco, Portugal

Artur Gil
CE3C - Centre for Ecology, Evolution and Environmental Changes, Azorean Biodiversity Group, Faculty of Sciences and Technology, Department of Biology, University of the Azores, 9501-801 Ponta Delgada, Portugal

Joana A. Paulo and Margarida Tomé
Centro de Estudos Florestais, Instituto Superior de Agronomia, Universidade de Lisboa, Tapada da Ajuda, 1349-017 Lisboa, Portugal

Pasquale A. Marziliano, Giuliano Menguzzato, Angelo Scuderi, Clemente Scalise and Vittoria Coletta
Agraria Department, University "Mediterranea" of Reggio Calabria. Loc. Feo Di Vito - 89060 Reggio Calabria, Italy

Nooreddin Noorian, Shaban Shataee-Jouibary and Jahangir Mohammadi
Forestry Department, Gorgan University of Agricultural Sciences, and Natural Resources, 386, Gorgan, Iran

Francisco Rodríguez y Silva and Juan R. Molina
Univesity of Córdoba,Dept. Forest Engineering, Edificio Leonardo da Vinci-Campus de Rabanales, 14071 Córdoba, Spain

Mercedes Guijarro, Javier Madrigal and Carmen Hernando
INIA, Forest Research Center, Dept. Silviculture and Forest Management, Forest Fire Laboratory, Ctra de La Couña km 7,5, 28040 Madrid, Spain
iuFOR, Sustainable Forest Management Institute UVa-INIA, Avda. Madrid 44, 34071 Palencia, Spain

Enrique Jiménez and José A. Vega
Xunta de Galicia, Lourizán Forest Management Research institute, 36080 Pontevedra, Spain

Ricardo Vélez
Spanish Society of Forest Sciences, 36080 Pontevedra, Spain

Ferhat Kara
Faculty of Forestry, Kastamonu University, Kastamonu, Turkey, 37100

Edward F. Loewenstein
School of Forestry and Wildlife Sciences, 602 Duncan Drive, Auburn University, Auburn, AL, 36849, USA

Dale G. Brockway
Southern Research Station, USDA Forest Service, 521 Devall Drive, Auburn, AL, 36849, USA

Manuel A. Valbuena
Education Department of the Basque Government, IES Murguia BHI. C/Domingo de Sautu s/n. 01130 Murguía (Araba-Álava), Spain

Esperanza Mateos
University of the Basque Country UPV/EHU, Department of Chemical and Environmental Engineering, University of the Basque Country UPV/EHU. Paseo Rafael Moreno Pitxitxi, 3, 48013 Bilbao, Spain

Francisco Rodríguez
Fora Forest Technologies SLL. C/Oreste Camarca 4. 42004 Soria, Spain

Ana Martín
Sustainable Forest Management Research Institute University of Valladolid-INIA, Avda de Madrid 44, 3004 Palencia, Spain

Brigite Botequim
Centro de Estudos Florestais (CEF), Instituto Superior de Agronomia, Universidade de Lisboa (UL). Edificio Mário de Azevedo Gomes Tapada da Ajuda, n/s, 1349 - 017, Lisboa, Portugal

Tiago M. Oliveira
The Navigator Company, Forest Protection Department, Apartado 55, 2901-861 Setúbal, Portugal

Alan Ager
USDA Forest Service, Rocky Mountain Research. 5775 Highway 10 West, Missoula, MT 59808 USA

Francesco Pirotti
Università degli Studi di Padova, Dipartimento Territorio e Sistemi Agro-forestali. Viale dell'Universita, 16, 35020, Padova, Italy

Begoña Abellanas and Simón Cuadros
Departamento de Ingeniería Forestal. Universidad de Córdoba. Campus de Rabanales. Córdoba. Spain

Manuel Abellanas and Dolores Lodares
Dpto. Matemática Aplicada. Facultad de Informática. Universidad Politécnica de Madrid. Campus de Montegancedo s/n. Boadilla del Monte. Spain

Arne Pommerening
Swedish University of Agricultural Sciences SLU. Faculty of Forest Sciences. Department of Forest Resource Management. Skogsmarksgränd. Umeå. Sweden

Fernanda M. G. Ramalho and Paulo R. G. Hein
Universidade Federal de Lavras, Dept. de Ciências Florestais, 37200-000, Lavras, Brazil

Jéssica M. Andrade
Universidade Federal de Lavras, Dept. de Engenharia, 37200-000, Lavras, Brazil

María Quirós-Segovia and Karel Drápela
Department of Forest Management and Applied Geoinformatics, Faculty of Forestry and Wood Technology, Mendel University in Brno, Zemědělská 3, 61300 Brno, Czech Republic

Sonia Condés-Ruiz
Department of Natural Systems and Resources, School of Forestry, Technical University of Madrid, Ciudad Universitaria s/n, 28040 Madrid, Spain

Amélia Palma
Instituto Nacional de Investigação Agrária e Veterinária. Av. República, Quinta Marquês, 2780-159 Oeiras, Portugal

João Miguel Pereira and Paula Soares
Universidade de Lisboa, Instituto Superior de Agronomia, Centro de Estudos Florestais. Tapada Ajuda, 1349-017 Lisboa, Portugal

Robert M. Brook and Neal Hockley
Bangor University; College of Natural Sciences; School of Environment, Natural Resources and Geography; Deiniol Road, Bangor, Gwynedd, LL57 2UW. United Kingdom

Salma N. Talhouk
Bangor University; College of Natural Sciences; School of Environment, Natural Resources and Geography; Deiniol Road, Bangor, Gwynedd, LL57 2UW. United Kingdom

Arbi J. Sarkissian
Bangor University; College of Natural Sciences; School of Environment, Natural Resources and Geography; Deiniol Road, Bangor, Gwynedd, LL57 2UW. United Kingdom
American University of Beirut; Faculty of Agricultural and Food Sciences, Dept. of Landscape Design and Ecosystem Management; AUB Riad El Solh, Beirut 1107 2020. Lebanon

Fernando Pérez-Rodríguez, Luís Nunes, Ângelo Sil and João C. Azevedo
Escola Superior Agrária, Instituto Politécnico de Bragança, Campus de Santa Apolónia, 5300-253, Bragança, Portugal
Centro de Investigação de Montanha, Campus de Santa Apolónia, 5300-253, Bragança, Portugal

Marcela Arguedas, María Rodriguez-Solis, Róger Moya and Alexander Berrocal
Instituto Tecnológico de Costa Rica, Escuela de Ingeniería Forestal, Apartado 159-7050, Cartago, Costa Rica

Pedram Attarod
Forestry and Forest Economics Department, Faculty of Natural Resources, University of Tehran, Iran
Key Laboratory of Water Cycle and Related Land Surface Processes, Institute of Geographic Sciences and Natural Resources Research, Chinese Academy of Sciences, China

Qiuhong Tang and Xingcai Liu
Key Laboratory of Water Cycle and Related Land Surface Processes, Institute of Geographic Sciences and Natural Resources Research, Chinese Academy of Sciences, China

John T. Van Stan II
Department of Geology and Geography, Georgia Southern University, USA

Maria Alló
Universidad de A Coruña. Facultad de Economía y Empresa. Dept. Economía. Campus de Elviña, s/n, 15071 A Coruña. Spain

Maria L. Loureiro
Universidad de Santiago de Compostela. Facultad de C. Económicas. Dept. Fundamentos da Análise Económica. Avda do Burgo das Nacións, s/n, 15782 Santiago de Compostela. Spain

Marian Drăgoi and Ionut Barnoaiea
University of Suceava, Suceava, 13, Universitatii str., 720229, Romania

Jaime Coello, Míriam Piqué, Pere Rovira and Carla Fuentes
Forest Science and Technology Center of Catalonia (CTFC), Carretera de Sant Llorenç de Morunys, km 2; ES-25280 Solsona, Spain

Aitor Ameztegui
Forest Science and Technology Center of Catalonia (CTFC), Carretera de Sant Llorenç de Morunys, km 2; ES-25280 Solsona, Spain
Department of Agriculture and Forest Engineering (EAGROF), University of Lleida, Lleida, Spain

Index

www.ingramcontent.com/pod-product-compliance
Lightning Source LLC
LaVergne TN
LVHW081017210926
848485LV00004B/886